FUTURE DEVELOPMENTS IN TELECOMMUNICATIONS

A James Martin BOOK

Prentice-Hall
Series in Automatic Computation

AHO, ed., *Currents in the Theory of Computing*
AHO and ULLMAN, *The Theory of Parsing, Translation, and Compiling,* Volume I: *Parsing;* Volume II: *Compiling*
ANDREE, *Computer Programming: Techniques, Analysis, and Mathematics*
ANSELONE, *Collectively Compact Operator Approximation Theory and Applications to Integral Equations*
AVRIEL, *Nonlinear Programming: Analysis and Methods*
BENNETT, JR., *Scientific and Engineering Problem-Solving with the Computer*
BLAAUW, *Digital System Implementation*
BLUMENTHAL, *Management Information Systems*
BRENT, *Algorithms for Minimization without Derivatives*
BRINCH HANSEN, *The Architecture of Concurrent Programs*
BRINCH HANSEN, *Operating System Principles*
BRZOZOWSKI and YOELL, *Digital Networks*
COFFMAN and DENNING, *Operating Systems Theory*
CRESS, et al., *FORTRAN IV with WATFOR and WATFIV*
DAHLQUIST, BJÖRCK, and ANDERSON, *Numerical Methods*
DANIEL, *The Approximate Minimization of Functionals*
DEO, *Graph Theory with Applications to Engineering and Computer Science*
DESMONDE, *Computers and Their Uses,* 2nd ed.
DIJKSTRA, *A Discipline of Programming*
DRUMMOND, *Evaluation and Measurement Techniques for Digital Computer Systems*
ECKHOUSE, *Minicomputer Systems: Organization and Programming (PDP-11)*
FIKE, *Computer Evaluation of Mathematical Functions*
FIKE, *PL/1 for Scientific Programmers*
FORSYTHE, MALCOLM, and MOLER, *Computer Methods for Mathematical Computations*
FORSYTHE and MOLER, *Computer Solution of Linear Algebraic Systems*
GEAR, *Numerical Initial Value Problems in Ordinary Differential Equations*
GILL, *Applied Algebra for the Computer Sciences*
GORDON, *System Simulation*
GRISWOLD, *String and List Processing in SNOBOL4: Techniques and Applications*
HANSEN, *A Table of Series and Products*
HARTMANIS and STEARNS, *Algebraic Structure Theory of Sequential Machines*
HILBURN and JULICH, *Microcomputers/Microprocessor; Hardware, Software, and Applications*
HUGHES and MICHTOM, *A Structured Approach to Programming*
JACOBY, et al., *Iterative Methods for Nonlinear Optimization Problems*
JOHNSON, *System Structure in Data, Programs, and Computers*
KIVIAT, et al., *The SIMSCRIPT II Programming Language*
LAWSON and HANSON, *Solving Least Squares Problems*
LORIN, *Parallelism in Hardware and Software: Real and Apparent Concurrency*
LOUDEN and LEDIN, *Programming the IBM 1130,* 2nd ed.
MARTIN, *Communications Satellite Systems*
MARTIN, *Computer Data-Base Organization,* 2nd ed.
MARTIN, *Design of Man-Computer Dialogues*
MARTIN, *Design of Real-Time Computer Systems*

MARTIN, *Future Developments in Telecommunications,* 2nd ed.
MARTIN, *Principles of Data-Base Management*
MARTIN, *Programming Real-Time Computing Systems*
MARTIN, *Security, Accuracy, and Privacy in Computer Systems*
MARTIN, *Systems Analysis for Data Transmission*
MARTIN, *Telecommunications and the Computer,* 2nd ed.
MARTIN, *Teleprocessing Network Organization*
MARTIN and NORMAN, *The Computerized Society*
MCKEEMAN, et al., *A Compiler Generator*
MEYERS, *Time-Sharing Computation in the Social Sciences*
MINSKY, *Computation: Finite and Infinite Machines*
NIEVERGELT, et al., *Computer Approaches to Mathematical Problems*
PLANE and MCMILLAN, *Discrete Optimization*
POLIVKA and PAKIN, *APL: The Language and Its Usage*
PRITSKER and KIVIAT, *Simulation with GASP II: A FORTRAN-based Simulation Language*
PYLYSHYN, ed., *Perspectives on the Computer Revolution*
RICH, *Internal Sorting Methods Illustrated with PL/1 Programs*
RUDD, *Assembly Language Programming and the IBM 360 and 370 Computers*
SACKMAN and CITRENBAUM, eds., *On-Line Planning: Towards Creative Problem-Solving*
SALTON, ed., *The SMART Retrieval System: Experiments in Automatic Document Processing*
SAMMET, *Programming Languages: History and Fundamentals*
SCHAEFER, *A Mathematical Theory of Global Program Optimization*
SCHULTZ, *Spline Analysis*
SCHWARZ, et al., *Numerical Analysis of Symmetric Matrices*
SHAH, *Engineering Simulation Using Small Scientific Computers*
SHAW, *The Logical Design of Operating Systems*
SHERMAN, *Techniques in Computer Programming*
SIMON and SIKLOSSY, eds., *Representation and Meaning: Experiments with Information Processing Systems*
STERBENZ, *Floating-Point Computation*
STOUTEMYER, *PL/1 Programming for Engineering and Science*
STRANG and FIX, *An Analysis of the Finite Element Method*
STROUD, *Approximate Calculation of Multiple Integrals*
TANENBAUM, *Structured Computer Organization*
TAVISS, ed., *The Computer Impact*
UHR, *Pattern Recognition, Learning, and Thought: Computer-Programmed Models of Higher Mental Processes*
VAN TASSEL, *Computer Security Management*
VARGA, *Matrix Iterative Analysis*
WAITE, *Implementing Software for Non-Numeric Application*
WILKINSON, *Rounding Errors in Algebraic Processes*
WIRTH, *Algorithms + Data Structures = Programs*
WIRTH, *Systematic Programming: An Introduction*
YEH, ed., *Applied Computation Theory: Analysis, Design, Modeling*

FUTURE

SYSTEMS AND TELECOMMUNICATIONS

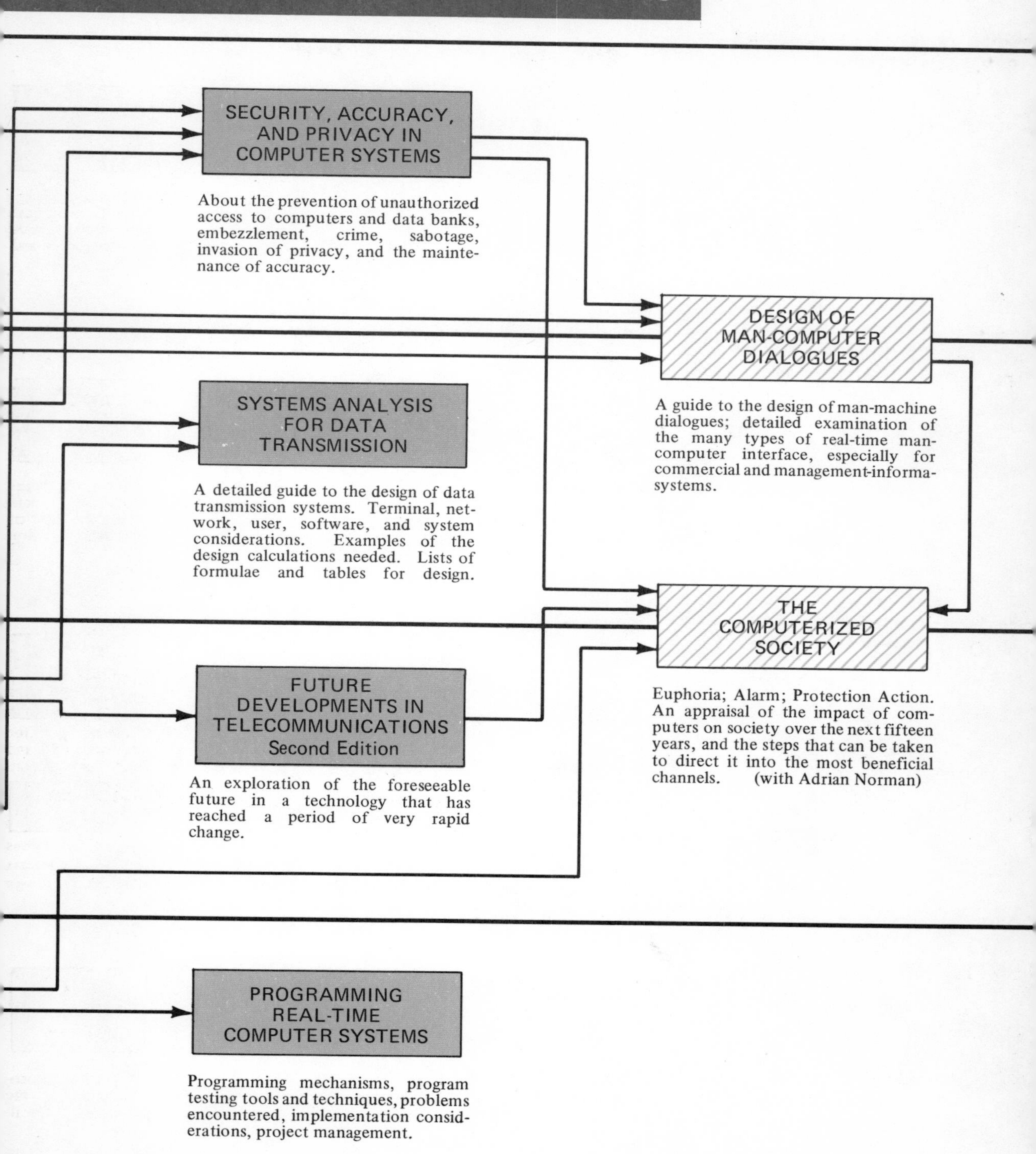
SECURITY, ACCURACY, AND PRIVACY IN COMPUTER SYSTEMS
About the prevention of unauthorized access to computers and data banks, embezzlement, crime, sabotage, invasion of privacy, and the maintenance of accuracy.
DESIGN OF MAN-COMPUTER DIALOGUES
A guide to the design of man-machine dialogues; detailed examination of the many types of real-time man-computer interface, especially for commercial and management-informa-systems.
SYSTEMS ANALYSIS FOR DATA TRANSMISSION
A detailed guide to the design of data transmission systems. Terminal, network, user, software, and system considerations. Examples of the design calculations needed. Lists of formulae and tables for design.
THE COMPUTERIZED SOCIETY
Euphoria; Alarm; Protection Action. An appraisal of the impact of computers on society over the next fifteen years, and the steps that can be taken to direct it into the most beneficial channels. (with Adrian Norman)
FUTURE DEVELOPMENTS IN TELECOMMUNICATIONS
Second Edition
An exploration of the foreseeable future in a technology that has reached a period of very rapid change.
PROGRAMMING REAL-TIME COMPUTER SYSTEMS
Programming mechanisms, program testing tools and techniques, problems encountered, implementation considerations, project management.

DEVELOPMENTS IN TELE-COMMUNICATIONS

Second Edition

JAMES MARTIN

The views expressed in this book are those of the author and do not necessarily reflect those of the IBM Corporation.

PRENTICE-HALL, INC., Englewood Cliffs, New Jersey 07632

Library of Congress Cataloging in Publication Data

MARTIN, JAMES (date)
Future developments in telecommunications.

(Prentice-Hall series in automatic computation)
Includes bibliographical references and index.
1. Telecommunication. I. Title.
TK5101.M325 1977 621.38 76-40103
ISBN 0-13-345850-4

Future Developments in Telecommunications, 2nd edition
James Martin

10 9 8 7 6 5 4

Printed in the United States of America

PRENTICE-HALL INTERNATIONAL, INC., *London*
PRENTICE-HALL OF AUSTRALIA PTY. LIMITED, *Sydney*
PRENTICE-HALL OF CANADA, LTD., *Toronto*
PRENTICE-HALL OF INDIA PRIVATE LIMITED, *New Delhi*
PRENTICE-HALL OF JAPAN, INC., *Tokyo*
PRENTICE-HALL OF SOUTHEAST ASIA PTE. LTD., *Singapore*
WHITEHALL BOOKS LIMITED, *Wellington, New Zealand*

TO CHARITY

THE STRUCTURE OF THE BOOK

A reader without a technical background can read this section of the book, without PART III.	PROLOGUE. PART I: TELECOMMUNICATIONS AND ITS USES. PART II: SYNTHESIS.
More detailed technical discussion is saved for PART III.	PART III: TECHNOLOGY.
The glossary gives the meaning of most telecommunications terminology.	GLOSSARY.

CONTENTS

Preface *xv*

Index of Basic Concepts *xvi*

1 **PROLOGUE**

2 Communications Revolution *3*

The Crystal Ball *9*

PART I **TELECOMMUNICATIONS AND ITS USES**

3 Categories of Telecommunication Links *27*

4 Analog versus Digital Transmission *45*

5 Speeds of Channels *57*

6 Intelligent Exchanges *71*

7 Pushbutton Telephones and Voice Response *95*

8 Video Telephones *113*

9 Cable Television and Its Potential *133*

10 Switched Data Networks *153*

11 Data Radio 171

12 Mobile Radio Transceivers 181

13 Satellite Antennas on the Rooftops 211

14 High Velocity Money 239

15 Electronic Mail and Messages 259

16 New Breeds of Carriers 271

PART II SYNTHESIS

17 Corporate Communications, An Unmanaged Resource 289

18 Computerized Networks 305

19 Terminals in the Home 313

20 The Wired City 327

21 Global Reach 339

22 The Law and Politics 347

23 A Future Scenario 379

PART III TECHNOLOGY

24 Radio Channels 405

25 Cables & Waveguides 445

26 Communications Satellites 473

27 Digital Channels and PCM 501

28 **Signal Compression** *523*

29 **Packet Switching** *539*

30 **Time-Division Switching** *559*

31 **TDMA** *575*

32 **Packet Radio** *585*

33 **Options in Network Design** *601*

Glossary *631*

Index *655*

BOX A Uses for Telecommunications Links

	S = Switched L = Leased line B = Broadcast	One-way or two-way?	C = Continuous channel B = Burst	Real-time?	Typical bandwidth (kilohertz)	Typical bit rate (thousands of bits per second)
Telephone	L,S	2	C	Yes	4	20 to 56
Picturephone	L,S	2	C	Yes	1,000	6.3
Sound broadcasting (low fidelity)	B	1	C	Yes	7.5	50 to 200
Sound broadcasting (high fidelity)	B	1	C	Yes	40	200–800
Television broadcasting	B	1	C	Yes	4,600	40,000 to 92,500
Large wall screen television	L,B	1	C	Yes	20,000 to 50,000	20,000 to 100,000
Closed circuit telephone intercom	L	2	C	Yes	4	20 to 56
Closed circuit television intercom	L	2	C	Yes	4,600	40,000 to 92,500
Still picture video telephone	L,S	2	C: voice B: picture	Yes	4 or higher	20 or higher
Still pictures on television screen	L,B	2*	B	Yes	40 or higher	400 or higher
Text on television screen	L,B	1 or 2	B	Yes		
Telegraphy	L,S	2	B	No		
Mailgram	S	1	B	No		
Facsimile	L,S	1	B	No		
Tele-photograph transmission	L,S	1	B	No		
Electronic mail delivery	L,S	1	B	No		
Voicegrams (voice messages)	L,S	1	B	No		

Electronic fund (cash) transfer	L,S	1	B	Yes or No		
Automatic meter reading (utilities)	S	1	B	Yes or No	Low	Low
Music library	L,S	2*	C	Yes or No	40	200–800
Movie selection in the home	L,S	2*	C	Yes or No	4,600	40,000 to 92,500
Stored television	L,S	1	C	No		
Radio paging	B	1	B	Yes	Variable	
Citizen's band radio	B	2	C	Yes	10	
Radio dispatching	B	2	C	Yes	25–50	20 to 56
Radio telephone	B	2	C	Yes	25–50	20 to 56
Vehicle location monitoring	L	1	B	Yes	0.1	0.1
Vehicle traffic control	L	2	B	Yes	4	2 to 10
Radar		2	C	Yes	500	
Interconnection between computers	L,S	2*	B	Yes or No	Any	Any
Batch data transmission	L,S	2*	B	No		
Access to time-shared computers	L,S	2	B	Yes	0.1 to 4	.1 to 10
Real-time systems such as airline reservations	L,S	2	B	Yes	0.1 to 4	.1 to 10
Fast alphanumeric man-computer dialogue	L,S	2	B	Yes	4	2.4 to 10
Man-computer dialogue with graphics (line drawings)	L,S	2*	B	Yes	4 to 50	2.4 to 56
Man-computer dialogue with still TV images	L,S	2*	B	Yes	4 to 100	4 to 100
Man-computer dialogue with moving TV images	L	2*	3	Yes	4,600	40,000 to 92,500
Man-computer dialogue with voice answerback	L,S	2*	B	Yes	4	20–56
Data collection systems	L,S	1 or 2	B	Yes or No	Low	Low
Voting by the public	L,S	1	B	Yes or No	Low	Low
Alarms (fire, burglar, system failure)	L	1	B	Yes	Low	Low

*(Transmission in one direction is at a much lower rate than in the other.)

(Continued over)

BOX A *Continued*

	S = Switched L = Leased line B = Broadcast	One-way or two-way?	C = Continuous channel B = Burst	Real-time?	Typical bandwidth (kinohertz)	Typical bit rate (thousands of bits per second)
Person identification systems (for security)	L,S	2*	B	Yes	Low or high	Low or high
Time transmission (exact time of day)	S,B	1	B	Yes	Low	Low
Wristwatch data receivers	B	1	B	Yes	Low	Low
Data broadcasting	B	1	B	Yes		
Pocket calculator terminals	S,B	1 or 2	B	Yes	Low	Low
Telephone conferencing	S,L	2	C	Yes	4	20 to 56
Voice conferencing	S,L	2	C	Yes	4,600	40,000 to 92,500
Interactive television (shopping advertising, games)	S,L	2*	C	Yes	4,600	40,000 to 92,500
					Low reverse channel	Low reverse channel
Computer assisted instruction	B,L,S	2	B	Yes	Wide range	Wide range
Library searches	L,S	2	B	Yes	4	2 to 10
Remote operation of switches	L,S	1 or 2	B	Yes	0.1	0.1
Remote control of machines	L,S	2	B	Yes	Wide range	Wide range
Emergency communications	S	2	C	Yes	Up to 4	Up to 56

PREFACE

A handful of entrepreneurs have become telecommunications millionaires in the 1970s. More will follow. The telecommunications industry is in a period of revolution for two reasons. First, the new technologies described in this book are explosive in their implications, especially for an industry that has been dominated by analog terrestrial voice circuits with mechanical switching. Second, in the United States, the shackles of regulation which prevented competition have been partially removed. The scope for innovation is immense. Few technologies can have a more profound effect on the future of society. Many countries, however, do not have the level of telecommunications innovation or expenditure that the more advanced countries have. There are major opportunities for third world countries which do not now have such vast commitments to already obsolete systems.

JAMES MARTIN

INDEX OF BASIC CONCEPTS

The basic concepts, principles and terms that are explained in this book are listed here along with the page on which an introductory explanation or definition of them is given. There is a complete index at the end of the book.

ALOHA protocol 585
Analog transmission 29, 45

Bandwidth 28–9
Broadcast data 172, 217
Burst modem 23–4
Burst multiplexing 156
Burst traffic 36

Cable T.V. 133
  two-way 141–42
Carterfone decision 351
CB radio (Citizen's Band radio) 182
C.C.I.T.T. 59
  modem and data transmission
    recommendation 623–25
  PCM transmission
    recommendation 509–12
Cellular mobile radio 193
C.E.P.T. 502
Circuit switching 168
Coaxial cable 445
Codec 5
Communication pulse 389
Communication satellites 211, 473–74
Compandor, companding 526
Compression, signal 523
Computer data transmission
    (requirements) 159
Computerized switching 74
Concentrators 566
Contention 578
Cream skimming 368

Data base, data bank 5
Data broadcasting 172, 217
Data network, switched 164
Dataphone Digital Service
    (DDS, from AT&T) 59, 163
Data radio 171
Delay, satellite 223–24
Delta modulation 524
Demand assignment 575
Demand Assignment Multiple Access
    (DAMA) 576
Demand Assignment Multiple Access
    equipment 233
Differential Pulse Code Modulation
    (DPCM) 524
Digital Radio systems
    (DR 18, from AT&T) 416
Digital transmission 29, 45, 233
Direct Access Arrangement (DAA) 363
Distributed switching (CATV) 333

Duplex transmission, full and half 35

Earth segment, of satellite transmission 219, 223
Earth stations, low cost 487
Elastic store, elastic buffer 567
Electronic Fund Transfer Systems (EFT, EFTS) 239, 277
Electronic mail 264
ESS #101 (Bell System Switch) 564

Frame grabber 148
Frequency Division Multiple Access (FDMA) 483
Frequency Division Multiplexing (FDM) 34, 41
Full duplex transmission 35

Grade of service 37
Guard band 34

Half duplex transmission 35
Helical waveguide 451
Hertz 29
Hotline 306
Hybrid communications 354
Hybrid data processing 353

Infrared transmission 423
Interconnect industry 350
Item switching 166

Large Scale Integration, LSI 27–8
Laser 463
Least-cost routing 90, 290
Light-Emitting Diode (LED) 423
Local loops 165, 273
Low-cost satellite earth stations 487

Message switching 540
Microcomputers 4
Microwave congestion 229–30
Millimeter wave radio 409
Minimum cost networks 290
Mobile radio 188
Modem 46
  burst 234
Multiple access 575
Multiplexing 34
  burst 156
  continuous channel 158
  Frequency Division (FDM) 34, 41
  Time Division (TDM) 34, 53

Office of Telecommunications Policy (OTP) 351
On-line, real time computers 4
"Open Skies" Policy 215, 354
Optical fibers 455
Optical transmission 423

PABX, PBX 71, 75, 564
Packet 539
Packet radio 177, 585
Packet switching networks 168, 539
Peak-to-average transmission ratio 156
POTS (Plain Old Telephone Service) 158
Pulse Amplitude Modulation (PAM) 49
Pulse Code Modulation (PCM) 50
Pushbutton telephone sets 95

Quantizing 50

Radio, Citizen's Band (CB) 182
Radio, data 171
Radio dispatching 188
Radio paging 188
Radio spectrum 181–82, 407–10
  shortage of 427–29, 435
Real time transmission 37
Repeater, regenerative 507

Satellite antennae, lowcost:
  broadcast capacity 216–17
  carrier 280
  communications 211, 473–74
  delay 223–24
  earth segment, grand segment 219, 223
  earth stations, lowcost 487
  links 235–36
  space segment 219
  transmission capacity 218
Session switching 166
Signal-to-noise ratio 48
Skin effect 446
Specialized Common Carriers 162, 272
Spectrum, electromagnetic 181–82, 406–10
Spectrum engineering 435
Switching:
  burst 156
  circuit 168
  continuous channel 156

Switching *(cont.)*
item 166
packet 168
session 166
time division 560
Symphonie Satellite 382

T-1 carrier, Bell System 53, 501–2, 505
T-2 carrier, Bell System 55, 501–2, 505
T-3 carrier, Bell System 501–2, 505
T-4 carrier, Bell System 501–2, 505
TASI (Time Assigned Speech Interpolation) 35, 224, 535
Telecommunication networks 31
design optimization 601
Teletext 172
Time Division Multiple Access (TDMA) 483, 575
Time Division Switching (TDS) 560
Transparency, of network 551
Transponder 211

Value added common carriers, value added value added networks 162, 280
Video telephones 113–14
Virtual call circuits 623
Vocoder 528
Voice answerback, voice response 97

Wideband links 59
"Wired" city 327

PROLOGUE

Telecommunications in New York, 1838. This semaphore station on Staten Island was an intermediate between Sandy Hook and Manhattan, seen in the distance.

1 COMMUNICATIONS REVOLUTION

Communications technology is in a period of revolutionary change. The new inventions and developments listed in Box 1.1. are involved. Any one of them has enormous potential. Taken in combination they will change the entire fabric of society.

Few technologies could have as profound an effect on the human condition as the full development of these inventions, and certainly additional inventions in telecommunications are yet to come, some perhaps of even greater impact. The last quarter of the twentieth century will be remembered as the era when man acquired new communication channels to other men, to libraries of film and data, and to the prodigious machines.

The new means of communication have potential both for great good, and for great evil. They will forge links between people, raise productivity, and make the best of man's culture available to all. On the other hand, they will also make the worst of man's culture available to all. They provide new techniques of tyranny. They make possible Orwell's telescreen and at the same time make it naive, for Orwell did not envision computers.

In the industrialized nations, these devices will provide the means of education needed to keep up with the ever-increasing rate of change—to combat the "future-shock" of too rapid an evolution. In the underdeveloped countries, they will provide the first hope of literacy and afterward a headlong plunge into the technology that made the northern white nations rich. However, along with education and culture, the changes can also bring propaganda, cruelty, and distortion.

International corporations, in the new era, will flourish with the help of worldwide computer networks. Human talent, the scarcest resource in the coming cybernetic age, will be made available worldwide. Communications are essential to a growing world economy. Business and societal patterns will change as screen-to-screen communication proves more efficient than traveling. In the

BOX 1.1 Major inventions and developments which, in combination, have the potential of changing the fabric of society.

The communication satellite. Satellites have provided telephone and television links to the underdeveloped world. The satellite antennae in some underdeveloped countries stand next to fields ploughed by oxen. Now satellites have the potential of revolutionizing corporate communications both nationally and internationally.

Low-cost satellite earth stations. Planar microwave circuits make it possible to mass-produce satellite receiving equipment at very low cost. Satellite receivers cheap enough for home purchase have been used in Canada, Japan, and India.

Demand-assigned multiple-access equipment. Satellite or high-capacity channels can be shared by multiple goegraphically dispersed users in a highly flexible manner, portions of channel capacity being allocated to users according to their instantaneous needs.

The helical waveguide. A pipe, now operating, that can carry 250,000 or more simultaneous telephone calls or equivalent information, in digital form, over long distances.

The laser. This means of transmission has the theoretical potential of carrying many millions of simultaneous telephone calls or their equivalent. It is being used with optical fibers to carry several thousand.

Optical fibers. A thin flexible fiber made of extremely pure glass which can carry a thousand times as much information as a copper wire pair. Optical communication fibers are now on the market; some are in use carrying public telephone calls. Many thousands of such fibers can be packed into one flexible cable.

Large-scale integration (LSI). A form of ultraminiaturized computer circuitry that probably marks the beginning of mass production of computerlike logic circuitry. It offers the potential of extremely reliable, extremely small, and, in some of its forms, extremely fast logic circuitry and memory. If large-enough quantities can be built, this circuitry can become very low in cost.

On-line real-time computers. Computers capable of responding to many distant terminals on telecommunication lines at a speed geared to human thinking. They have the potential of bringing the power and information of innumerable computers into every office and eventually every home.

Microcomputers. Mass-producible miniature computers of low cost.

Video telephones. Telephones with which subscribers see as well as hear each other or can see still images.

BOX 1.1 *Continued*

Large TV screens. TV screens that can occupy a wall if necessary.

Cable TV. A cable into homes with a potential signal-carrying capacity more than one thousand times that of the telephone cable. It can be used for signals other than television.

Voice answerback. Computers can now assemble human-voice words and speak them over the telephone. Voice answerback and the pushbutton telephone set, makes every such telephone a potential computer terminal.

Millimeter-wave radio. Radio at frequencies in the band above the microwave band can relay a quantity of information greater than all the other radio bands combined. Chains of closely spaced antennas will distribute these millimeter-wave signals.

Cellular mobile radio. A system organization that will permit many radio telephones or other mobile radio devices in a city.

Packet radio. Radio systems for computer terminals that will make pocket terminals, or other small mobile terminals practicable.

Data broadcasting. Information can be broadcast in digital form at VHF or UHF frequencies for reception on home TV sets, special terminals, or portable devices.

Pulse code modulation. All signals, including telephone, Picturephone, music, facsimile, and television, can be converted into digital bit streams and transmitted, along with computer data, over the same digital links. Major advantages accrue from this.

Codecs. Circuits which convert signals such as speech, music, and television into a bit stream, and convert such bit streams back into the original signal. Codecs will become increasingly inexpensive and efficient.

Computerized switching. Computerized telephone exchanges are coming into operation offering many new services, and computerlike logic can be employed for switching and "concentrating" all types of signals.

Data banks. Electronic storage for huge quantities of information that can be manipulated and indexed by computers and that can be accessed in a fraction of a second.

Packet switching networks. One way of building generalized switched data networks interconnecting computers and terminals. A widely accepted standard CCITT x .25 exists so that packet switching networks of different countries will be interconnectable.

future, traveling will be more for pleasure than for business. Certain types of people may work at home much of the time, dialing computers and participating in teleconferences. In addition, they may shop at home, receive education at home, be entertained at home, and interlink each others' homes with video devices. Money transactions will generally be handled without cash or check, and man's credit rating will mean more than his pedigree. Furthermore, our concepts of privacy may change in a world of interconnecting computers.

The satellites and screens will undoubtedly open windows between nations, and wise use of them will do much to forge links and understanding between men. The result may well be worldwide diffusion of knowledge, skills, and culture. Perhaps the best form of foreign aid rich nations can give to poor ones is mass education and training in agricultural and technological skills. The new communication links make this help possible. Thousands of villages in India—many without schools, many without radio or telephone, many without light from sunset to sunrise, and isolated from the world by the inability of their people to read had a village television set receiving educational programs from a NASA satellite. But television can also increase unrest; it can spread and often amplify chaotic situations. To what extent is the unrest in the United States and Japan amplified by television? What will be the effect on multimillions of previously isolated people in the underdeveloped nations of seeing worldwide riots and American affluence on their screens?

There is little doubt that in the United States the new communication technology will enrich society enormously. It will have a great effect on education; it will increase industrial productivity and make possible many new ventures in industry. It seems likely that in some countries the technology described here will be far ahead of that in others. This fact, along with computers, will further enhance the competitive edge of the technically advanced nations. The investment in new forms of education alone will produce high dividends.

As industrial automation grows, a higher proportion of people will work in service industries, and there will be an ever-increasing amount of leisure. Many service industries will be dependent on or based on the techniques discussed in the chapters ahead. Education for leisure will be vitally important, for it will be a moribund society where leisure becomes equated with passive television watching. The communication links will be vital for education and computer-dependent leisure activities, and the media in the home must be interactive rather than passive.

To the accepted stages of economic growth of a society, a new one will be added in time. Societies will modify their behavior patterns when a state of "saturation of affluence" is reached. Eventually, as the Gross National Product per capita rises, through automation, the mass striving for more money and material goods must give way to other drives. A bigger electric can opener does not give life a richer texture. Music, opera, education in the arts and literature, informed conversation, leisure facilities to explore an intellectual world, and entertainment and hobbies using the new media, do. The Protestant ethic will have given way, perhaps to attitudes closer to those of Athens in its prime, perhaps to new attitudes unclassifiable in terms of past history. Man will have the opportunity to once again become civilized in a "Postindustrial" society.

In an era of "Limits to Growth" in the conventional trappings of affluence and limits to growth in transportation, because of world's rapidly declining resources and rapidly growing population and pollution, there are no limits to growth in telecommunications or culture. We are entering an era of staggering growth in knowledge and the ability to disseminate it, in entertainment and nonpolluting electronic technology.

To provide the new services, new telecommunication links are needed. When the links are there, the services will grow. The greater the usage of the links, the lower their cost, for the new technology brings great economies of scale. On the other hand, some services will not be initiated *until* the costs of the links are low. To some extent we have a chicken and egg argument. Which comes first, the telecommunication network or the services using it?

Physical communications like railways and canals had a major effect on the growth of the Industrial Revolution. To a large extent, they made it possible. The same is true in the electronic revolution, and unavailability of communication links will impede progress. In my work in the design of computer systems, I have seen many splendid schemes fall by the wayside because the telecommunication links needed were not available or were too expensive. As will be illustrated, incomparably better data transmission facilities than those now available could be provided, and the result would have a major effect on industry.

Recently I returned from a developing island that as yet has no doctor and almost no roads. While there, I heard a conversation between a sage local black leader of the community and a well-meaning American on the subject of which should come first: a doctor or a road across the island. The American insisted that a doctor was all important, and the local man said there was no point in having a doctor until the road existed so that he could get to his patients. The latter view will prevail. The island cannot develop until it has its road.

Lack of adequate telecommunication links or suitably priced tariffs suppress the development of applications and their marketing. Such links will have a major effect on productivity, economic growth, and the lives of people. They are the catalyst of the electronic revolution. There are few better investments that an advanced country could make at this point in time than extensive spending on a data transmission network and other such facilities described in the chapters ahead.

One of the most difficult sections of the American coast-to-coast telegraph line was built over the Sierra Nevada Mountains in 1861. On the early telephone lines, 40 years later, it became possible, amazingly, to speak across a distance of 2000 miles, from New York to Denver, without amplifiers (because the vacuum tube had not been invented). *(Courtesy The Western Union Telegraph Company.)*

2 THE CRYSTAL BALL

The intent of this book is to take all the facts that seem relevant to telecommunications, explain them to the reader, and piece by piece build a picture of where we are going and what is likely to be achieved. Before embarking on this it would be as well to see what success others have had at predicting the directions of technology, and establish some ground rules.

There have been some spectacular failures.

When Thomas Edison announced that he was working on an incandescent lamp, gas securities dived and the British Parliament set up a committee to investigate. The committee reported that Edison's ideas were "good enough for our trans-Atlantic friends . . . but unworthy of the attention of practical or scientific men." The chief engineer of the British Post Office called it *ignis fatuus.*

The airplane, the telephone, the space rocket, the atomic bomb, the computer, and radio and television broadcasting all met similar derision shortly before becoming practical realities. The derision usually came from leading scientists or engineers of the day and often from committees of them.

In 1956 the British Astronomer Royal Dr. Woolley announced to the press that "Space travel is utter bilge" [1]. This remark was made *after* President Eisenhower had announced the United States satellite program. The very next year the Russians launched Sputnik I. Lack of forecasting accuracy does not seem to matter; Dr. Woolley later became a leading member of the committee advising the British government on space.

Although several thousand German V2 rockets had blitzed London and Belgium in World War II, causing great loss of life, Dr. Vannevar Bush advised the United States Senate in December 1945 that a 3000-mile-range bomb-carrying rocket was "impossible." He said, "I say, technically, I don't think anyone in the world knows how to do such a thing . . . I think we can leave it out of our thinking. I wish the American public would leave it out of their thinking." [2]

Winston Churchill's scientific adviser, Professor F. A. Lindemann (Lord Cherwell), had given the House of Lords the same misinformation months earlier. He had not learned from the fact that nine months before the V2s rained down on London, he had advised Churchill that he "discounted the probability of the use of large rockets" [3]. He had refused to believe that the German V2 could fly, although there was photographic evidence that it could.

FAILURES OF NERVE

Arthur C. Clarke, the science-fiction author, divides forecasting failures into two classes: "failures of nerve" and "failures of imagination" [1]. The former case occurs, he says, when, *given all of the relevant facts* the would-be prophet cannot see that they point to an inescapable conclusion. He often refuses to believe that anything fundamentally new can happen. "Failures of nerve" are frequently accompanied by substantial emotion. The more spectacular illustrations, such as the preceding ones, often come from older persons who are more reluctant to change preconceived ideas.

Robert U. Ayres points out in his book *Technological Forecasting* [4] that committees for forecasting are often prone to failure of nerve. The first major forecasting effort by committee was conducted in 1937 by the U.S. National Research Council [5]. It produced a sober and responsible document, which began "In an age of great change, anticipation of what will probably happen is a necessity for the executives at the helm of the Ship of State" and which missed virtually every major development of its decade, including antibiotics and radar (both of which had existed for ten years), jet engines (which had been designed in theory), and atomic energy (which had been much speculated about). In 1940 a National Academy of Sciences Committee was set up in the United States to evaluate the proposed gas turbine [6]. The committee concluded that the turbine was quite impractical because it would have to weigh 15 lb/hp (pounds per horsepower) as opposed to 1.1 lb/hp for internal combustion engines. One year later a gas turbine was in operation in England and it weighed 0.4 lb/hp. One finds corporate strategy reports today that are almost as erroneous.

The presumed advantage of a committee is that interactions between the members will produce a creative synergism and that individual biases and hobbyhorses will be averaged out. In fact, however, overconservatism usually seems to result. In a ten-year study of economic forecasting, the U.S. National Bureau of Economic Research concluded that individuals do consistently better than consensus polls, at least on economic issues [7]. Ayres comments "All revolutionaries are aware of the Leninist warning that the most dangerous threat to its leadership always comes from the radical left. On technical committees, the unspoken rule is generally to avoid being outflanked on the right (i.e., by the more conservative elements)!" [4]

We are all, in differing degrees, prisoners of familiarity. This fact is often true of the technologist viewing his own discipline and the industrialist viewing his own product. The technologist and industrialist understand in such fine detail how things are done today that they cannot imagine the sweeping changes that will come tomorrow; the detail of these is not yet known. They focus on the limitations set by the current state of the art.

FAILURES OF IMAGINATION

Arthur Clarke's second category, failures of imagination, applies with a vengeance to the data processing and communication industries. When the computer came into existence at the end of the 1940s, confounding all the forecasts that said it was a technical impossibility, very few people had the imagination to see how it could be used.

Forecasters estimated a commercial market of not more than 12 machines in the United States. "Only ten or a dozen very large corporations will be able to take profitable advantage of the computer" was a view expressed in 1948. Sometime later IBM (International Business Machines) made an historic decision not to market the computer because it would never be profitable. The problem was that people failed to see how the machine would be used; they lacked the imagination to think of suitable applications. They thought of the machine as doing only scientific calculations and could only visualize a small number of calculations that were big enough. As the computer industry has grown, forecast after forecast has suffered from a similar, if less spectacular, failure of imagination. When disk memory had been available for a few years, a leading computer salesman in England assured me that it was an American gimmick that would soon disappear from the marketplace. (This person now has a high management position.) When real-time systems were working in airlines and savings banks, it was a commonly held view that there would be no application for real-time methods other than in airlines and savings banks. Similarly, when data transmission first became available, one heard repeatedly that nobody in his right mind would use it because the mail was cheaper.

The same failure to foresee applications has plagued the industries concerned with transmitting information. When broadcasting was first proposed earlier this century, a man who was later to become one of the most distinguished leaders of the industry announced that it was very difficult to see uses for public broadcasting. About the only regular use he could think of was the broadcasting of Sunday sermons, because that is the only occasion when one man regularly addresses a mass public. Late in the last century the minutes of a meeting set up by Western Union contained the following passage about Alexander Graham Bell, the inventor of the telephone:

> "Bell's profession is that of a voice teacher . . . yet he claims to have discovered an instrument of great practical value in communication, which has been overlooked by thousands of workers who have spent years in this field.

> "Bell expects that the public will use his instrument without the aid of trained operators. Any telegraph engineer will at once see the fallacy of this plan. The public simply cannot be trusted to handle technical communications equipment . . . when making a call, the subscriber must give the number verbally to the operator who will have to deal with persons who may be illiterate, speak with lisps or stammer, or have foreign accents or who may be sleepy or intoxicated when making a call.
>
> "Bell's instrument uses nothing but the voice, which cannot be captured in concrete form . . . we leave it to you to judge whether any sensible man would transact his affairs by such a means of communication.
>
> "In conclusion the committee feels that it must advise against any investment whatever in Bell's scheme."

About the same time in London it was pronounced that "the telephone may be appropriate for our American cousins, but not here because we have an adequate supply of messenger boys."

At many times in telephone history, engineers have considered their latest high-capacity channel the ultimate, and have failed to imagine how the public could use greater capacity. Consequently, new applications have often failed to materialize because of bandwidth shortage. One still sees forecasts for data transmission that apparently regard it as an extension of telegraphy, and fail to realize that data transmission will have fundamentally new uses. The corporate strategy reports in the information industries today seem virtually guaranteed to be failures of imagination.

TECHNOLOGICAL SURPRISES

In the preceding categories, we discussed failures to forecast when all the facts are known. In addition, technology has surprises in store for us that are likely to be entirely unpredictable. Some of these surprises contradict the most cherished views of their time. In this century several laws that were regarded as the most fundamental foundation stones of physics have been proven wrong. My dictionary still defines an atom as "a particle of matter so minute as to admit of no division." At school I was taught "matter can neither be created nor destroyed." The Heisenberg uncertainty principle, the general theory of relativity, quantum mechanics, the discovery of mesons, the quasar, and perhaps the pulsar, have all cracked the foundation stones of their day. A few years ago I would have argued vehemently that nothing could travel faster than light. Now science journals discuss the possibility of faster-than-light particles.

However, a technological surprise does not need to crack the foundation stones in order to play havoc with the best-laid plans of mice, men, and telephone companies. Although he might possibly have realized that they were conceivable in theory, a reasonable forecaster in 1940 would not have predicted the computer; in 1945 he would not have predicted the transistor; in 1950 he would not have predicted the laser; in 1955 he would not have predicted the use of pulse code modulation, large-scale integration, solid state switching, computer time sharing, on-line real-time systems in commerce, direct-access data banks or synchronous communication satellites; in 1960 he would not have predicted holography or satellite antennas on the rooftops; in 1965 he would not have predicted the hand-held calculator or the spread of microcomputers. All of these inventions have a major effect on the story we have to tell.

Today's computers would have been quite inconceivable in 1940. The idea would have been laughed at as the wildest fantasy. Yet by 1990 the data processing industry in the United States is likely to exceed the U.S. Gross National Product of 1940.

The rate of developing new technology is increasing constantly. We can expect the number of technological surprises in the two decades ahead to be greater than the number in the past two decades.

UNDERESTIMATING DEVELOPMENT TIME

On the other side of this argument, an important lesson to learn from technological history is that the appearance of a surprise invention does not immediately make existing equipment obsolete. In fact, there is quite a lengthy time between the conception of a complex invention and its use in public systems. The first working laser was developed in 1959. Devices using it were being sold in the second half of the 1960s, but public telecommunication channels based on it are unlikely to be working before the early 1980s.

In the provision of highly complex and expensive systems, such as new communication channels, we can expect a time lag of some years because of three factors.

1. Several years of refining a basic invention are needed before it is suitable for such systems, and several years more are needed to perfect manufacturing processes and bring down costs. With many new technologies—with the laser in fact—more than a decade of research is needed before it becomes a suitable basis for system development.
2. Very large blocks of capital are needed for the equipment and its manufacturing processes. There is a reluctance to commit these until the economics of the new techniques are proven.
3. Particularly important in telecommunications, there is a vast investment in present-day plant, which cannot simply be discarded when a new technology appears. A problem today is that one technology is following another at an increasingly rapid rate. Telecommunications plant has traditionally been designed to have a 40-year lifetime. Fortunately, the growth rate is sufficiently fast that this lifetime does not prevent the introduction of new types of system, but it does mean that the old equipment must co-exist with the new.

The development lead time (the time taken from the start of development to the first commercial application) is sufficiently great that surprise inventions coming this year would almost certainly not find their way into public communication channels and switching during the next decade.

UNDERESTIMATING THE COMPLEXITY

If excessive conservatism and failures of nerve prevent predictions of change, on the other hand, once a new concept is grasped would-be forecasters often let their enthusiasm run away with them. Not only is the development lead-time forgotten, but there is also an almost inevitable tendency to underestimate the complexity of implementation. The press and popular books about the future are full of this kind of oversimplification. The gee-whiz school of forecasting suffers from a failure to work out the details. A simple-seeming end product such as Picturephone or the ability to dial up movies for showing on the home television screen can be grasped without appreciation of the detailed difficulties that are entailed.

One can look back ten years at the forecasts then made for computers. Today's applications are much more diverse than these predictions (failure of imagination). Today's hardware is faster, more reliable, cheaper, and has more on-line storage than was forecast. The machines make a major unforecast use of telecommunication links and terminals. However, in one aspect they entirely failed to live up to the forecasts—that is, in the imitation of intelligent human functions. We have not come close to the predictions made for language translation, speech recognition, intelligent robots, and other forms of "artificial intelligence." The gee-whiz forecasts for computers writing Western movie scripts have not been fulfilled. The reason is that these are all functions that we can imagine easily because they are human functions, but we totally underestimated how complex they really are. This complexity was painfully discovered during the years of trying to program them.

We must make a cautious assessment of the complexity of schemes eligible for discussion in this book.

LEGAL AND POLITICAL PROBLEMS

Many worthy schemes have foundered on political and legal rocks, and the waters we navigate in discussing communications are filled with such rocks.

Multibillion dollar vested interests are at work. Those operating terrestrial telephone lines have no intention of losing a vast revenue to satellites. The television networks see grave dangers from cable TV. The computer companies live in fear of the Federal Communications Commission (FCC) or other government regulation. The term "computer utilities" is taboo in some circles. The heavily regulated common carriers have no intention of allowing small companies to move in and skim the cream of the business using new equipment.

Some of these factors will be discussed later, but generally it is so uncertain who will win the battle of politics that most of the book concentrates on what is *technically* and *economically* possible. As will be discussed in Chapter 22, it seems sadly probable that although great strides will be made, the development of this vital technology will be far from optimal because of political scuffling and legal entanglements.

FAILURE TO FORECAST MARKET CONSTRAINTS

Another failure in forecasting is caused by market constraints. Sometimes a viable and interesting product is never placed on the market. This may be due to lack of entrepreneurial interest. It may be because it cuts across a large corporation's existing market, or because the profit margin has been judged not high enough.

Sometimes inventions related to telecommunications can only be marketed by a large common carrier like AT&T. This fact would be true, for example, for innovations in city telephone switching. There have been many ideas for improving the service of the telephone network which have never reached the marketplace. It would be useful, for instance, and not difficult technically, to have automatic message recording at the local central office (telephone exchange). If a subscriber were out when called, his caller could leave a spoken message, which would be automatically recorded, and that subscriber could retrieve it when he returned, for a small fee. Although AT&T did not implement this relatively simple scheme, they did implement the Picturephone® service, which was exceedingly difficult technically, very expensive, and some say, of dubious market potential. You never can tell.

Similarly, some of the ideas discussed in this book may never reach the marketplace; on the other hand, the majority of them probably will. One could certainly invent many ingenious gadgets that use telecommunications but that would be unlikely to be marketed. I would like to have a grand piano, for example, capable of being played either by a pianist or by a stream of digital impulses coming over a telephone line—a teleprocessing piano. The user could dial a computer and request Rubinstein playing Mozart on his home piano, or he might play a duet with the computer when he is learning, the machine taking the left hand, for example. He could prerecord certain trills or phrases and make the machine play these by pressing a single key. He could listen on his own piano to distant friends playing or could play duets with them. The teleprocessing piano has many possibilities. The computer-minded reader might like to work out how he would organize the bit stream and buffer, assuming no more than 4800 bits per second on the telephone line.

It is virtually certain that the teleprocessing piano will not come to the marketplace, however, because too few people would be prepared to pay for it. When other possible services are discussed in the chapters ahead, reader should ask himself: Is this something that can be sold, or is it just another teleprocessing piano?

® Registered Trade Mark.

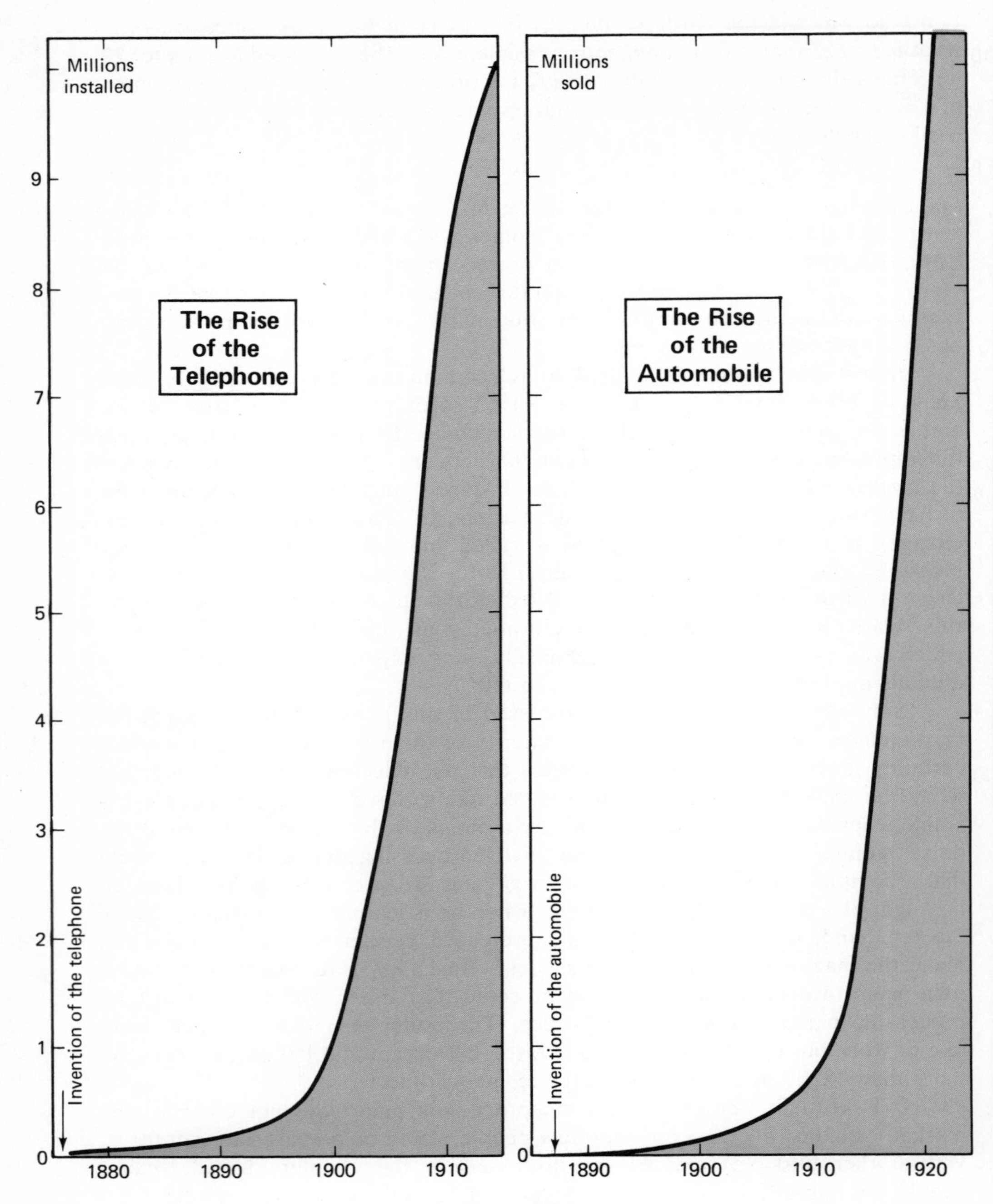
Millions installed
9
8
7
6
5
4
3
2
1
0
The Rise of the Telephone
Invention of the telephone
1880
1890
1900
1910
Millions sold
0
The Rise of the Automobile
Invention of the automobile
1890
1900
1910
1920

Figure 2.1

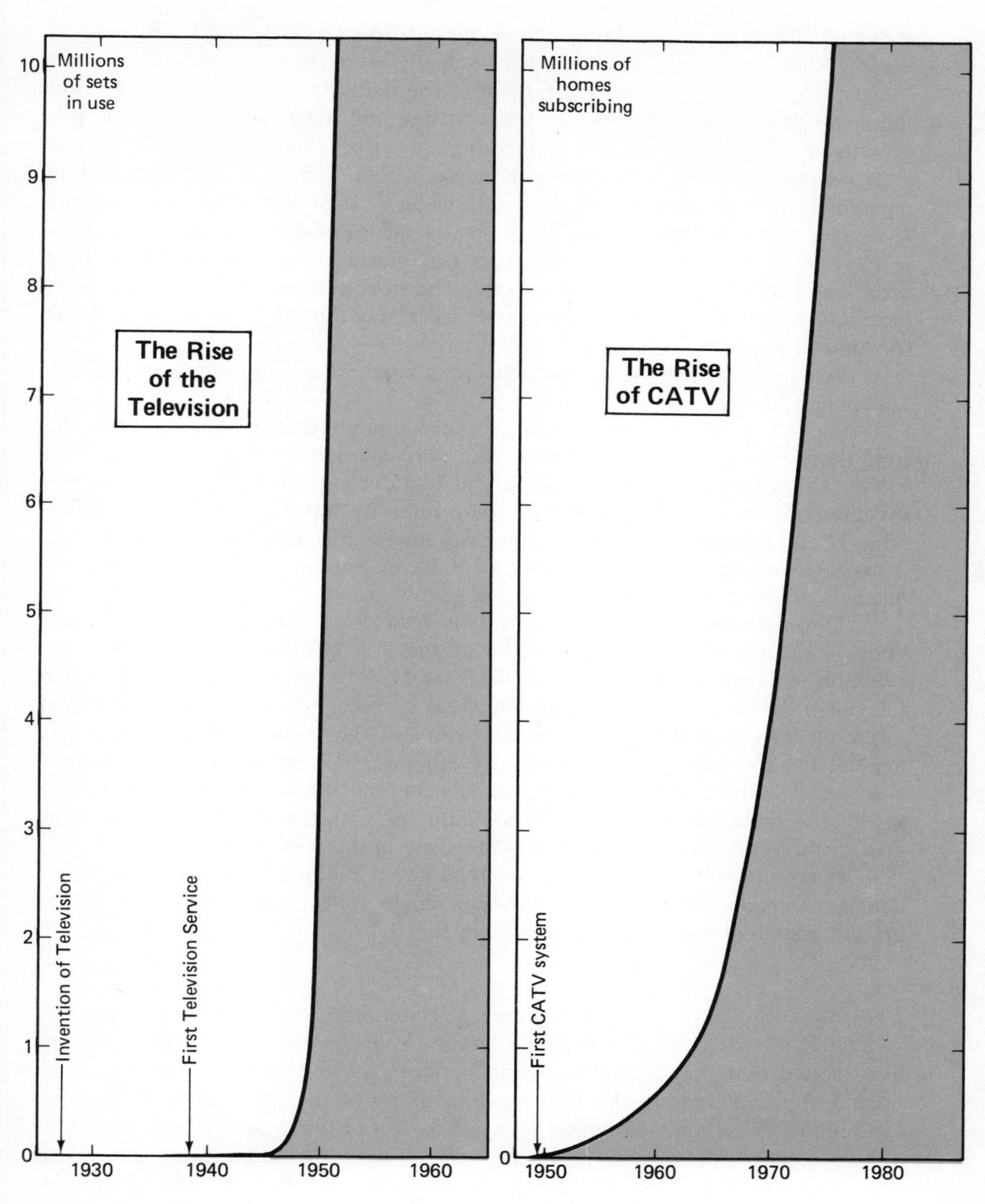

Figure 2.1 (cont.)

MARKET GROWTH RATES

Most telecommunication services which are entirely new in nature have been slow to take off. It is a characteristic of the industry that new ways to communicate gain market acceptance slowly at first and then build up to a massive growth rate. The reason is that pioneering subscribers have few people to call. Few people want to have a Picturephone set if there are very few other sets in existence—there is almost nobody to talk to on it. It is only when other people have sets that the Picturephone becomes useful. Similarly, few people wanted to have television when the service was first started. The industry for making programs had not built up, and so, once the novelty had worn off, there was little worth watching. Once many people have sets then the money is available for making programs.

Again we have the telecommunications chicken-and-egg argument—which comes first, the service or its users?

Fig. 2.1 shows the rate of rise of some communication services from the time they were invented to the time of a mass market buildup. The telephone curve swings upwards spectacularly about two decades after the telephone was invented. At this time citywide and some intercity links were becoming available. The automobile (which in a sense is a device for communicating) also had a massive market rise two decades after its invention. Just as the telephone needed cables so the automobile needed suitable roads.

The rise of television also came two decades after its invention. It came about a decade after the introduction of the first television service, but was probably delayed somewhat by World War II. The slope of the curve when television did take off is amazingly steep. It is easier to build transmitters than intercity links, and the entertainment nature of television had high consumer appeal. CATV had a much slower growth rate. It needed expensive cables, was not at first providing anything really new, and it had an uphill battle with the giant establishment of the television industry. Citizens' Band radio had a slow growth for years and then took off explosively in the mid-1970's.

In the chapters ahead when you read about Picturephone, Teletext, interactive television, and other services you might reflect upon how their future growth could compare with the curves in Fig. 2.1.

CAPITAL

Nationwide telecommunication facilities require enormous capital investments. AT&T management have stated that they intend to spend between $9 and $10 billion per year on capital improvements to the Bell System. Many of the other schemes we discuss require vast sums of capital to progress from prototype systems to nationwide systems.

The American economy may be characterized for the next ten years by extreme capital shortage. This may restrict part of the growth that would otherwise occur in telecommunications. It may create a climate in which large A-rated corporations can obtain capital but small corporations, which are often the most innovative, cannot. Such a climate would impede the growth of the specialized and value-added common carriers (discussed later).

LONG-TERM TRENDS

In many complex technologies, a long-term trend can be plotted as a relatively smooth curve in spite of "surprises" that occurred on the way. Often a growth rate in capability close to an exponential curve can be traced.

If we can take *one class* of invention, a measure of its performance often follows a curve like that of Fig. 2.2. The speed of the automobile and the power of vacuum-tube computers followed such a curve, for example. After a slow start, technical improvements gather momentum until saturation eventually sets in and it is either technically or economically unfeasible to increase the performance of the invention further.

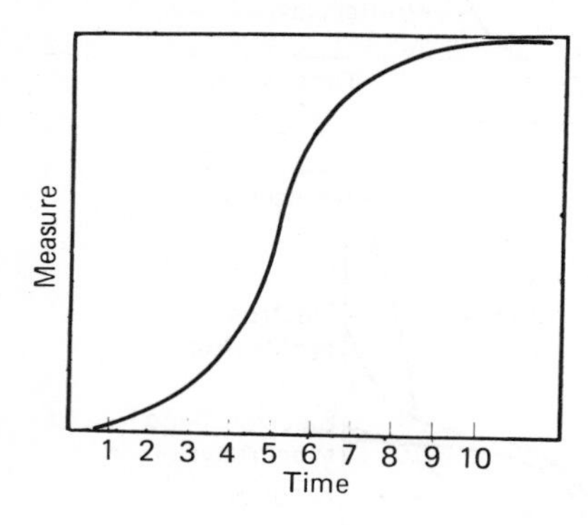

Figure 2.2

In many technologies, however, new classes of invention replace earlier ones for performing the same function. In looking at the performance of one class by itself, we find that it follows a curve like that in Fig. 2.2, but in studying the change as a whole, we must consider the effect of many such curves superimposed as each class of invention replaces the previous one. This situation is illustrated in Fig. 2.3, where four generations of technology or classes of invention replace one another. Each, by itself, is limited. Developments in engineering carry its performance so far, but then the natural restrictions of the technique place a limit on further improvement of performance. Meanwhile, however, a new technology comes into being. The performance curve for the industry as a whole is the envelope of the component performance curves, which tends to maintain exponential growth until there becomes limited value in further development.

This process is happening in many industries. In the computer field, one "technology" has been replacing the previous one about every six years, and the situation seems likely to continue. In the first generation, price and perfor-

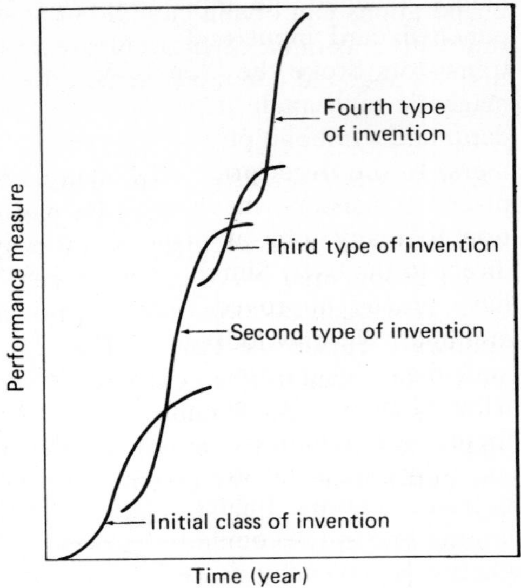

Figure 2.3 Because one type of invention or technology replaces another the growth in performance continues for a sustained period. This happened in computers, for example, as one "generation" has replaced another.

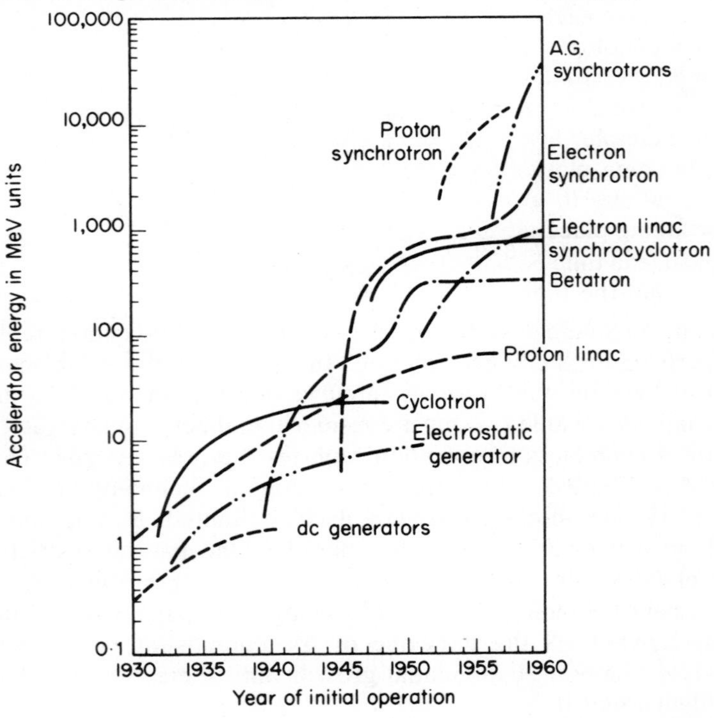

Figure 2.4 The rate of increase of operating energy particle accelerators. *(From M. S. Livingston.)*

mance reached the limits imposed by vacuum tubes, delay line memories, punched card input and output, and bit-by-bit programming. In the second, transistors broke the "reliability barrier," magnetic cores gave fast memories, magnetic tape made it possible to feed the machine as fast as it could digest data, and symbolic programming languages raised the productivity of programmers. In the third, price dropped another tenfold, "solid logic technology" replaced transistors, machine organization raised throughput and allowed attention to many jobs at once, and higher-level languages brought the machine direct to the user. Simultaneously, communications between men and machines have greatly improved. Such a pattern of development will almost certainly continue. We now look forward to higher levels of large-scale integration, solid-state files, a high degree of parallelism, better data base management systems, distributed processing computer network architectures, satellite communications, increasingly intelligent terminals, and other new ideas that change the technology.

Figure 2.4 illustrates a similar sequence of developments in nuclear particle accelerators. Figure 2.5 shows that such a sequence has been going on for a long time in telecommunications. In this way, the capability of computers, the energy achieved by nuclear accelerators, and the capacity of telecommunication highways have each followed an overall growth curve close to exponential. Many other technologies have had a similar exponential growth rate.

Although we cannot anticipate the surprise inventions, it does, in some cases, seem reasonable to extrapolate the exponential performance curves. Figure 2.5, for example, has been approximately a straight line for so long (the vertical scale being plotted logarithmically) that it seems reasonable to assume that it will continue to grow in this way for some time ahead, and basic research now taking place yields credence to this view.

A typical lead time between a basic invention being made in Bell Laboratories or elsewhere and the time when this invention is being used as part of a public service is about 20 years. This lead time has applied all the way up the curve in Fig. 2.5 for the past 100 years. For some inventions like microwave, spurred on by World War II, the lead time has been a little faster; for some it has been a little longer than 20 years. It seems reasonable to assume a 20 year lead time for future technology. This assumption enables us to estimate whether the curve in Fig. 2.5 will continue to go up as a straight line. We can look at the inventions of the last 20 years, estimate their potential information-carrying capacity and apply a 20-year lead time. The result suggests that the growth will continue and indeed the straight line will bend upwards as shown. The capacity of the optical fiber cables now being developed is very high if we assume that multiple glass fibers are used in one cable. There is not likely to be enough *voice* traffic to fill such cables, but perhaps by the time they are available there will be enough *video* traffic to fill them. Perhaps because of shortage of traffic they will be deployed later than shown on Fig. 2.5, thus avoiding the sharp upward swing of the curve.

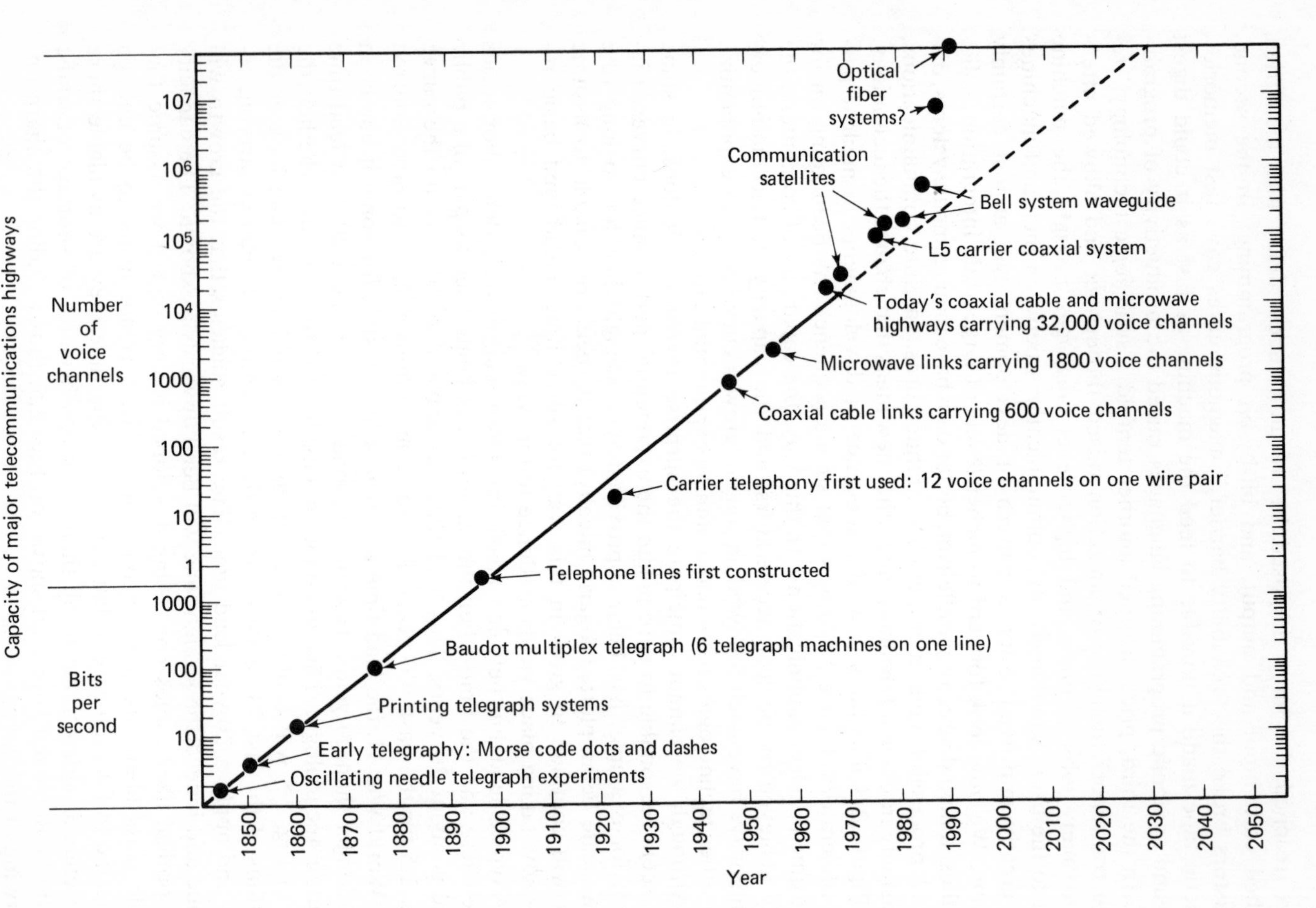
Capacity of major telecommunications highways
Number of voice channels
Bits per second
Year
Optical fiber systems?
Communication satellites
Bell system waveguide
L5 carrier coaxial system
Today's coaxial cable and microwave highways carrying 32,000 voice channels
Microwave links carrying 1800 voice channels
Coaxial cable links carrying 600 voice channels
Carrier telephony first used: 12 voice channels on one wire pair
Telephone lines first constructed
Baudot multiplex telegraph (6 telegraph machines on one line)
Printing telegraph systems
Early telegraphy: Morse code dots and dashes
Oscillating needle telegraph experiments

Figure 2.5 The sequence of inventions in telecommunications.

GROUND RULES We are now in a position to lay down some ground rules for our discussion of future developments in telecommunications.

First, because of the development lead time in this field, we can expect that all types of channels in service in the next ten years will be based on a technology that is now known. They will be engineered differently from today in some cases, but the potential improvements in engineering will not invoke major "surprises." On the other hand, such surprises *will* appear at an increasing rate, and thus the types of channels and services available in twenty years time and beyond will probably employ revolutionary new technology. The technology of 1999 is as unpredictable in 1979 as the transistorized computer with mass on-line storage (1960) was before World War II.

Much of this book discusses the facilities likely to be available during the next ten years. In that discussion we may be safe from major technological surprises, although many exciting new uses of communications are planned for the next decade. We will not be safe from "failures of imagination" and can be certain that the channels will be used in ways we have not yet thought of.

Parts of the book wander beyond the next ten years. Chapter 23 discusses the potential of the next 25 years and some of this potential must include techniques entirely beyond the knowledge of today's engineers.

Because our legal, political, and societal structure is extremely slow to change, it is of value to project the technology three decades into the future if we can. Only by looking well ahead can we hope to direct our turbulent evolution into channels that will make man's world a better place to live in.

We can make one firm prediction. The change in communications during the next two decades will have a major effect on the life of every individual.

REFERENCES

1. Arthur C. Clarke, *Profiles of the Future* (London: Victor Gollancz Ltd., 1962).

2. Vannevar Bush, testimony before the Special Committee on Atomic Energy, December 1945.

3. Winston S. Churchill, *The Second World War. Closing the Ring* (Boston: Houghton Mifflin, 1951), p. 239.

4. Robert U. Ayres, *Technological Forecasting* (New York: McGraw-Hill, 1969).

5. W. C. Ogburn et al., "Technological Trends and Nation Policy," U.S. National Research Council, Natural Resources Committee, 1937.

6. Technical Bulletin No. 2. U.S. Navy, Bureau of Ships, January 1941.

7. Victor Zarnowitz "An Appraisal of Some Aggregative Short-term Forecasts," National Bureau of Economic Research, December 1964.

PART I TELECOMMUNICATIONS AND ITS USES

"Mr. Watson, come here; I want you"—the first articulate sentence ever spoken over an electric telephone by Alexander Graham Bell, March 10, 1876, spoken after he spilled on his clothes some of the acid that was part of this transmission apparatus. *(Courtesy Western Electric.)*

3 CATEGORIES OF TELECOMMUNICATION LINKS

In the years ahead we will have telecommunication channels of enormous capacity, and logic circuitry of high complexity capable of manipulating what happens on them. These channels will change the way men communicate.

Behind much of the development in telecommunications that will take place in the next ten years, two fundamental changes are occurring. These changes are so powerful that it will be necessary to rethink almost all aspects of the technology. The engineer who rests upon his long experience of telecommunications technology is having the rug pulled out from under him.

The first of these dominant trends is a continuing and rapid increase in the capacities of the channels in use. The television cables now being laid into homes have a thousand times the information-carrying capacity of the existing telephone cables into the home. In 1940 long-distance cables carried 60 telephone conversations. Today AT&T's L5 carrier carries 108,000, and the helical waveguide will soon carry 230,000. Experimental optical fiber systems have a much higher capacity. Satellites offer the prospect of taking very high-capacity channels into user locations.

The other dominant trend is in the use of computers and computerlike logic. Directly accessible computer storages are increasing in size at a staggering rate. Computer logic is rapidly increasing in speed, decreasing in size, and increasing in reliability. Most important, it is fast dropping in cost, as we can observe in the cost of pocket calculators.

We have now entered a new era in logic technology—the era of *Large-Scale Integration, LSI.* What this means in economic terms is that if many thousands of a logic circuit are needed, it can be manufactured at low cost even if it is an exceedingly complex circuit. The process etches the circuit onto a silicon chip in a way that somewhat resembles lithograph reproduction. The process for producing one logic circuit is expensive to set up because the "masks" for etching have to be carefully drawn and debugged. However, once it is set up, additional quantities can be mass-produced like newspaper printing at little extra cost. LSI circuits will be mass-produced with increasing quantities of transistors on one chip or wafer. Increasingly complex and powerful microprocessors and memory modules will be made in this way.

It is a little like making plastic elephants. The first plastic elephant is expensive because the molds must be made and the machinery set up. But once the first is made, plastic elephants can be turned out by the million at little cost. We have now reached the plastic elephant era in logic circuitry.

To make effective use of the new mass-producible logic circuitry, it is often necessary to rethink and restructure traditional methods of operating. It takes great imagination and inventiveness to perceive that the best way to do something is fundamentally different from that before LSI.

In the story that unfolds in the chapters ahead we shall frequently discuss converting signals into a digital form and manipulating the resulting bit stream, sometimes in a complex manner. This process would have been unthinkable when today's telephone networks were architected. From now on digital transmission, storage, and manipulation of all types of telecommunications signals will be increasingly economically viable.

Increasing bandwidth and decreasing logic cost are a potent combination. Both changes are happening fast. The technical press calls it a "revolution," but we have barely begun to grasp the implications.

We should no longer think of telecommunications as meaning telephone and telegraph facilities. Almost every means of communication known to man, except love-at-first-sight, can be converted to an electronic form.

Box A at the front of the book lists some major types of uses for telecommunications. There are many items on the list that have not been part of traditional telecommunications. The new potentials arise to a large extent from the use of computers, low-cost logic circuitry, and the employment of channels of higher information-carrying capacity.

Telecommunication channels differ greatly in their capacity just as some water pipes can carry more water than others. Some of the signals we transmit require high capacity channels and others low capacity. Television transmission, for example, needs more than a thousand times the channel capacity of telephone voice. Telephone voice needs more than ten times the capacity of telegraphy.

BANDWIDTH AND BITS

The traditional way of quoting channel capacity is in terms of *bandwidth*. Bandwidth is the *range* of frequencies that a channel can transmit. If the lowest frequency in the range is A and the highest frequency is B, the bandwidth of the channel is B–A.

The amount of information that a channel can carry is proportional to its bandwidth. If an AM radio reproduces sound frequencies between 100 and 7500 hertz (cycles per second)†, we would say that it has a bandwidth of 7400 Hz. On the other hand, as any hi-fi enthusiast knows, to reproduce music perfectly one needs to transmit frequencies from about 30 to 20,000 Hz (even if the top frequencies are only heard by passing bats!), which is a bandwidth of about 20 kHz. Color television needs a much higher bandwidth—about 4.6 MHz. Telephone voice uses a bandwidth of about 3.0 KHz. Telegraph signals are transmitted with bandwidths below 200 Hz. Box A shows typical bandwidths used for the signals listed.

There are different ways of organizing the information to be sent—ways which differ widely in their efficiency. One way, which plays a major role in the story we have to tell, is to convert the signal into a series of bits (one bit is an *on* or *off* pulse—for example, the presence or absence of current) that are transmitted digitally, as in the flow of data in computer circuits. Any of the items in Box A could be sent this way, as will be discussed in Chapter 4. On the other hand, data that are basically digital in nature, such as that from a computer, could be (and usually are) sent over nondigital telecommunication channels, by appropriate conversion.

Signals needing a large bandwidth need a large number of bits per second to carry them. However the number of bits per second needed is not necessarily proportional to the bandwidth needed for different signal types, as we will explain later.

MAJOR TRENDS

Figure 3.1 is a diagram of the transmission process showing the major trends that are changing the way channels are used.

Related to the increasing channel capacity, the cost of long-distance bandwidth is dropping substantially.

Almost all signals used to be sent in an analog fashion, i.e., as a continuous range of frequencies. Now it has become economical to use digital transmission on many links, and the mechanisms for encoding the information to be transmitted employ computer-like processes. Because the cost of digital logical circuits is dropping rapidly, the complexity of the encoding mechanisms is likely to increase. There is a trade-off between the complexity of encoding and the channel bandwidth, which we discuss later in the book.

† Today the term "cycles per second" is no longer used.
1 hertz (Hz) means the same as 1 cycle per second.
1 kilohertz (kHz) is one thousand cycles per second (kilocycles)
1 megahertz (MHz) is one million cycles per second (megacycles)
1 gigahertz (GHz) is one billion cycles per second.

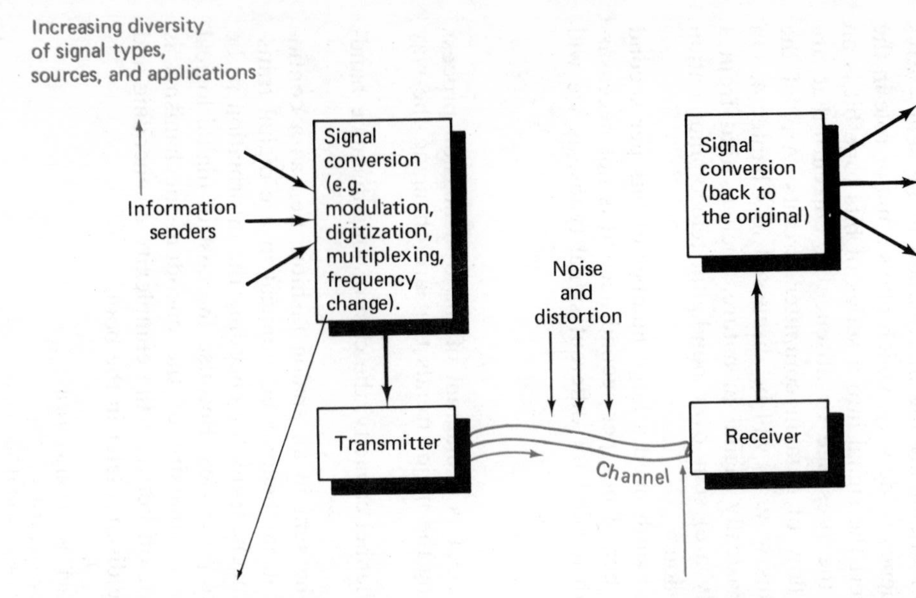

- Higher levels of multiplexing.
- Increasing complexity of control mechanisms permit different signals to be transmitted together.
- Conversion to digital rather than analog form increasingly used.
- More complex encoding, especially for signal digitization.
- Greater degrees of signal compression and data compaction so that signals can be sent on a given channel.

- Increasing channel capacities.
- Increasing transmission frequencies.
- Decreasing cost of long-distance channels.
- Larger numbers of signals occupying the same physical channel.
- Channels increasingly designed for digital rather than analog transmission.
- New types of channels, e.g. satellite.
- Channels with higher levels of noise and distortion used for cost reasons, with more complex signal encoding to protect the signals from this noise and distortion.

Figure 3.1 The communication process and major trends effecting it.

FOUR PARTS TO TELECOMMUNICATION NETWORKS

There are four major parts to telecommunication networks such as the telephone network.

1. *Instruments,* the devices the subscriber uses to originate and receive the signals. The vast majority of instruments are telephone handsets. Today, however, an endless array of other devices are being attached to the telephone lines, including the terminals used for data transmission.
2. *Local loops,* the cables from the subscriber to his local switching office. Telephone and telegraph loops today consist of wire-pair cables. Every subscriber has his own pair of wires to the central office, and nobody else uses these unless he is on a party line. Coaxial cables are now being laid into people's homes for the distribution of television, and these cables have many potential uses other than television.
3. *Switching,* the facilities which permit users to be interconnected on demand. Most switching is done by the world's vast network of telephone exchanges, though today important new types of switching are being developed for computer data and other new forms of telecommunications.
4. *Trunks,* the channels which carry calls between the switching offices, sometimes over long distances. One trunk can usually carry many calls simultaneously.

COST CHANGES

In the Bell System the cost breakdown between these areas is approximately [3]:

Instruments:	23%
Local loops:	15%
Switching:	45%
Trunks:	17%

New technology is changing these costs. The fastest change is in long-distance transmission. The capacity of such systems is increasing greatly without a proportionate increase in cost. Techniques in the laboratory today portend much greater economies of scale in the years ahead. The investment cost of adding a channel mile to the Bell System is dropping rapidly as the channel capacity of the links used increases. Satellites and new terrestrial technology will probably lower the cost of long-distance bandwidth to one tenth of its present cost. On the other hand the traffic volumes are rising and some signals, mainly video signals, require much more bandwidth than telephone signals.

The other costs are not dropping as fast. The largest portion of the telephone cost is that of switching, which contributes 45% to the cost of an average telephone call; the figure is 54% for long-distance calls. Only 28% of the cost of an average call is in the switching equipment; the remainder is in operator salaries [4]. The key to switching economy lies in computers. If used fully, they can reduce the cost in operator salaries. Computers and electronic circuitry, like computer circuitry, are beginning to replace electromechanical telephone exchanges. The cost of computer circuitry is dropping rapidly, and its maintenance costs will be much lower than those of the vast arrays of electromechanical switches. Unfortunately, electromechanical exchanges presently represent billions of dollars of investment and consequently will not be replaced quickly. The Bell System, however, is now installing computerized exchanges at the rate of one per day [5]. Strangely some operator functions remain which other countries have automated, for example listening to a coin box operator deposit coins.

The local loops in the telephone network are unlikely to come down much in cost, although the use of concentrators may drop their cost somewhat. A concentrator enables the signals from a number of subscribers to be sent over one pair of wires. It is possible that with digital transmission and time division multiplexing a quite different configuration of local loops could be built that would be lower in cost. However, the vast investment sums tied up in today's wiring inhibit too drastic a change.

A change that has great potentialities *is* occurring. The local telephone loops can be made to carry a *much greater bandwidth* than is required for the telephone. In addition other channels of enormous capacity are now being wired into homes to carry cable TV. These are not switchable as is a telephone channel; still, new forms of switching could be used in which many subscribers communicate at once over the same coaxial cable, perhaps gaining access to computerized education or music library services.

The cost of the telephone set (23% of the total) is the least likely to drop. Instead, many new forms of terminal, some of which are very expensive at the moment, are coming into use. These forms include Picturephone sets, facsimile machines, typewriterlike data terminals, televisionlike data terminals in which the user may have a keyboard or may point at the screen images with a "light pen," terminals using a keyboard and "voice answerback" response, terminals for checking credit cards or obtaining airline tickets, and so on. The cost of data terminals can be expected to drop as markets increase and logic-circuitry mass-production techniques are perfected. If a true mass market for data terminals develops, it is conceivable that they could drop to costs not much higher than the cost of today's television sets. (See Fig. 3.2). The domestic television set and data terminal may possibly become combined. Extensions of the telephone handset are being marketed for use as data terminals. These are inexpensive, capable of innumerable applications, and may come into very widespread use (Fig. 3.3). It is possible that a high proportion of telephone extensions will eventually have data capability.

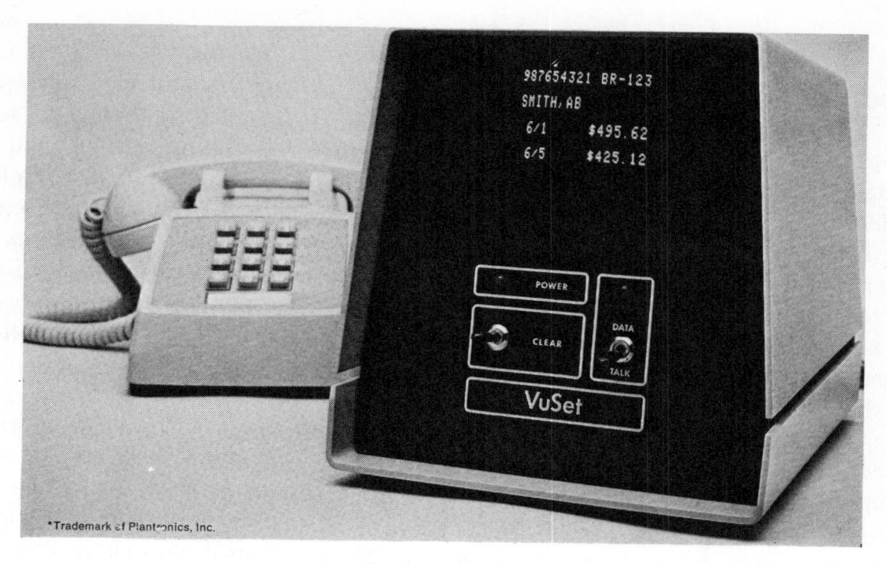

Figure 3.2 AT&T's VuSet©, an inexpensive desktop terminal linked to the Touchtone® telephone. It displays four lines of 16 characters each, using a 64-character set, and is designed to be highly reliable.

Figure 3.3 Data terminals which are an inexpensive extension of the telephone handset could come into very widespread use. It is possible that a high portion of telephones could eventually have data capability. Such devices can be used for innumerable applications, possibly with templates for labeling the keyboard like those illustrated in Figs. 7.5 to 7.10.

This figure shows AT&T's Transaction II telephone. It has an 8-character visual display and reads magnetic-stripe cards conforming to ABA standards.

MULTIPLEXING

A channel of a given capacity can be split up into a number of lesser channels. A given bandwidth or range of frequencies can be split up into smaller bandwidths by a technique called frequency-division multiplexing [1]. Similarly a bit stream can be subdivided to carry lesser bit streams with their bits interleaved—for example, every tenth bit in a high-speed bit stream may be a bit from one subchannel. The bit after that is from a second subchannel and so on, until ten subchannels are derived. The process is called time-division multiplexing [1]. The topic is covered in more detail later in the book. We shall frequently refer to the subdivision process, saying, for example, that one pair of wires carries 12 voice channels or that 100 music channels are sent over one television channel.

When frequency-division multiplexing is used, some space is needed between the channels to avoid interference; this is called the guard band. For telephone traffic, for example, the world's transmission media are divided into slices of 4 kHz; however, the usable bandwidth is closer to 3 kHz. If ten 20-kHz hi-fi channels are packed into a bandwidth of 200 kHz, only about 18 kHz of each channel will remain (more than enough for most human ears). The remaining of 2 kHz between each channel is required to separate the channels without causing appreciable distortion.

Time-division multiplexing also is less than 100% efficient, and requires some housekeeping bits among the information bits. Some sawdust is inevitable when a tree is cut into logs.

SHARING WITHOUT MULTIPLEXING

Some channels are shared but without multiplexing—that is, only one signal is sent at a time. Such is the case with a telephone *party line*. Several subscribers share the line but only one can use it at one instant. The same is true with many schemes for data transmission. A line may serve many terminals but only one can transmit at once.

SWITCHING

Another way to interconnect many users is by means of switching. The user has a line to a telephone exchange and there his line can be connected to that of a person he wishes to talk to. The switching may take place in the network facilities, for example in a telephone exchange, or it may take place on the subscriber's premises enabling many to share a small number of access lines.

POINT-TO-POINT OR MULTIPOINT?

When a telecommunication path is set up it may connect two points only. Sometimes it connects several points as with a conference line, or a line connecting several terminal locations to a computer. Sometimes it connects one location to many, as in broadcasting or cable television. In the latter examples the path is a one-way path rather than the two-way path of telephone transmission. One-to-many transmission (i.e., broadcasting), is common. Many-to-one transmission is less frequently found but is sometimes used for data collection.

Two-way transmission can be *duplex* (full duplex) or *half-duplex*. Duplex transmission means that signals can go into both directions at the same time. Half-duplex transmission means that signals can go in either direction but only in one direction at once. A local telephone call is full duplex. Both parties can talk and be heard at the same time. On the other hand it is not absolutely necessary for a telephone connection to be full duplex. A perfectly satisfactory conversation can be had over a half duplex connection—it does not add much to the conversation when both parties talk at once.

A Picturephone line, unlike a telephone line, needs to be full duplex. You want to watch the other person's face both when you are talking and when he is talking.

Half duplex transmission is satisfactory for most transmissions between computers and terminals. However, many lines are constructed to be fully two-way channels and the data transmission machines may as well take advantage of this fact. In the United States use of a full duplex line often costs 10% more than use of a half duplex line. It is therefore more efficient to organize the transmission so that it goes in both directions at once.

DYNAMIC CHANNEL ALLOCATION

Where a telephone line is long and expensive, as with a trans-Atlantic line, channels may be allocated dynamically. The link is subdivided into many channels in both directions. When a user talks, a circuit detects that he is talking and allocates a channel to him. When he stops talking the channel is made available for other users. When the other party replies he is given a channel in the opposite direction. The channel allocations are done at electronic speed within a tiny fraction of a second of the start of the speech. The technique is referred to as TASI, Time Assigned Speech Interpolation, and can increase the effective capacity of lines.

When the speech travels in a digitized form, digital channel capacity can be allocated to it as required, by computer-like circuits. When a speaker says a word, he is given channel space, and when he becomes silent the channel space is taken away from him. With digital circuitry this process can occur at high speed. Channel capacity is allocated dynamically to speech just as memory blocks are allocated dynamically to users in a computer. Most telephone speakers spend about 55% of their time pausing or listening. Dynamic channel space allocation done very rapidly can more than double the capacity of a high-capacity channel. The front of some words of speech will be removed as the channel allocation cannot occur instantaneously. Usually only a few milliseconds will be removed, but occasionally many milliseconds will be removed because, by chance, most persons are speaking in one direction and there is no free channel in that direction until one of them pauses briefly. If about 50 milliseconds are chopped off the front of every word spoken, most speech is still quite intelligible and recognizable.

This technique is called *digital speech interpolation,* DSI.

CONTINUOUS OR BURST TRAFFIC

Some telecommunications users need a continuous unbroken signal. Television and music are unbroken signals. So is telephone speech in conventional transmission, although it can be fragmented with techniques like digital speech interpolation.

Other users transmit bursts. Telegraph messages or data messages require bursts of transmission. When a person communicates with a distant computer, bursts of data go between the two with substantial pauses between them. Much machine communication is in burst form rather than continuous form. Switching and multiplexing techniques designed for burst transmissions can be different from those for continuous transmissions, the bursts being interleaved like traffic on a highway system. As we will see later, it can be very inefficient to send burst traffic over facilities designed for continuous traffic, and yet this has been commonly done because the ubiquitous telephone facilities use continuous-channel multiplexing and switching. New types of facilities are developing to handle burst traffic.

REAL-TIME OR NONREAL-TIME?

It is important to distinguish between real-time and nonreal-time usage of telecommunications channels.

Real-time usage of a channel means that the receiving party or machine receives the signal at approximately the same time that it originates. In practice there will always be a small transmission delay. This delay should be sufficiently small that the channel is used effectively as though there were zero delay.

Thus telephone transmission is real-time. The receiving party reacts to the signal as though there were no delay in its transmission. Telegram transmission is nonreal-time. The receiving party sees the telegram hours after it was sent. Live television of a ball-game is real time; a broadcast of a program recorded the previous night is nonreal-time. Batch data transmission for computers is nonreal-time; transmission permitting interactive use of a computer terminal (as with an airline reservation system) is real-time. When a terminal operator uses a distant computer, there could be a transmission delay of several seconds and we would still refer to it as "real-time" because the operator uses the terminal in effectively the same way as he would if there were no delay.

Box 3.1 summarizes these categories of usage of telecommunication links.

SLOW SCAN

When real-time transmission is needed the bandwidth of the channel used must be at least as great as the bandwidth of the signal. When the transmission is nonreal-time the channel bandwidth could be smaller, and the signal must be stored until it is used.

Real-time television transmission in America, for example, needs a bandwidth of 4.6 MHz. Prior to satellites the television links across the Atlantic or Pacific—the suboceanic cables—did not have channels of that bandwidth. Television was therefore slowed down and transmitted nonreal-time over smaller-bandwidth channels. It was stored on videotape machines and then broadcast. This technique is called slow-scan transmission.

Hi-fi music could be transmitted over ordinary telephone lines in a similar manner. It would be slowed down, for example, by playing a 15 inches-per-second tape at $1\frac{7}{8}$ inches per second, and raised in frequency by about 400 Hz to fit into the central part of the telephone bandwidth.

Any nonreal-time signal can be slowed down or speeded up to fill the channel it is transmitted over.

GRADE OF SERVICE

When switching is used as on the public telephone network, or when many users *share* a smaller number of channels, there is a certain probability that a user will not be able to obtain a connection when he needs one. All channels are busy. When this happens on the telephone network he receives a "network busy" signal (an "engaged" signal in British parlance), which he may distinguish from a called-party busy signal because it is faster.

The probability of receiving a network busy signal is referred to as the "grade of service" of the network. A telephone system may be engineered to give a grade of service of 0.002, i.e., 99.8% of all calls made will be connected to the called telephone (which may itself be busy). Some congested areas have a grade of service which is not this good. Corporate tie-line networks are usually engineered more frugally with a grade of service between 0.01 and 0.05.

A good grade of service can only be achieved with real-time signals such as telephone if the average utilization of the channels is substantially less than 100%. For a given number of channels and a given average utilization the grade of service can be calculated [2]. Figure 3.4 plots the channel utilizations of a group of channels which give grades of service of 0.05, 0.01, and 0.001. The acceptable channel utilization increases as the number of channels increase. For 10 channels with enough traffic to utilize 5 of them on average, the grade of service is 0.018. To achieve a grade of service of 0.001 with that traffic, fourteen channels are needed. Similarly with real-time data transmission a mean channel utilization substantially lower than 100% is used in order to achieve an acceptably fast response time. The situation is more complex on a mesh-structured network with alternate routing.

BOX 3.1 Categories of communication links.

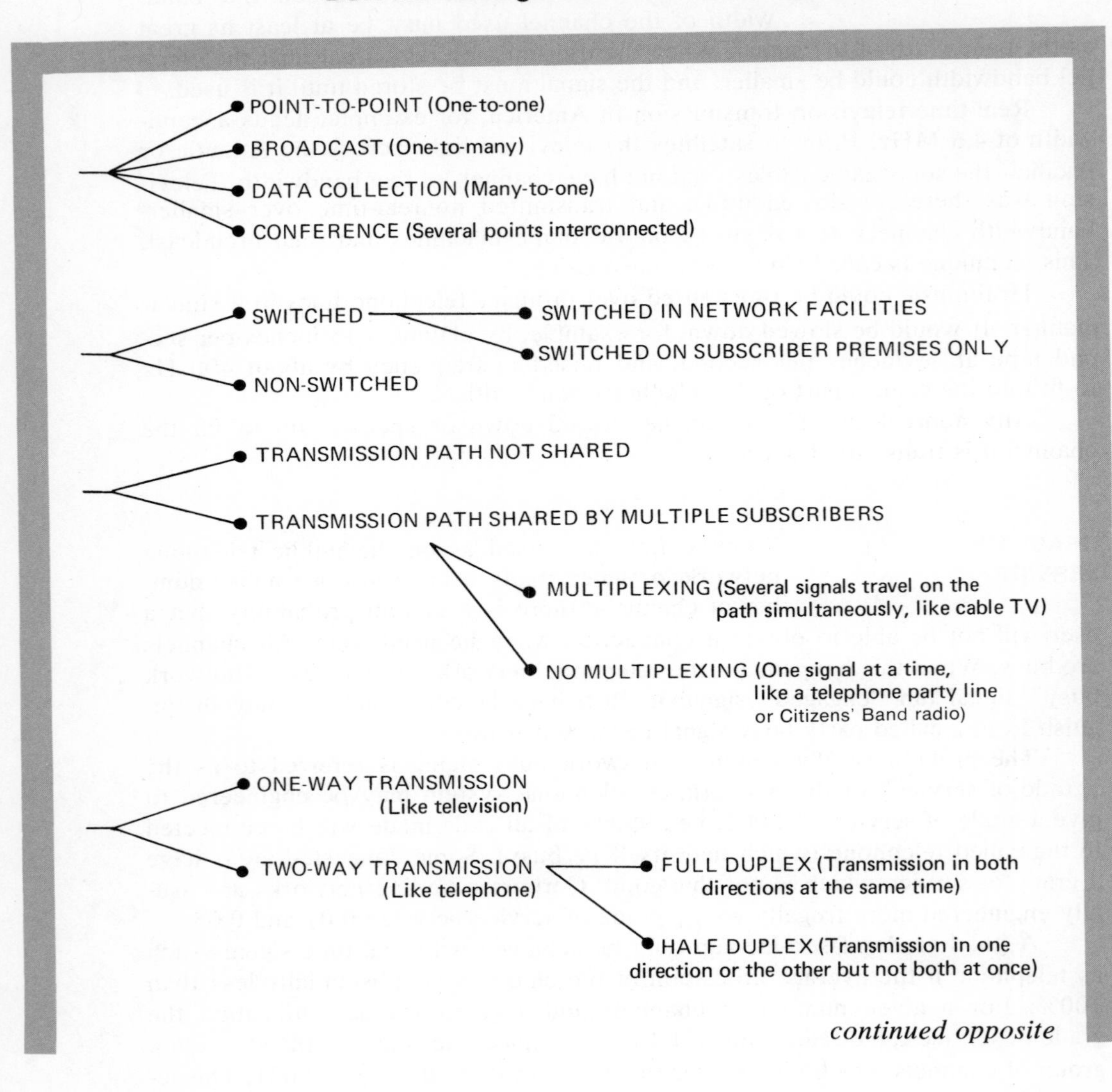
POINT-TO-POINT (One-to-one)
BROADCAST (One-to-many)
DATA COLLECTION (Many-to-one)
CONFERENCE (Several points interconnected)
SWITCHED
SWITCHED IN NETWORK FACILITIES
SWITCHED ON SUBSCRIBER PREMISES ONLY
NON-SWITCHED
TRANSMISSION PATH NOT SHARED
TRANSMISSION PATH SHARED BY MULTIPLE SUBSCRIBERS
MULTIPLEXING (Several signals travel on the path simultaneously, like cable TV)
NO MULTIPLEXING (One signal at a time, like a telephone party line or Citizens' Band radio)
ONE-WAY TRANSMISSION (Like television)
TWO-WAY TRANSMISSION (Like telephone)
FULL DUPLEX (Transmission in both directions at the same time)
HALF DUPLEX (Transmission in one direction or the other but not both at once)

continued opposite

BOX 3.1 *Continued*

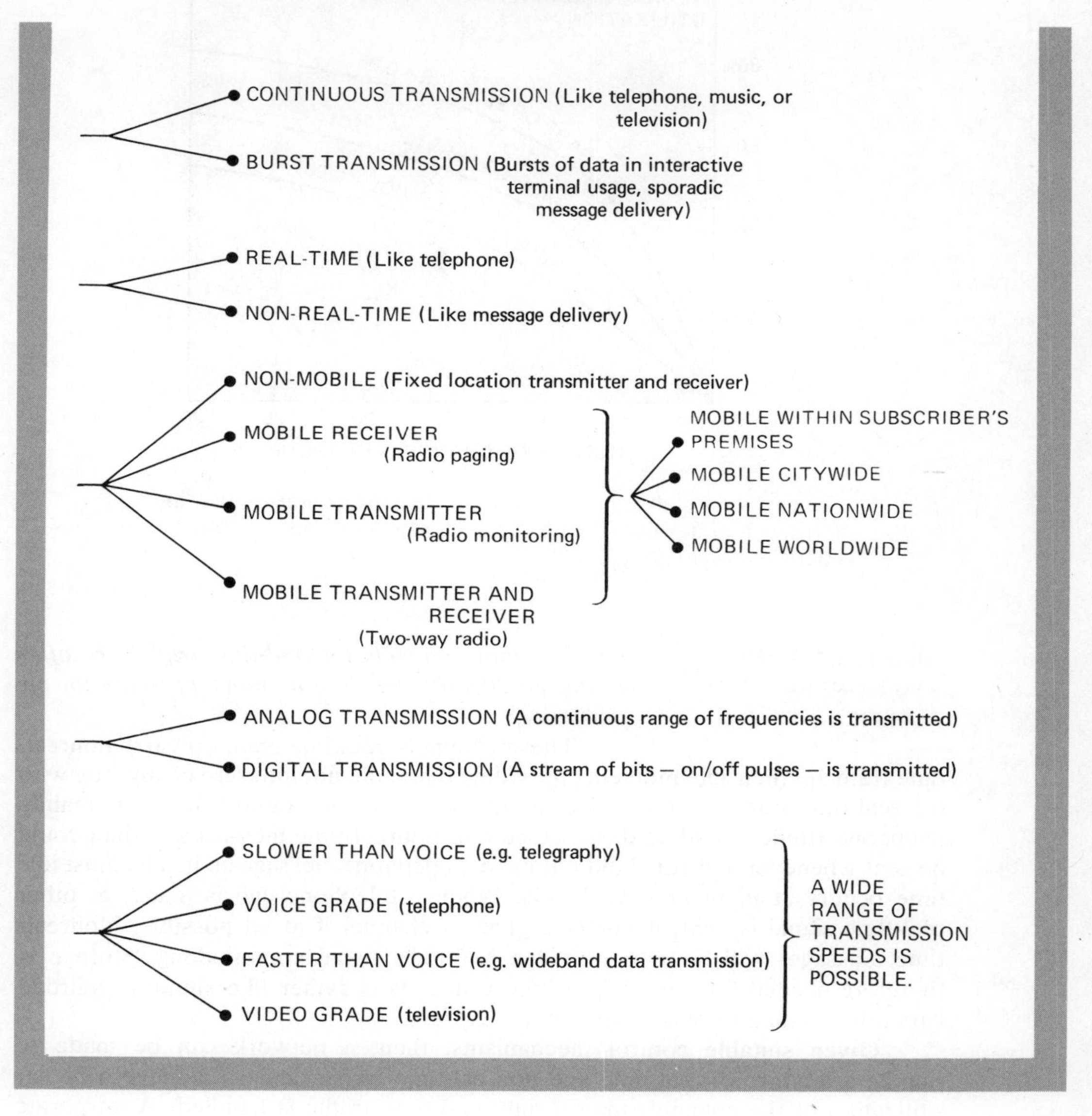

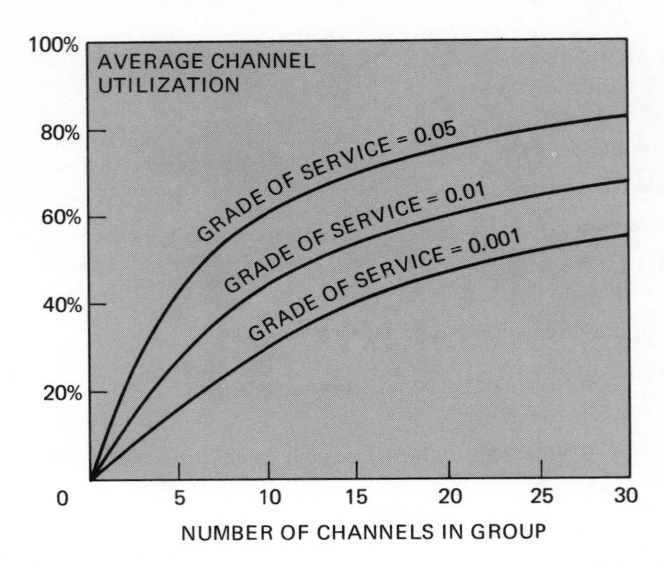

Figure 3.4 To give an acceptable grade of service with real-time signals (such as telephone), the average channel utilization must be substantially lower than 100%

MIXED REAL-TIME AND NONREAL-TIME TRAFFIC

The fact that channels for real-time traffic are often unutilized 30% of the time or more presents an opportunity.

These channels could be made to carry nonreal-time traffic in their idle moments providing that it did not interfere in any way with the real-time traffic. A corporate tie-line network, for example, handling mainly telephone traffic, could be designed so that nonreal-time messages or data could be sent whenever a channel did not have a telephone message on it. The most idle time occurs at night and weekends. When a telephone call is made, or other real-time signal is sent, it must be given a channel if at all possible. Nonreal-time messages will have to wait if a real-time signal comes along. Storage is therefore needed for the nonreal-time traffic. It is rather like shunting railroad cars into a siding to make way for the express.

Given suitable control mechanisms, then, a network can be made to handle a mixture of real-time and nonreal-time traffic, and so achieve a higher utilization of the channels than if only real-time traffic is handled. A corporate communication network has much idle channel time, especially at night, and nonreal-time messages such as cables, letters, and computer data could be sent without needing more channels. As logic and storage costs drop there will be various examples of nonreal-time traffic "piggybacking" on the gaps in real-time traffic.

Box A at the front of the book indicates which of the signals it lists are real-time and which are nonreal-time. Nonreal-time traffic includes:

Telegraph messages.

Facsimile transmission of documents.

Electronic mail.

Mailgram (a Western Union service in which telegraph messages are transmitted to a post office for hand delivery).

Electronic fund (cash or check) transfer.

Voicegram—delivery of spoken one-way telephone messages.

Electronic transmission of news photographs.

Batch data transmission.

Data collection signals.

Computer data entry.

Computer message generation.

Video presentations for use in corporate training.

MOBILE DEVICES?

When the transmitter or receiver is mobile it must be connected to the transmission network by radio. New uses of radio are growing including computer terminals in vehicles, the paging of individuals via small receivers they carry on their belt, and portable telephones. The spread of mobile radio applications depends, as we will discuss later, on improved regulation of how the radio spectrum is used.

EMERGENCIES

Modern telecommunications offer superb potential for automatically reporting emergencies. Fire warnings or vital machine breakdowns can be signalled instantly to monitoring computers that sound alarms and display required details at fire and breakdown stations. The crime rate is rising at such an alarming rate in the United States (and in other countries with an extensive use of modern communications media) that strong action is becoming increasingly vital. Eventually we may find burglar alarms in most homes, coupled to the telephone (or other) cables. The triggering of such alarms could be detected by computers, which again would display relevant information in police stations and would perhaps automatically contact squad cars. If crime in the streets of cities like New York continues its exponential growth, one can imagine pedestrians carrying radio alarms that will enable a computer to dispatch the nearest prowl car to a victim in the street.

Emergency communications or other high-priority traffic may have the capability to preempt normal communications. A designated authority can override normal network controls with an appropriate signal, forcing his way through occupied parts of the network that would otherwise have given him a "busy" signal. He can preempt all or part of the network for emergency purposes. On some future systems several levels of priority will be used.

ARCHITECTURE Given the diverse new applications and structures of telecommunications, given mass-producible logic circuitry, given new technologies as dramatically different as communication satellites, it behooves us to ask: how can all of these be put together to provide the facilities that would be most valuable for a nation, or a city, or a large corporation, or a small planet? The answers to that question suggest that the optimal telecommunication architectures for the future are very different from those of the past.

REFERENCES

1. James Martin, *Telecommunications and the Computer, Second Edition* (Englewood Cliffs, N.J.: Prentice-Hall, 1976), Chap. 12.
2. James Martin, *Systems Analysis for Data Transmission* (Englewood Cliffs, N.J.: Prentice-Hall, 1970), Fig. 3.2 was plotted using Table 11 in the Appendix of that book.
3. Eugene V. Rostow, "The President's Task Force on Communication Policy," Staff Paper No. 1, Washington, D.C., 1969.
4. *Ibid.,* p. 44.
5. Bell System Tech. J., October 1969. A complete issue on ESS No. 2.

Figure 3.5 Many changes in telecommunications devices will result from the use of microminiature computers. This photograph shows Intel's 8748 single-chip microcomputer. Twenty-seven input/output channels (top right) and 8000 bits of programmable read-only memory (right) are on the same chip as the microprocessor.

The receiving device from which Mr. Watson first heard, to his great surprise, Alexander Graham Bell's voice. Bell was not trying to invent the telephone at the time. *(Courtesy AT&T.)*

4 ANALOG VERSUS DIGITAL TRANSMISSION

Basically, there are two ways in which information of any type can be transmitted over telecommunication media: analog or digital.

Analog means that a continuous range of frequencies is transmitted. The sound you hear and the light you see consist of such a continuous range. Sound, as any hi-fi enthusiast knows, consists of a spread of frequencies from about 30 to 15,000 Hz or, for people with very good ears, 20,000 Hz. It cannot be heard below 30, and it cannot be heard above 20,000. If we wanted to transmit high-fidelity music along the telephone wires into your home (which is technically possible), we would send a continuous range of frequencies from 30 to 20,000. The current on the wire would vary *continuously* in the same way as the sound you hear.

The telephone companies, conscious of costs, transmit a range of frequencies that may vary from about 300 to 3000 Hz only. This is enough to make a person's voice recognizable and intelligible. When telephone signals travel over lengthy channels, they are packed together, or *multiplexed,* so that one channel can carry as many such signals as possible. To do this your voice might have been raised in frequency from 300–3000, to 60,300–63,000 Hz. Your neighbor's voice might have been raised 64,300 to 67,000. In this way, they can travel together without interfering with one another; but both are still transmitted in an *analog* form—that is, as a continuous signal in a continuous range of frequencies.

Digital transmission means that a stream of on/off pulses are sent, like the way in which data travel in computer circuits. The pulses are referred to as *bits.* It is possible today to transmit at an extremely high bit rate.

Figure 4.1 shows an analog signal and a digital signal. A transmission path can be designed to carry either one or the other. As we shall see in the chapters ahead, this fact applies to all types of transmission paths—wire pairs,

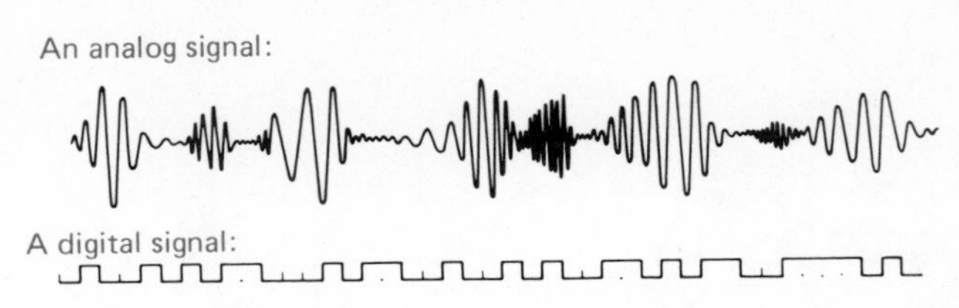

Figure 4.1 Any information can be transmitted in either an analog or a digital form.

high-capacity coaxial cables, microwave radio links, satellites, and the new transmission media, such as waveguides and lasers.†

In order to follow the arguments in the chapters ahead, it is important to understand that *any type of information can be transmitted in either an analog or a digital form.*

The telephone channel reaching our home today is an analog channel, capable of transmitting a certain range of frequencies. If we send computer data over it, we have to convert that digital bit stream into an analog signal using a special device known as a *modem*. This converts the data into a continuous range of frequencies—the range of the telephone voice. In this way, we can use any of the world's telephone channels for sending digital data.

On the other hand, where digital channels have been constructed, it is possible to transmit the human voice over them by converting it into a digital form. Similarly, *any* analog signal can be digitized for transmission in this manner. We can convert hi-fi music, television pictures, temperature readings, the output of a copying machine, or any other analog signal into a bit stream. High-fidelity music would need a larger number of bits per second than telephone sound. Television would need a much higher bit rate than sound transmission. The bit rate needed is dependent on the bandwidth, or range of frequencies, of the analog signal, as well as the number of different amplitude levels we want to be able to reproduce.

Almost all the world's telephone plant today grew up using analog transmission. Much of it will remain so for years to come because of the multi-billions of dollars tied up in such equipment. However, digital technology is rapidly evolving, and major advantages in digital transmission are beginning to emerge. It is probable that if the telecommunication companies were to start afresh in building the world's telecommunication channels, they would almost entirely be digital with the possible exception of the local "loops" between a subscriber and his nearest switching office. A new and different form of plant would be installed using a technique called *pulse code modulation,* in which the voice and other analog signals would be converted into a stream of bits looking remark-

† These different types of transmission media are described in the author's *Telecommunications and the Computer,* 2nd Edition, Chapter 9.

ably like computer data. Some developing countries with a less-massive investment in old equipment are installing pulse code-modulated systems, and lines of this type are already in extensive use in the United States, Canada and Japan.

If digital streams form the basis of our communication links, then computer data will no longer need to be converted into an analog form for transmission, as it is today. However, analog information, such as the sound of the human voice, needs to be coded in some way so that it can be transmitted in the form of pulses and then decoded at the other end to reconstitute the voice sounds. This is already done on many short-haul telephone trunks. The circuits of the future will be designed to transmit very high speed pulse trains into which voice, television, facsimile, and data will all be coded and sent in a uniform manner. Instead of manipulating data so that they can be squeezed into channels designed for voice, the voice will be coded so that it can be sent over channels that are basically digital.

ECONOMIC FACTORS

The economic circumstances favoring digital transmission stem from two main factors. First, it is becoming possible to build channels of high bandwidth—that is, high information-carrying capacity. Indeed, it is now appreciated that many existing wire-pair channels, which represent an enormous financial investment, could be made to carry much more traffic than they are currently doing. However, a high level of multiplexing is needed to make use of high-capacity channels. In other words, many different transmissions must be packed together to travel over the same channel. When such packing occurs, the circuit-mile cost for one signal, such as a telephone transmission, drops greatly. But the cost of packing and unpacking them, plus switching them, remains high.

When analog signals are multiplexed together, each must occupy a different range of frequencies within the overall range that is transmitted. The frequency range (bandwidth) available is divided up and allocated to the separate signals. This process, known as *frequency-division multiplexing,* uses fairly expensive circuit components such as filters. When thousands of telephone conversations travel together over coaxial cable or microwave links, they must be demultiplexed, switched, and then multiplexed together again at each switching point. While there is great economy of scale in the transmission, there is not in this multiplexing and switching operation. As the channel capacities increase, so the multiplexing and switching costs assume a greater and greater proportion of the total network cost.

Digital circuitry, on the other hand, is dropping in cost at a high rate. With the maturing of large-scale integration techniques, it will drop even more. Where digital rather than analog transmission is used, this increasingly low-cost circuitry will handle the multiplexing and switching. The telecommunication networks will become in some aspects like a vast digital computer.

There is one other major advantage in using digital techniques for transmission. In analog transmission, whenever the signal is amplified, the noise and distortion is amplified with it. As the signal passes through its many amplifying stations, so the noise is amplified and thus is cumulative. With digital transmission, however, each repeater station regenerates the pulses. New, clean pulses are reconstructed and sent on to the next repeater, where another cleaning-up process takes place. Therefore the pulse train can travel through a dispersive noisy medium, but instead of becoming more and more distorted until eventually parts are unrecognizable, it is repeatedly reconstructed, and thus remains impervious to most of the corrosion of the medium. Of course, an exceptionally large noise impulse may destroy one or more pulses so that they cannot be reconstructed by the repeater stations.

A distinctive characteristic of digital transmission is that a much greater bandwidth is required. In order to send a given quantity of telephone conversations, for example, we would need a much higher bandwidth than with the analog systems in use today. However, because the signal is regenerated frequently, the pulse code modulation signal can operate with a lower signal-to-noise ratio. Thus there is a trade-off between bandwidth and signal-to-noise ratio in the transmission of a given quantity of information. If a given pair of wires is used, for example, a wider range of frequencies can be employed for transmission because of the frequent regeneration of the signal, and because only two states of a binary signal need to be detected—not a continuous range of amplitudes as in an analog signal.

An additional economic factor is the rapidly increasing use of data transmission. Although data transmission still employs only a small proportion of the total bandwidth in use, it is increasing much more rapidly than other uses of the telecommunication networks. Data are transmitted much more economically over a digital circuit than an analog circuit. With the present facilities almost ten times as much data can be sent over a digital voice line as over an analog voice line with typical modems.

Thus several factors are swinging the economics in favor of digital transmission.

1. The trend to much higher bandwidth facilities.
2. The decreasing cost of logic circuitry, which is used in coding and decoding the digital signals and in multiplexing and switching them.
3. The increase in capacity that results from the use of digital repeaters at frequent intervals on a line.
4. Improvements in codec design enabling speech to be encoded into a smaller number of bits.
5. The rapidly increasing need to transmit digital data on the networks.

In terms of the immediate economics of today's common carriers, pressed for capacity, digital transmission is appealing for short-distance links because with relatively low-cost electronics it can substantially increase the capacity of existing wire pairs. This is particularly important in the congested city streets.

An important long-term advantage is the fact that all signals—voice, television, facsimile, and data—become a stream of similar-looking pulses. Consequently, they will not interfere with one another and will not make differing demands on the engineering of the channels. In an analog signal format, television and data are much more demanding in the fidelity of transmission than speech and create more interference when transmitted with other signals. Eventually, perhaps, there will be an integrated network in which all signals travel together digitally.

We will say more about the detailed economics in the chapters which follow.

PULSE AMPLITUDE MODULATION

In order to convert an analog signal such as speech into a pulse train, a circuit must sample it at periodic intervals. The simplest form of sampling produces pulses, the amplitude of which is proportional to the amplitude of the signal at the sampling instant (see Fig. 4.2). This process is called *pulse amplitude modulation* or PAM.

The pulses produced still carry their information in an analog form; the amplitude of the pulse is continuously variable. If the pulse train is transmitted over a long distance and subjected to distortion, it may not be possible to reconstruct the original pulses. To avoid this we employ a second process, which converts the PAM pulses into unique sets of equal amplitude pulses so that we need only detect the presence or absence of a pulse (bit), not its size. As we shall see in Chapter 30, the PAM pulses themselves are used in certain switching equipment in which the switching is done by electronically controlling the flow of PAM pulses.

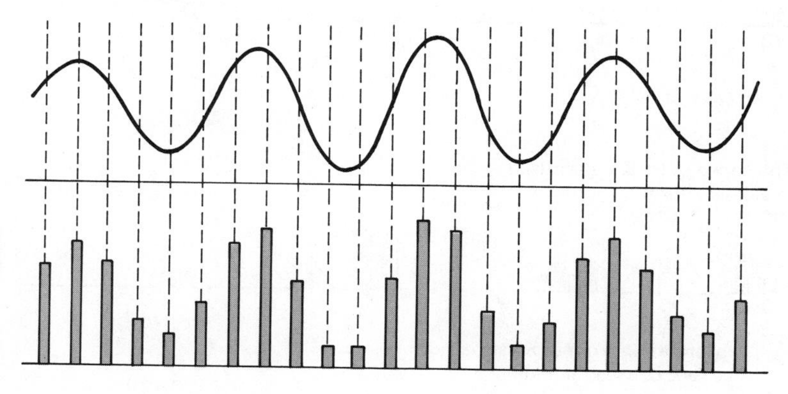

Figure 4.2 Pulse amplitude modulation (PAM).

PULSE CODE MODULATION

The amplitude of the PAM pulse can assume an infinite number of possible values ranging from zero to its maximum.

It is normal with pulse modulation to transmit not an infinitely finely divided range of values but a limited set of specific discrete values. The input signal is *quantized*. This process is illustrated schematically in Fig. 4.3. Here the signal amplitude can be represented by any one of the eight values shown. The amplitude of the pulses will therefore be one of these eight values. An inaccuracy is introduced in the reproduction of the signal by doing this, analo-

1 The signal is first "quantized" or made to occupy a discrete set of values

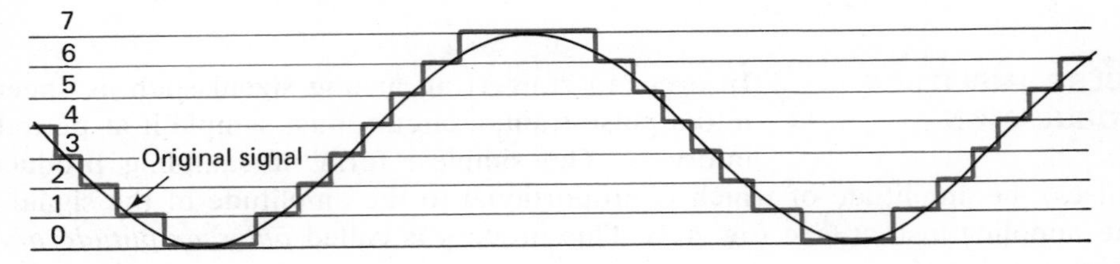

2 It is then sampled at specific points. The PAM signal that results can be coded for pulse code transmission

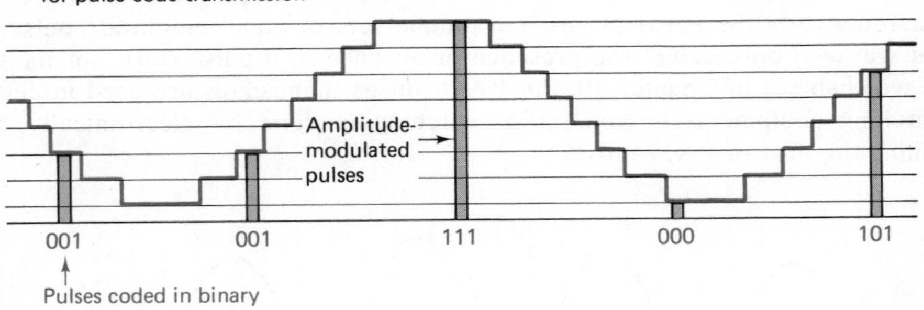

3 The coded pulse is transmitted in a binary form

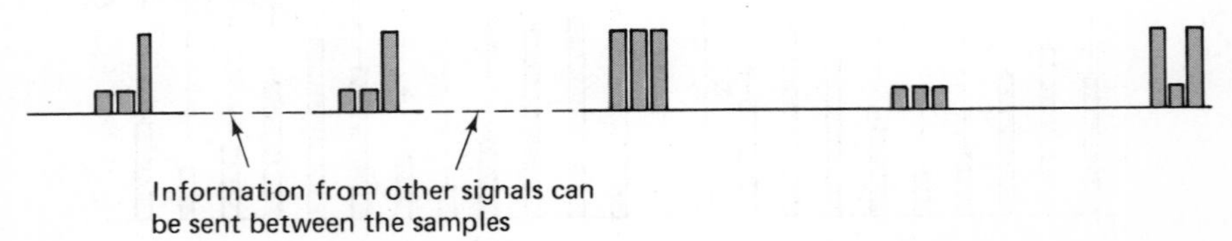

Figure 4.3 Pulse code modulation (PCM).

gous to the error introduced by rounding a value in a computation. Figure 4.3 shows only eight possible values of the pulse amplitude. If there were more values, the "rounding error" would be less. In systems in actual use today, 128 pulse amplitudes are used, or 127 to be exact for the zero amplitude is not transmitted.

After a signal has been quantized and samples taken at specific points, as in Fig. 4.3, the result can be coded. If the pulses in the figure are coded in binary, as shown, three bits are needed to represent the eight possible amplitudes of each sample. A more accurate sampling with 128 quantized levels would need seven bits to represent each sample. In general, if there were N quantized levels, $\log_2 N$ bits would be needed per sample.

The process producing the binary pulse train is referred to as *pulse code modulation*. The resulting train of pulses passes through frequent repeater stations that reconstruct the pulse train, and is impervious to most types of telecommunications distortion other than major noise impulses or dropouts. The mere presence or absence of a pulse can be recognized easily even when distortion is present, whereas determination of pulse magnitude would be more prone to error.

On the other hand, the original voice signal can never be reproduced exactly, because of the quantizing errors. This deviation from the original signal is sometimes referred to as *quantizing noise*. It is of known magnitude and can be reduced, at the expense of bandwidth, by increasing the number of sampling levels; 128 levels, needing seven bits per sample, are enough to produce telephone channels having a signal-to-noise ratio comparable to that achieved on today's analog channels.

HOW MANY SAMPLES ARE NEEDED?

The pulses illustrated in Figs. 4.2 and 4.3 are sampling the input at a limited number of points in time. The question therefore arises: How often do we need to sample the signal in order to be able to reconstruct it satisfactorily from the samples? The less frequently we can sample it, the lower the number of pulses we have to transmit in order to send the information, or, conversely, the more information we can transmit over a given bandwidth.

Any signal can be considered as being a collection of different frequencies, but the bandwidth limitations on it impose an upper limit to these frequencies. When listening to a violin, you hear several frequencies at the same time, the higher ones being referred to as "harmonics." You hear no frequencies higher than 20,000 Hz, however, because that is the upper limit of the human ear. (The ear has a limited bandwidth like any other channel.) When listening to a full orchestra, you are still hearing a collection of sounds of different frequencies, although now the pattern is much more complex. Similarly, other signals that we transmit are composed of a jumble of frequencies. A digital signal can be analyzed by Fourier analysis into its component frequencies.

It can be shown mathematically that *if the signal is limited so that the highest frequency it contains is W hertz, then a pulse train of 2W pulses per second is sufficient to carry it and allow it to be completely reconstructed.*

The human voice, therefore, if limited to frequencies below 4000 Hz, can be carried by a pulse train of 8000 PAM pulses per second. The original voice sounds, below 4000 Hz, can then be *completely* reconstructed.

Similarly, 40,000 samples per second could carry hi-fi music and allow complete reproduction. (If the samples themselves were digitized, as with PCM, the reproduction would not be quite perfect because of the quantizing error.)

Table 4.1 shows the bandwidth needed for four types of signals for human perception, plus the digital bit rate used or planned for their transmission with PCM.

In telephone transmission, the frequency range encoded in PCM is somewhat less than 200 to 3500 Hz. 8000 samples per second are used. Each sample is digitized using seven bits so that $2^7 = 128$ different volume levels can be distinguished. This gives $7 \times 8000 = 56{,}000$ bits per second. High-fidelity music with five times this frequency range and a need for finer quantizing needs more bits per second.

In videophone encoding, a smaller number of bits are used to code each sample. It is not necessary to distinguish as many separate levels of brightness. The ratio between the bit rate and the bandwidth used is therefore smaller. On high-fidelity services such as network color TV, however, a larger number of bits are employed to minimize quantizing noise.

Table 4.1 Bandwidths and equivalent PCM bit rate for typical signals

Type of Signal	Analog Bandwidth Used (kilohertz)	Number of Bits per Sample	Digital Bit Rate Used or Needed (thousand bits/second)
Telephone voice	4	7	$7 = 56$
High-fidelity music	20	10	$20 \times 2 \times 10 = 400$
Picturephone	1000	3	$1000 \times 2 \times 3 = 6000$
Color television	4600	10	$4600 \times 2 \times 10 = 92{,}000$

MULTIPLEXING

As noted, $4000 \times 2 \times 7 = 56{,}000$ bits per second are needed to carry a telephone voice. However, all the transmission facilities that this bit stream is likely to be sent over can carry a much higher bit rate than this. A thin pair of wires, such as those laid under city streets for ordinary telephone distribution, can be made to carry two million bits per second or more by using digital repeaters sufficiently closely spaced in manholes. Coaxial cables and microwave radio carry much more than this.

It is therefore worthwhile to send more than one telephone conversation over one pair of wires. This is done by interleaving the "samples" that are transmitted. If four voice signals are to be carried over one pair of wires, the samples are intermixed as follows:

Sample from speech channel 1
Sample from speech channel 2
Sample from speech channel 3
Sample from speech channel 4
Sample from speech channel 1
Sample from speech channel 2
Sample from speech channel 3
and so on

This is illustrated in Fig. 4.4. By sampling the signals at the appropriate instants in time, a train of PAM pulses is obtained; these pulses are then digitally encoded. For simplicity, only a four-bit code is shown in the diagram. Each PAM pulse is encoded as four bits. The result is a series of "frames," each of 16 bits. Each frame contains one sample of each signal.

In order to decode the signal, it is necessary to be sure where each "frame" begins. The signals can be reconstructed with this knowledge. The first four bits relate to speech channel 1, the second four to channel 2, and so on. A synchronization pattern must also be sent in order to know where each frame begins. This, in practice, can be done by the addition of one bit per frame. The added bits, when examined alone, form a unique bit pattern that must be recognized to establish the framing.

This process is called *time-division multiplexing*. It takes place at electronic speeds in computerlike logic circuits. The circuit components for digital multiplexing of this type are much lower in cost than those for frequency-division multiplexing. In the latter process, the range of frequencies available for transmission is divided up into smaller ranges, each of which carries one signal.

THE BELL SYSTEM T1 AND T2 CARRIERS

The most widely used transmission system at present with time-division multiplexing is the *Bell System T1 carrier*. This carrier uses wire pairs with digital repeaters spaced 6000 ft apart to carry approximately 1.5 million bits per second. Into this bit stream 24 speech channels are encoded, using pulse code modulation and time-division multiplexing. Eight thousand frames per second travel down the line, and each frame contains 24 samples of eight bits. Seven bits are the encoded sample; the eighth forms a bit stream for each speech channel, which contains network signaling and routing information.

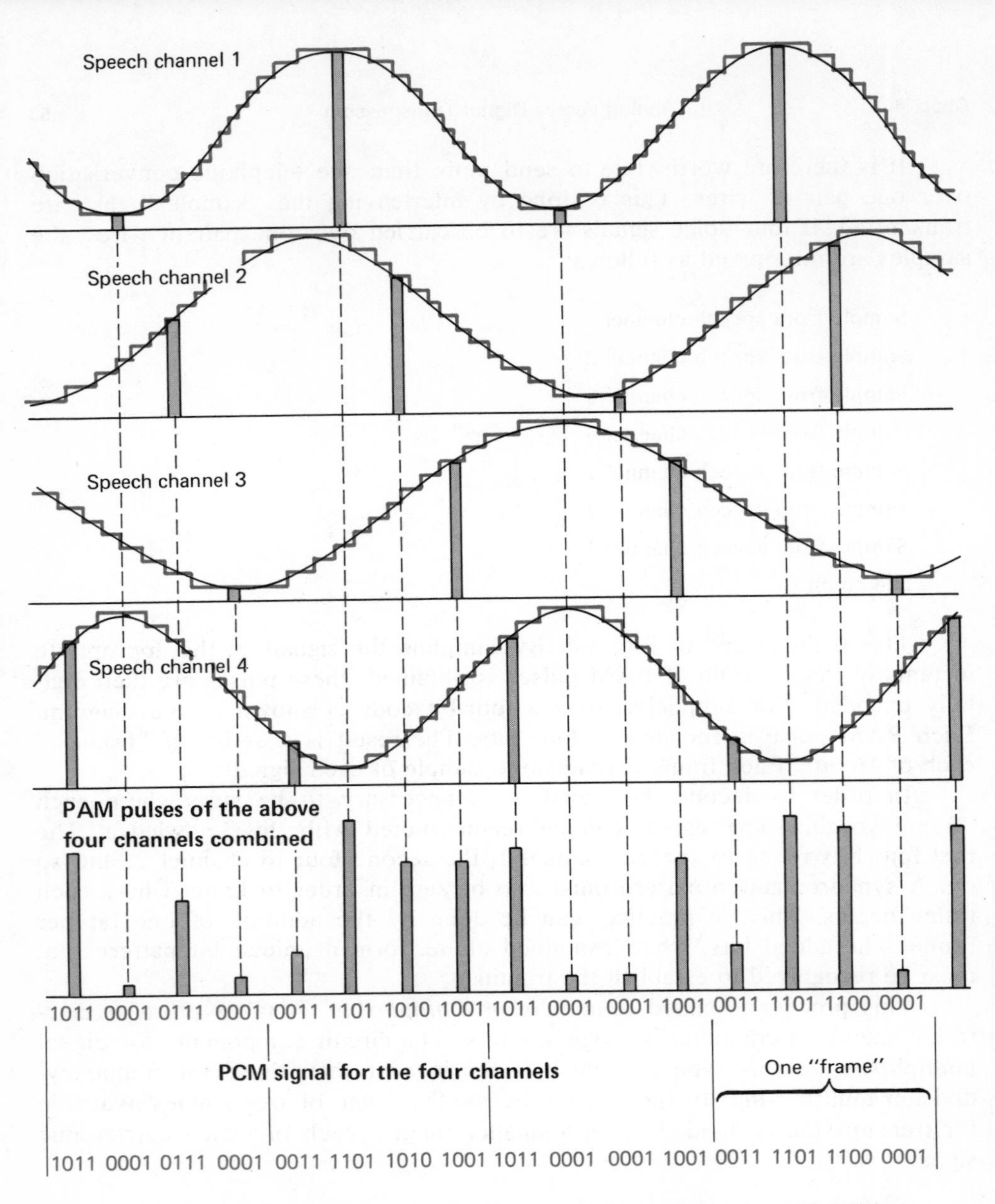

Figure 4.4 A simplified picture of time-division multiplexing with PCM transmission.

The T1 carrier will be discussed in more detail in Chapter 27. Meanwhile, let us note that it is likely to be the basic building block of many of our future telecommunication networks. Other such systems will be built by other telecommunication companies and in other countries, but the ability to encode the basic telephone channel into 64,000 bits per second will be fundamental to them. International standards are in existance for this [3, 4].

This technology is attractive for taking advantage of the vast quantities of wire-pair circuits that exist. The telephone companies have an enormous investment in these circuits, which span rural areas and are laid beneath the cities. The possibility now arises of making them carry 24-voice channels or more rather than the single channel that most of them now carry. Frequency-division multiplexing is also used on wire pairs, but this process normally gives 12-voice channels. Furthermore the cost of time-division multiplexing will increasingly tend to be less. Time-division multiplexing is also becoming attractive in the switching technology, as we shall see in Chapter 30.

When digital data are sent over a private analog voice channel, speeds of about 4800 bits per second are typical although 2400 is often used still. 9600 can be achieved with the penalty of a higher error rate. A rate of 4800 bits per second is used over a public channel with network signaling on it although here 1200 is still common. Some common carriers, particularly in countries other than the United States and Canada, permit transmission at speeds of only 1200, sometimes even less, on their public voice lines. These low speeds indicate the difficulty of designing *modulation* equipment to convert the data into a suitable form for traveling over the analog telephone line because of the high level of distortion on such lines.

The conversion to digital lines is good news for the data processing men. With synchronous transmission of seven-bit characters, the 56,000 bits per second of the PCM voice line give about 7500 characters per second, with a powerful error-detecting code. This is more than ten times the speeds in conventional use today.

Data processing specialists, who have had reason to complain that the printer on the other end of a telephone line is slow, can be encouraged by the thought that over a PCM telephone line they could print at about 4000 lines per minute, or fill a very large screen full of data in a second. To do so, however, the high-capacity link would have to go into their premises. At present it commonly ends at the local telephone office.

The T1 carrier and corresponding international facilities are only the beginning. A hierarchy of interlinking digital channels is planned for the Bell System. The next step up is the T2 carrier, which is designed to take the signals from four T1 carriers or, alternatively, to carry one Picturephone signal. This signal operates at 6.312 million bits per second. Higher still are the other T carriers which will carry hundreds of megabits per second over broadband transmission facilities. An equivalent hierarchy, operating at a different set of speeds, is coming into use in countries outside North America.

More than 50 million channel miles of T1 carrier are already in operation and are proving very successful. A few million channel miles of T2 carrier are in operation. The higher T carriers are under test but not yet in production. Some problems in the use of the higher-speed channels are yet to be resolved.

One of the major needs of the 1970s is a public-switched network of channels for data transmission of widely varying speeds and flexibility. The T1 and T2 carriers are clearly candidates for the transmission links for this, although as we shall see there are alternative candidates.

The first commercial telephone, developed by Bell, went into service in 1877 when a Boston banker leased two instruments. The user placed his mouth and ear to the opening alternately. England had nine telephones installed by the end of 1879. *(Courtesy AT&T.)*

5 SPEEDS OF CHANNELS

Telecommunications channels of a wide variety of transmission rates, or bandwidths, are desirable. In the past most channels were 3 KHz channels designed for telephone speech. In the future, with CATV cables, data networks, private microwave, and satellites, a wide variety of channel speeds will be used.

HUMAN CHANNEL RATES

Different transmissions for human consumption require different rates. The eye can absorb about a thousand times as much information as the ear, when measured in terms of signal bandwidth. Similarly measured, high-fidelity stereophonic sound contains about ten times as much information as telephone sound. "High-fidelity" wall-screen television would require ten times the bandwidth of today's television—and today's television more than one hundred times that of high-fidelity sound.

Fig. 5.1 illustrates the human capacity for communication and storage of information. A young person's ear can hear about 20 KHz of sound which could be encoded into about 200,000 bits per second. With elaborate encoding, 10,000 bits per second can give intelligible speech input with the speaker's voice being recognizable. The eye requires far more. Even with highly elaborate coding more than 100 million bits per second would be needed to represent the image we see—for example to encode a wide-screen movie. Nevertheless an interesting visual image can be created with 1 million bits per second.

The sensory input is processed and sifted before being stored in the brain's long-term memory. It appears to be stored in a highly compacted form with remarkable retrievable capabilities that we could not imitate today with electronics. The brain has a limited capability to retain facts and figures as it receives them, and uses a short term buffer forthis purpose which is part of its

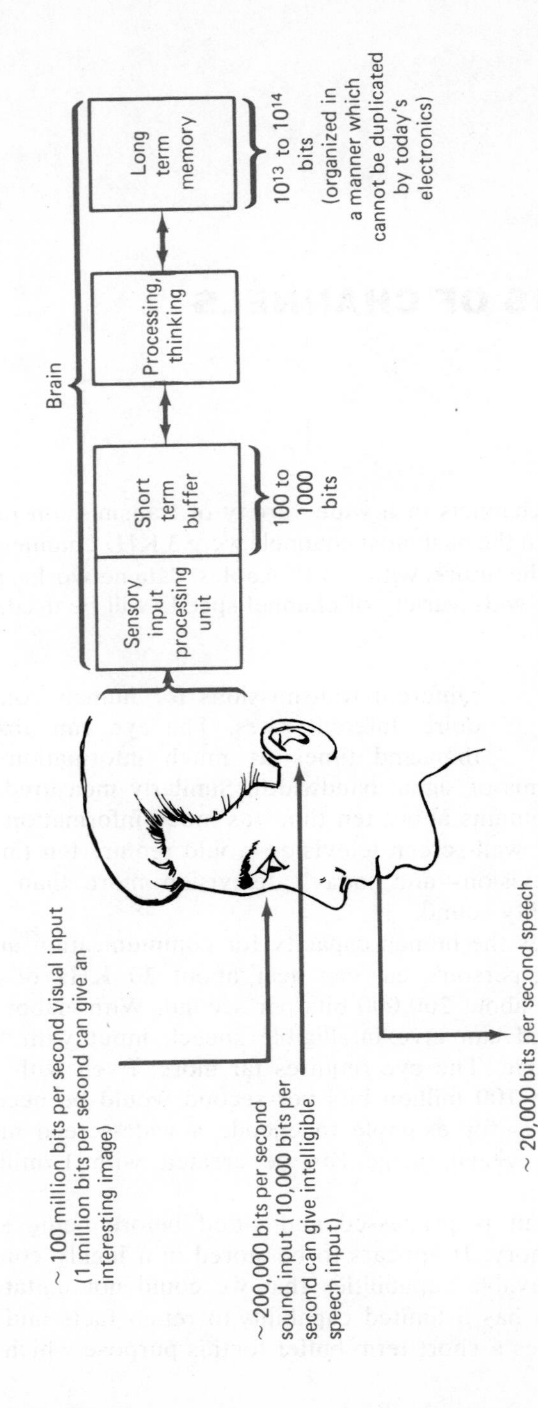
~ 100 million bits per second visual input
(1 million bits per second can give an
interesting image)
~ 200,000 bits per second
sound input (10,000 bits per
second can give intelligible
speech input)
~ 20,000 bits per second speech
Brain
Sensory
input
processing
unit
Short
term
buffer
100 to
1000
bits
Processing,
thinking
Long
term
memory
10^{13} to 10^{14}
bits
(organized in
a manner which
cannot be duplicated
by today's
electronics)

Figure 5.1

sensory input apparatus. We use this short-term buffer heavily in conversation with humans and dialog with machines [1].

A given piece of information is conveyed to humans, sometimes using a low transmission capacity and sometimes a large capacity.

If we watch five minutes of a television talk show a message may be conveyed to us. To transmit television digitally (with PCM encoding) requires 92.5 million bits per second. The five minutes therefore take 5 × 60 × 92.5 million = 27.75 billion bits. More compact signal encoding can reduce this somewhat.

The same message may be conveyed in speech in five minutes of conversation over a telephone line. If the line is digital (PCM), operating at 56,000 bits per second, this uses 5 × 60 × 56,000 = 16.8 million bits. In fact it uses twice this because a typical PCM telephone transmits in both directions together so that the parties can both talk or hear each other.

Human conversation is highly redundant, and if the person with the message to convey had been better organized he might have spoken it in 120 words, taking one minute. This would have needed 3.36 million bits.

The 120 words could have been sent by data transmission. Using the coding that is typical today, this would need 4800 bits. If a tighter code had been used (5-bit Baudot code instead of 8-bit ASCII) 3000 bits would have sufficed. If message compaction techniques had been used, the message could have been sent in less than 1000 bits.

The different ways of sending the same information will be important in considering the usage of various transmission media later in the book, for example the usage of television channels, communication satellites, or the limited capacity of mobile radio channels.

In tomorrow's communication world, many different types of signals are going to be transmitted. By their nature they require widely differing transmission rates, or bandwidths.

Data transmission has primarily used a fairly restricted range of speeds so far. The most common speeds are: the telegraph channel speeds 45, 50, 75, and 150 bits per second, subvoice grade lines outside North America at 200 bits per second, and speeds derived from voice lines, 600, 1200, 2400, 4800, 7200, and 9600 bits per second. Wideband data links operating at 19,200, 40,800, 48,000, and 50,000 bits per second are becoming more widely used, and there is occasional use of higher speeds. The AT&T Dataphone Digital Service, DDS, transmits bit rates of 2400, 4800, 9600 and 56,000 bits per second.

As mentioned earlier, analog and digital signals are interchangeable. We should keep this interchangeability in mind in discussing the speeds of future communication links. Computer data at 4800 bits per second can be converted into an analog form to travel over telephone channels of 3 KHz. On the other hand, telephone signals can be digitized and represented by a stream of 56,000 bits per second (hence the top DDS transmission rate).

CCITT, the international standards organization for telephony and telegraphy, has recommended [2] that data transmission speeds over telephone circuits should be 600 N bits per second, where N is an integer between 1 and 18. In making the recommendation it recognizes that in some countries, a data rate of 2000 bits per second is in common use and does not conform to the recommendation. Of the recommended speeds, certain are classed as "preferred" speeds: 600, 1200, 2400, 3600, 4800, 7200, and 9600. Standardization of higher data rates is now necessary and should relate to the new hierarchy of PCM channels. CCITT has standardized the PCM sampling rate at 8000 samples per second, giving a 64,000 bit-per-second channel. AT&T derives 56,000 bits per second from its digitized speech channel, the reaminder being used for synchronization, signaling, and control.

CCITT has standardized two PCM transmission rates of 1.544 and 2.048 million bits per second [3 & 4]. It has been suggested that higher capacity PCM channels should transmit at 2.048×2^N bits per second, where N is an integer. AT&T has standardized its own digital channel speeds to be:

Level 1: 1.544 m b/s (T1 Carrier)
Level 1C: 3.2 m b/s (T1C Carrier)
Level 2: 6.3 m b/s
Level 3: 45 m b/s
Level 4: 274 m b/s (T4 Carrier)

These speeds will probably become standard for North America.

In the wide range of speeds at which transmission will take place, certain points stand out as being of particular importance. Some of these factors relate to human needs, such as the desire to make telephone sound intelligible and the caller recognizable with a low expenditure of bandwidth; some relate to machine needs, such as the maximum speed of mechanical printers; and some relate to channel properties, such as the maximum capacity of a wire pair.

Let us note the significant points in our range of speeds, as they indicate requirements for communication channels. Eventually it will be desirable to have all these channels available on a dial-up or on-demand basis.

1. Very Slow Transmission

Some types of machines transmit *conditions* to each other and no bulk of data is transmitted; for example, a remote burglar alarm transmits a simple yes/no condition signal. A vehicle detector in the street transmits simple pulses as cars go over it. When you pick up or replace your telephone handset, this fact is transmitted to the relevant switching locations. Signals of these types require only a very small bandwidth.

2. Typewriter Speeds

The input speed on a typewriter is the maximum speed at which a human being can type. Fifteen characters per second is enough for the most nimble-fingered. The output speed of a typewriter-like terminal need not necessarily be the same; however, this is the speed of a typewriter printing mechanism today. The price of a faster printer is higher. There will be many requirements for channels of this speed, say 150 bits per second, in the foreseeable future. Today's telegraph channels of 75 bits per second and below are a little too slow for the fastest keyboard operators and for typical electric typewriter mechanisms. It is convenient for many purposes to have printing terminals with speeds higher than 15 characters per second.

3. Human Reading Speed

If you are a fast reader, you might be able to read this page in a minute—in other words, at a speed of about 250 characters per second. If information is being displayed to you on a screen unit, it is desirable that it be transmitted as fast as you can read it. Doing so is a requirement of an efficient man-machine interfacee. Display screens operating today at 2400 bits per second (300 characters per second) on typical commercial applications do seem to provide an effective form of man-computer communication. Computer output that is substantially slower than this rate can be frustrating for the user. On computer-assisted-instruction systems, it has been commented that a lower-speed output is like having a teacher with a speech impediment.

Sometimes a speed slightly higher than 300 characters per second seems desirable—for example, when the "page" is being skip-read, and the terminal user quickly flashes on the next page. This practice is likely to be common in browsing or searching operations. In looking through a telephone directory, for example, you do not read every line. When tables are displayed also, a fast operator can handle speeds somewhat higher than 300 characters per second. Many screen devices that can operate at 4800 bits per second (600 characters per second) are likely to be used today on analog voice lines. This rate seems to be a generally useful speed for such display terminals. It may well become the most commonly used speed on digital data networks.

It should be noted that although the terminal operator can usefully absorb 4800 bits per second, he certainly cannot respond at this speed (except by voice). A 150 bit-per-second return channel is adequate for his response with foreseeable mechanical devices, such as keyboards and light pens.

4. Telephone Channels

The channel we are all most familar with was designed with the important economic constraint that the bandwidth used should not be larger than necessary. The maximum number of telephone calls can then be multiplexed together over long-distance links or links between offices. The result is a bandwidth of about 3 KHz, which is enough for you to recognize the voices of your callers and comprehend what they say. There is no need for high fidelity.

The systems we can design today are largely dominated by this bandwidth. It is highly desirable now to break away from this domination as quickly as possible.

5. Machine Printing and Reading Speeds

A typical high-speed computer printer operates at about 1200 lines per minute—that is, about 20,000 bits per second. The highest speed card reader also reads cards at a speed close to 20,000 bits per second. It would be of value to have switched telecommunication lines interconnecting these machines with magnetic tape and disk units, and computers. Perhaps because of mechanical improvements, the input/output speeds will increase even more. Xerographic printing can give speeds that are higher than 20,000 bits/second. Optical document reading may also give higher speeds. Standard broadband channels in increasing use in the United States and elsewhere operate at 56,000 bits per second.

6. A PCM Voice Channel

A voice channel using pulse code modulation transmits at about 56,000 bits per second. It is probable that many of the world's voice channels will use this form of transmission in the not-too-distant future. This rate is close to the speed requirement for high-speed printers and readers, and thus it would be valuable to have a switched public network of this speed. AT&T's Dataphone Digital Service (DDS) provides leased channels of up to 56,000 bits per second, derived in part from PCM telephone channels.

A typical typewritten page can be transmitted in a digitized facsimile form at 56,000 bits per second in about 4 seconds. Channels of this speed could thus form the basis for an interactive information retrieval system.

7. A High-Fidelity Sound Channel

Telephone channels are restricted to a bandwidth of 3000 Hz. A bandwidth of 20,000 is needed for full high-fidelity transmission, and twice that amount is required if the transmission is stereophonic. The wires entering the home are capable of carrying this range of frequencies; therefore domestic distribution of high-fidelity music is technically feasible, and could be on a dial-up basis if desired. An analog channel engineered for high-fidelity transmission could carry about ten times as much information as a telephone channel.

When a pulse code modulation channel is designed for high-fidelity transmission, more detectable signal levels (quantizing levels) are used than in speech transmission. Ten bits per sample may be used instead of seven bits for speech. The total bit rate is about 320,000 bits per second. Stereophonic transmission can be carried by about 640,000 bits per second instead of 56,000 for telephone speech.

Using more complex means of digitally encoding the stereo signal a smaller number of bits can be used to carry music. This trade-off between bit rate and logical complexity of the digital encoding applies to most other types of digital transmission also. The T1 carrier could be made to carry the highest quality quadraphonic hi-fi.

8. The Maximum Capacity of a Wire Pair

Multibillions of dollars worth of wire pairs are laid down under the streets and along the highways. In recent years the wire pair has been made to carry a bit rate of a few million bits per second. Digital repeaters at appropriately close intervals on the line, which regenerate, reshape, and retime the bits being transmitted, have been used. These repeaters are the basis of the Bell T1 carrier, and the CCITT G.733 Recommendation [3], which send 1,544,000 bits per second over wire pairs with repeaters every 6000 ft. These are normally used to transmit 24 telephone conversations simultaneously. This is likely to be a standard for decades to come. A T1 carrier has been used for data transmission at 1,344,000 bits per second. Such speeds are appropriate for tape-to-tape or disk-to-disk transmission. This facility or a similar one seems likely to play a major role in data transmission because it makes the most effective use of the ubiquitous wire pairs.

Higher bit rates can be sent over a twisted wire pair with digital repeaters. The T2 carrier transmits 6.312 million bits per second and this is close to the maximum rate practical on a wire pair. The CCITT G.732 Recommendation is for 2.048 million bits per second [4].

9. Picturephone

When Bell System Picturephone signals are encoded in a digital form, six million bits per second are required. These bits can travel over a Bell T2 carrier. It is doubtful how widespread the Picturephone service will become; however, the T2 carrier will become one of the main short-haul trunks of the United States and will be used for many different types of transmission. Four T1 signals are multiplexed together to travel over one T2 channel.

10. Television

The next step in speed relating to human communication is television, which requires about 4-6 MHz when transmitted over analog circuits. Television is transmitted over analog trunks along with large groups of telephone calls. In the future, television, like every other type of signal, will probably be carried by digital pulse streams, and a channel operating at 92.5 million bits per second will be used for television. Such a channel is planned in the Bell System future hierarchy of digital channels.

The television screen itself may eventually become larger and of higher fidelity. If the number of lines on the screen is doubled, then four times the bandwidth will be needed. If we have a 5-ft wall screen for television, or perhaps eventually a screen that occupies most of one wall, the resolution per inch will probably not need to be quite as good as that on today's small sets. Ten times the present bandwidth will probably be enough for even the most spectacular home screens. This would give approximately the resolution I obtain from my 35-mm slide projector, which has a 6-ft screen in my apartment and seems sharp enough for all practical purposes. Such transmission would probably not use ten times the *bit rate* of today's television because powerful signal compression techniques can be used with digital transmission.

The ultimate requirement of visual transmission outside ultralarge-screen theaters may, then, be about 50 MHz. The Bell System T4 carrier of 274 million bits per second could probably be made to carry this video signal. Coaxial cables like those laid into homes today to carry CATV *could* be engineered to transmit such signals, but more likely such bit rates will await the deployment of optical fibers.

In all probability computers will find uses for this high transmission speed eventually. Signals to and from many terminals will be multiplexed onto one such channel. Responses to terminals on some systems will be in the form of pictures as well as alphanumeric responses. Data banks will be remote from the machines using the data. Time-sharing systems of immense versatility will be able to call in programs from remote locations. Small machines will be able to handle highly elaborate applications with graphics by use of data networks employing such channels.

11. High-Speed Digital Channels

The Bell System T4 carrier transmits 274 million bits per second. This is implemented in Canada as the LD4 coaxial system. The Bell WT4 waveguide system transmits 60 of the T4 bit streams through a single pipe, each modulating a different sine wave carrier, and giving a total throughput of 16 billion bits per second. Designs for satellites have been discussed with total satellite throughputs of billions of bits per second, but it is not yet economical to launch them. Optical fibers have been made to transmit a billion bits per second and many such fibers could be packed into one cable.

It is thus clear that physical facilities are emerging which are capable of transmitting much higher bit rates than those in common use today.

SUMMARY

Box 5.1 summarizes the main landmarks in this wide range of usable bit rates. On the left are bit rates corresponding to human transmissions of various types, such as typing, tele-

BOX 5.1 A comparison of the speeds of physical channels with the speeds of signals employable by humans.

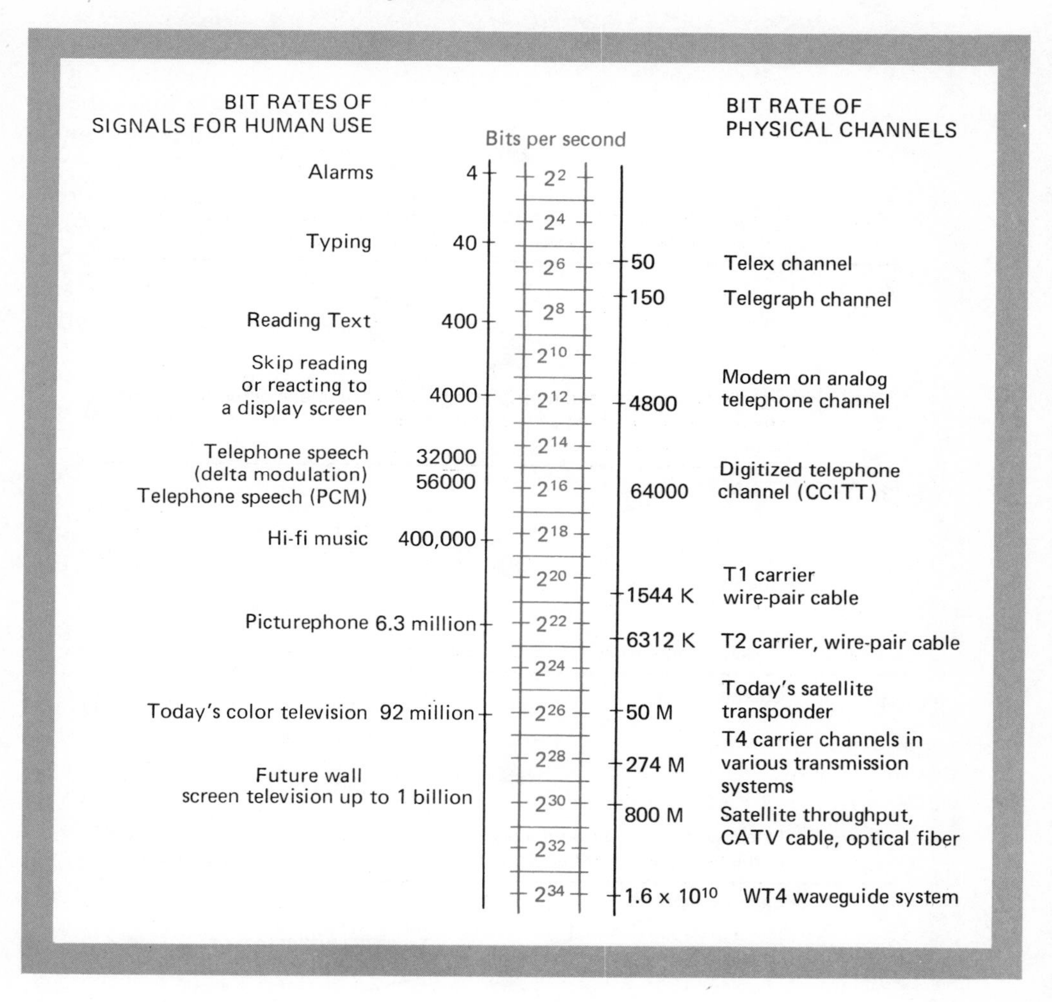

phone, hi-fi music and television. On the right are bit rates of physical channels, ranging from the slow rate of telex to the high rate of AT&T waveguide.

TRADEOFFS

A telecommunications system designer can exercise a number of tradeoffs in his use of channel capacity. These tradeoffs are illustrated in Box 5.2. Some of them could (and should) have major effects on the architecture of telecommunications systems.

BOX 5.2 Some trade-offs in telecommunication system design to channel capacity.

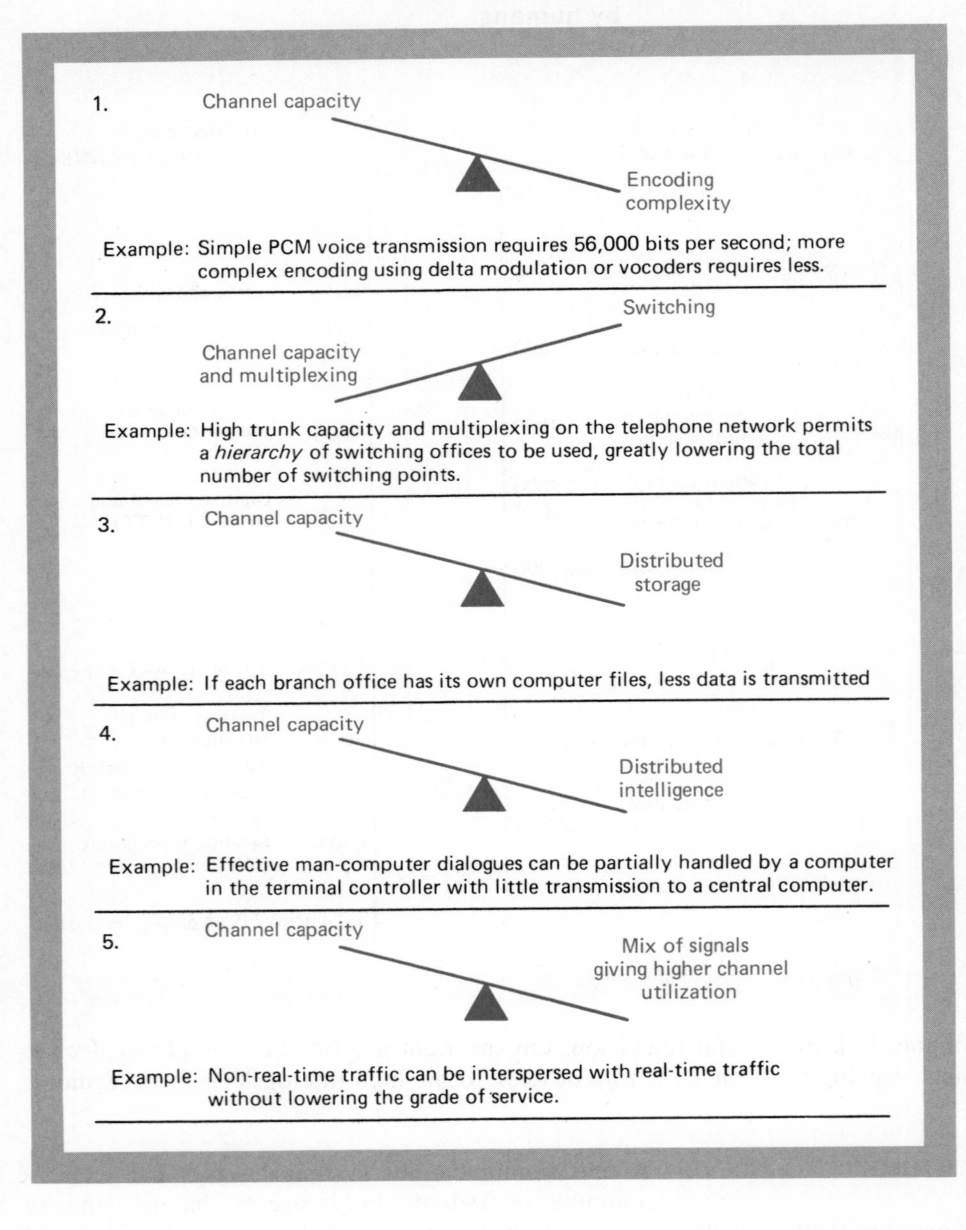

1.
Channel capacity
Encoding complexity
Example: Simple PCM voice transmission requires 56,000 bits per second; more complex encoding using delta modulation or vocoders requires less.
2.
Switching
Channel capacity and multiplexing
Example: High trunk capacity and multiplexing on the telephone network permits a hierarchy of switching offices to be used, greatly lowering the total number of switching points.
3.
Channel capacity
Distributed storage
Example: If each branch office has its own computer files, less data is transmitted
4.
Channel capacity
Distributed intelligence
Example: Effective man-computer dialogues can be partially handled by a computer in the terminal controller with little transmission to a central computer.
5.
Channel capacity
Mix of signals giving higher channel utilization
Example: Non-real-time traffic can be interspersed with real-time traffic without lowering the grade of service.

1. Complexity of Encoding

First there is a tradeoff between the number of signals that can be packed into a given channel and the complexity of their encoding. Telephone companies digitize the telephone voice today, for example with simple PCM encoding. This requires 56,000 bits per second with Bell System standards (T1 Carrier, etc.) and 64,000 bits per second with some CCITT standards (CCITT Recommendations G732 and G733). Delta modulation techinques use more complex encoding, which we discuss in Chapter 27, and can achieve the same telephone sound quality with half the number of bits per second. Again, dynamic allocation of the bit stream may be used to double the number of telephone calls transmitted, as described in Chapter 3. Using delta modulation and dynamic channel allocation a 4-wire T1 carrier circuit (1.544 bits per second full duplex) can be made to carry 96 telephone calls instead of the 24 it carries today. Telephone speed can be compressed much further with more elaborate techniques. The price of doing so lies in the complexity of the encoding electronics.

Similar arguments apply to the transmission of television over satellite channels, the compaction of digitized facsimile documents, and the compaction of data sent to and from computers.

In general, a channel can carry more signals with complex encoding than with simple encoding. The cost of complex encoding is dropping as the technology of large-scale-integration circuitry develops.

2. Switching versus Multiplexing

If a network connects many points and little use is made of multiplexing, then extensive use of switching is needed.

Large networks usually have more than one level of switching office. The higher levels in the hierarchy, for example the toll switching offices in a telephone network, have high capacity trunks interconnecting them with a high degree of multiplexing. This greatly reduces the total channel mileage and the total number of switching offices needed.

The local loops of a telephone network connect each subscriber, with no multiplexing, to his local telephone office. This results in a very large cable mileage. If small concentrator switches were used in apartment buildings or dense subscriber groupings, and multiplexing were used on the cables, a great reduction in cable mileage could result. The cable television cables now being laid in many locations could be employed with concentrators and multiplexing to greatly reduce telephone cable mileage—however in industrial countries the telephone loops are already in place.

3. Channel Capacity versus Distributed Storage

Much of the data transmission used in computer systems is designed to obtain information from distant data storage. The paperwork files in branch offices are replaced with communication links to a centralized computer with a data base. The cost of small storage units is dropping fast and their reliability is increasing; consequently it is becoming economic to store some data in peripheral units and avoid the data transmission.

There is a tradeoff between data transmission and distributed storage.

4. Channel Capacity versus Distributed Intelligence

Man-computer dialogues designed for computer terminal users who cannot program need to be psychologically appropriate for the user. This requires a large number of bits going to and from the computer which performs the dialogue processing. A trend in the computer industry is to use increasingly "intelligent" terminals which can carry out some of the dialogue processing at the terminal location. Terminals or their controllers are being designed to contain microcomputers.

The greater the level of programmed function at the peripheral locations, the less the data which need to be transmitted to the central computer. There is a trade-off between distributed intelligence and channel capacity. Much of the motivation in the computer industry for using distributed microprocessors has resulted from the fact that the data transmission links required for good interactive system design were too expensive, unavailable, or insufficiently reliable.

5. Mix of Signals

An objective of telecommunications design ought to be to achieve a high utilization of the channels. Much of the difficulty of achieving high utilization stems from the fact that most users want to employ the channels at random times—times of their own choosing—and then want the channel to be available to them almost immediately when they request it. The probability of a user being refused a channel when he requests it (the "grade of service") must be set at a low level, for example 0.01 or lower.

A variety of mechanisms is employed to allocate channels to users when they need them. In general higher channel utilization will be achieved if more channels can be grouped together for allocation purposes, as illustrated in Fig. 3.2. Higher grouping is possible if a larger number of users share the group. Rather than constructing data networks separately for each application, it would be more economical to combine the data networks, and more economical still to combine a corporation's voice and data networks.

As we commented in Chapter 3 a particular opportunity for improved channel utilization exists in the use of priority mechanisms so that real-time and nonreal-time traffic fills in the gaps between the telephone calls and other real-time calls. Much data traffic can be deferred until hours when there is little real-time traffic.

REFERENCES

1. James Martin, *Design of Man-Computer Dialogues,* Chapter 19, Prentice-Hall, Inc., Englewood Cliffs, N.J., 1973.

2. CCITT Recommendation V.22 bis. "Standardization of Data-Signalling Rates for Synchronous Data Transmission on Leased Telephone-Type Circuits." CCITT Green Book, Vol. VIII, Data Transmission, International Telecommunications Union, Geneva, 1973.

3. CCITT Recommendation G.733 on PCM multiplex equipment operating at 1544 KB/S. CCITT Green Book, Volume III, Line Transmission, International Telecommunications Union, Geneva, 1973.

4. CCITT Recommendation G.732 on PCM multiplex equipment operating at 2048 KB/S. CCITT Green Book, Volume III, Line Transmission, International Telecommunications Union, Geneva, 1973.

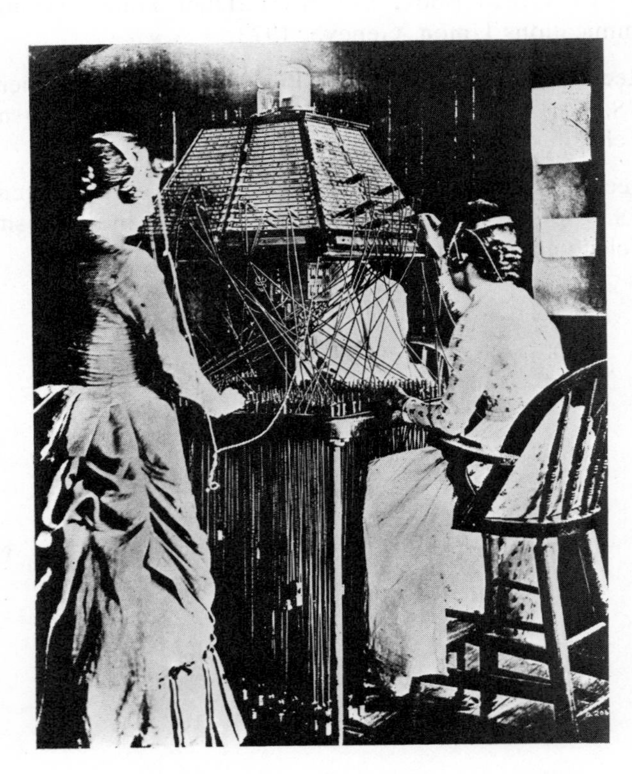

One of the earliest private branch exchanges (PBXs), a pyramid switchboard, 1881. *(Courtesy AT&T.)*

6 INTELLIGENT EXCHANGES

To be fully useful to society, telecommunication networks must be able to interconnect a very large number of users. Switching facilities have been a vital part of telecommunications since its earliest days. Until recently, all automated switching offices were built out of electromechanical components like those in Fig. 6.1. Large switching offices have an amazing quantity of electromechanical equipment clicking away, routing the calls.

Today computers are coming into common use for switching. In advanced computerized switches the calls flow through solid state circuitry under computer control, and there are no moving parts. Computer-like devices, dropping rapidly in both cost and size, offer the possibility that switching can be done in many small machines distributed close to user locations, rather than in large centralized offices with vast numbers of cables going to them. Switching, like computing, can be done in mini- and micromachines of high reliability, as well as in the traditional large centralized machines.

Furthermore computers can carry out many functions associated with switching that would have been too complicated to be worthwhile with electromechanical devices. Just as we talk about *intelligent* terminals when they have microcomputers under their cover, so we can talk about *intelligent* switching mechanisms or *intelligent* exchanges. Computerized mechanisms for switching and control change the nature of what is possible in the linking together of facilities to form new telecommunications systems. Corporate satellite systems, electronic mail systems, radio telephone systems, and others, require the sophisticated forms of control which computers permit.

There is an amazing difference between the PABX† (Private Automatic Branch Exchange, Fig. 6.1) installed on typical corporate locations and the

† PABX is the international abbreviation for a private automatic branch exchange. In the United States the abbreviation PBX is used rather than PABX.

Figure 6.1 Before computers elaborate electromechanical switches were used. Figure 6.2, opposite, shows banks of these strowger switches in a typical small telephone exchange. Most of the world's telephone exchanges are of this type.

Figure 6.2 Prior to the late 1970s, most PABXs consisted of electromechanical step-by-step switches such as those shown here. This equipment is now being replaced by computers.

Figure 6.3 AT&T's *Dimension* PABX, a computer controlled system with solid-state, time-division switching. The flexibility of stored program control makes it possible to give such a PABX many features not practicable on electromechanical systems.

computerized device with which it is being replaced. The telephone company PABXs existing in most organizations prior to the late 1970s are rooms full of electromechanical switches, such as that in Fig. 6.2. Figure 6.3 shows AT&T's *Dimension* PBX† which replaces the equipment in Fig. 6.2, a small elegant-looking minicomputer-controlled machine with no moving parts in the switches. The switching is performed by interleaving streams of pulses at high speed in computer-like LSI circuitry, in a fashion described in Chapter 30. The computer-controlled machine is extremely flexible compared with its clicking mechanical predecessors. It can be programmed to perform all manner of functions not practicable with mechanical switching.

BASIC PRINCIPLES

The essential elements of a computerized switching system are shown in Fig. 6.4. The incoming lines all enter some form of switching network (1). On the latest systems this is in solid-state electronics—no relays or moving parts. There are many different paths through the switching network, and the switch controls for setting up different paths are operated by the computer (2).

There is a mechanism for constantly scanning the activity on the lines (3) and informing the computer program of significant events. The program must be informed when you pick up your telephone handset; it must read the number you dial and must be informed when you replace your receiver after a call. Similarly, it must detect and interpret the signals from other switching offices that arrive on the trunks.

There are various signals that it must send down the lines, such as "busy" signals ("engaged" signals in British parlance), signals to make telephones ring, dialed numbers, and on-hook/off-hook signals, which it sends to other switching offices. The signaling equipment (4) might be regarded as one of the computer output units in this type of system.

†See footnote on p. 71.

Figure 6.4 Basic elements of a computerized switching system.

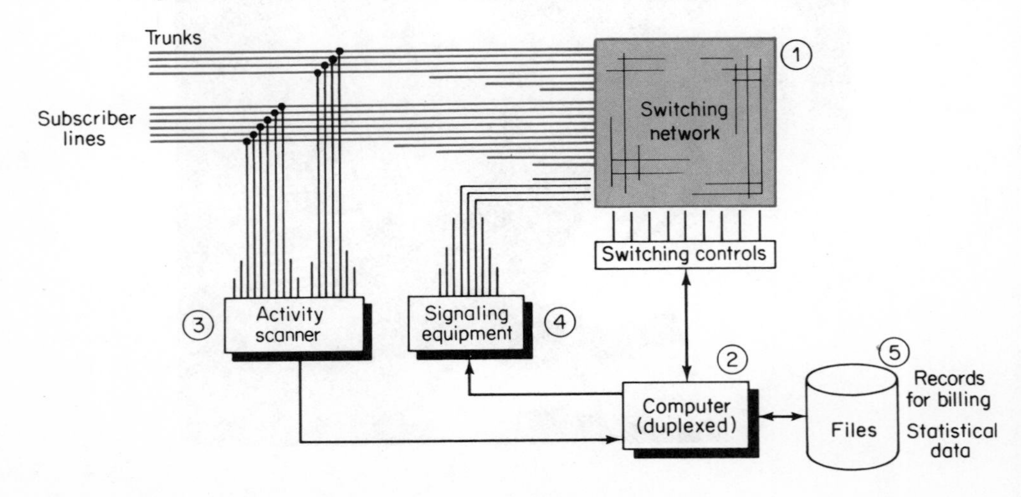

Information about calls must be gathered for billing purposes, and details about the uses of the network will be filed for statistical analysis. These will be recorded by the computer (5).

Because the switching is carried out by computer many special types of instruction can be sent to the PABX services.

PABX FUNCTIONS

A computerized PABX can allow the telephone users to carry out many of the functions that before only an operator could handle. They can set up conference calls, transfer calls, and dial other parties to add on to an existing call. Users need not waste their afternoon redialling when a call does not go through. They can instruct the exchange to redial. Users can be allocated different priorities to ensure that the important calls always get through. When users change their offices, it is no longer necessary to rewire their telephones in order that they keep the same number; all that is necessary is an instruction sent to the PABX.

In general, as soon as the switching is done by computer, all manner of options can be programmed. Box 6.1 lists some of the capabilities of computerized PABXs. The functions marked with crosses on that chart were done before computerization by the best of electromechanical PABXs.

Because of the flexibility of stored-program control a variety of different facilities and restrictions can be associated with each individual extension. Among the facilities that may or may not be assigned to any extension on the IBM 3750 PABX are

1. Outgoing calls permitted via the attendant.
2. Direct outward dialing of *local* calls by *day*.
3. Direct outward dialing of *local* calls by *night*.
4. Direct outward dialing of *national* calls by *day*.
5. Direct outward dialing of *national* calls by *night*.
6. Direct outward dialing of *international* calls by *day*.
7. Direct outward dialing of *international* calls by *night*.
8. *Intrusion allowed*—the extension may be interrupted for a pending call.
9. *Interrupt*—the extensive user may signal a busy extension that he wished to interrupt.
10. *Add-on third party*—the extension is allowed to initiate a three-way call.
11. *Paging*—the extension user is allowed dial a paging signal.
12. *Contact monitoring*—allows the extension to sense and operate a contact.
13. *Numeric data collection*—allows numeric data to be collected through an extension.
14. *Alphanumeric data collection*—allows alphanumeric data to be collected through an extension.

Many other functions are available on today's electronic PABXs.

BOX 6.1 Features of PABX Systems.

Features of Some Conventional PABXs	Features of Bell System PABX packages. Series 100	 Series 200	 Series 300
• ATTENDANT'S CONSOLES Operators at one or more consoles assist with the switching of calls. Varying numbers of switching functions are automatic.	X	X	X
• PUSHBUTTON STATION SELECTION The attendant has a status light and button for each station she controls.		X	X
• STATION TO STATION DIALING A station can dial any other station attached to the PABX without assistance from the attendant.	X	X	X
• DIRECT OUTWARD DIALING A station can dial a call on the public network without assistance from the attendant.	X	X	X
• STATION HUNTING There are several lines to the PABX location. When the first line dialed is busy, the others are tried.	X	X	X
• CALL TRANSFER BY ATTENDANT A station receiving a call can signal the attendant and request that that call be transferred to another station.	X	X	X
• CALL STACKING ("CAMP-ON") BY ATTENDANT When one or more calls are received for a station which is already busy, the attendant can put the calls in a waiting state until the station is free.		X	X
• CALL WAITING ("CAMP-ON") SIGNAL A person receiving a call can be signaled to inform him that another call is waiting.		X	X

BOX 6.1 ***Continued***

	Features of Bell System PABX packages.		
	Series 100	Series 200	Series 300
• CONFERENCE CALLS SET UP BY ATTENDANT Calls between more than two telephones (sometimes not more than three) can be connected by the attendant.	X	X	X
• NIGHT SERVICE TO A FEW DESIGNATED STATIONS The external lines serving the PABX can each be switched through to a certain station when there is no attendant.	X	X	X
• POWER FAILURE TRANSFER Certain prearranged stations can continue operation when there is a commercial power failure.	X	X	X
• STATION RESTRICTION FROM OUTGOING CALLS Designated stations are prevented from making outgoing calls.	X	X	X
• SECRETARY CONSOLES A secretary can have a console for handling the calls of one or more employees.	X	X	X
• KEY TELEPHONE STATIONS LINKED TO SECRETARIES An employee can signal his secretary and vice versa. Secretaries can intercept calls.	X	X	X

Other Features Possible on Computerized Exchanges

- EACH STATION HAS ITS OWN EXTERNAL TELEPHONE NUMBER
  Possible on Centrex systems in which the switching is done at the central office location.

BOX 6.1 *Continued*

Feature	Series 300
• DIRECT INWARD DIALING Stations can be dialed from the outside without intervention by the attendant.	
• CALL TRANSFER BY INDIVIDUAL STATIONS A station can transfer a call from the outside to another station without assistance of the attendant.	X
• CONSULTATION HOLD BY INDIVIDUAL STATIONS A user can place a call on "hold" while he dials another station; he can then return to the interrupted call.	X
• THIRD PARTY ADD-ON BY INDIVIDUAL STATIONS A user receiving a call can dial another station thereby setting up a three-way conference call.	X
• NIGHT ANSWERING SERVICE When the attendant console is unoccupied, any station may answer incoming calls by dialing a code. Attendant-seeking calls may be automatically reassigned to selected stations.	X
• LISTING OF CHARGES INCURRED BY STATIONS A listing is provided for each station of the nonlocal calls dialed giving the number, duration and cost of each call.	
• RECORDING OF LOCAL CALL INFORMATION Local call message unit information is automatically recorded for each extension. This is used by hotels for billing customers.	

BOX 6.1 *Continued*

- AUTOMATIC CALL WAITING FACILITIES
  Calls to a busy station are automatically held. The calling party is notified that he is on hold, and the busy party is signalled that a call is waiting.
- AUTOMATIC CALL FORWARDING
  A person may go to a different location and inform the PABX of its extension number; the PABX will forward his calls to that number.
- AUTOMATIC CALL STACKING
  When calls arrive for a station which is busy, they will be automatically queued, possibly with a spoken "wait" message being played to the caller.
- AUTOMATIC CALL DISTRIBUTION
  When calls may be answered by any of a group of stations the calls are automatically distributed to the first free station.
- AUTOMATIC CALL BACK
  When a user places a call and the number is busy, the user may instruct the PABX to call him back when it is free.
- EXTERNAL NUMBER REPETITION
  When a user dials a long number and it is busy, the user may instruct the PABX to remember the number so that it can repeat the dialing.
- EXTERNAL CONFERENCE CALLS DIALED BY INDIVIDUAL STATIONS
  A user can set up an external conference call with the help of the attendant.
- USERS WHO MOVE CAN RETAIN THEIR NUMBERS
  The number can apply to a new station without rewiring, merely by changing the tables used by the PABX.
- ABBREVIATED DIALING
  Commonly used lengthy numbers are replaced by 2-digit numbers.

BOX 6.1 *Continued*

- INTRUSION SIGNAL
A signal is automatically sent to a user if a third party comes on the line (for privacy protection).

- AUTOMATIC CALL TRANSFER
Incoming calls to a busy station are automatically transferred to another designated station.

- CALL PICKUP
A user can dial a code to pick up calls to other extensions within a preset pickup group.

- ALARM CLOCK CALLING
Users can register in the PABX a time at which they wish to be called.

- PAGING BY ATTENDANT
When a subscriber cannot be located, the attendant can page him (possibly by a radio pager). He dials his own number and is connected to the party trying to contact him.

- PAGING DIALED BY STATIONS
A paging operation can be initiated automatically by any user.

- DO-NOT-DISTURB FACILITY
A user may dial a code requesting that no telephone calls be sent to him.

- SELECTIVE DO-NOT-DISTURB FACILITY
Certain extensions are designated high priority. A user may dial a code requesting that only calls from high priority extensions be put through to him.

- STATION-TO-STATION RESTRICTIONS
Certain extensions can be controlled so that they cannot receive calls from certain other extensions. A hotel manager, for example, may prevent his phone from receiving calls from guests.

BOX 6.1 *Continued*

- CALL CHAINING
  When a user makes many successive calls to a remote location he is not disconnected at the end of each call so that he has to redial, but is automatically transferred to the remote operator or PABX.
- TRAFFIC MONITORING AND MEASUREMENT
  Traffic is continuously monitored and a manager can obtain a traffic report at any time.
- CORPORATE NETWORK CONNECTIONS
  The PABX makes connections to and from a corporate network including tie-lines, CCSA, foreign exchange lines, WATS lines and specialized common carrier facilities. Network users can call public numbers and public telephones can make calls via the network.
- PRIORITY ACCESS TO CORPORATE NETWORK FACILITIES
  A priority structure is used so that certain subscribers are given priority access through corporate facilities and do not normally receive network busy signals.
- TRUNK GROUP WARNING INDICATORS
  The console attendant receives a visual indication when only a preset number of trunks remain in a trunk group.
- MULTITONE TELEPHONE BUZZER
  The telephone extension can "ring" with different sounds to indicate to the user where the call is from before he picks up his telephone. The user may wish to pick up only certain calls. "Priority" calls may make a different sound.

BOX 6.1 *Continued*

- HOT LINE SERVICE
  Allows callers from predesignated stations to place calls automatically without dialing, i.e., whenever they pick up the phone the call is directed to a given extension or trunk.
- FACILITIES FOR INTERCONNECTING COMPUTERS AND TERMINALS
  The PABX is designed to handle data traffic, possibly from terminals without modems. The PABX may be directly compiled to a data processing system.
- DATA COLLECTION FACILITY
  The PABX automatically scans data collection devices, and assembles the data for retransmission.
- CONTACT MONITORING AND OPERATION
  The PABX automatically monitors contacts, e.g., for fire or burglar protection or for process control applications; and may operate certain contacts.
- SECURITY FEATURES
  Security provisions are provided to help ensure privacy.

Additional Features Which Can Lower Telephone Costs:

- FACILITY FOR DIALING PERSONAL CALLS
  Users may dial personal calls by prefacing them with a code; the users are then billed for their personal calls.
- AUTOMATIC MINIMUM-COST ROUTING ON CORPORATION NETWORKS
  A call is routed to a number accessible via a corporate network by whatever is the cheapest route, e.g., first choice: tie-line; second choice: WATS line; third choice: Direct Distance Dialing, applicable only to calls of certain priorities.
- REMOTE ACCESS FOR CORPORATE NETWORK FACILITIES
  The corporate network may be accessed from certain telephones outside the corporation to obtain lower cost long distance calls.

BOX 6.1 *Continued*

- AUTOMATIC MONITORING OF CHARGES INCURRED BY USERS
  Charges incurred by all users are continuously monitored and may be inspected by a manager at any time.
- RESTRICTIONS ON WHAT NUMBERS A STATION CAN DIAL
  Specified extensions are prevented from making trunk calls, or from calling certain area codes or office codes.
- TIME RESTRICTIONS ON STATION USAGE
  Specified extensions can be used only at specified times (e.g., 9 A.M. to 5 P.M.). Time restrictions and code restrictions may be combined.

Additional Features With Voice Recording Devices:

- PRERECORDED MESSAGES
  Spoken messages can be automatically played to callers kept waiting, and for other conditions. Some systems play music or advertisements to waiting callers.
- AUTOMATIC TELEPHONE ANSWERING
  When a user does not answer the PABX can play a prerecorded message from him, possibly requesting that a message be left.
- AUTOMATIC MESSAGE RECEPTION
  Messages for a user who does not answer can be spoken to the PABX which records them for that user.
- REMOTE READING OF MESSAGES
  A user who is traveling can dial his PABX and instruct it (using an appropriate security code) to play back the messages that have been left for him.
- DICTATION SERVICES
  Users may dictate memos to the system, to be typed by a typing pool.
- REMINDER MESSAGES
  Users may speak reminder messages to the system to be played back to a given extension at a given time.

NUMBERING SCHEME

In order to carry out functions other than the straightforward dialing of telephone numbers a user must have some means of informing his PABX what he wants to accomplish. He usually has only the telephone dial (or touchtone keys) to use for this purpose. Special meaning must therefore be associated with the dialing of certain digits. An advanced PABX will normally have an instructional brochure telling its users how to employ the special dialing codes. An alternative is to provide many extra keys on the telephone handset for carrying out special functions as is done (for example, on the key telephones of Executone, Inc).

A common example is the dialing of a digit (often 9) to obtain an outside line. Sometimes an 8 is dialed preceding the dialing of any number on the corporate tie-line network. A 0 is dialed to reach the attendant. On some systems a 1 is dialed for manager-to-secretary or secretary-to-manager call. Dialing 20 before an outside number indicates, on some systems, a personal call and will be billed to the individual in question.

Some dialing codes such as the preceding one are used before a call is placed. On computerized PABXs another set of dialing codes are designed to be used in the middle of an established call. For example on a typical IBM 3750 system a 2 is dialed, followed by an extension number to transfer a call to another extension. A 2 can be dialed to place an existing call on "hold" while its recipient dials and talks to another subscriber. Dialing 4 returns to the "held" call, and dialing 3 would set up a conference call between the three parties. Dialing 0 in the middle of a call connects its recipient to the operator while putting the caller on "hold."

Some dialing codes carry out special operations. (For example extensions so authorized may dial 22 to bar all incoming calls and unbar them again by dialing 23.) Extensions so authorized may interrupt busy extensions by dialing 6. This code places a soft "camp-on" signal on the busy extension, requesting its user to put his call on "hold" for a moment, or terminate it.

If a user dials a lengthy number and receives a "busy" signal he may instruct the PABX to store that number, by dialing 7. When he tries to reach the number again he dials 27, and the PABX dials the stored number. If he dials an extension and the person is not there, he may dial 5 to initiate a paging operation. If the person is wearing a small radio paging unit he will hear a paging signal. (Alternatively loudspeaker paging could be used.) The paged person goes to the nearest telephone and dials 25 followed by his own extension number, and is automatically connected to the extension which paged him.

Many other dialing codes are used on some systems. The use of programmed control gives a PABX planner the type of flexibility that data processing designers have in the choice of codes. An example of a dialing code scheme is given in Box 6.2.

BOX 6.2 An illustration of a numbering plan that could be used with a computerized PABX having many new functions.

Function	Code
To dial an extension:	
Numbers available for extensions	3xxx 4xxx 5xxx 6xxx
To dial an external number:	9 + number
To dial a number on the corporate tie-line network:	8 + tie-line code + number
Abbreviated dialing of an external or tie-line number:	70 to 79
To dial the attendant:	0
Manager-to-secretary or secretary-to-manager call:	1
To dial personal calls to an outside number:	20 + number
To transfer a call being received to another extension:	2 + extension number
To place a caller on "hold" and consult with another extension:	3 + extension number
To place a caller on "hold" and consult with the attendant:	0
To return to the "hold" caller:	4
To add a third party to a call:	8 + extension
To disconnect a third party from the call:	4
To attract the attention of a busy party by sending a "CAMP-ON" tone to an extension already dialed:	6
To accept a "CAMP-ON" call, ask the existing caller to hold and dial:	6
To return to the "held" party:	4
External number repetition:	
i. Instruct the PABX to store the last number dialed:	7

continued

BOX 6.2 *Continued*

ii. Instruct the PABX to dial the stored number:	27
Paging:	
To page a person after dialing his extension:	5
To respond to a paging signal (from any extension):	25 + your own extension
To obtain the time of day:	21
To bar incoming calls (DO NOT DISTURB):	221
To unbar incoming calls:	222
To enter a reminder (alarm clock) message to be generated on date DD at time TTTT:	23DDTTTT
To instruct the system to forward all calls to extension xxxx:	241xxxx
To instruct the system to stop forwarding calls:	242
To enter a message for dictation to extension xxxx:	25xxxx
To leave a prerecorded message for all callers who call when the extension is unattended:	26
To instruct the system to read messages that have been left:	28 + Security code
To leave a spoken message for an extension which does not respond:	9

Figure 6.5 illustrates a 12-key pushbutton telephone labled to indicate PABX functions. It shows some of the functions of the Rohm PABX. To give the keys a special meaning, pressing them is preceded by pressing * or #.

ADD-ON FACILITIES

Many users, however, retain a common carrier PABX without many of the functions in Box 6.1. In some countries they do not have a choice. A variety of devices are coming onto the market which operate *in conjunction* with an

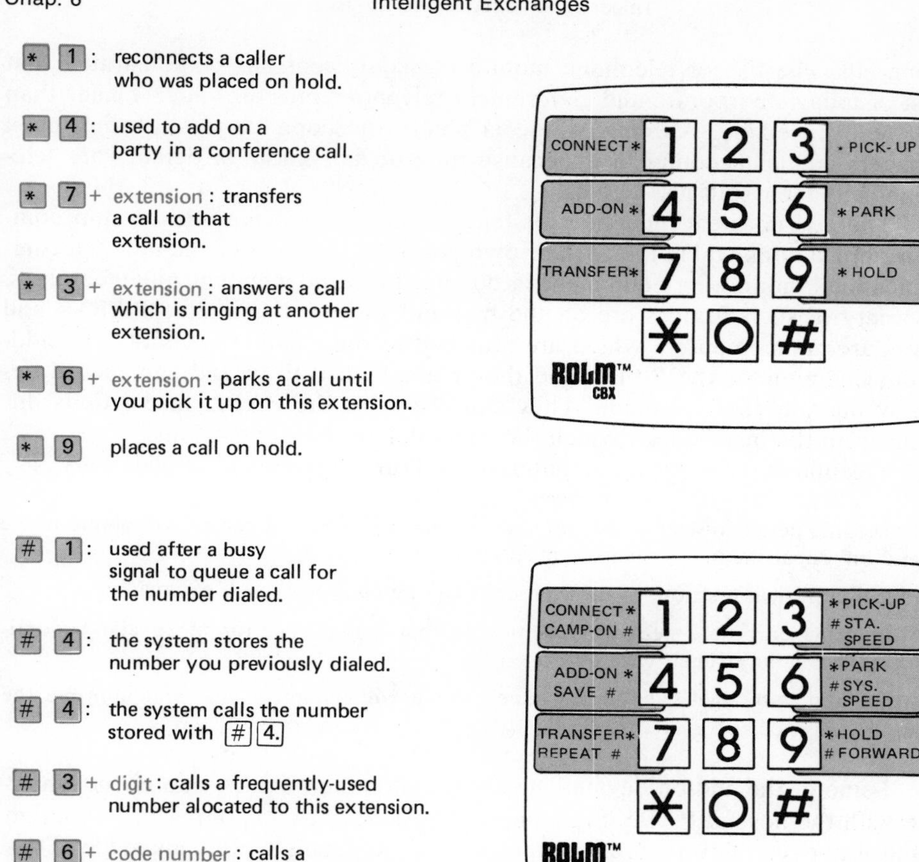

* 1 : reconnects a caller who was placed on hold.
* 4 : used to add on a party in a conference call.
* 7 + extension : transfers a call to that extension.
* 3 + extension : answers a call which is ringing at another extension.
* 6 + extension : parks a call until you pick it up on this extension.
* 9 places a call on hold.
CONNECT*
ADD-ON*
TRANSFER*
* PICK-UP
* PARK
* HOLD
ROLM™ CBX
1 : used after a busy signal to queue a call for the number dialed.
4 : the system stores the number you previously dialed.
4 : the system calls the number stored with # 4.
3 + digit : calls a frequently-used number alocated to this extension.
6 + code number : calls a frequently-used business number allocated to the system.
9 + extension: diverts a call to another extension.
CONNECT * CAMP-ON #
ADD-ON * SAVE #
TRANSFER* REPEAT #
* PICK-UP # STA. SPEED
* PARK # SYS. SPEED
* HOLD # FORWARD
ROLM™ CBX

Figure 6.5 Labels on a pushbutton telephone used to indicate the functions of the Rohm PABXs.

established PABX to provide additional features. Such features include traffic monitoring, automatic call distribution, paging, automatic telephone answering, least-cost routing, automatic call accounting, alarm clock features, scanning data collection terminals, and other teleprocessing features. In some installations one finds an electromechanical PABX connected to a small computer which provides add-on functions.

LOWERING THE TELEPHONE BILL

One category of computer function which might be of particular concern to today's telecommunications manager is that which can lower the overall telephone bill. The use of telephone monitoring equipment has demonstrated that most of today's corporate and government telephone bills are much higher than they really need be. Although there is plenty of scope for cost cutting most managers cannot accomplish it because they do not know in detail what telephone calls their staff are making.

There is no great incentive for the telephone companies to develop computer applications which lower their own revenue, so much of the new telecommunications monitoring equipment is coming from independent manufacturers. A variety of such devices are on the market. Some are complete PABXs and others are minicomputers which are attached to the existing PABXs. The add-on devices monitor the PABX and the lines attached to it, and can record details of all calls that are made. They can also give the PABX instructions different from the instructions which the users dial at their extensions.

Telephone monitoring computers have four categories of applications:

1. Recording details of calls made by users so management can control telephone usage and bill departments or clients appropriately.
2. Placing physical restrictions on what calls can be placed by each extension.
3. Monitoring corporate network facilities so that network performance can be optimized and failures dealt with quickly.
4. Automatic routing of calls in a manner which will minimize cost and improve the service to specified users.

Some of the add-on devices merely monitor the circuits, and do not interfere with the normal PABX functioning. These devices do not perform functions 2 and 4 above. They record on tape or disk the details of all calls made. These data are processed by a conventional computer to provide all manner of reports for management.

One type of report lists the calls made and the costs incurred by each extension. The extension user may then be asked to indicate which were personal calls, and to explain particularly expensive telephone usage. The telephone expenses are broken down by department, and managers are often astonished by them. In many organizations there is a high level of telephone abuse and the telephone bills drop as much as 40% as soon as the calls are monitored. Service organizations who bill their phone calls back to clients, for example lawyers, architects, consultants, accounts, etc., can have the computer prepare these bills. Most such organizations handle telephone charges very loosely, often billing their clients only 55% to 65% of the billable calls [1], the remainder being absorbed by the partnership as a cost of doing business. Some monitoring devices make the telephone user key in a number with all calls to indicate the subject matter, for billing or budget control.

Another type of report relates to the utilization of the corporate network. A corporate network can use many different types of lines and tariffs—tie-lines,

Figure 6.6 A telephone system for small business which prints out a record of employees' calls so that the office manager can immediately control misuse of the telephone. *(Manufactured by Tele/Resources, Inc.)*

WATS line, foreign exchange lines, specialized common carrier circuits, telpak, satellite circuits, as well as direct distance dialing. There is much scope for cost reduction in selecting the optimum mix of facilities. Cost minimization needs a computer program and a detailed knowledge of the traffic. The telephone monitoring computer can provide reports on network utilization and recommendations about where to modify the network—reduce trunks, add a WATS circuit, a foreign exchange line, and so on. The monitoring reports will summarize calls which were not successfully connected as well as the successful calls. This is valuable because most telecommunications managers have no measure of user frustration.

Often, corporations have no means of knowing whether their expensively leased tie-lines are all working correctly. The lines are arranged into groups of trunks; when one appears busy the next is automatically selected. A faulty trunk may appear permanently "busy" and the equipment always bypasses it. Sometimes the user fails to obtain a connection and he simply redials. Monitoring equipment can be programmed to detect trunks which are permanently "busy" or otherwise faulty. When such equipment was first used in the mid 1970s the results were a shock to many corporations. A 20% trunk outage rate was not uncommon [1], i.e., one in five of the trunks that some corporation had been paying for had been faulty with the faults undetected. The users had merely encountered somewhat more busy signals than usual.

Intelligent monitoring equipment reports faulty trunks almost as soon as they go faulty. Users encountering tie-line problems can dial a code (on some systems they dial the word "BAD"); the system then examines the circuit and makes appropriate trouble reports.

CALL INTERCEPTION

Devices added to PABXs which intercept the calls the users dial are more expensive, but can control telephone costs in a more direct fashion. Each extension can have restrictions placed on it which prevent it making trunk calls, prevent it making international calls, allow it to dial only between 9 A.M. and 5 P.M., allow it to call only certain area codes, and so on. Some systems force a user key in a billing or subject matter code before they complete any trunk call.

A more elaborate capability which can save more money in a large corporation is *least-cost routing*. The potential savings from least-cost routing are likely to become more diverse and more widespread. PABXs of sophisticated corporate users can connect to any of the variety of different types of trunks that corporations use. These trunks all have different costs and hence the desirability of minimum cost routing decisions. The user placing a call is not motivated to route his call at minimum cost—he merely wants his call to go through quickly—neither is he able to select the minimum cost path because he does not know the tariffs or the current call activity.

A minimum-cost routing facility will try to route each call by the cheapest route first—perhaps a WATS line, perhaps a specialized common carrier trunk, perhaps in the future a satellite circuit. The lowest-cost circuits to that destination may all be occupied. Then a decision is necessary: does the facility place the call by a more expensive route or does it make the caller wait? Calls for high executives may be put through immediately, even if the most expensive routing is necessary—direct distance dialing. Lesser callers may be made to wait, but not more than 30 seconds; if no cheap trunk becomes free in 30 seconds they are given a more expensive route. The lowest level callers may be permitted to use only the cheapest trunks.

It will thus be observed that the question of caller priority enters into call routing. The highest priority callers may be given excellent service and almost never receive network *busy* signals. Lower priority callers may be given busy signals or made to wait a short time until a low cost circuit becomes available. Route selection mechanisms thus not only lower the telephone bill, but provide better service to those persons whose time is deemed valuable. To an executive plagued by an excessive number of busy signals on his corporate network, the improvement in service can be a godsend.

Box 6.3 summarizes the effect of installing a WATSBOX, an add-on minicomputer marketed by Action Communication Systems, Inc., in six large corporate users. The number of calls that can be handled by a given corporate network is increased yet the total costs are decreased [2].

BOX 6.3 The effect of connecting an Action Communication Systems, Inc. WATS-BOX to existing PABXs in six typical installations [2].

Customer	Long Distance Costs	Cost Per Call	Total Calls Per Month
A	Down 32%	Down 44%	Up 36%
B	Down 32%	Down 54%	Up 54%
C	Up 4%	Down 28%	Up 47%
D	Down 6%	Down 19%	Up 15%
E	Down 14%	Down 18%	Up 38%
F	Down 20%	Down 35%	Up 54%

An application of private switching devices which has been particularly economical in some organizations permits calls to be placed to or from telephones not on the premises of the organization in question. Calls can be placed from employees' homes to a local PABX or add-on device which then permits the employees to set up long-distance or conference calls using the corporate facilities. An executive at home in New York can call an executive at home in San Francisco at night for the cost of a local call. The call is routed over otherwise idle trunks, leased in some cases from a specialized common carrier. Similarly customers in a distant city can be called on corporate tie-lines to a PABX in that city.

When, in the future, corporations find it economical to have some of their employees working at home, it will be valuable to link their homes in this way into the corporate telephone networks.

An add-on device manufactured by the North Electric Company, called a *communications extender,* gives dialed access from external telephones to a corporate PABX and hence enables remote locations to take advantage of it. A user dials the telephone number of the device and then dials a security code (to prevent unauthorized use). The communications extender is connected directly to a PABX and permits the calling telephone to do whatever could be done from a PABX extension. In particular it may be connected to corporate network facilities so that the user is given direct access to specialized common carrier or leased trunks.

Figure 6.7 The PABX features discussed in this chapter, and other features such as *direct inward dialing* and *cell tracing*, will be provided by computerized central offices such as Bell ESS (Electronic Switching Systems). One central office system can provide the features to many small users. *(Photograph courtesy of AT&T.)*

CENTRAL OFFICE SYSTEMS

The PABX functions we have described in this chapter could be performed by central office equipment. One central office can provide them in a time-shared manner for many subscribers. If the central office is a stored program computer, new functions can be provided by modifying its programming. Central office provision of the functions has two advantages. First, the users do not have to worry about purchasing or leasing PABX equipment. Very small users could make use of central office features. Many of the functions we have discussed can be provided for *home* use. Second, a central office can provide *direct inward dialing;* i.e., each extension has a separate external telephone number and hence can be dialed directly on the telephone network (*Centrex* service in North America).

AT&T is planning a set of Centrex ESS (Electronic Switching System) user packages. Features such as those listed in this chapter are packaged into "Generic Programs." There are some features which can only be done by central office operation, such as *call tracing,* i.e., establishing a local caller's telephone number.

Less than five per cent of central offices are computerized as yet in the U.S.. and a lower proportion in most other countries. So central office programming will not eliminate the need for intelligent PABXs.

FUTURE SYSTEMS

Computerized switching will be a vital part of many future networks. The advent of this technology makes complex schemes practical incorporating radio telephones, satellites, and other new technology. Microminiaturized solid state switches will be a part of many future systems. Telecommunications has at last escaped the domination of electromechanical switching.

REFERENCES

1. Harry Newton, "Computerized Telephone Accounting Explodes," *Telecommunications,* May 1975.
2. Action Communication Systems, Inc., a publicity description of usage of their WATSBOX, Suite 3005, 122 E. 42 St., New York, N.Y., 10017, 1975.

Boy operators were used in the first offices, following the pattern of the telegraph industry which had always used boys as operators and messengers. Girls later proved to be more capable as operators and by 1890 girls were operating practically all Bell System switchboards during the daytime. They did not operate them at night however because it was considered improper for a girl to take a job that involved being away from home after dark. *(Courtesy AT&T.)*

7 PUSHBUTTON TELEPHONES AND VOICE RESPONSE

The cheapest way to converse with a remote computer today is by means of a conventional telephone. The telephone user first dials the computer, and when the connection is established, uses the same instrument for sending data. This operation is facilitated by the use of a telephone with a pushbutton keyboard such as the AT&T *Touchtone*® telephone.

CCITT has recommended that pushbutton telephones have the 16 buttons shown in Fig. 7.1, or a subset of them. Each button generates two fre-

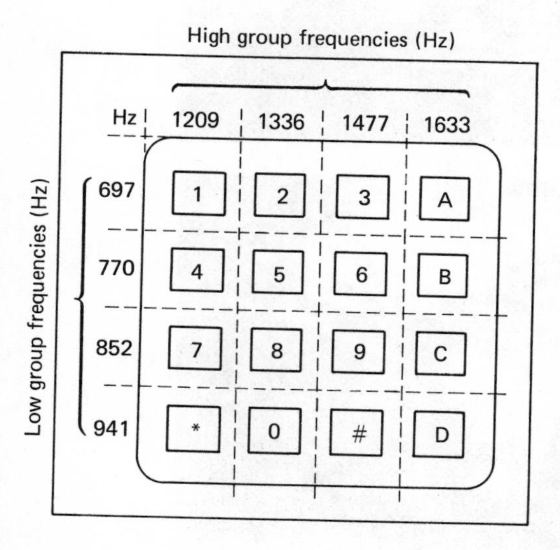

Figure 7.1 CCITT recommendation Q23 for the buttons and frequencies on a pushbutton telephone set. Pressing a button generates two frequencies, one from each group. The North American Touchtone telephones conform to the recommendation but do not use the A to D keys.

® Registered Trade Mark.

quencies in the voice band as shown. Most North American telephones have 12 buttons, and exclude the A to D buttons. For data transmission at least one button is needed in addition to the 10 numeric buttons to condition the meaning of these 10 buttons. Pushbutton telephones are spreading fast in North America. Touchtone telephones are cheaper to manufacture than rotary-dial telephones which need an electromechanical mechanism with many precise components. For areas which do not have pushbutton telephones, a small extra keyboard may be added to a dial telephone, as in Fig. 7.2.

Figure 7.2 (a) The Touchtone telephone keyboard has twelve keys rather than ten. The two nonnumeric ones, "✻" and "#", facilitate its use in many data transmission applications. (b) Where pushbutton telephones are not yet in service, a small separate keyboard can be added to the telephone, as shown.

(a)

(b)

In this chapter we discuss how the pushbutton telephone can be used as a computer terminal without additional equipment, except at the computer center. As we will see, several types of interesting and useful applications can be constructed. The availability of such services will enable the public to dial a computer from their own homes and use it at very little cost.

A variety of computer terminals have also used voice answerback because it provides the cheapest method of giving a versatile response. There are many types of dialogue in which this is satisfactory although for some a written or displayed response would be much better.

VOICE ANSWERBACK MACHINES

A variety of voice answerback machines exists on the market. Several are inexpensive and have a vocabulary of a small number of fixed words. Some are parts of computer systems with potentially very large vocabularies.

There are basically three ways in which the human voice can be stored in machines:

1. *In an analog form.* It can be stored as on a phonograph disk, magnetic disk, or recording tape, perhaps in fixed-length words. A simple electronic circuit can transfer it to the telephone line.
2. *In a directly digitized form.* Telephone sound can be converted directly into a digital bit stream as in PCM telephony in which one second of sound is represented by 56,000 bits. These bits could be stored on computer disks or other storage devices. A computer program could select and put together different variable-length strings. A digital-to-analogue operation converts them back into sound.
3. *In encoded digital form.* Rather than storing digitized sound, the words themselves can be encoded. A sequence of digits represents the timing, energy, and frequency associated with the spoken words, in a compact form. Special circuits must then reconstruct the sound of the word in order to speak it. Spoken words can be encoded in various ways. Typical schemes would require less than one-twentieth of the bits direct digital conversion requires. They would give clearly intelligible words but would not reproduce inflections exactly or necessarily permit the speaker's voice to be recognized as would direct digitization. However, the whole human vocabulary could be stored on a computer disk.

An inexpensive example of the first method is the Cognitronics "Speechmaker." This uses a 31-word vocabulary in the form of a photographic film drum. A five-bit code is used to select one of the 31 words, or silence. The drum is easily interchangeable. Several drums with prerecorded vocabularies are available, and drums with new words can easily be made. The sound is generated by a light beam shining through the film onto a photoelectric cell. The output of the cell is amplified and put onto a telephone line. The drum rotates every 625 milliseconds, and each word or segment of speech lasts 600 milliseconds. The drum generates a timing pulse once per revolution, and in the next 25 milliseconds the switching to the selected sound track takes place. The circuit remains switched for the 600 milliseconds of playing time.

For many applications it is desirable to have some longer responses—for example, complete sentences like "Please key in your account number" or "Consignment of automotive parts shipped to Amsterdam on Pan American Flight Number 2 on Saturday November 18." Such sentences can be composed from variable-length segments of speech on machines like the Periphonics voice response units. The Periphonics machines can store a large vocabulary of variable-length speech segments on a magnetic disk. The phrases can be assembled by a computer and spoken over telephone lines using conventional input/output instructions.

An example of the third approach is the IBM 7772 [2]. With this machine, the words are stored on computer storage in compacted digital form. Table 7.1 shows the lengths of typical words and the number of eight-bit bytes of storage they require. One second of sound requires about 2400 bits of stor-

Table 7.1 The storage requirement for typical spoken words in the IBM 7772 voice answerback system [2]. This is only 4% of the bits used for PCM telephone speech.

Spoken Word	Length (in milliseconds)	Number of 8-Bit Bytes Used for Storage
One	466	139
Two	447	137
Three	472	145
Four	554	122
Five	686	154
Six	567	148
Seven	524	129
Eight	456	157
Ten	461	148
Accrue	676	229
Accrued	772	262
Action	536	223
Actual	600	233
Affiliation	923	259
Allot	581	207
Allowable	779	210
Allowed	642	186
Analysis	776	233
Appropriate	746	303
Appropriation	950	391
Asked	601	190
Assistance	703	259
Assumed	703	211

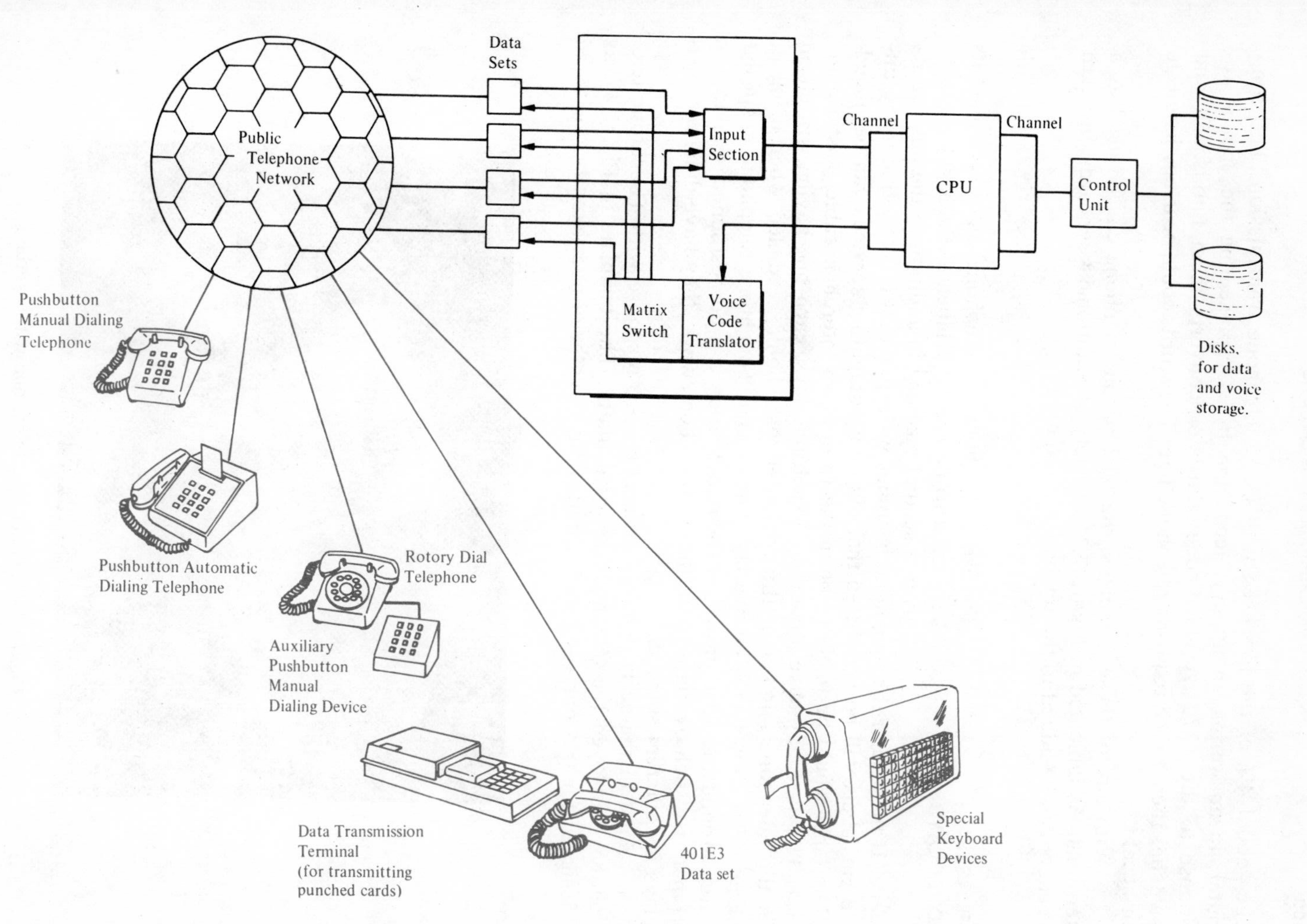

Figure 7.3 Devices used with voice response systems.

age—about 4% of the bits used with PCM telephone transmission. The computer, in answering an inquiry, feeds the bytes representing the spoken response to the 7772. A voice-code translator converts them into sound and switches them to the requisite telephone line. The words could be in any language or accent.

With any of these machines, the pushbutton telephone can be used as a terminal. As indicated in Figure 7.3, a variety of additional keyboards and card readers can be added to the telephone if required.

PLASTIC OVERLAYS

To use a pushbutton telephone as a terminal, its keys need to be specially labeled. Each key must be given a second meaning in addition to the digit on its face. If this use of the telephone becomes widespread, it is likely that many different labels will be needed so that different computing services can be dialed. The label would be likely to be in the form of an interchangeable plastic overlay. The user may have a booklet of such overlays, each one having written on it the telephone number of the computer that should be dialed when using it. Figure 7.4 shows such an overlay. The normal Touchtone telephone has a flat area around the keys on which such an overlay can be positioned. Children in Japan use such overlays to do their school homework; they have different overlays for different subjects.

Some of the early demonstrations of voice response with Touchtone telephones employed the telephone as a calculator. The template in Fig. 7.4 was designed for this purpose. Such applications have been overtaken by the rapid

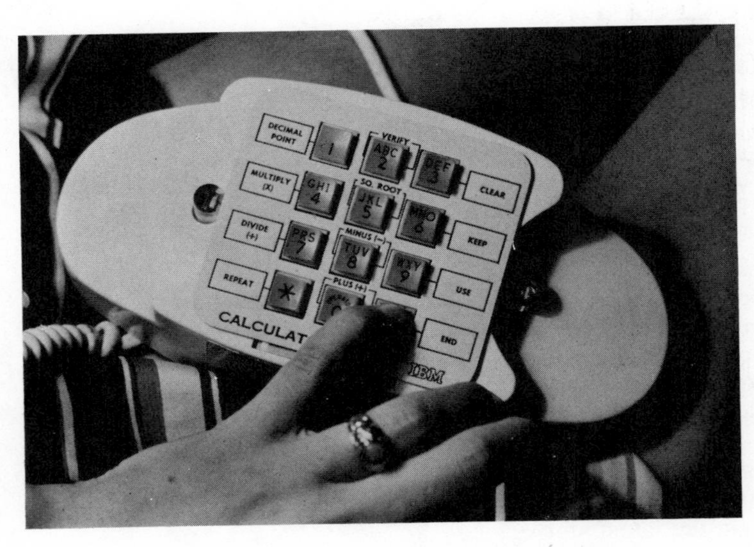

Figure 7.4 Touchtone telephone with overlay.

spread of pocket calculators. The more ingenious schemes included ways to write programs and enter them with telephone pushbuttons [2].

The spread of pocket machines, including programmable ones, implies that most viable applications of voice response will be to computers which provide information or give a service which cannot be performed with pocket machines or other microcomputers.

VARIETIES OF APPLICATIONS

Voice answerback has a large number of potential applications both for commercial and scientific uses:

Manufacturing Industries
Salesmen inquiries
Order entry
Job status
Parts inquiries
Inventory control

Retail Stores
Credit inquiries
Stock status
Merchandise control
Order inquiries
Credit and collection inquiries

Hotels and Motels
Reservations inquiries

Hospitals
Laboratory report inquiries
X-ray results
Patients' characteristics
Patients locations

Insurance Companies
Policy inquiries

Educational Institutions
Substitute teacher assignments

Banks and Financial Institutions
Deposit accounting inquiries
Installment and commercial loans
Savings account inquiries
Brokerage inquiries
Customer credit information

In addition to private commercial applications, voice answerback could have a variety of uses that could appeal to the domestic user. For example, he could obtain the latest stock prices and ask for other information about a stock that might interest him. His home telephone would become the equivalent of the terminal that a stock broker uses. Similarly, he could request movie and theater information—perhaps making reservations directly from his home. He could ask for plane, train, and bus schedules. New transportation schemes have been proposed that involve computerscheduled buses that pick customers up from their homes or offices with completely flexible routing. The customer dials for the bus, and could communicate with its scheduling computer using a pushbutton telephone.

Diet planning and menu selection by means of a voice answerback computer have been proposed. New employment could be sought. A voice answerback system could be used for obtaining hotel rooms, domestic help, library books, magazine subscriptions; in fact it could be used for almost anything from attempting to buy a second hand car to real-time computer dating.

The applications can become quite complex. North American Aviation has used a scheme to allow engineers to inquire about engineering drawings. A file of 75,000 or so drawings was searched, and the latest changes listed. The Equitable Life Assurance Society has used voice answerback in the processing of insurance applications.

In Detroit a real-estate system was developed by the Realatron Corporation, whereby several thousand participating brokers could enter information about the location, price range, style, number of bedrooms, and up to fifteen other characteristics of houses in which they were interested. After receiving this data, the computer then searches its files. A variety of credit-checking and banking operations has been performed with pushbutton telephones.

In New York and Illinois, Touchtone telephones have been used for teaching. The New York City Board of Education has used a system that can drill up to 2,000 students, in grades 2 through 6, in arithmetic. The student uses the Touchtone telephone at home. The computer poses voice questions to the student. He answers at his own rate and the computer notes the responses. In Oak Park River Forest High School, near Chicago, a computer instructs and drills telephone students on subjects as varied as history, biology, languages, business education, mathematics and the physical sciences. The instruction is from audio lessons taped on continuous loops of one-inch tape. Each loop contains 32 different sound tracks of up to 15 minutes in length.

Where a number of plastic overlays are used to enable a user to carry out different applications at his terminal, it would be useful to standardize some of the operations. Figure 7.5 shows a blank overlay; Figures 7.6 to 7.9 show overlays marked for several applications that might apply to a home user. Each overlay contains the telephone number of the computer that must be dialed with it. An "application number" tells the computer which overlay is on the telephone. The number is sent when the computer requests it.

Figure 7.6 shows a template labeled for obtaining stock market information. Figure 7.7 shows one for sports information. In the former, the user must know the three alphabetic characters that are used to reference each stock. In the second, he must have a listing of numbers referring to teams and players.

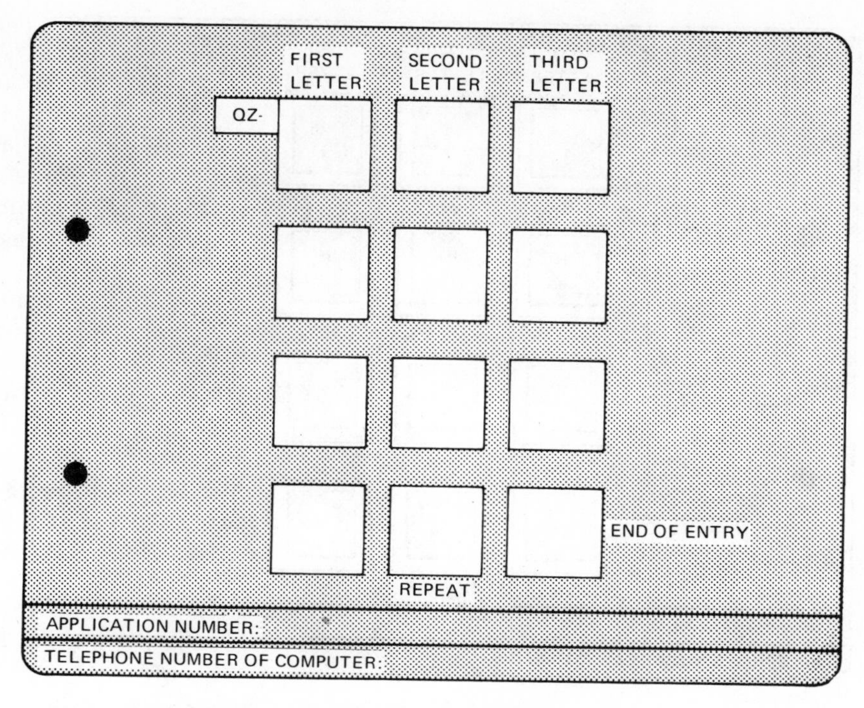

Figure 7.5 Blank overlay for labeling pushbutton telephone for diverse applications (see Figs. 7.6 through 7.10).

Figure 7.6 Overlay for labeling pushbutton telephone for stock market enquiries.

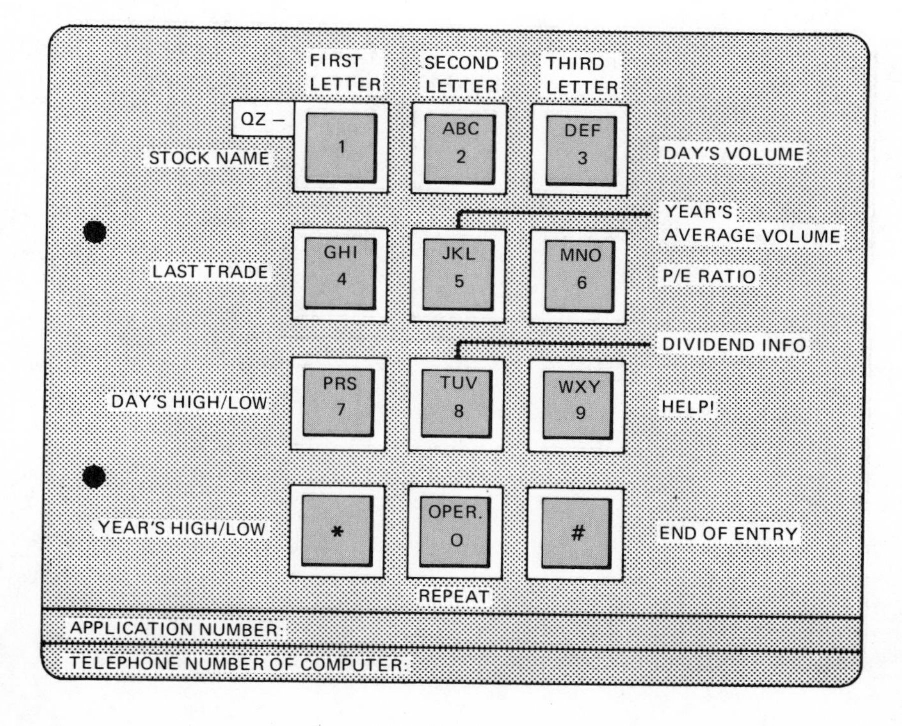

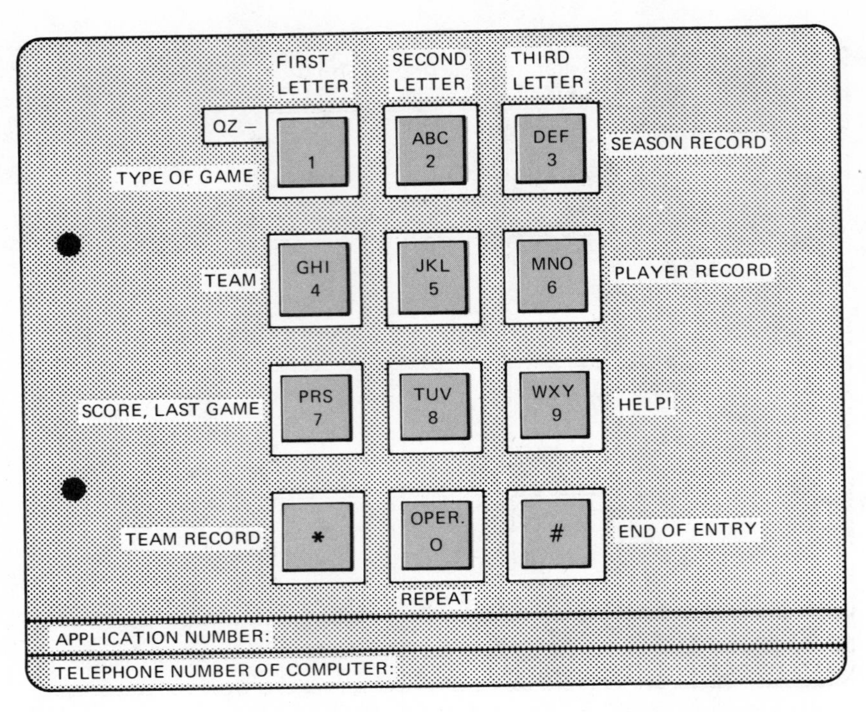

Figure 7.7 Overlay for labeling pushbutton telephone for obtaining sports information.

Figure 7.8 Overlay for labeling pushbutton telephone for theater/stadium inquiry/booking service.

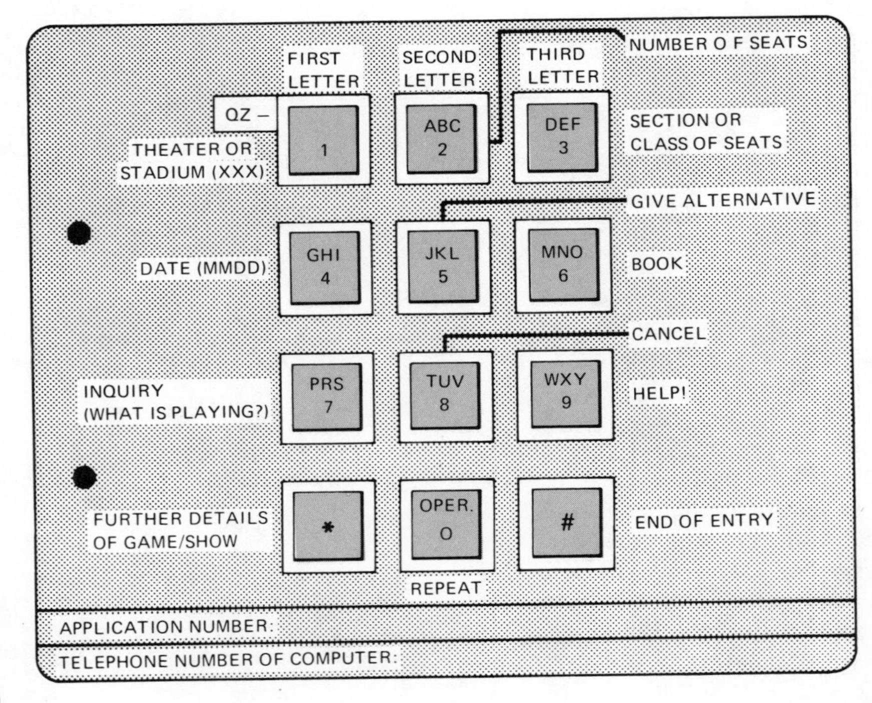

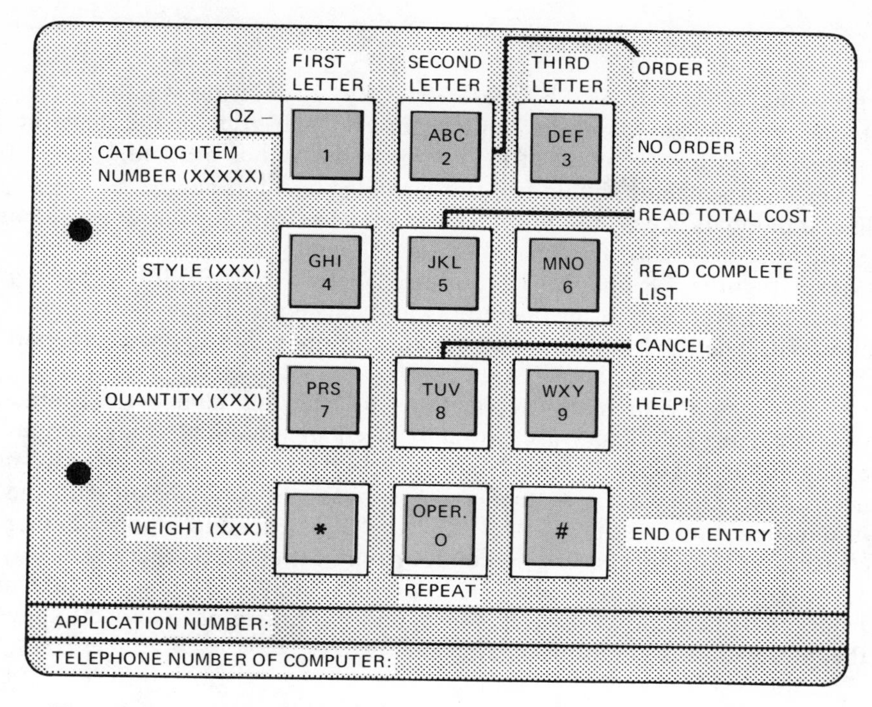

Figure 7.9 Overlay for labeling pushbutton telephone for home catalogue ordering service.

The question arises, how will a user be billed for these services? He could be automatically billed by the telephone company every time he dials the number of the computer in question. Alternatively, he might pay once a quarter, perhaps, for the privilege of using the service and would then be given a self-checking user number that could be registered in the computer. Information could then only be obtained after keying in the user number. A better method would be to make each telephone set automatically generate a unique signal number after the key labeled "User Number" has been pressed. This is done on some terminals.

The next two figures show templates in which the user is requesting a service. Here the billing problem does not arise because the user is billed only for the service he receives that normally involves something mailed or delivered to his address. Mere inquiries are free, although again a customer number is required. Figure 7.8 shows a template for inquiring about theater programs, concerts, sports events, and for making reservations. Such a service might be provided by a theater agent. The user will key in the number of seats he wants and the section or class. The machine will read out appropriate seat numbers and the user will inform it if he wishes to reserve these. He can request alternatives and can cancel seats if he wants. If time permits, the tickets will be mailed to the customer; otherwise he will pick them up at the office or theater.

Figure 7.9 shows a similar scheme for ordering goods. Such a service might be provided by a merchandising organization. For each item number keyed in, the computer will read back what it is as a check and will ask the user what quantity or weight he requires. It will then quote a price. The user will press the "ORDER" or "NO ORDER" key to indicate whether or not he wants the item. At any point, he can ask for the total price of the items he has ordered so far or for a readout of the complete list he has ordered.

Any such schemes should use self-checking account numbers to lessen the likelihood of bogus access to the system. It would be of value on some systems to have other security schemes for identifying the caller.

Both Figures 7.8 and 7.9 have a key labeled "HELP." Pressing this key gives the caller access to an operator who can help overcome any difficulties encountered. The "#" key either terminates an operation or terminates an entry.

It is necessary in many applications to transmit alphabetic characters. Three alphabetic characters are normally used in obtaining stock quotations, for example. On a Touchtone keyboard these have to be entered in a two-character form. An easy way to do this is to first press the key labeled with the letter in question, and then press "1," "2," or "3," depending upon whether the user wishes to transmit the first, second, or third letter on that key. Thus pressing "5" and then"2" would mean "K." Unfortunately, "Q" and "Z" are missing from the Touchtone keyboard. These, and "–" have been added to key "1" for this illustration, and are marked on the overlay.

CORPORATE SERVICES

In a corporation a wide variety of applications can be built around the use of pushbutton telephones. Pushbutton dialogues have been used for shop-floor data collection. A system could be designed for locating employees. A variety of information services could be built.

Figure 7.10 shows a template for providing information about products. The user can press any labeled key and then "END-OF-ENTRY," or he can follow the pressing of the labeled key with the entry of a number of a qualifier. Thus if he presses MODEL, END-OF-ENTRY, the machine will say what models exist of the machine in question. If he presses MODEL, 2, END-OF-ENTRY, the machine will speak details of Model 2.

To be useful the voice answerback unit must have the capability to play back variable-length segments of speech. It may, for example, have the capability to scan a section of magnetically recorded sound very quickly and select the required response. A variety of devices have this capability.

A dialogue with such a device might proceed as follows:

TELEPHONE USER: MACHINE NUMBER, 3, 3, 3, 0, END-OF-ENTRY
VOICE ANSWERBACK: 3330 LARGE CAPACITY DISK STORAGE
TELEPHONE USER: HIGHLIGHTS, END-OF-ENTRY
VOICE ANSWERBACK: 1. DATA RATE
2. ACCESS TIMES
3. CAPACITY

Figure 7.10 Overlay for labeling pushbutton telephone for providing information about products in a corporation.

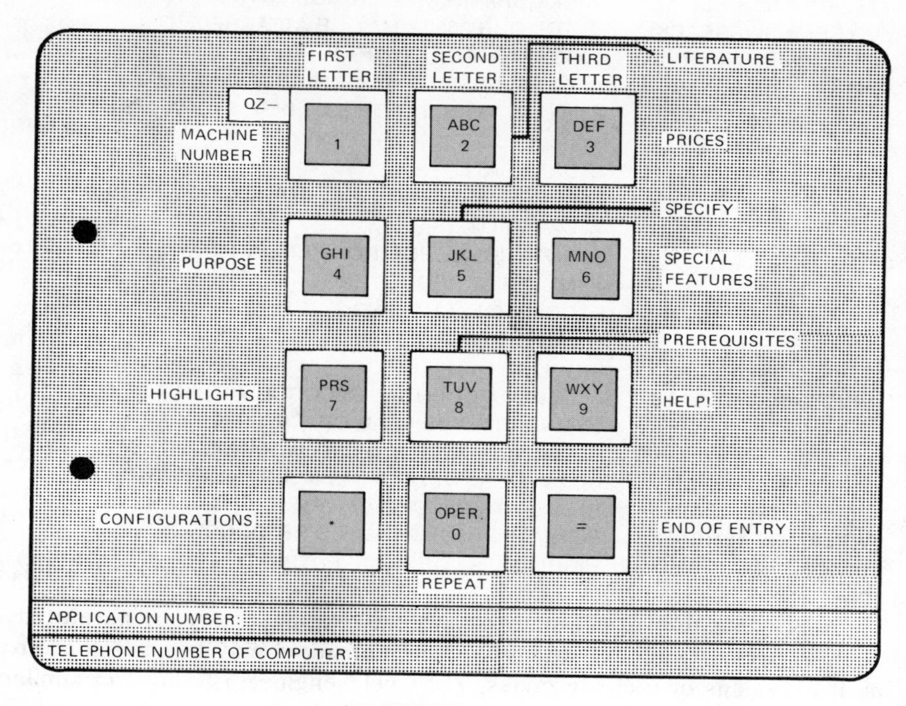

	4. CYLINDERS
	5. POWER DRAWERS
	6. WRITE INHIBIT SWITCH
	7. DISK PACKS
TELEPHONE USER:	HIGHLIGHTS, 1, END-OF-ENTRY
VOICE ANSWERBACK:	DATA RATE 806 THOUSAND BYTES PER SECOND
TELEPHONE USER:	HIGHLIGHTS, 2, END-OF-ENTRY
VOICE ANSWERBACK:	AVERAGE ACCESS TIME: 30 MILLISECONDS
	MINIMUM ACCESS TIME: 10 MILLISECONDS
	MAXIMUM ACCESS TIME: 55 MILLISECONDS
TELEPHONE USER:	HIGHLIGHTS, 3, END-OF-ENTRY
VOICE ANSWERBACK:	TRACK CAPACITY: UP TO 13,030 BYTES
	CYLINDER CAPACITY: UP TO 247,570 BYTES
	DISK PACK CAPACITY: UP TO 100,018,280 BYTES
	UNIT CAPACITY: UP TO 800,146,240 BYTES
TELEPHONE USER:	REPEAT, END-OF-ENTRY
VOICE ANSWERBACK:	REPEATS THE PREVIOUS REPLY
TELEPHONE USER:	LITERATURE, END-OF-ENTRY
VOICE ANSWERBACK:	MANUAL NUMBER GA22-6822
TELEPHONE USER:	CONFIGURATIONS, END-OF-ENTRY
VOICE ANSWERBACK:	1. MINIMUM CONFIGURATION
	2. MAXIMUM CONFIGURATION

TELEPHONE USER:	CONFIGURATIONS, 2, END-OF-ENTRY
VOICE ANSWERBACK;	UP TO FOUR 3330's MAY BE ATTACHED TO A 3830 STORAGE CONTROL
TELEPHONE USER:	MACHINE NUMBER, 3, 8, 3, 0, END-OF-ENTRY
VOICE ANSWERBACK:	3830 CONTOL UNIT FOR 3330 DISK STORAGES
TELEPHONE USER:	HIGHLIGHTS, END-OF-ENTRY
VOICE ANSWERBACK	1. DATA RATE 2. ERROR DETECTION AND CORRECTION CAPABILITIES 3. ROTATIONAL POSITION SENSING 4. MULTIPLE REQUESTING 5. COMMAND CHAINING 6. RECORD OVERFLOW
TELEPHONE USER:	HIGHLIGHTS, 5, END-OF-ENTRY
VOICE ANSWERBACK:	ALLOWS SEQUENTIAL RECORDS WITHIN A CYLINDER TO BE READ OR WRITTEN BY A SEQUENCE OF CHANNEL COMMANDS WITHOUT ROTATIONAL DELAYS BETWEEN RECORDS
TELEPHONE USER:	HIGHLIGHTS. END-OF-ENTRY
VOICE ANSWERBACK:	REPEATS THE HIGHLIGHTS RESPONSE
TELEPHONE USER:	HIGHLIGHTS, 3, END-OF-ENTRY

It is possible that a salesman or systems engineer may be permitted to dial such a service from a customer's office. Outsiders may be prevented from dialing it by means of security codes. IBM field engineers today can similarly dial a voice answerback service to assist them in diagnosing teleprocessing faults.

An application such as that on the previous page needs a voice response unit with a large vocabulary and the ability to speak sequences varying in length from very short to very long. A number of applications are already in successful operation which have these needs. One is used by Emery Air Freight in Connecticut.

Emery wanted to give customers the capability to enquire about the status of their shipments from an ordinary telephone or from an inexpensive terminal. Voice response seemed to be the best way to do this but a very large vocabulary was needed and it was necessary to handle a substantial number of incoming calls simultaneously.

A typical "conversation" with the Emery system goes as follows:

1. The customer dials the (unlisted) telephone number of the computer.
2. An appealing voice replies in natural-sounding speech: "You have contacted Emery's audio response system. Please enter your Emery shipment number."
3. The customer keys the number on his airbill into either a Touchtone telephone or a small portable terminal which operates like an expanded version of a Touchtone telephone and makes it easier to key in the alphabetic characters in the shipment number. The terminal can be attached to any telephone by means of its acoustical coupler.

4. The voice says "Emery shipment 70216 is being traced." There is then a pause while the control computer, an IBM 370, reads the shipment record and assembles the response. This pause is usually only a second or two because the response time of the system is good. The voice continues "The shipment was delivered on October 19th at 11:25 a.m. Total transit time was 16 hours. The destination tariff point is Birmingham, England. There are three pieces, consisting of automotive parts. The shipment entered Emery service at 1435 hours on October 18th. Thank you for calling Emery.

The voice response unit could have been replying to as many as 93 other enquirers at the same time.

The size of the vocabulary needed is made large by the inclusion of words such as "Birmingham, England" and "automotive parts." The vocabulary must, in fact, include all the cities to which Emery ships and descriptions of all the goods that are used. The vocabulary must be easily expandable because more city names and goods descriptions are likely to be added.

The following questions should be asked in the selection of a voice response unit:

1. How many responses can it give simultaneously?
2. How many lines can it handle (if this is not the same as the answer to the previous question)?
3. What is its vocabulary size?
4. Can the vocabulary be changed easily in the field?
5. Can the vocabulary and number of simultaneous responses be expanded?
6. Are the sound segments it "speaks" fully variable in length?
7. Is the speech natural sounding and of good fidelity?
8. What types of terminals can it support?

In addition, questions about the software are important, such as:

1. Can it be programmed for in the main computer in a simple, problem-free manner?
2. Can the programs be in a high-level language such as COBOL?
3. Will the manufacturer of the voice response unit provide complete software support so that the user has no programs to write other than his application programs in COBOL or the equivalent?
4. Will the manufacturer's support handle the input from telephone or keyboards as well as the spoken output?
5. Will the software edit the incoming messages?
6. Will it detect operator errors in input and automously instruct the operator to correct them?
7. Will it provide a security check on input?
8. Will it handle functions needed for controlling the telephone lines, such as disconnect and dialing out?

In the Emery case, the software is programmed by the manufacturer of the voice response device, Periphonics Inc., and resides in a peripheral PDP 11 computer. It does handle the above functions and the peripheral computer is programmed to emulate a conventional tape unit. The application programs can then be programmed in the main computer in COBOL (or any other high-level language) and the input/output needs merely the GET and PUT and other instructions used when programming for tape units. Using simply structured control words, the COBOL programmer can select a segment of speech that starts and ends at any position on a voice track.

OTHER PUSHBUTTON DEVICES

Keyboards other than pushbutton telephones are used with voice response devices. Often the CCITT recommended frequencies of Fig. 7.1 are employed. 255 different characters could be encoded with these 8 frequencies, or 127 if odd-even parity checking is used. A four-out-of-eight code, in which each character must have exactly four frequencies, can give

Figure 7.11 Small portable pushbutton devices may be used with voice response from telephones anywhere. Salesmen, for example, may send details of orders to a computer. Techniques are in use for building an appropriate level of security in such transmission.

somewhat better protection from errors than odd-even parity checking, and permits 70 possible characters.

Voice response techniques permit a wide variety of applications using all manner of different inexpensive devices. Figure 7.11 shows such a device being used in a public telephone booth.

REFERENCES

1. *Vocabulary File Utility Program for the IBM 7772 Audio Response Unit,* Manual C27.6924-2. IBM Corp., Kingston, N.Y., 1969.

2. Albert Newhouse and Robert A. Sibley, Jr., "On the Use of Very Low Cost Terminals," Document AD 691398. Houston University, Clearinghouse for Federal Scientific and Technical Information, U.S. Department of Commerce, National Bureau of Standards, 1969.

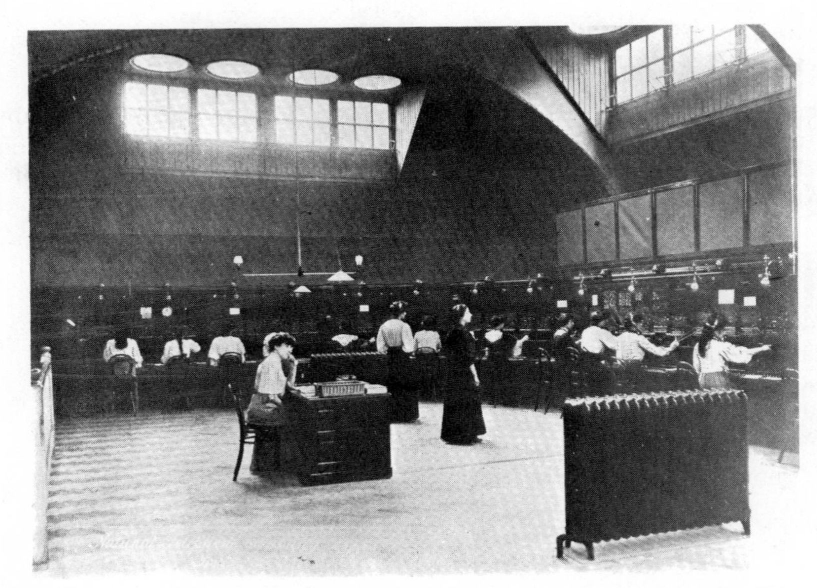

A typical telephone exchange in England in 1906. This exchange handles 4600 lines. *(Courtesy The Post Office.)*

8 VIDEO TELEPHONES

A video telephone is one which transmits and receives an image on a screen. It may be a moving image of a person talking, as with the Bell System Picturephone ® service. It may be a still picture which can change periodically, as with the RCA Videovoice ® set (see Fig. 8.7).

The Picturephone set now being marketed in certain cities is shown in Fig. 8.1. The Picturephone service will be integrated into the voice facilities so that customers will be able to dial and use private-branch-exchange facilities and Centrex facilities in the same way as with their telephone.

THE SUBSCRIBER EQUIPMENT

The Bell Picturephone System design is an impressive piece of modern electronic engineering. The subscriber uses three components (Fig. 8.1). First, there is a conventional *Touchtone telephone,* with 12 keys as in Fig. 7.1. Second, there is the *Picturephone set* with its screen, camera, and loudspeaker; and third, there is a *control unit* which contains a microphone and which permits the user to adjust his picture and speaker volume. There is also a separate control unit containing the power supply and line interfacing electronics. This unit is attached to a wall often far away from the Picturephone set so that it does not clutter the user's room. The user does not need to hold a telephone handset while talking and so can talk naturally and is free to move.

The main characteristics of the system are listed in Table 8.1. The display unit stands on a nonslip ring and can be turned through almost 360 degrees on it. It can be adjusted for tilt. The screen displays 30 frames per second, composed of about 250 lines per frame. However, the lines are interlaced, with odd-numbered lines on one image and even-numbered lines on the next so as to give a flicker rate of 60 times per second, which makes the flicker virtually un-

Figure 8.1 The Picturephone subscriber uses these three components: (1) a conventional Touchtone telephone, (2) the Picturephone set with its screen, camera, and loudspeaker, (3) a control unit that permits adjustment of the picture, and speaker volume. The control unit also contains a microphone.

noticeable (as in television). In a face-to-face conversation we look into each other's eyes part of the time. The Picturephone user will usually look at the eyes on the screen rather than at the camera lens, and thus will appear to be looking away slightly. In order to minimize the annoyance of this, the camera lens is placed just above the screen, which is as close as possible to the eyes of

Table 8.1 Picturephone standards

Bandwidth	1 MHz
Screen size	5 1/2″ × 5″
Frames per second	30
Number of lines per frame	250
Number of picture elements per line	211
Normal viewing distance	36″
Normal area of view	17 1/2″ × 16″ to 28 1/2″ × 26″
Close-up area of view	5 1/2″ × 5″

the person on the screen. The displacement will make a user appear to be looking down slightly, which we often do in normal conversation.

The camera lens has a viewing arc of about 53 degrees. It is slightly less than 12 inches above the desk, which gives a natural viewing angle of the face of a person sitting at the desk. If the person is 3 feet away, the normal field of view will be 17 1/2 in. × 16 in., which gives the person freedom to move from side to side. The control unit, however, has a SIZE knob that can expand from field of view up to 28 1/2 in. × 26 in., which would permit more than one person to be in view. This "zooming" is done electronically, not by lens movement.

The knob above the camera lens gives a distance setting. It can be set to 3 ft, 20 ft, or 1 ft. At the 20-ft setting, it can view a room or a blackboard. At the 1-ft setting, it views an object lying on the desk underneath it, thus enabling it to project documents or pictures. (In the early system installed at the Westinghouse Electric Corporation, one of the users used to put a photograph of the Westinghouse president in this position whenever the Picturephone rang!) The viewing area under the set is 5 1/2 in. × 5 in. To view this, a swivel-mounted mirror automatically swings in front of the lens, and the picture scanning is appropriately reversed. The top of the field of view is the edge of the ring the set stands on.

The lens iris adjusts automatically to the lighting conditions. The strength of the video signal acts as an "exposure meter," but the upper and lower quarters of the picture are not included in the measure so as to avoid false readings from ceiling lights or white shirts. The iris is a unique friction-free mechanism designed like all of this equipment to be as reliable as possible and not incur maintenance expenses. When the iris is fully open, an automatic gain circuit takes over and the set can be used in quite dim lights. The picture quality, however, is better in good light, and the depth of focus is greater when the iris of the f/2.8 lens is not fully open.

OPERATING THE EQUIPMENT

A Picturephone call is initiated by the telephone set in exactly the same way as a telephone call except that the # key is pressed prior to the keying of the telephone number. (The # key is the bottom right-hand one of the Touchtone keyboard.) The telephone number keyed will be the regular telephone number of the person in question. If a subscriber is called who does not have a Picturephone, his telephone will ring and an indication will be displayed on the screen of the calling party that he does not have Picturephone. The calling party can talk to him, and may comment on his lack of Picturephone equipment. It was thought that this might put social pressure on subscribers in affluent areas to obtain Picturephone sets when they have become an accepted status symbol.

The telephone of the person called makes a distinctive Picturephone ring. The person called picks up the telephone handset and the pictures appear on

Figure 8.2 Picturephone in use in a Washington office.

the screen. The user can adjust his picture with the SIZE, HEIGHT, and CONTRAST knobs. When doing so, he can if he likes look at his own image on his screen by pressing the VU-SELF key. If he does not want to appear on the caller's screen, he can press the DISABLE key. He can mute his microphone if he wants with the ON/OFF keys and can adjust the loudspeaker volume of his set with the VOLUME knob.

All the facilities possible with a normal telephone can be used, such as card dialing, key telephone, secretary extensions with or without a picture set, Centrex and private-branch exchange facilities.

PICTUREPHONE PSYCHOLOGY

Picturephone has been heralded by AT&T as a major social innovation. Julius P. Molnar, executive vice president of the Bell Telephone Laboratories, wrote the following [2]:

> Rarely does an individual or an organization have an opportunity to create something of broad utility that will enrich the daily lives of everybody. Alexander Graham Bell with his invention of the telephone in 1876, and the various people who subsequently developed it for general use, perceived such an opportunity and exploited it for the great benefit of society. Today there stands before us an opportunity of equal magnitude—Picturephone service.

He goes on to say in an article in the *Bell Laboratories Record* [2]:

> Most people when first confronted with Picturephone seem to imagine that they will use it mainly to display objects or written matter, or they are very much concerned with how they will appear on the screen of the called party. These reactions are only natural, but they also indicate how difficult it is to predict the way people will respond to something new and different.
>
> Those of us who have had the good fortune to use Picturephone regularly in our daily communications find that although it is useful for displaying objects or written matter, its chief value is the face-to-face mode of communication it makes possible. Once the novelty wears off and one can use Picturephone without being self-conscious, he senses in his conversation an enhanced feeling of proximity and intimacy with the other party. The unconscious response that party makes to a remark by breaking into a smile, or by dropping his jaw, or by not responding at all, adds a definite though indescribable "extra" to the communication process. Regular users of Picturephone over the network between the Bell Telephone Laboratories and AT&T's headquarters building have agreed that conversations over Picturephone convey much important information over and above that carried by the voice alone. Clearly, "the next best thing to being there" is going to be a Picturephone call.

Not everyone shares this enthusiasm. The London *Economist* in a special issue on telecommunications [3] described the use of the Picturephone set as "a social embarrassment" and said that "talking into it was like talking to a mentally defective foreigner." A more normal reaction is that users like it and enjoy playing with it, but are not yet ready to pay the cost of the service.

Use of the Picturephone does create a feeling of closeness between the parties which is absent on the telephone. There is slight distortion of the face which is disturbing to some users. The camera lens is above the screen and so the eyes of the person you talk to do not appear to look directly into your eyes, but rather to a point below your eyes. There is some parallax distortion when the face is too close to the camera. The camera is sensitive in the infrared part of the spectrum and the human skin is partially transparent to infrared. This causes dark whiskers beneath the skin of a clearly shaven man to be visible, especially when the light is dim.

In talking to my wife on the Picturephone, the visual distortion seems to cause minor emotional distortion. The face talked to does not look quite as it should and we tend to be self-conscious about the unattractive rendering of our own faces which we can see by pressing the VU-SELF key. My wife thinks she needs Picturephone makeup and has a tendency to pull faces. Nevertheless, the call is more intimate and enjoyable than a telephone call.

Another disturbing aspect of Picturephone conversation is that callers tend to stare at each other's eyes. In most normal conversation people look only occasionally into the eyes of the person they are talking to. Lovers and salesmen may stare more constantly. Normally there is an unconscious and elaborate eye ritual during conversation. A Picturephone correspondent tends to stare at you like a television announcer, and this sometimes makes conversation uncomfortable.

Eye contact ritual is not the only aspect of conversation that is changed by Picturephone. I asked a journalist if he could interview people effectively by Picturephone and his reply was emphatically negative. He said that there are many "body language" cues which he receives during an interview which are missing on Picturephone.

After extensive use of Picturephone people become accustomed to it and more relaxed with it. Staff at the Bell Laboratories who have had a Picturephone for some years are remarkably casual in its use often glancing at the set only occasionally. One man has a habit of initiating calls standing up, revealing only his torso to the party he calls.

Picturephone was one of AT&T's most expensive development projects. The research and development cost several hundred million dollars [4] not counting work on the T2 carrier transmission facility (described later) which is needed for Picturephone. Considering the expense, we might have expected substantial research into the relative effectiveness of this over other means of communication. According to Reid [5], however, research justifying video telephones was not done. He comments:

> Although the study of human interactions occupies a central place in a number of disciplines, the human aspects of person-person telecommunications have been curiously ignored. The well-organized platoons of the established disciplines have marched around, rather than across, this area of research. . . . It is necessary to make the negative point that in all this huge body of experimental social psychology research, the question of communication by any media other than face-to-face has been virtually ignored.
>
> . . . despite their obvious relevance to each other, the fields of social psychology and telecommunications engineering have made little contact.

Research into the effectiveness of other new means to communicate is also scant. In the very complex area of men communicating with computers, for example, there is a surprising absence of research. People in many walks of life will interact with computer terminals in the world we are building, yet most man-terminal dialogues are designed very ineffectively. Organizations with enormous financial stakes in this form of communication are charging ahead without studying the psychology of the communication process.

COST AND MARKETING

Curiously, *market* research also seems to have been absent in the development of Picturephone. Many persons at the 1964 World Fair were asked if they liked the Picturephone on display; almost all said "yes," but none were asked "Would you buy it at such-and-such a price?"

The cost in Chicago in the early 1970s was $50 per month for a Picturephone line (which included 30 minutes of usage per month) and $25 per month for each set. The charge for usage after 30 minutes was $0.15 per minute. This charge was promotional, however, and in Pittsburgh it cost twice as much.

In the late 1960s American Telephone and Telegraph forecast that for 1% of the domestic telephones in service, there would be a Picturephone set by 1980, and the figure would be 3% for business telephones [1]. This would mean that for 1.5% of all telephones there would be a Picturephone by 1980. AT&T revised the forecast in 1973 to half a million sets installed by 1980. The initial marketing of Picturephone, however, even at Chicago prices, was not a success. Even if the cost of Picturephone drops substantially with the use of new technology, it seems unlikely to have high sales in the 1970s.

For the 1970s it seems likely that Picturephone cost will remain at least ten times that of telephone. Later it may drop because of widespread use of digital lines, much lower cost of logic circuitry, and compaction techniques which encode the Picturephone signal so that much less information is sent.

As we will explain in more detail later, there are two keys to marketing video telephones. The first is great reduction of the bandwidth (or bit rate) required. The second is giving the user more functional capability.

Probably the severest criticism from communications men in industry is that the money used for Picturephone development could have been better spent. Today's voice channels are overloaded. A grave need for better data communication facilities exists. If the same talent, bandwidth, and capital had been spent on a flexible dial-up data network, its effect on the United States economy would have been incomparably greater.

Indeed, it is possible that the major economic benefit of the Picturephone service may come from its provision of dial-up lines between major business locations that can carry a million bits of data per second or more. This could revolutionize the way we use computers.

TRANSMISSION ON LOCAL LINES

A major tenet of Picturephone development has been to take maximum advantage of existing Bell System facilities. For this reason, a conventional telephone is used for establishing the calls. The control of the network and the signals used for its control and for dialing are basically the same as for the conventional telephone. What must be added are wires to carry the picture in parallel with the existing local loops for telephones into the subscribers' premises, new switching facilities operated in parallel with the telephone switching but under the same control, and trunks capable of carrying the Picturephone signals as well as today's telephone speech. All these factors have been worked out in a fashion that minimizes both their cost and the consequent upheaval in existing telephone plant.

A bandwidth of 1 MHz (1 million cycles per second) is needed to carry

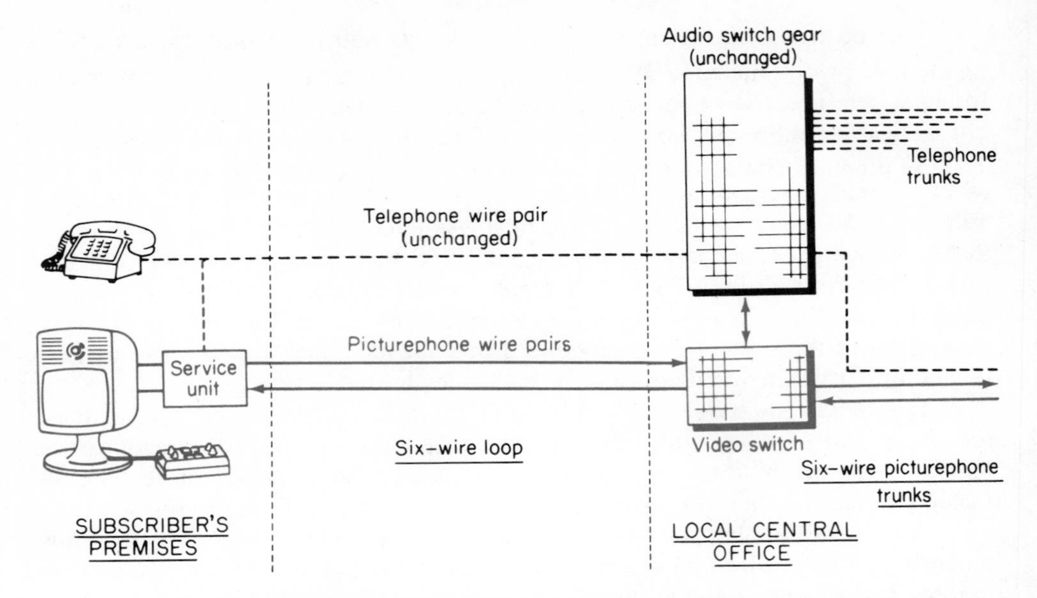

Figure 8.3 Three wire pairs connect the Picturephone equipment to the local central office.

the Picturephone image. This, surprisingly perhaps, is accomplished on wire-pair lines like telephone lines. It was achieved by placing equalizers on the telephone wire pairs about every 6000 ft to give a gain equal and opposite to the attenuation on the wire pair. The line must be equalized over a frequency band from 1 Hz to 1 MHz. Telephone lines now have load coils (inductances) at intervals of about 6000 ft, and the equalizers will replace these. In this way, Picturephone can be carried by modifying today's telephone connections rather than by replacing them. This is an important factor in making the service feasible.

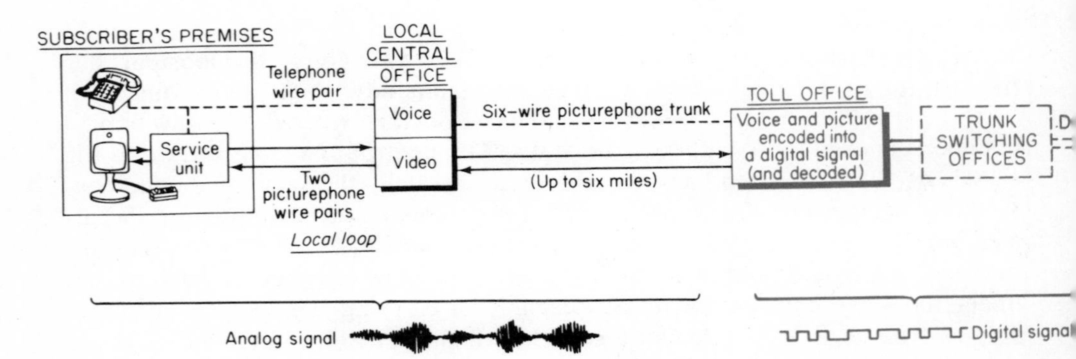

Figure 8.4 Picturephone network links.

The Picturephone set requires two such wire pairs, one for each direction of picture transmission; in addition, the connection to the telephone will remain unchanged. Thus there are six wires from the subscriber's equipment to his local switching office. The equalizers on the picture wires must be carefully adjusted; otherwise distortion accumulating through successive equalizers will produce ghost images caused by echoes in the picture.

Figure 8.3 shows the configuration of these lines.

LONG-DISTANCE TRANSMISSION

As discussed in Chapter 4, telecommunication signals can be carried in either an analog or a digital form. Picturephone signals can be encoded into a bit stream of 6.3 million bits per second. Eventually all Picturephone transmission between offices (except those less than 6 miles apart) will be done in this digital form. Once digitally encoded, the signals will remain in that form until they reach their destination office or the closest office to it that can decode them (Fig. 8.4).

Transmission in this form has two major advantages: first, it prevents the cumulative distortion that transmission over a variety of analog facilities would cause; second, it is substantially less costly.

Two types of transmission media are used today to carry telephone signals across long distances: microwave radio systems consisting of chains of line-of-sight antennas on the hill tops and coaxial cable systems consisting of high-capacity "pipes" buried or strung between poles. Initially the Bell TD2 microwave system and the L4 coaxial cable system will be used to carry the digital Picturephone signal. In terms of the capacity it uses, however, the cost of this long-distance transmission is high. In the microwave system, one Picturephone channel occupies a capacity that could carry 400 telephone channels. In the coaxial system, it displaces 300 telephone channels.

The cost will eventually become less because systems designed for digital transmission will be installed—the T2 and T4 carriers described later on. With

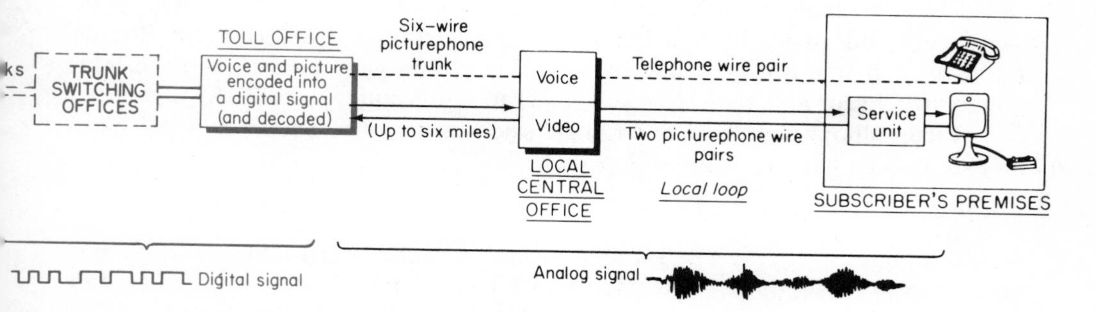

these carriers, the Picturephone channel will displace only 96 telephone channels.

The trade-off in *local* Picturephone distribution will remain far better than that for its transmission between cities. In the local network, only two extra telephone wire pairs are needed and must be modified. The cost of a long-distance call occupying 96 telephone channels will be high; however, certain techniques we will discuss in later chapters promise a dramatic drop in long-distance transmission.

PICTUREPHONE SWITCHING

Just as the transmission of Picturephone signals requires no fundamental change in today's circuits, so in the switching of them maximum use must be made of what is already in existence. It is the intention of AT&T that all the public dial-up capabilities and private branch exchanges available to telephone subscribers should also be available to Picturephone subscribers. This capability must be built into the framework of the existing telephone plant.

A separate and relatively small switching unit designed to switch the broadband four-wire Picturephone signals will be added to existing exchanges. The telephone signal will be switched in the normal manner by the old equipment, and the new broadband switch will be operated in parallel with it to switch the video signal (Fig. 8.3). No additional common control mechanism will be needed. The existing control mechanism for telephone calls will also control the video switches. When the special prefix #, indicating a Picturephone call, is dialed, this step causes the video switch to be operated in parallel with the telephone switch.

Picturephone switching equipment may similarly be added to a private branch exchange, as shown in Fig. 8.5. A separate switch unit that employs reed relays is used.

The attendant's console may have a Picturephone set or not. If it does not, a fixed pattern, of the user's choice, will be transmitted until the desired person is reached. Both voice and picture go through the new section of the exchange. If a number is dialed at one of the Picturephone stations and it is preceded by a #, then the line circuit switch will connect it to the Picturephone switchgear and give a busy indication to the telephone PBX (Private Branch Exchange). If it is not preceded by a #, this switch will connect it to the telephone switchgear and give a "busy" signal to the Picturephone PBX.

Picturephone switching and transmission can thus be accommodated ingeniously into an existing telephone plant.

IMPROVEMENTS TO PICTUREPHONE

Bell Laboratories have been hard at work on improvements to Picturephone. A set with a color picture has been developed. Better methods of encoding the present picture are being worked upon.

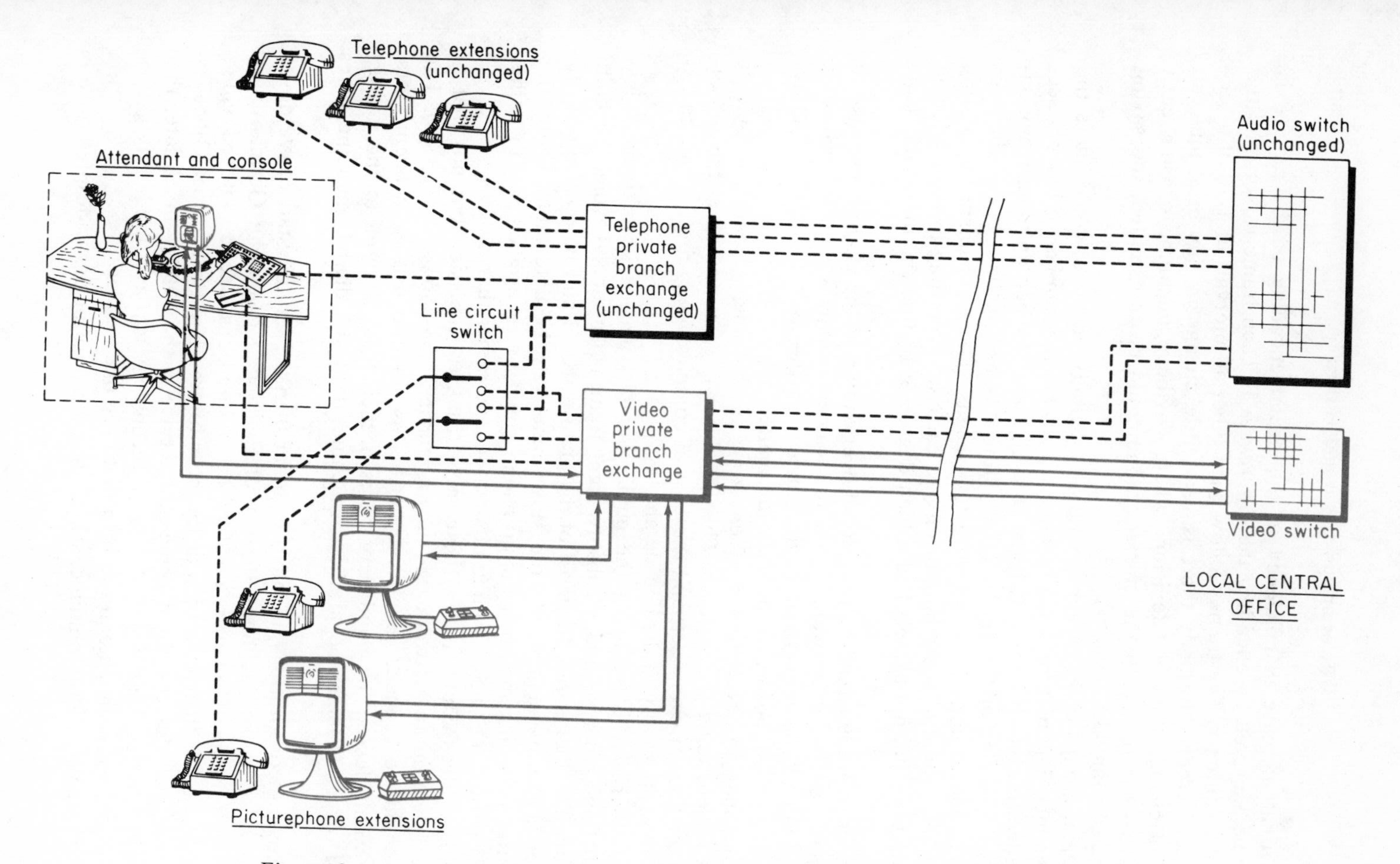

Figure 8.5 The addition of Picturephone equipment to a private branch exchange. (Redrawn from reference 2).

A great increase in efficiency in the digital encoding of the picture is likely to be achieved by encoding the *differences* in each frame rather than the total frame contents. The frames follow each other at intervals of 1/30 of a second on the screen. Normally there is very little change in this brief interval, much less change than in television, where cowboys might be chasing each other across the screen. If the changes alone could be transmitted, the saving would be substantial.

Again, groups of Picturephone signals, digitally encoded, could share the same channel. The picture could be encoded so that the number of bits per second needed to carry it varied with the rate of change. When an almost motionless face is looking at the set, relatively few bits are needed, but if the user turns his head, a large burst of extra bits results. The mechanism for sharing the channels would allocate bit capacity to each set according to its needs and would assume that not all sets had a high degree of picture motion at the same time.

In either there would be occasional overloading of the channel because of momentary excessive movement of the camera subjects. The overloading would result in the movement being blurred. In the uses for which Picturephone is intended, however, this may not matter.

The snag in these forms of encoding is that highly complex logic is needed for encoding the pictures. This logic, however, would be at the toll office rather than at every set, and the cost of complex logic circuits can be expected to drop greatly in the decade ahead.

Another way of reducing the number of bits needed is to transmit less than 30 frames per second. Probably eight frames per second would be quite adequate. The flicker rate for most of the picture would still be 60. Only in the areas of fast movement would the slow frame rate be detectable. Using this and better encoding methods Picturephone could be redesigned for T1 rather than T2 carrier (i.e. 1.544 million bits per second rather than 6.312).

These ideas could reduce the cost of trunk transmission but not that of the local distribution network (the analog transmission shown in Fig. 8.4). The latter could be reduced greatly, nevertheless, by the use of concentrators, which would lessen the number of wires needed from an area to the nearest central office. It might be at these machines that the digital encoding takes place.

Several corporations in countries other than the United States have developed a video-phone, including Plessey of England, Siemen's of Germany, and Toshiba of Japan. However, none of them foresee a public-switched videophone service in their countries in the near future. The British Post Office expects that such devices might come into service in the 1990s—much later than the United States. This is not so much a technology gap as a credibility gap. Is it worth paying so much extra to be able to see one's caller?

Eventually, Picturephone cost will drop because of the widespread use of digital lines, the lower cost of logic circuitry, and signal compaction. Possibly it can be sold even at this price. Modern American advertising seems able to sell almost everything. If it saves me business trips on appallingly overcrowded airways or if it enables me to see my grandchildren, it may be a good buy. Particularly attractive may be other attachments on the dial-up Picturephone lines, for example, computers, data bases, high-speed facsimile printers, etc.

BETTER COMMUNICATIONS IN BUSINESS

There is no question that better communications are needed within corporations. Many managers and professionals have the feeling that they could do their job better if they communicated better, but the people they need to talk to are far away. To travel by plane to find out information is very time-consuming and many people hate flying. We now have the capability to forge new links within corporations, to enable man to communicate with man better, and develop the information sources he needs without long-distance travel. What should the new links be?

Picturephone Model II does not seem to be the answer. It has two serious drawbacks. First, it consumes prodigious bandwidth and so is expensive for transmission other than that within the same building. Second, normal typewritten documents are not readable over it. When corporation men communicate they usually want to be able to see documents, contracts, invoices, program listings, insurance policies, line drawings, block diagrams and so on.

Both of these objections can be overcome in a video-telephone which transmits still rather than moving images. As we have commented, Picturephone Model II transmission requires a bit stream of 6.3 million bits per second. A system delivering a Picturephone-quality still image in 4 seconds, using good digital compaction techniques, would require a channel of about 56,000 bits per second for those four seconds, and the bit rate could be used for other transmission in the long pauses between still images being sent. 56,000 bits per second is the speed of AT&T digitized voice channels.

If high resolution images are needed, the number of bits would be greater. On the other hand, the number of bits can be reduced by using black and white images rather than pictures with shades of gray. A normal-size densely-typed page can be reproduced with clarity using 200,000 bits. To transmit this in 4 seconds would need 50,000 bits per second. The bit rate of a digitized telephone line can thus carry either Picturephone-quality photographs, or high-resolution black-and-white images, sufficiently fast for them to be used as a supplement to human conversation. Given appropriate equipment, persons talking by telephone could pause for four seconds when appropriate to transmit an image of a document, a face, an object, or a finger pointing to something on a diagram.

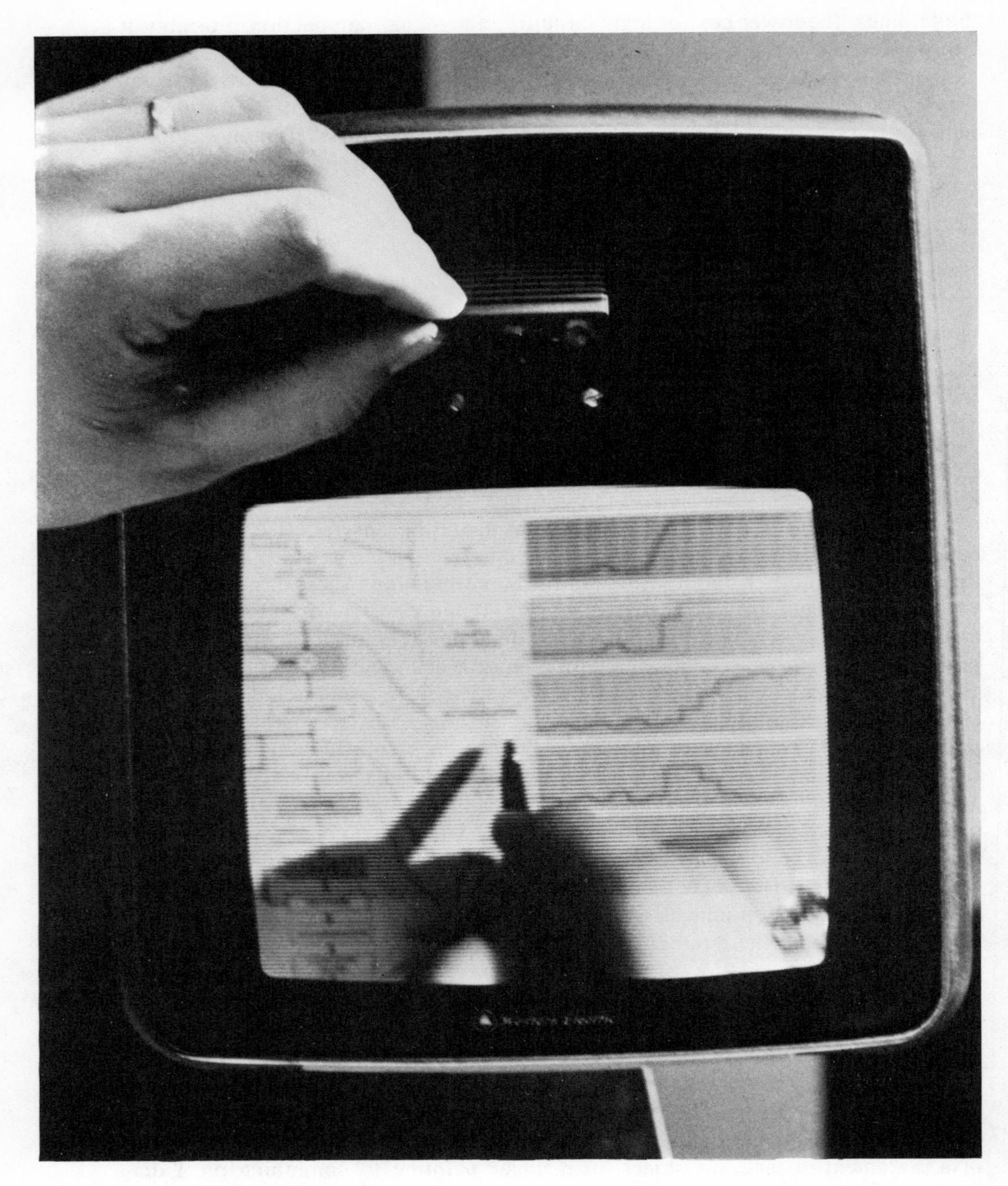

Figure 8.6 250 lines per frame give insufficient resolution to read typed documents or to examine detailed diagrams.

SLOW SCANNING If Picturephone is going to be a substitute for some business traveling and is going to enable certain people to work at home, a sharper or bigger picture will sometimes be advantageous. Bell Laboratories is working on a slow-scan mode using the same transmission facilities. This will permit high-resolution images to be sent, but it will not be possible to follow rapid movement. The terminal used would have a large screen. Possibly a means for printing documents in this mode will be provided.

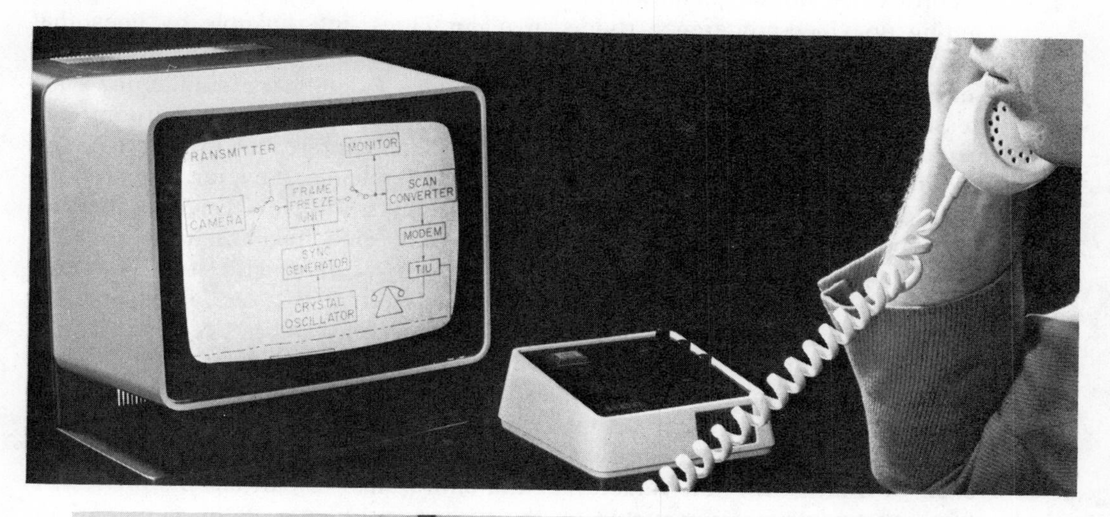

Figure 8.7 The RCA Videovoice set which transmits still black-and-white television pictures over the telephone circuit used for voice conversation. The camera is at the bottom left. Often it sits on top of the screen unit.

VIDEO CONFERENCES

Picturephone in the home is likely to remain too expensive for many would-be users. Public Picturephone booths in the main cities may become popular for talking to far-away friends and relatives.

Higher-quality video equipment will be still more expensive and thus is unlikely to be connected to switched public lines for some time. Instead, it will be used on leased lines, and again city facilities may be available. Some countries that do not plan switched video-phone services in the near future *are* setting up video-conference rooms in major cities, interlinkable by lines that carry full-quality television pictures. The British Post Office has developed a service called Confravision, in which it provides video-linked studios for businessmen to hold meetings in. If successful, a Confravision meeting room may be built in many cities, all interconnectable. Video conferences in industry over private leased lines are also likely to prove valuable. Several corporations and government agencies have set up video links. In New England there is a video network interconnecting hospitals. One such link has been used for some time in the Bell Laboratories between Murray Hill and Holmdel, New Jersey (35 miles). The link is a particulary interesting one, for it uses automatic switching between the conference participants and has features for the transmission of printed material and drawings [6].

Figure 8.8 shows the layout of the conference rooms in this system. Up to nine people sit at a curved table at each conference room location. Three microphones and cameras are used, with up to three persons within range of each. A fourth overview camera has the entire table of participants in view. A fifth camera, which is mounted on the ceiling and points at the center of the tabletop, is used to transmit documents, handwriting, small objects, or drawings, possibly with a speaker's finger pointing out features on them. Cameras can be placed in a variety of other positions, and often a camera giving a close-up of a speaker's face is used.

During normal operation, automatic switching will occur between the cameras. The logic of camera switching has been designed to imitate as closely as possible what a person does in a face-to-face conference. When a person speaks, the camera trained on his face is automatically switched on. The participants at the distant location will thus see his face on the screens in front of them. When another person speaks, voice-activated switching will switch the transmission to his face. If nobody speaks, the camera will remain on the last speaker for several seconds and the transmission will then be switched to the overview camera.

The screen in front of each participant normally displays the image being transmitted from the distant location. If someone is speaking then the image will be that of the speaker's face. If no one is speaking, it will be the overview picture. When a person is speaking himself, he will thus see the entire group at the far location. When somebody responds there he will see a close-up of that person. At the same time, the higher screen in Fig. 8.8 is displaying the image that is being sent to the far location.

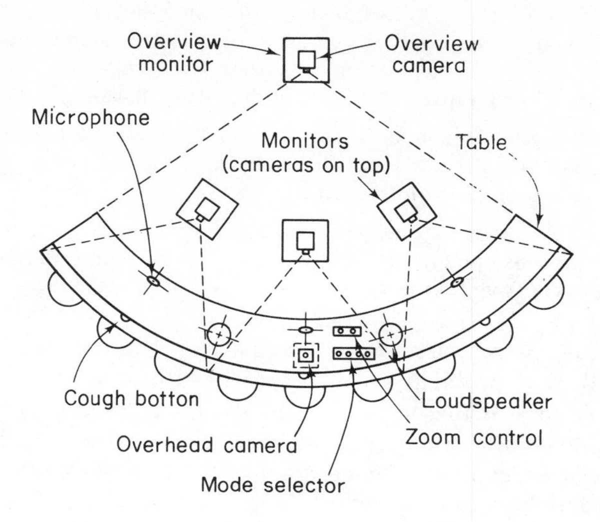

Figure 8.8 The Bell Laboratories video conference system transmits between distant conference locations using ordinary television transmission facilities. In "normal mode" operation the cameras automatically switch to the person speaking.

This system can be switched by the conference leader into one of three modes other than the normal mode just described.

1. *Locked graphic mode*. Here the outgoing transmission and the screens at the sending location are locked to the overhead graphics camera. A remote zoom control can make this camera show images of objects from 4 1/2 in. × 6 in. to 18 in. × 24 in. The speaker can control the zoom to obtain the clearest image.
2. *Leader graphics mode*. This mode is the same as the locked graphics except that the voice-actuated switching is not suppressed. Sustained speech of a remote participant will cause his face to appear on the screen.
3. *Leader mode*. This mode is used when the conference is being addressed by the leader and the screens contain his face, except when he is interrupted for a sustained period by a speaker from the remote location.

The reaction of the participants to this teleconference scheme was generally very good. One problem had to be dealt with; when persons coughed or sneezed, the loudness of this noise caused switching to the camera trained on them. One can imagine a conference during the annual flu epidemic of the locality in which the screens were filled most of the time with the contorted faces of the various coughers and sneezers. The problem was solved by placing a "cough button" within reach of each participant. Pressing this button prevented the system switching to them when they coughed, sneezed, hic-cupped, or belched.

Teleconference studios may become a common feature of the future.

Figure 8.9 AT&T's Intercity Visual Conferencing Service uses Picturephone equipment to provide video conferences for up to six participants in each city.

They would enable countries who cannot afford a switched Picturephone service to offer a glamorous facility employing their existing television transmission facilities between cities.

REFERENCES

1. Figures taken from "The President's Task Force on Telecommunications Technology," Staff Paper No. 1, Appendix A, p. 16, Washington, D.C., June 1969.
2. *Bell Laboratories Record,* May–June 1969, introduction to a special issue on Picturephone. Bell Telephone Laboratories, Inc., Murray Hill, N.J.
3. *The Economist,* London, August 9, 1969, a special section on telecommunications.
4. Edward M. Dickson & Raymond Bowers, *The Video Telephone, a Preliminary Technology Assessment,* prepared for the National Science Foundation, Cornell University, Ithaca, N.Y., 1973.
5. Alex Reid, *New Directions in Telecommunications Research,* a report prepared for the Sloan Commission on Cable Communications, Telecommunications Headquarters, British Post Office, London, 1971.
6. D. Mitchell, "Better Video Conferences," *Bell Laboratories Record,* January 1970.

The Bell telephone laboratories gave what was claimed to be the first demonstration of television in 1927. The screen consisted of 50 neon-filled tubes, each divided into small segments. *The New York Times* stated that television's commercial value was "in doubt."

9 CABLE TELEVISION AND ITS POTENTIAL

Television is the most influential communications medium in the history of man.
—*Robert Sarnoff, then Chairman of R.C.A.*

Television is a low-brow medium. This is its social role, and there is no sense in attempting to improve it.
—*Federal Communications Commissioner Lowinger 1967.*

Few media have a greater potential for changing the culture of a society than the coaxial cables being laid into homes by the cable television companies. There are, however, severe obstacles in the way of this potent medium developing to the full benefit of society.

The cable has an information-carrying capacity roughly a thousand times that of the telephone cable, and yet is laid into homes at less than half the cost because one cable serves many homes. Fig. 9.1 illustrates the difference between telephone and television cabling into the home. Telephone and Picturephone cables connect one subscriber only (except with party lines) to a switching office. The expense of such a local distribution network is high because large numbers of separate channels are needed and all are highly underutilized. Cable television wiring links large number of subscribers to the same cable, but does not provide public switching facilities at the head of the cable.

With modern electronics cable TV can be used in ways quite different from those originally intended, and the potentialities are exciting.

THE SPREAD OF CATV

Cable television was started in 1949 by a local radio dealer in an Appalachian village, a man who was oblivious to its potentialities. He realized that the

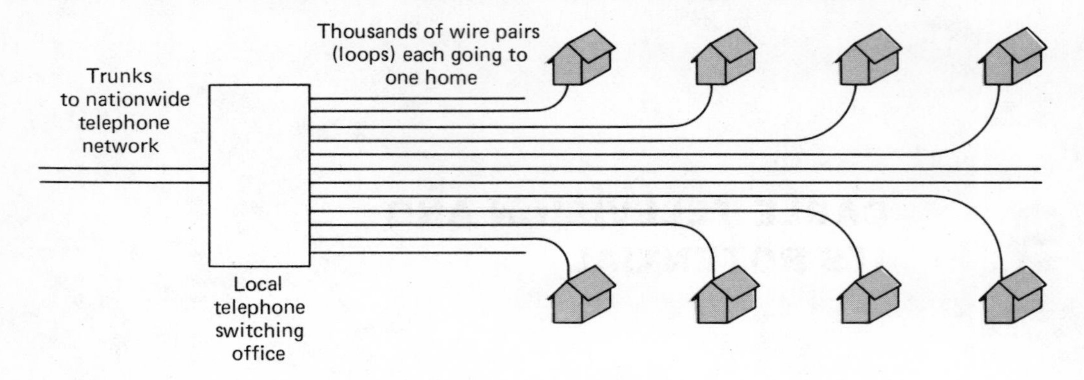

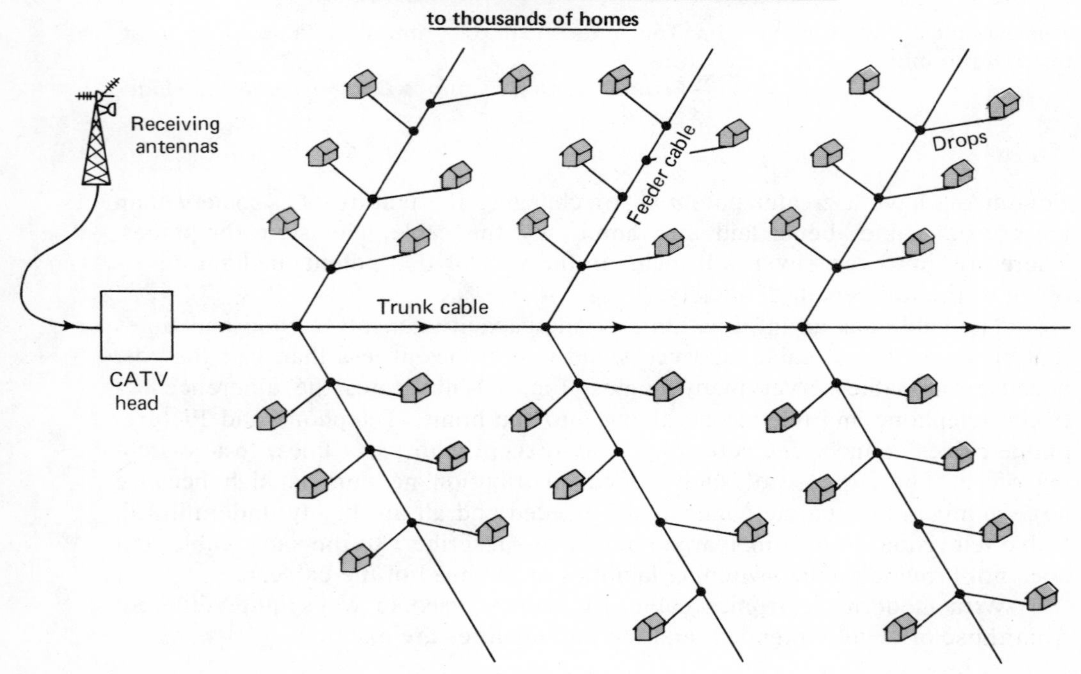

Figure 9.1 The difference between telephone and television cables into the home.

surrounding mountains were spoiling the reception of television signals from Philadelphia, which meant that he would be unable to sell many sets, so he set to work on an enterprising scheme. He erected a tall antenna on a suitable mountain where the reception was good and ran coaxial cable from it into the homes of people willing to pay a small fee. In this way, television came to an Appalachian village, and a technique came into use that should, if allowed, revolutionize home electronics.

Cable television spread steadily in the 1950s and 1960s hampered by FCC regulation and by lobbying from the big three networks who feared its competition. Its initial spread in the United States was in rural areas where the reception was poor. In Canada it spread in the cities, largely because cable television operators were allowed to import distant signals and hence provide more programs. In 1968 a U.S. Justice Department antitrust decision urged the FCC to allow cable television to develop as a competitive medium with its own program origination and advertising. With much catching up to do, cable television systems were then built in U.S. cities.

Cable television is referred to by the initials CATV (Community Antenna Television). However, with many systems today there is no community antenna; the cable goes directly into the office from which the programs are distributed.

The main sales arguments today are the provision of advertisement-free programs, sports broadcasts, first-run movies, news and stock market reports, and the improvement of reception. As the pamphlets mailed to New York apartments say:

> Ghosts, unsteady pictures, weird psychedelic colors and assorted eye-jarring interference have plagued TV reception in this area since commercial television first became a reality. . . .
>
> The tall buildings between your set and the broadcasting tower deflect the local TV signals. The signals bounce off these buildings before reaching your antenna. The result: ghosts. And that's just the beginning.
>
> TV signals ricocheting off planes flying overhead are subjected to electrical interference ranging from automobile ignitions to your neighbor's vacuum cleaner. They finally reach your set much the worse for these interferences. You get way-out colors, flip-flopping pictures, snow, herringbone patterns, and assorted other "stray" signals in addition to the signal sent out by the station.
>
> . . . As a cable TV subscriber, you, too, can discover the pleasure of clear, sharp, ghost-free, snowless jitterless television in brilliant color or crisp bright black and white [1].

Cable television serves more than ten million homes in the United States at the time of writing, and the number is increasing rapidly as indicated in Fig. 9.2. It has been estimated that by the early 1980s half the homes in this coun-

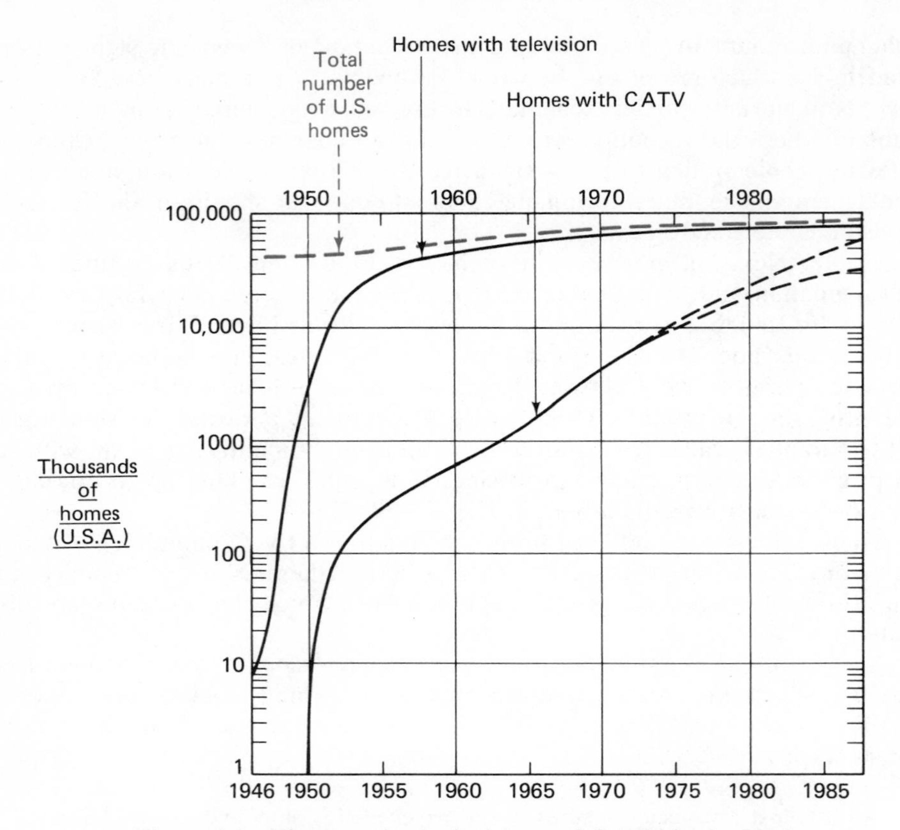

Figure 9.2 The growth of television and CATV in the USA.

try will be wired for it. There is, however, a wide divergence of opinion as to the rate at which it will spread, as well as its eventual penetration. One body of opinion holds that *all* broadcasting *should* reach the home by cable, thereby freeing the airwaves for more important uses. It is doubtful whether this situation will occur for several decades because it is not economical to operate cables in sparsely populated districts. 10 or 20 per cent of American homes are likely to remain out of reach of the cables but will probably be reached by other television technologies eventually, such as satellites.

MANY CHANNELS

Fig. 9.3 shows a CATV cable passing through a Manhattan apartment. It passes through all the apartments in the building. This little cable can carry 20 television channels, some not used as yet. Other installations offer 36 and 40 channels. The President's Task Force on Communication Policy estimated that an 81-channel system would cost about $215 per subscriber location, and a low monthly fee could be charged for such a system [2]. With today's technology, this would need two cables. One cable could carry 50 channels.

The American public's appetite for television is enormous. *Business Week* quotes one satisfied cable subscriber as saying "I can see Perry Mason at 180 lb, Perry Mason at 205 lb, and Perry Mason at 230 lb. It just depends on when he made the shows they're rerunning" [3]. The average American home has the television switched on for 6 hours per day. However, the public is hardly likely to need 81 broadcast channels. It is in the alternative uses of the cable that the most interesting potentialities lie.

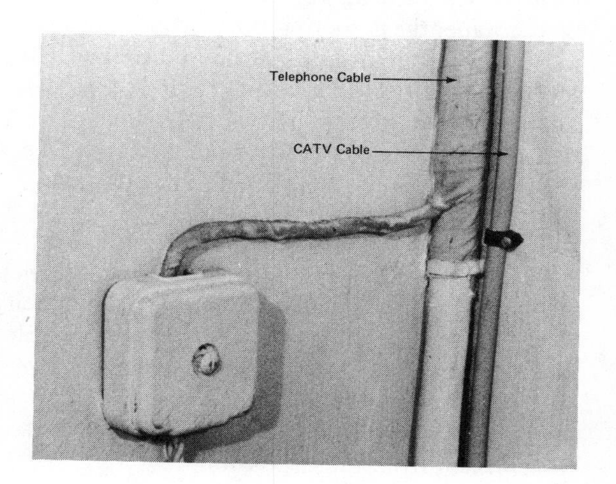

Figure 9.3 The CATV cable in a Manhattan apartment Many of the television channels in such cables are today unused, leaving a very high bandwidth that could carry other services into the home.

HOW SHOULD THE CHANNELS BE USED?

Ours is an age dominated not by culture but by technology and lawyers, and the burgeoning CATV industry seems to be having an identity crisis concerning how it should fill its many channels. Consumers are indicating that they want more value for their CATV fee.

Many of the cable systems are using some of their channels for services such as weather reports, news headlines, time signals, and stock market reports. Most of these services are provided very inexpensively with a single fixed camera. Some cable television companies run their own films and produce their own programs, in some cases serving a local rather than a statewide need. Local ball games, local shopping information, politicians whose constituency is numbered in a few thousands . . . these seem a natural for cable TV.

A television program need not be expensive. Some CATV companies have a very simple studio with little equipment other than a camera and lights. The studio can be used at little cost by any group who wants it. Local school boards have used this facility to discuss matters such as upcoming bond issue referendums. Only a simple camera is needed for many local events, and a person can learn to operate such a camera very quickly. City council and board meetings, school events, local sports, amateur entertainment productions are shown on the local cable. One system in Manhattan transmits Columbia University's basketball games. Thus cable television has been claimed to be a local medium rather than McLuhan's "global" one.

How, then, should the extra channels be filled? Worldwide there are enormous quantities of excellent program material that has not found its way onto the cables. The BBC archives, for example, are gigantic and filled with past television programs, on all manner of subjects, too good to deserve a fate of permanently gathering dust. Many countries make good childrens' programs to which different-language soundtracks could be added. Japan spends much money on TV production. Britain's University of the Air creates very high quality teaching courses for television. The world would surely be a richer place if the spare CATV channels could be filled with quality programming from different nations and different cultures.

DIVIDING UP THE CHANNEL

A television channel has a potentially high information-carrying capacity—a thousand times higher than a telephone speech channel. (Like any channel, it can be subdivided into channels for signals of smaller bandwidth, which could be used for voice, high-fidelity music, still pictures on the screen, slow-scan images, data or facsimile.) A variety of proposals for future domestic products have arisen from these possibilities.

A guard band is needed to separate the channels, which, on a cable, may be up to 25% of the signal bandwidth. A signal with a bandwidth of 35 kHz could provide the highest quality stereophonic music. One television channel in a cable could therefore carry 100 hi-fi stereo music channels.

Of the many television channels, perhaps one could be divided up for classical and one for pop music. The classical music channel could broadcast 100 symphonies, operas, and concertos simultaneously, each occupying a different frequency slot within one television channel. The user would "tune" to the frequency slot he wants. Perhaps a third channel could be divided into non-stereophonic sound channels with a bandwidth of 6 kHz for speech programs. One television channel could carry 600 such sound channels simultaneously (allowing for appropriate channel spacing). They might be used for continuous news broadcasts, weather forecasts, a time channel (why use a whole TV channel for telling the time?), continuous sports reports, stock market reports, community information reports such as theaters and movies, and shopping news. More than 500 sound channels would still be left for radio plays, talks, poetry, language teaching, and other programs, which might be repeated continuously for each day. Seventy-eight channels on an 81-channel system would remain for Perry Mason and other television! Such is the power of coaxial cable wiring.

As with all we have to say about cable TV, it is far from certain that the full capacity will be utilized. It is far from being utilized today. Moreover, it is uncertain who would pay for such a proliferation of programs, although the programs need not be expensive. There could be frequent reruns of old programs, as well as music without broadcasters. Foreign programs could be relayed via satellite. It is worthwhile calculating how the capacity *could* be used. With Picturephone there is a great shortage of bandwidth; with cable TV there is, at least for today, an embarrassing excess of it.

MUSIC LIBRARY

One appealing consumer service would be a CATV music library.

Let us suppose that a music "library" that plays an extremely large number of pieces of music continuously is set up. A tape deck, like a videotape machine, might be used to play 100 pieces of stereophonic high-fidelity music continuously. Forty such machines might be used with different playing times to cater to the different lengths of the works which would mean that 4000 music tracks were playing simultaneously. The music would include every major symphony, opera and concerto, and a vast amount of jazz, pop, and show music. The cost of the multichannel tape deck might be $4000, giving a cost for the entire facility of perhaps $250,000. One coaxial cable trunk could carry all 4000 sound channels to locations where they would be appropriately switched on to different CATV systems. If one such facility served many thousands of CATV subscribers, the cost per subscriber could be low.

The majority of sound channels on the cable could carry music determined by the programming authorities. The majority of subscribers, most of the time, would be content to listen to what was fed to them, especially if a large number of channels existed, as indicated earlier. At any one time, a minority of subscribers would be playing a symphony or other piece of music *requested* from the library.

Let us suppose that there are 3000 subscribers on a cable, and not more than half have their sets switched on in the peak hour. Of these subscribers, not more than 20% are likely to be listening to sound-only programs, and of those that are, only a quarter are listening to *requested* music; the rest are listening to the preselected channels. That could give a total of 75 request channels and 25 preselected channels of music on *one* television channel.

Let us suppose that the music library is connected to several cables, and hence serves a total of 100,000 subscribers. (In a big city and suburban area it could serve a million.) Let us suppose that the subscribers are automatically billed for using the music library service, at a rate of 25 cents per hour, and that the subscribers listen to the music for one hour per week on average. This adds up to $1.3 million per year revenue from the 100,000 subscribers. Such a system would seem to be economically feasible for cities with high CATV penetration. In smaller cities, use would depend on the cost of long-distance transmission of the television channel from a remote music library.

SELECTION OF OTHER PROGRAMMING

The idea of being able to request one's programs instead of merely accepting broadcasts, which are usually planned for a mass audience rather than a selective audience, is appealing. With coaxial cables going into a large proportion of homes, this step seems economically feasible with sound-only programs. It does not appear feasible with TV programs, given today's technology.

Another possibility that has been suggested is to send movies to the home on request at off-peak viewing hours and store them automatically on videotape. Suppose that 20 television channels are used for this purpose during 16 hours of the day (thus avoiding an 8-hour peak-viewing period). If the average length of a movie is 2 hours, then 160 movies per day could be transmitted to subscribers whose tape machines would be automatically switched on when the movie began.

Home selection from a large movie library does not appear likely for the next ten years. Selection of movies by *institutions* for CATV viewing *is* feasible. CATV cables are used in some schools for on-request transmission of educational movies from a manually operated movie library.

PAY TV

While even 40-channel cables do not have enough capacity to allow their many subscribers to select movies freely, they could permit a limited degree of selection. In one proposed system subscribers select programs from a guide and inform the system what they want to see at a given time. The selection is such that most subscribers pick certain items such as currently popular movies and sports events, hence most requests can be met. The most popular requests are determined by a minicomputer. Some subscribers are disappointed, told that their selection can be scheduled at another time, and offered a second choice.

This scheme is one variant of a much discussed cable service—Pay TV. The CATV industry is searching for cable services that would increase cable revenue. Most market studies on the subject put *Pay TV* at or near the top of the list. Pay television, also called *Pay Cable, Subscription Television,* and euphemistically, *Premium Television,* would attempt to provide programs of an exceptionally desirable nature, and charge specifically for those programs. Such programs would include sports events and first-run full-length feature movies without advertising. A service called Home Box Office is now a success in many U.S. cities. The subscriber to this service pays a monthly fee for advertisement-free movies and other programs on a number of channels.

REVERSE CHANNELS

Many of the interesting new applications of cable TV require reverse-direction channels on the cable. The early cables were all one-way cables for the passive distribution of television. In 1972 the FCC ruled that U.S. CATV systems should be constructed with reverse-channel capability.

A two-way cable permits services in which the subscriber can respond to the signals that reach his home via the cable. His responses can be interpreted by a computer at the cable head or elsewhere, and innumerable new uses of the medium become possible. Pay TV and sound-library selection are two such services. Others include news and financial services, computer assisted education, interactive advertising, automatic collection of television ratings, access to encyclopedic sources of data, and so on. All manner of computer applications could be made available in the living room. The CATV industry refers to such uses of the cable as *Subscriber Response Services (SRS).*

Some representatives of the CATV industry believe that there is a gigantic future revenue potential in subscriber response services. Hubert Schlafly, when Executive Vice President of Teleprompter, wrote that they would have a greater impact on the public than television: "two-way, mass entry, substantially real-time, data communication between individually selective addresses, and a nation full of computer processing potential, provides a flexibility of use that exceeds the imagination." [5]

There are three main types of methods for making the cable carry signals back to the cable head—the top three illustrations of Fig. 9.4.

First, the cable could be in the form of a loop. Signals originating at a subscriber location travel in the same direction as the other signals, but are carried back to the cable head. This would be effective if the subscribers were strung together in a linear fashion, but they are not. They are connected in the tree-structured fashion shown in Fig. 9.1, which makes looped cables uneconomical.

Second, the cable could have two-way amplifiers and electronics which permit the reverse-channel signal to occupy a different frequency band to the forward-channel signals. This is done on some cables: it provides a means of converting existing one-way cables to two-way operation. The forward, "downstream," signals are at conventional television frequencies. The reverse, "upstream," signals are below the television frequencies—in North America, below Channel 2 (54 Mz)—in the 5 to 30 MHz range. A filter is used at the amplifiers to separate out the upstream signals and feed them through an upstream amplifier.

Third, two cables can be used, one downstream and one upstream. Most of the cost of the CATV wiring is not in the cable itself but in digging up the streets, installing the cable, obtaining permission, and in administrative expenses. A two-cable system is generally less than 20% more expensive than a one-cable system, and has more flexibility for future use. Some of the far-sighted CATV companies have installed two cables to each subscriber in recent years, but initially attached no electronics to the upstream cable. This "shadow" cable awaits whatever markets may develop.

It is the intention of CATV organizations to provide more channels than those received by a conventional noncable set. Where two cables are used, one

1. A cable which loops back to the cable head

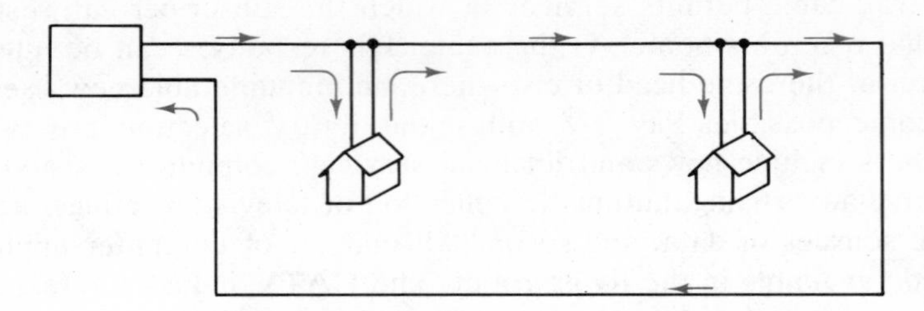

2. Two directions of transmission on one cable

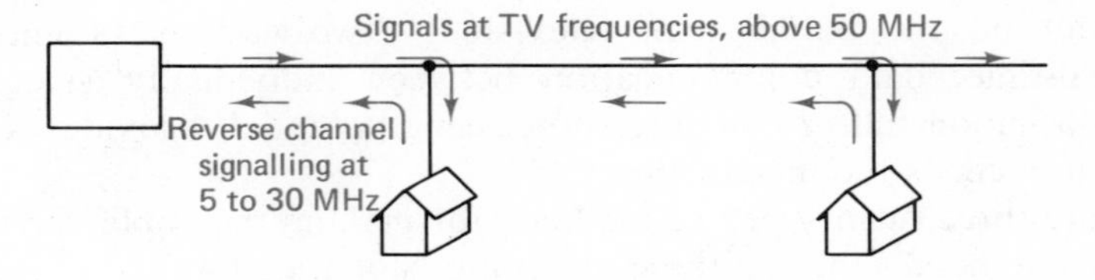

3. Two separate cables

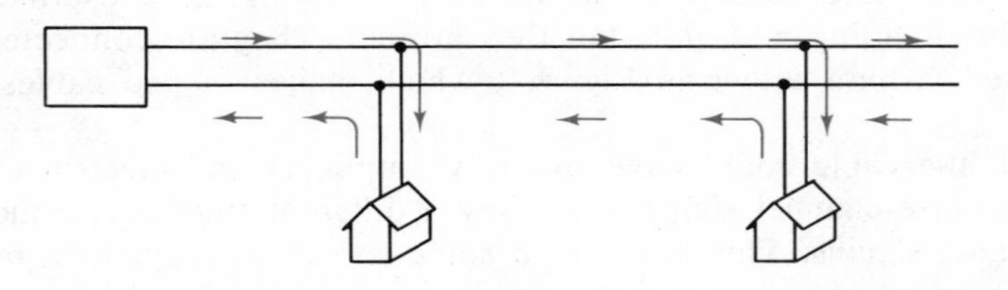

4. One conventional cable and one two-way cable

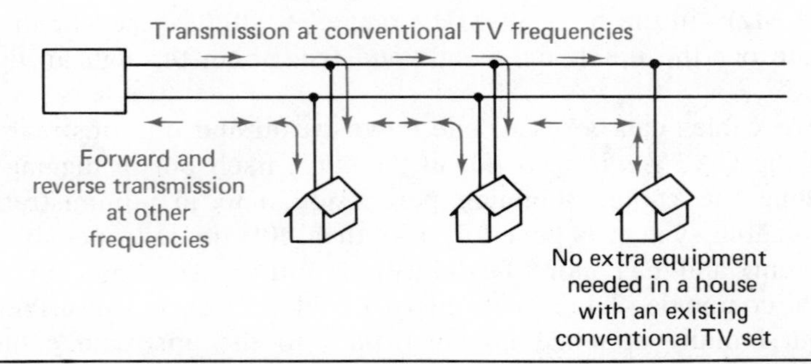

Figure 9.4 Four techniques for reverse-channel signalling.

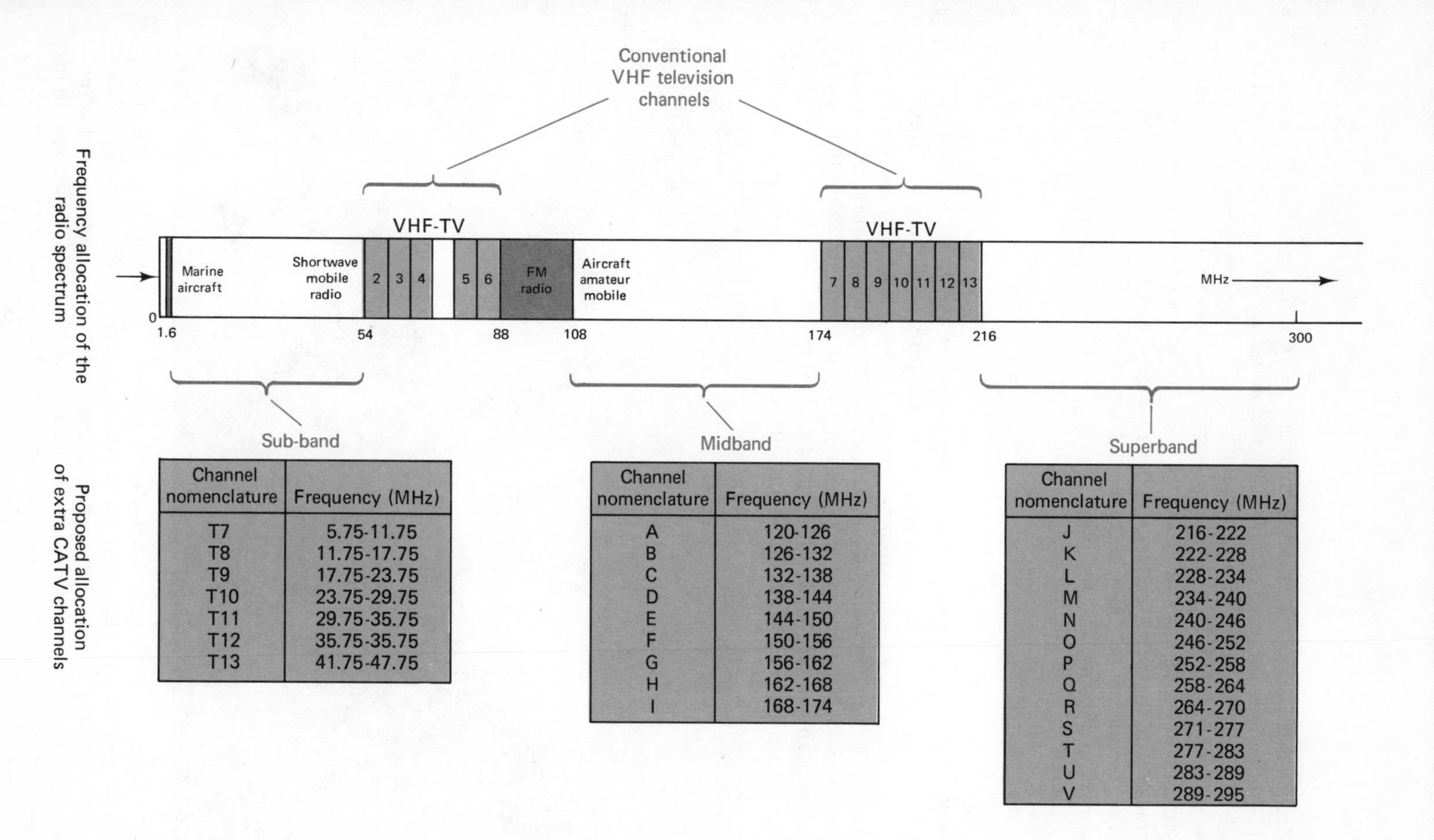

Channel nomenclature	Frequency (MHz)
T7	5.75-11.75
T8	11.75-17.75
T9	17.75-23.75
T10	23.75-29.75
T11	29.75-35.75
T12	35.75-35.75
T13	41.75-47.75

Channel nomenclature	Frequency (MHz)
A	120-126
B	126-132
C	132-138
D	138-144
E	144-150
F	150-156
G	156-162
H	162-168
I	168-174

Channel nomenclature	Frequency (MHz)
J	216-222
K	222-228
L	228-234
M	234-240
N	240-246
O	246-252
P	252-258
Q	258-264
R	264-270
S	271-277
T	277-283
U	283-289
V	289-295

Figure 9.5 Channel allocations on CATV.

may be designed to carry conventional channels to existing sets. Fig. 9.5 shows the VHF frequencies of conventional television channels and the new frequencies which CATV uses. The new frequencies could all be used for passive television, or some could be used for new services or reverse-channel signaling. In order that no extra equipment is needed by a subscriber with a conventional set, channels using the new frequencies, both forward and reverse, may be carried on the second cable. The bottom diagram of Fig. 9.4 shows one cable serving conventional sets, and the other carrying forward and reverse signals at new frequencies.

EVOLUTION OF CATV

Figure 9.6 shows stages in the evolution of CATV. It began as merely a means of picking up television transmission at a better location than the houses it served. Later, the cable systems acquired facilities for their own program origination, local television studios, and links for shipping in programs other than those broadcast through the air. Microwave links were employed for carrying signals across cities to the cable head. First-run movies were broadcast and in some

Figure 9.7 5-meter and 10-meter satellite earth stations are used in the U.S.A. to distribute CATV programs so that many towns can have services like *Home Box Office*. Satellites permit cable TV systems to be built in remote areas. Persons in isolated or rural towns can then enjoy first-run movies, sports events, and other CATV offerings. In the future they might be linked into interactive facilities. *(Photograph courtesy Scientific-Atlanta.)*

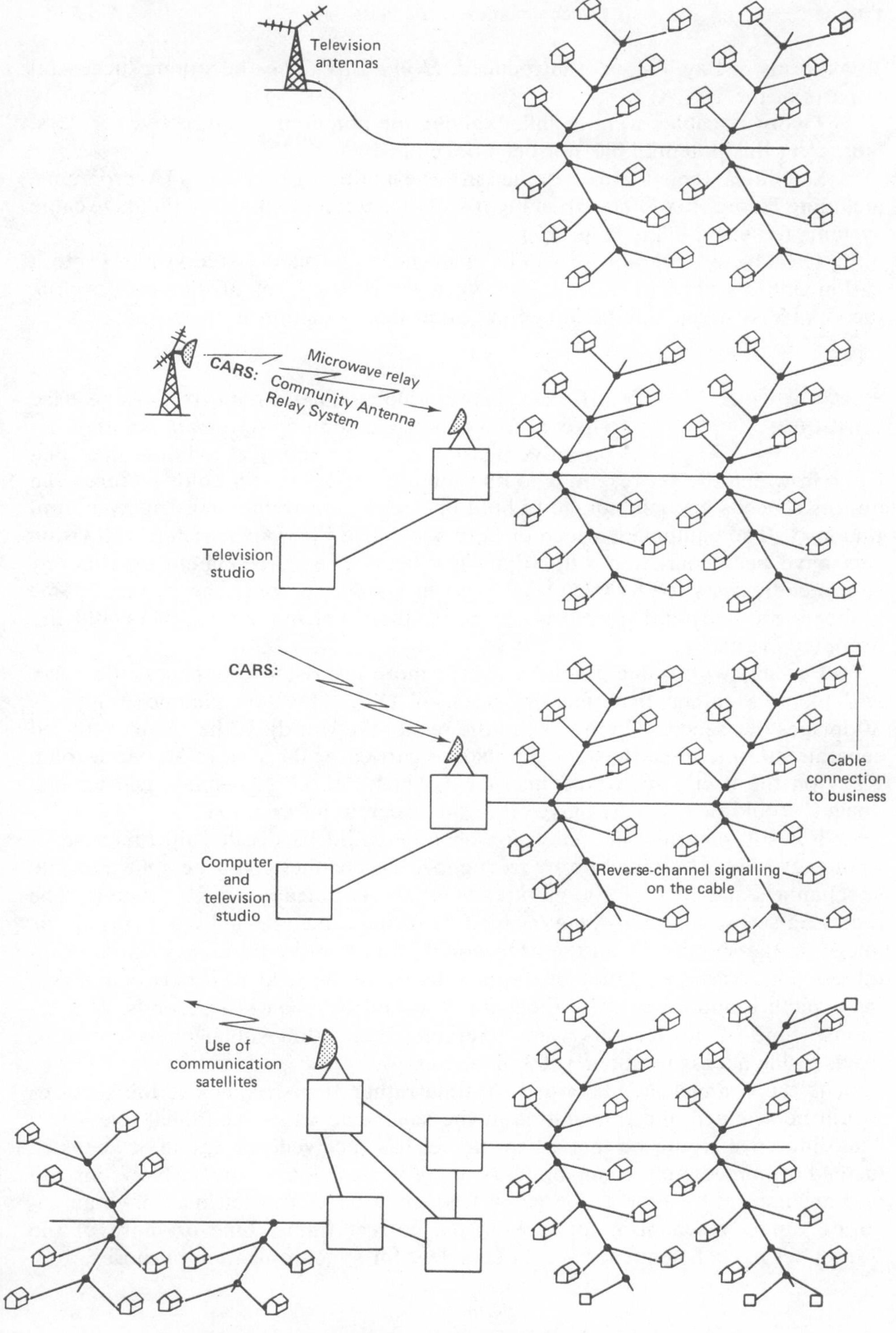

Figure 9.6 Phases in the development of the CATV industry.

areas forms of Pay TV were introduced, *Home Box Office* becoming successful in parts of the U.S.A.

Two-way cables were installed, giving the potential of interactive services, but as yet this potential has not been developed.

Satellite antennas were installed in large numbers to carry CATV programs, including Home Box Office, to many towns. Very remote towns could have cable systems fed via satellite (Fig. 9.7).

The future evolution of CATV may hook up many cable systems into a nationwide distribution facility, and when interactive uses of the sets develop, the CATV systems may be linked in nationwide computer networks.

FRAME GRABBERS

If a set displays alphanumeric data it could be used like a conventional computer terminal. Another attractive possibility is to employ a computer-like dialogue in which the set responds to its user by displaying still color pictures. To do this it needs enough storage to hold one television frame and display it continuously. The equipment which does this is called a *frame grabber*. Television sets have been marketed with frame grabbers. They are sometimes used by sports enthusiasts for freezing a moment in a game to study the player. Frame grabbers are substantially more expensive than alphanumeric buffers, but are dropping in cost.

For sets with frame grabbers one or more television channels could carry *still* pictures rather than moving ones. A U.S. television channel transmits 30 images per second (25 in most of the rest of the world). If the channel carried separate still pictures, these could also be carried at the rate of 30 per second. Keys on the user's set would instruct it which "page" to display, and enough "pages" could be carried to give all of the program information.

For still pictures, the television channel could be divided up timewise or frequency-wise. If divided up by frequency, the channel could be split into 200 subchannels and each of 200 pictures would be scanned every 10 seconds. The receiving set would have some means for storing a picture and regenerating the image on the screen. Again the user would "tune" to the frequency band he required. The "response time" or the time he would have to wait between selecting a picture and completely receiving it would be about 10 seconds. If a response time of 50 seconds were tolerable, then 1000 subchannels could be used, giving access to about 1000 pictures.

If the channel is subdivided by time rather than frequency, the pictures would be scanned and transmitted at the same rate as conventional television. The difference would be that when the set has received one frame, it keeps it, instead of immediately changing it, as today's television set does. Now, instead of a mechanism for tuning to a given frequency band, the set must have an accurate timing mechanism for picking the correct frame. One or more of the frames set in each sequence would be solely for synchronization purposes.

The still-picture facility could have many uses. News services, weather forecasts, and stock market reports could use some of the frames in the channel. One's daily newspaper in the future may occupy one such television channel. Local shops, theaters, and movies may use such a channel in color, and many other local advertisers could employ it. Programmed instruction sessions may be scheduled, with the student "branching" from one page to another according to his level of knowledge or success in answering questions.

It has been proposed that facsimile machines should become part of such equipment in the future, and the user would have the facility to copy pages he wanted.

DIALOGUES WITH PICTURES

Many appealing forms of dialogue can be imagined in which the responses come in the form of a color picture on the TV screen rather than in the form of a digital response (with letters and digits).

Consider the following case as an illustration of the possibilities. A woman wants to continue planning her vacation, or perhaps she dreams about a vacation she is unlikely to have. She uses her set to contact the local travel-agent computer, having been instructed how to do so by advertisements on the other channels. The travel-agent computer has her account record, which states how previous dialogues have concluded. It displays the following message on her screen:

GOOD EVENING MRS. SMITH.
WOULD YOU LIKE A CONTINUED PRESENTATION
ON ONE OF YOUR PRIME SELECTIONS:
1. PERU?
2. BOLIVIA?
3. HAWAII?
4. EASTER ISLAND?
IF SO PLEASE KEY THE ABOVE NUMBER

She presses the 4 key on her Touchtone telephonelike keyboard. The "4" travels over the digital input channel, via the cable computer to the travel-agent computer. The travel-agent computer responds:

EASTER ISLAND.
WHICH OF THE FOLLOWING WOULD YOU PREFER:
1. SCENERY?
2. ENTERTAINMENTS?
3. HOTEL INFORMATION?
4. SHOPPING?
5. TIME-TABLE?
6. TO MAKE RESERVATIONS?

Figure 9.8 Warner Brothers' QUBE cable television system provides many channels of pay (Premium) television and free television, and viewers can interact with some programs using this small cable-connected console.

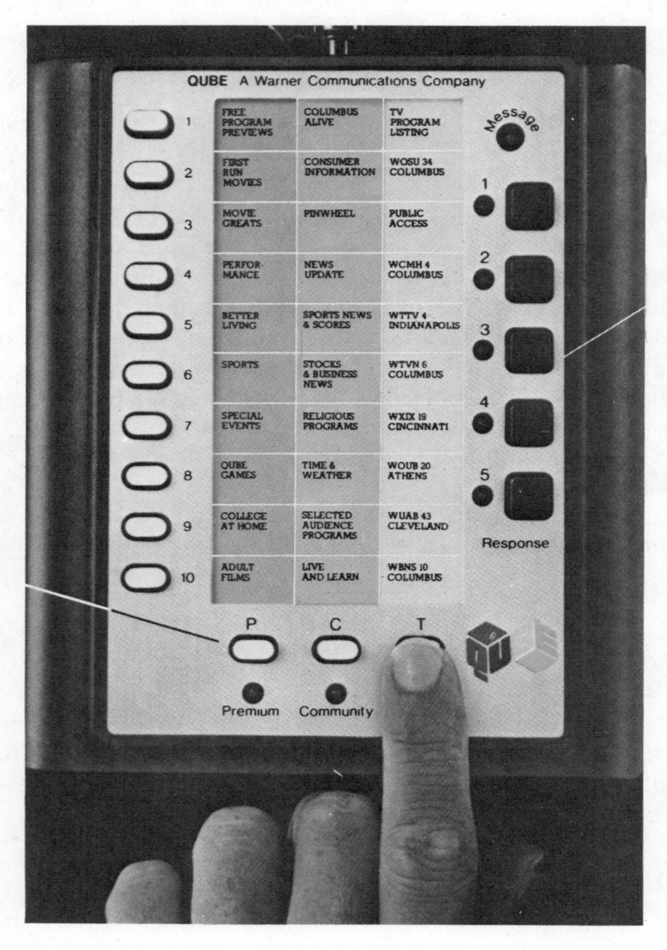

Each of these buttons activates the column of channels above it:
P: Pay television channels.
C: Ten community channels.
T: Conventional television and public access channels.

A message light

Lights and buttons used for viewer responses e.g., to play a game, cast a bid in an auction, or speak out on an issue.

Qube gets you into the action.

1. Field goal? 2. Quarterback sneak? 3. Quick short pass? 4. Fullback off tackle? 5. Halfback sweep?

Eight seconds left. Time for one play. The ball is on the 2-yard line. Your team is behind by 3 points. What do you think the coach should do?

Go for a tie with a field goal? Or go for the win? And how? What would you do?

Touch the button and let your opinion be known...in seconds. You'll see how your decision compares with everyone else's.

Columbus Votes	
Field Goal	7%
Quarterback Sneak	15%
Quick Short Pass	21%
Fullback Off Tackle	30%
Halfback Sweep	27%

Now sit back and watch what play the coach himself calls.

PLEASE KEY THE ABOVE NUMBER.
She presses the 1 key and the screen says:
SCENERY.
PLEASE PRESS THE # KEY TO CHANGE SLIDES AND BEGIN PRESENTATION.

She presses the # key. There is a pause of about 6 seconds, and then a magnificent aerial view of Easter Island appears on the screen in color. The woman looks at breathtaking pictures for the next half hour and then presses a key labeled END. A message appears on the screen trying to sell her a booking. She responds negatively and switches the set off.

The pictures may come from 35 mm slides in a large random-access file, or from frames on microfiche or film cartridges in units controlled by the travel-agent computer. The image is scanned and passed over a communication line to a switch unit at the head of the TV cable and from there is switched to the appropriate subchannel.

All manner of applications for dialogue with pictures can be imagined. It may be particularly valuable for teaching, especially with children. Madison Avenue probably would be the richest source of material for such a channel initially, but other longer-lasting "programs" would steadily amass like books in a library.

Again, there are a variety of ways in which such a channel could be organized. Perhaps the simplest way to organize it would be to assign a still-picture subchannel permanently to a subscriber for the duration of his picture dialogue.

Let us again assume that one cable is limited to 10,000 subscribers. At the peak-viewing period, not more than 40% of the subscribers are watching their sets. Most subscribers still spend most of their time with "passive" channels rather than with those demanding a response. In many homes the passive channels are a half-perceived background to daily living. During the peak viewing period, less than 5% of the viewers, say, will use the still-picture channel—say 200 viewers per cable.

One television channel could be used for 200 viewers and give a response time of 10 seconds with frequency-division multiplexing and from 0 to 10 seconds with time-division multiplexing. If two channels were used for the 200 viewers, these times would be halved.

Such a system would have a small but finite probability that all 200 subchannels are busy when a subscriber attempts to obtain a still-picture service. In this case, he will obtain a "busy" indication, just as he sometimes does from his telephone. If the average viewer watches the still-picture channel for no more than half an hour at a time, another channel will become free, on the average, every 6 seconds. If the subscriber attempts to obtain a channel again after the busy signal he will probably succeed.

If the demand for still-picture service is in fact greater than the preceding figures indicate, more than one of the many television channels may be used for it.

The response times in practice might be much lower than the above figures, and the probability of delay very low, because fewer than 10,000 sets would be attached to each cable.

INTERCONNECTIONS TO THE CABLE

At the cable head, the various channels will be connected to other forms of transmission media. The television signals may come from local cables to television studios or from long-distance links such as those provided by the Bell System for today's television distribution. They may come from satellite links, perhaps from other countries. The sound channels may go through switching facilities to sound studios or a music library. They may also be switched to long-distance links or obtained from radio antennas.

The digital channels, when used, will enter and be transmitted from a computer at the cable head, which, in many cases, would act as a means of switching the data to links with other distant, special-purpose computers. The computer would permit a terminal on the cable to "dial" computers anywhere and make the connections over the public network.

REFERENCES

1. Advertisement from Manhattan Cable Television.
2. President's Task Force of Communications Policy, Staff Paper No. 1, Eugene V. Rostow, Washington, D.C., 1969.
3. *Business Week,* November 22, 1969.
4. John DeMercado, *Video Networks,* International Computer/Telecommunication Conference, Lake Maggiore, Italy, 1973.
5. Teleprompter Corporation brochure on CATV, 1973: "The Computer in the Living Room" by Hubert J. Schlafly.

Under the Special Patronage of Her Majesty

And H. R. H. Prince Albert

GALVANIC AND MAGNETO

ELECTRIC TELEGRAPH,

GT. WESTERN RAILWAY.

The Public are respectfully informed that this interesting & most extraordinary Apparatus, by which upwards of 50 SIGNALS can be transmitted to a Distance of 280,000 MILES in ONE MINUTE,

May be seen in operation, daily, (Sundays excepted,) from 9 till 8, at the

Telegraph Office, Paddington,

AND TELEGRAPH COTTAGE, SLOUGH.

***ADMISSION* 1s.**

" *This Exhibition is well worthy a visit from all who love to see the wonders of science.*"—MORNING POST.

Despatches instantaneously sent to and fro with the most confiding secrecy. Post Horses and Conveyances of every description may be ordered by the ELECTRIC TELEGRAPH, to be in readiness on the arrival of a Train, at either Paddington or Slough Station.

The Terms for sending a Despatch, ordering Post Horses, &c., only One Shilling.

N.B. Messengers in constant attendance, so that communications received by Telegraph, would be forwarded, if required, to any part of London, Windsor, Eton, &c.

THOMAS HOME, *Licensee.*

G. NURTON, Printer, 48, Church Street, Portman Market.

An advertisement of the Great Western Railway, England, 1845. (*Courtesy The Post Office, England.*)

10 SWITCHED DATA NETWORKS

For half a century telecommunications (other than broadcasting) has been dominated by telephone users. There have been a thousand times as many telephone subscribers as other types of telecommunication users. Consequent economics have dictated that telegraph and data traffic should be converted to a form in which they can travel over the telephone system.

In earlier decades it was a dream of the telephone companies that every telephone in a country would be able to dial every other telephone. Some even dreamed that all business offices and almost all houses would have such a telephone. In its day, this hope fired the imagination of the large numbers of people who made such a dream come true.

In today's dream we would like *machines* everywhere to be able to intercommunicate and would like fluency in man-machine communication. We would like to enter data into a terminal and dispatch the information at appropriate speeds to a data processing machine anywhere in the country or—and what is of rapidly increasing importance—anywhere in the world. We would like a burglar detector to send details to the police computer instantly. We would like machinery with an overheating bearing to dial for help. We would like a checkless society with electronic transmission of money. From our screen unit we would like to be able to browse in or query all kinds of computer files. Eventually, in a slightly more distant future, we would like to be able to set up a connection between our television set or wall screen and locations from which a picture can be received, cameras in friends' apartments or in work locations, and videotape machines.

In the 1970s new types of common carriers are beginning to emerge, some with a desire to build their own nontelephone networks. Separate data transmission networks for computer users are being built or discussed in most industrial countries. Some of these will operate by attaching new types of equipment to the telephone networks. Others will employ new transmission networks, physically separate from the telephone networks.

THE POTS MENTALITY

The majority of the revenue of telephone companies still comes from telephone service—sometimes referred to in such organizations as "POTS"—Plain Old Telephone Service. The telecommunication administrations in most countries outside North America receive more than 98% of their revenue from telephone service. Most of the engineering development, capital expenditure, and design philosophy is oriented to POTS. If digital, rather than analog, links are being built it is because this has become the best way to carry voice, not data; and the transmission and switching are designed for telephone traffic, not computer users.

It is possible to derive data channels of various types from the telephone facilities, and many countries are now using these to provide new types of data networks.

It cannot be stressed too strongly that computer users have fundamentally different characteristics and requirements from Plain Old Telephone Service. To force computer traffic to fit into telephone channels with telephone switch-

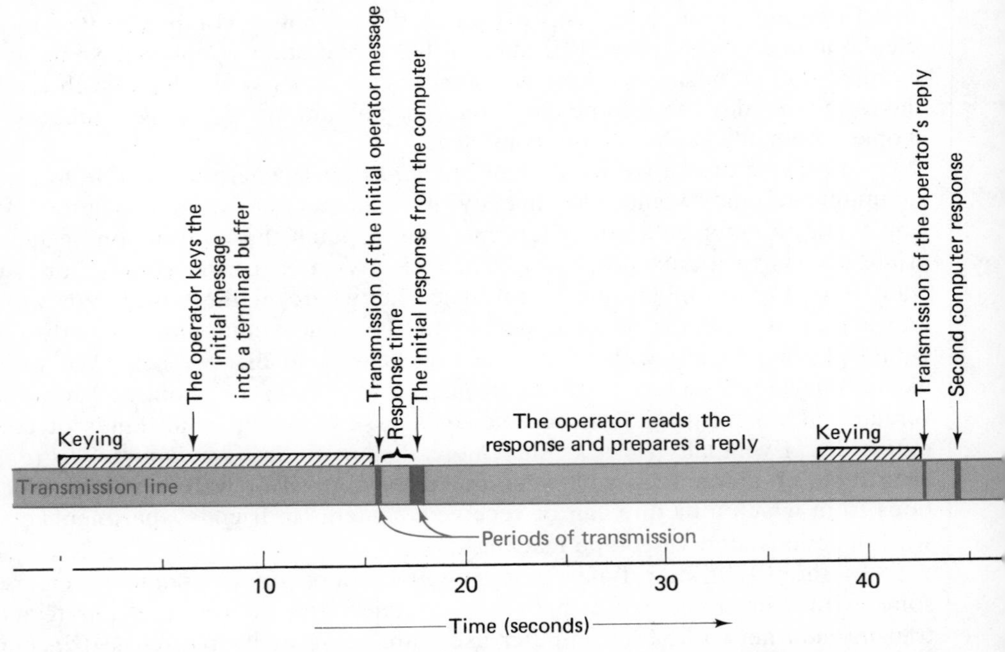

Figure 10.1 Most man-computer interactions make highly sporadic use of the transmission facilities.

ing both severely limits the potential of the computer and is likely to waste the channel capacity. New network architectures are being employed for data transmission and a whole new body of theory is emerging. Sometimes old-established POTS engineers have difficulty grasping that the needs of computer users are so different from telephone users.

PEAK TO AVERAGE RATIO

Perhaps the most important difference between terminal users and telephone users is that whereas telephone users talk or listen *continuously,* terminal users transmit and receive *bursts* of data with periods of silence between the bursts.

Figure 10.1 illustrates one terminal using a transmission line in a typical interactive fashion. The line is idle the majority of the time. When a telephone line is used by one interactive computer terminal the line is used extremely inefficiently because of the gaps between transmissions.

To overcome these disadvantages a network is needed which is designed to handle sporadic bursts of transmission rather than continuous telephone channels. As we shall see, the *transmission* facilities in existence for telephone traffic are well suited to such a network but the *multiplexing* and *switching* fa-

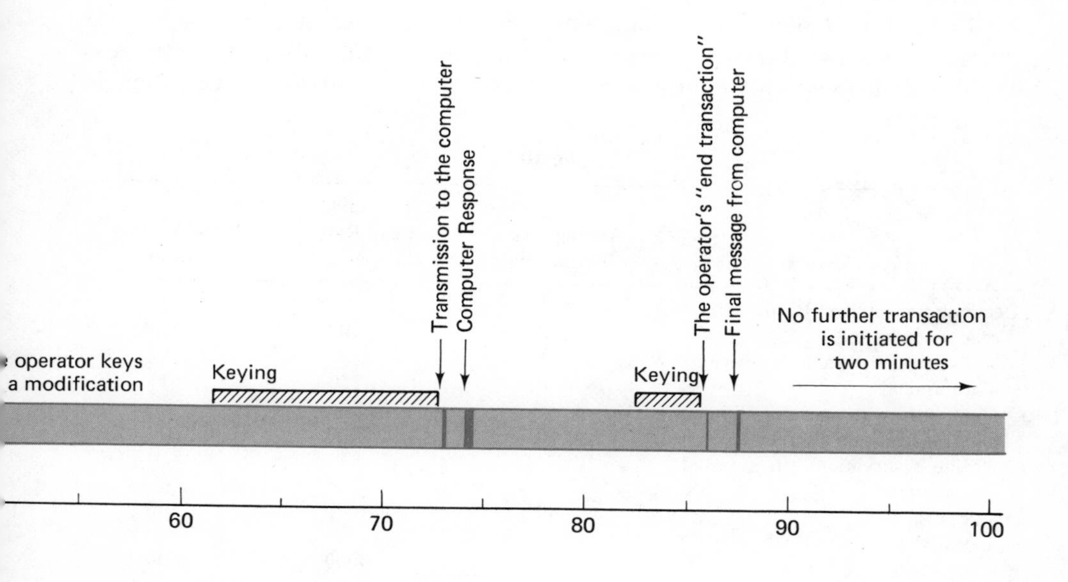

cilities need to be different. Such a network needs *burst multiplexing* and *burst switching* rather than continuous-channel multiplexing and switching. There are several different ways in which burst multiplexing and burst switching can be achieved.

How strongly *burst* switching and multiplexing is advantageous over *continuous-channel* switching and multiplexing depends upon how bursty the traffic is. A useful measure of this is the ratio of *peak to average* data rate of the transmission from one user. The peak-to-average ratio can be assessed for any specific application of data transmission, such as the dialogue of a person at a computer terminal. Table 10.1 shows examples for some typical dialogues [1].

The *average* bit rate of these dialogues is calculated from the total number of bits that pass to and from the terminal. The *peak* rate is calculated for the period of maximum desirable data flow to or from that terminal. The peak might occur, for example, when a screen is being filled with information. Some of the dialogues need a fast response time if they are to be effective and hence a screen must be filled with information in a second or two. Almost all man-computer dialogues have a peak-to-average ratio greater than 10. Most dialogues with a buffered visual display unit have a peak-to-average ratio greater than 100; some greater than 1000. The trend to "intelligent" programmable terminals tends to reduce the *average* data flow because more interactions can take place at the terminal location. Nevertheless the peak data flow often remains the same, hence the peak-to-average ratio is often greater with the newer intelligent terminals.

In the future new uses of graphics will come into existence, designed to enhance man's capability to grasp and manipulate information at terminals. Graphics dialogues will require a much higher bit rate to the screen than is

Table 10.1

Application	Average Rate Bits/Second	Desirable Peak Rate Bits/Second	Peak/Average Ratio
1. Calculation in BASIC on a teletype machine	8	100	12.5
2. Stock analysis on a typewriter-like terminal	10	150	15
3. Data entry with operator filling in multiple forms on a screen	10	2000	200
4. Airline reservations on a visual display unit	40	4000	100
5. Sales order			
i. With simple visual display terminal	2	4000	2000
ii. With programmable visual display terminal	15	1500	100
6. Circuit design with a graphics terminal	200	200,000	1000

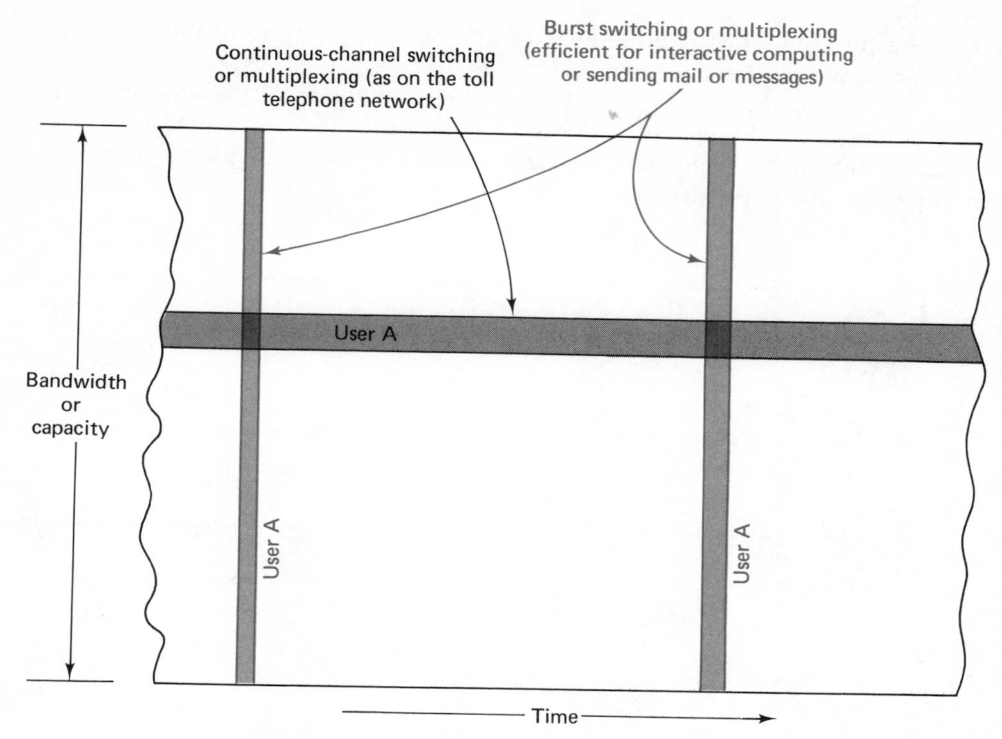

Figure 10.2 Two ways to divide up telecommunications capacity. Traditional telecommunication networks use *continuous-channel* multiplexing and switching. Switched data networks need burst multiplexing and switching.

common today—again in the form of high-speed bursts. Such bursts could either come from a local computer or from a communication link designed to transmit bursts.

The continuous-channel switching and multiplexing used for conventional telephone traffic thus has two disadvantages for computer terminals. First it causes inefficient utilization of the facilities because of the high peak-to-average ratio. Second the transmissions are restricted to the maximum speed of the telephone channels when in some cases faster bursts of transmission would be beneficial.

Figure 10.2 illustrates the two fundamentally different ways of dividing up telecommunications capacity. The horizontal line represents the traditional methodology; a portion of the bandwidth or capacity is allocated to one user for as long as he needs it. The vertical lines represent burst subdivision. A user is given the whole capacity for a brief period of time, and other users' bursts are interleaved. The bursts are assigned asynchronously whenever the user has the need for them. The two methods of operation apply both to the multiplexing process and to the switching.

The two methods of Fig. 10.2 are sometimes combined, as shown in Fig. 10.3. Some networks operate using continuous subchannels which are leased from the telephone companies, but the channels are subdivided using burst multiplexing and burst switching.

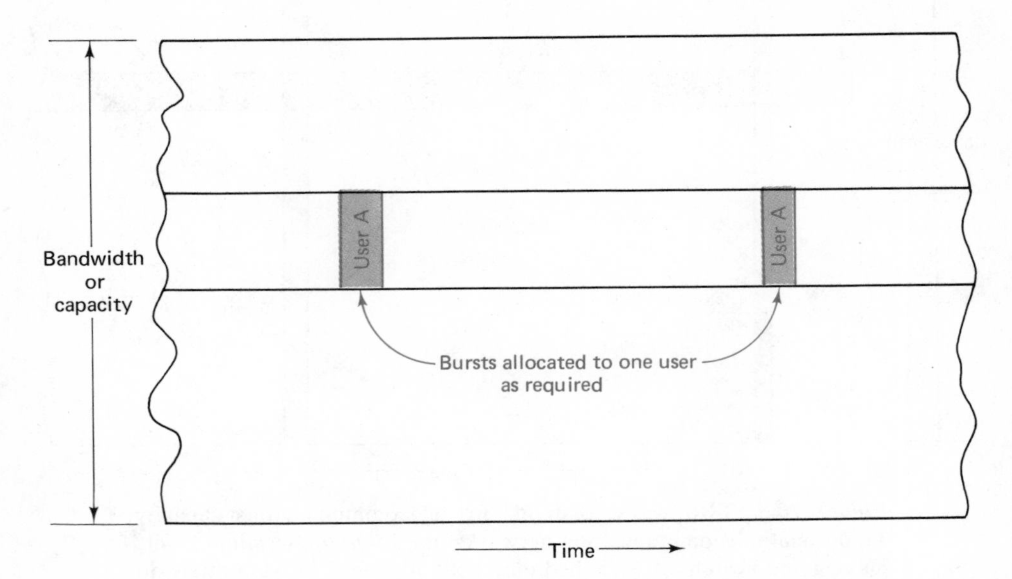

Figure 10.3 A combination of continuous-channel and burst switching or multiplexing. Burst switching may be used with leased continuous subchannels.

OTHER REQUIREMENTS OF DATA USERS

Box 10.1 shows other requirements of data users.

1. Rate of Transmission

The telephone operates at only one bandwidth. If data are sent efficiently over a dial-up telephone line, the speed is a few thousand bits per second, although lower speeds are often used. As we saw earlier, a wide variety of speeds ranging from very slow to exceedingly fast is needed. It is desirable that all these speeds of transmission eventually be available as switched public channels. Some corporations have installed their own switched data networks of high speeds, using leased lines to their own private switchgear.

BOX 10.1 Telephone traffic and computer traffic have characteristics so different that different network architectures are needed.

Telephone Users	*Computers and Terminals*
• Require a fixed capacity channel.	• Require a very wide spread of channel capacities ranging from a few bits per second to (ideally) millions of bits per second.
• Always carry out a two-way conversation.	• One-way or two-way transmission.
• Tolerant to noise on the channel.	• Data must be delivered without errors.
• Transmit or listen continuously until the call is disconnected.	• In a man-computer dialogue, transmission is in bursts.
• Require immediate delivery of the signal.	• In non-real-time data transmission the data can be delivered later, when convenient.
• The transmission rate is constant.	• In a man-computer dialogue the mean number of bits per second is usually low, but the peak requirement is often high. The peak to average ratio is often as high as 1000.
• The time to set up the connection can range from a few seconds to one minute.	• Sometimes it is desirable that the connection should be set up in a second or less.
• Switching is carried out only at the start of a conversation.	• Efficiency can be improved if the messages which constitute a dialog are individually switched with a very low switching time.
• Telephone dialing is manual.	• Setting up a connection between machines is often automatic.
• Telephone callers employ simple compatible instruments.	• Incompatible machines may intercommunicate giving a need for code or signal conversion.

2. Speed of Connection

Manual dialing is slow; typically it takes about 20 seconds to dial a 10-digit number. It is faster with pushbutton telephones but still slow by computer standards. After the number has been dialed, there is a delay before the connection is established. The delay is sometimes a few seconds, often as many as 10 seconds, and may be as long as 30 seconds on long-distance calls. The delay is inherent in the electromechanical switchgear used for making the connection and for searching for alternate routes. These time delays are acceptable for human telephone conversation, but with switched connections for other purposes, we must consider what interconnection times are desirable.

For nonreal-time data transmission, the end-to-end time can be quite long, as it is in some message-switching systems. For real-time work, a fast interconnection is needed, sometimes very fast. A request entered into a terminal needs a reply in a few seconds. Many airlines have specified that 90% of their transactions must have a response in 3 seconds or less, which requires a mean response time typically of about 1 1/2 seconds. If such transactions, or some of them, were finding their way through a switched network, very fast switching would be required.

It is possible to build extremely fast switching equipment. Once fast-switched networks are in existence, a pattern of data processing that will emerge in some applications will be one in which real-time actions initiated at terminals sometimes require processing by more than one machine. When processing a transaction, a computer will request information or assistance from other computers. When booking a vacation, a travel agent's computer will interrogate many such sources. Data requested from a management information system or other form of information system will often require the interrogation of data banks in other locations.

3. Speed Conversion

Terminals connected to a switched data network will operate at many different speeds. Some method is required for accommodating the different speeds without undue wastage of channel capacity. A switched analog telephone network cannot handle high-speed signals, and when low-speed signals are sent over it, as from the large numbers of typewriter-speed terminals thus connected, the available capacity is often less than 3% utilized. (Transmission at 15 characters per second over lines which can handle 600 is common practice in the data processing industry today.)

There are two ways of tackling this problem on a digital network. One is by means of time-division multiplexing in which the bits from the slow terminal are interleaved into bit streams from other devices so that full use can be made of channels between the multiplexing equipment. The other method is by means of digital storage built into the network and performing buffering operations.

4. Code Conversion

Different terminals use different transmission codes. An inexpensive terminal using, for example, five-bit Baudot code could be connected to a similarly inexpensive machine using seven-bit U.S. ASCII code, if the digital devices on the network performed the translation. Different terminals also use different line control procedures. The ability in a network of enabling incompatible terminals to communicate by automatic conversion of their character codes and line control procedures is referred to as "terminal transparency."

5. Data Conversion or Selection

It is sometimes advantageous to convert *data* as well as code for several reasons. They include cryptography enciphering and deciphering for security, error detection and correction, the compacting of information transmitted, the use of different language control messages in international transmission, and the structuring of responses that make the terminal easy to use.

SEPARATE OR INTEGRATED DATA NETWORK?

One of the decisions that the telecommunications organizations have to make is whether the switched data network and the voice network should be integrated or not. Should telephone and data traffic travel over the same switched network? Should they share the trunks and the switching mechanisms or should they have entirely separate networks?

The arguments changed entirely with the introduction of digital (PCM) circuits to carry telephone transmission. As we discussed in Chapter 4, a PCM channel designed to carry the human voice transmits 56,000 or 64,000 bits per second and could alternatively carry this rate of data. Many bit streams are interleaved (time-division multiplexing) between offices to give much higher-speed data channels. With suitable terminals, feeder channels, and switching, these bit streams could be used to carry all types of data. The switched data network and the telephone network could use the same trunks. However, as we have stressed, telephone switching is inefficient for data. A long and expensive voice line might typically be obtained for half an hour to connect a teletype machine to a distant computer. In this time the user, working at human speed, with many pauses to think, might cause a *thousand* characters to be transmitted. With efficient organization the same line could transmit a *million* characters in the same time. If it is a PCM line, it could transmit *ten million*. If the same line formed part of a network designed for data transmission, this inefficiency could be avoided by interleaving many separate transmissions.

The advocates of a separate data network claim that in addition to greatly improved efficiency it would have many special features, such as code and speed conversion which are not required in voice transmission. Furthermore, and this is a strong argument, the switched data network is needed *now* and can be built now by adding separate facilities to existing telecommunications channels.

ORGANIZATIONS WHICH PROVIDE DATA NETWORKS

Both separate and integrated networks are emerging in different parts of the world. They are being built by five categories of organizations:

1. Established Telephone Administrations

With the construction of PCM telephone links it seems natural that switched data networks should be derived from the telephone networks. In the United States AT&T is building its DDS (Dataphone Digital Service) network which provides data channels of speeds up to 56,000 bits per second (the speed of the AT&T digitized telephone channel). The network demonstrates that 56,000 bits per second user-to-user channels with a low error rate can be derived from the telephone network. (See Fig. 10.4.) DDS speed could be increased up to 1.544 million bits per second if the customers required the higher speeds. In many other countries switched networks have been planned or implemented by the telephone administrations.

2. Specialized Common Carriers

New common carriers, notably Datran in the United States, have built or planned switched data networks. Datran built a microwave system, with computerized switches, which is designed for data transmission only. It was subsequently taken over by Southern Pacific Communications. It is entirely separate from the telephone network except that many users can only access the system by means of telephone loops or trunks.

3. Value-added Carriers

A value-added data network is one in which existing transmission links are used in a new way which enhances their value to the end users. Value-added networks have been set up by new corporations, such as Telenet and Graphnet, in the United States, and by existing common carriers in other countries. Value-added carriers add computers to existing transmission links to provide new transmission services such as data networks which move bursts (packets) of data rapidly and inexpensively from one subscriber location to another.

4. Private Corporations

Large corporations have set up switched networks for their own use, leasing channels and adding switching equipment to them. Some of these networks cost many millions of dollars per year. They are likely to become more elaborate in the future with satellite channels, and computers for controlling switch-

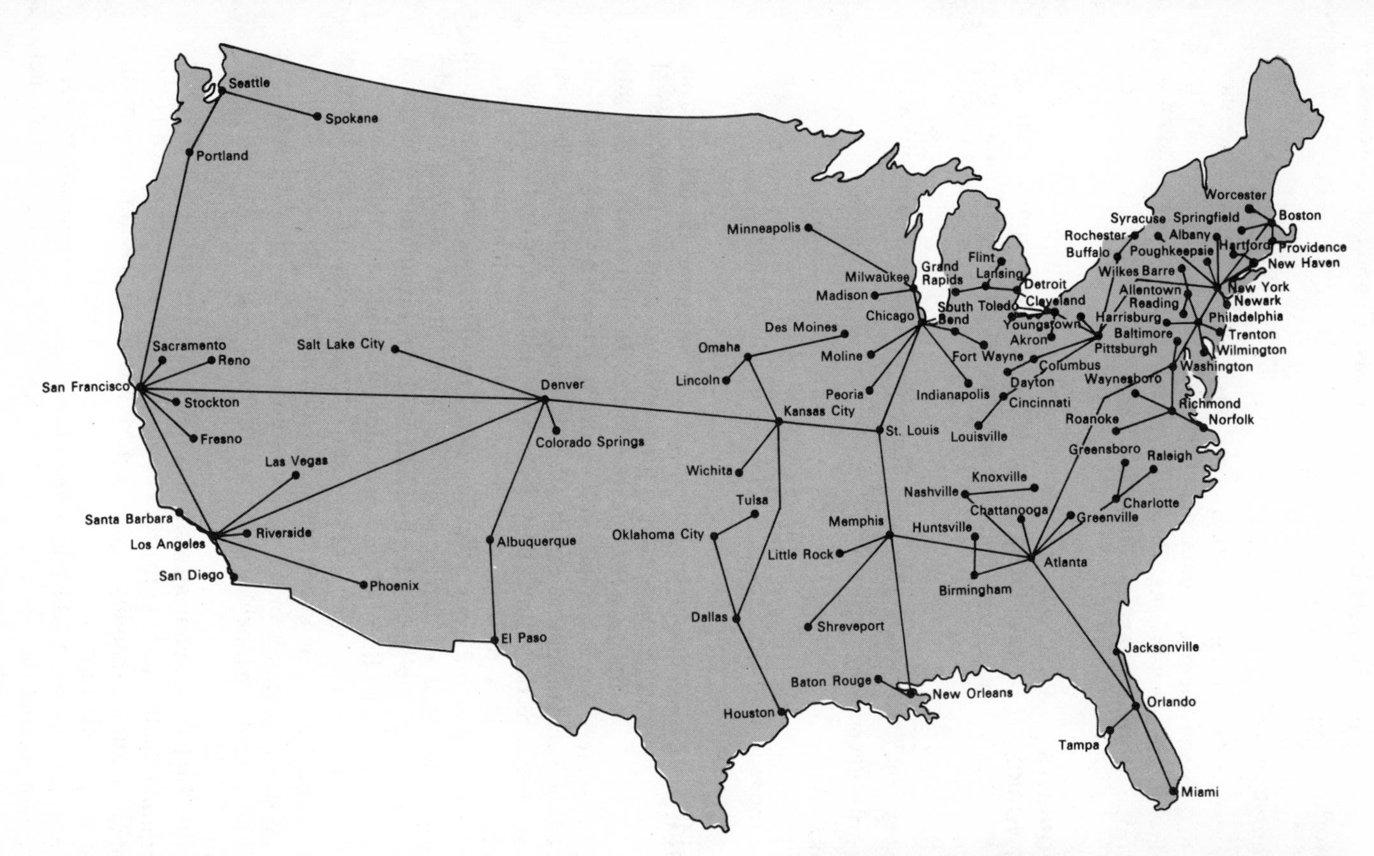

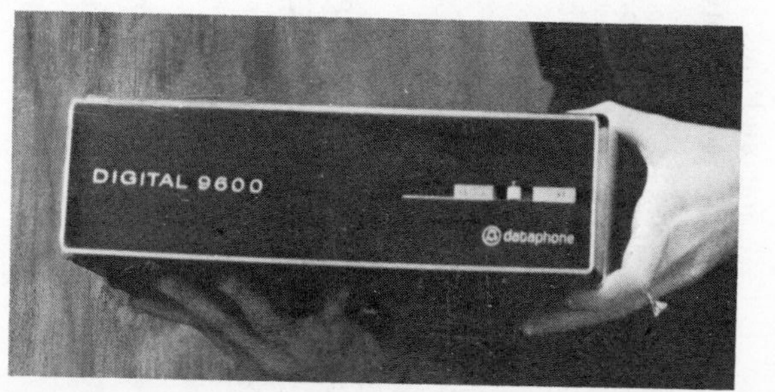

Figure 10.4 AT&T's planned DDS network giving point-to-point lines between the cities shown which transmit data at speeds of 2400, 4800, 9600, or 56,000 bits per second. These links are digital, not analog. Hence, they do not need a modem, but employ the service unit (right) which permits computers or terminals to be connected to the DDS line. (*Pictures courtesy AT&T.*)

ing and the assignment of channels on a demand basis. Chapter 17 discusses corporate networks.

5. Service Corporations

In some industries, groups of corporations have formed separate service organizations to provide them the data networks they need. The airlines use ARINC to relay airline reservations. Banks have formed corporations such as SWIFT, in Europe, to relay financial transactions and information.

THREE COMPONENTS OF DATA NETWORKS

Whatever the technology used for constructing a switched data network, three system components are necessary in order to provide end-to-end data communications:

1. The trunking system which provides the main highways between switching offices,
2. The local distribution network, and
3. The switching system.

Let us discuss these in turn.

THE TRUNKING SYSTEM

Very high bit transmission rates can be obtained using microwave radio circuits, coaxial cables, satellites, waveguides, or even wire pairs with sufficiently closely spaced repeaters. Such facilities will form the trunking system of future data networks. The transmission rate will generally be far too high for the users of the network; therefore the trunks must be made to carry the traffic of many users simultaneously. This process can be handled by synchronously interleaving the bit streams of different users through time-division multiplexing, in the same way that the Bell T1 carrier interleaves the bits of 24 digitized telephone signals. Alternatively it can be handled by asynchronously interleaving blocks of data, each of which carry their own identification and address.

In order to standardize and mass produce the channeling and terminal equipment, a basic operating speed may be selected for the main building blocks of the system. Channels of this speed will then be switched between the end offices. Many such channels will be multiplexed together over the high-capacity trunks.

The Southern Pacific data network has 4800 bits per second as its basic speed. Its trunking facilities are built from multiples of this speed, which is well suited to human reading capability and therefore apt to be the most commonly used speed for display terminals.

Other systems use the PCM telephone channel as the standard speed. A data network built around this speed can utilize the world's vast telephone networks, wherever they employ PCM channels.

The entire bit rate of the T1 carrier or its CCITT equivalents (1.544 or 2.048 million bits per second) may be the basic channel speed of the network. The bit rate of a satellite transponder may be employed between earth stations and subdivided.

Given the high bit rates of physical paths the trunking system should be able to move blocks of data quickly. An architecture should be selected that does not degrade the response time to the users.

LOCAL DISTRIBUTION NETWORK

There has been a great drop in cost in the long-distance trunks because of the higher bandwidth of microwave radio and coaxial cable. It is probable that solid state switching will reduce the cost of switching offices and increase their capability. The local loops from switching office to subscriber, however, have not dropped in cost, although multiplexers and concentrators using large-scale-integration circuitry will enable many devices to share one such link.

In general, the local loops are twisted wire pairs in multipair cable. Their loss-versus-frequency characteristics were designed for the requirement of telephone transmission only. It is desirable to use these wire pairs in the local data distribution network because of their wide availability, and the enormous capital that would be required to replace them with different facilities.

There are two types of approach to using the wire pairs for carrying a high bit rate. The first is to use equalization on the loop and design a modem that permits a wideband signal to be sent over it in an analog fashion. The Bell System 303-type data set transmits up to 230,400 bits per second over local loops. The second approach is to transmit the data in a baseband digital form without a modem. Most local loops can transmit at least 56,000 bits per second without modems, and this forms the basis of AT&T's DDS (Dataphone Digital Service). A *signal conditioning unit* is employed to ensure that the bit stream is sent at suitable amplitude with correct timing. In order to transmit wideband signals or high speed bit streams, the local loops must have no load coils. Most of today's loops, at least in North America, *are* unloaded.

Much higher bit rates can be transmitted over wire-pair telephone lines if digital repeaters are used at suitably frequent intervals. This is the basis of the T1 and T2 carriers which transmit at 1.544 or 6.312 million bits per second respectively, but not generally on local loops. Most local loops are not as thick as the typical short-haul wire-pair trunk used for T1 or T2 transmission. The wire sizes for local loops generally vary from 0.5 mm to 0.88 mm in diameter, whereas short-haul trunks are typically 1.25 mm. Nevertheless, with suitable repeater spacing most local loops could be made to carry 1.544 million bits per second.

In North America, *competition* is growing in long distance transmission, but it is proving more difficult to have competition in local distribution. Local distribution is the achilles heel of the specialized common carriers. After shipping their signals across the country the new carriers usually have to deliver them to customers via the local loops of traditional telephone companies.

Communication system designers would like alternatives to conventional analog telephone loops with modems for two reasons. The main reason is to obtain higher bandwidths—high data rates for computer users and hence in many cases faster response times, video channels, high-speed facsimile or information retrieval channels, etc. The second reason is to have an alternative means of transmission when the local cable or its connections fail. In some cities the old telephone loops are deteriorating and the advent of catastrophic failures (which sometimes last for days) is increasing.

Box 25.1 lists some possible alternatives to local telephone loops. Alternatives appropriate for data transmission include the use of CATV cables, packet radio (Chapter 32), and satellite antennas on user premises (Chapter 13).

SESSION-SWITCHING VERSUS ITEM-SWITCHING

For telephone users, switching times of 10 seconds or more are acceptable. They are acceptable also when large batches of data are transmitted. For interactive terminal use, however, long switching times cause inefficient use of the transmission facilities.

If the switching time is several seconds the circuit will be established, like a telephone circuit, before the dialogue starts and will not be disconnected until the dialogue has ended. The term "session" is used to describe the period of time a terminal user is connected through the network. Because only one terminal uses the switched path for the duration of the session the efficiency of transmission usage will be related to the *peak-to-average* ratio of the dialogue shown in Table 10.1. The majority of today's dialogues result in less than 10 bits per second being transmitted on average. Over a telephone operating at 4800 bits per second, this represents a transmission efficiency of 0.002. Over a PCM line of 56,000 bits per second, it will represent an efficiency of less than 0.0002.

If the switching is fast enough a fundamentally different approach can be used. Instead of switching the connection for the entire session, it can be switched for each individual data item that is sent, and then disconnected. With item-switching the transmission path need not be idle between items as it is in Fig. 10.1.

The transmission path is idle for the duration of the switching operation, and therefore high utilization of the transmission path can be achieved only by making the switching time short relative to the transmission time. If a message of 480 bits is transmitted at 4800 bits per second it occupies the channel for 1/10 second. The switching time therefore needs to be substantially less than the 1/10 second to give high channel utilization. If the same message is sent over a channel operating at 1.544 million bits per second it occupies the channel for 311 microseconds, and to give high channel utilization the switching should be performed in microseconds.

Figure 10.5 shows the maximum channel utilizations for different switching

FOR TRANSMISSION AT 4800 B/S:

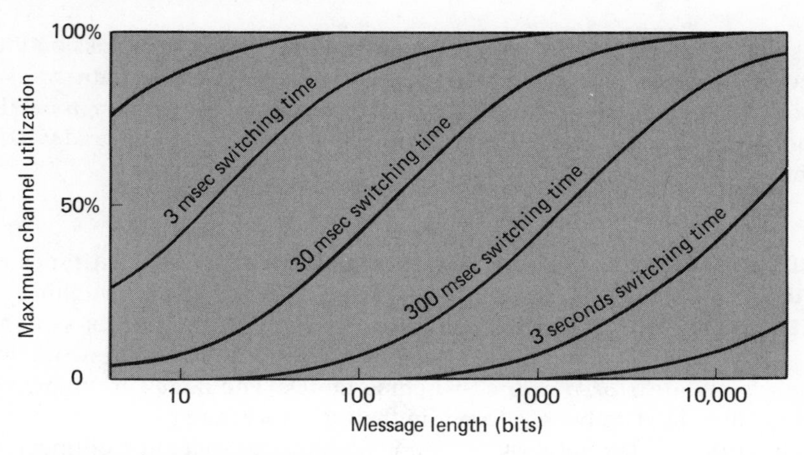

FOR TRANSMISSION AT 56,000 B/S:

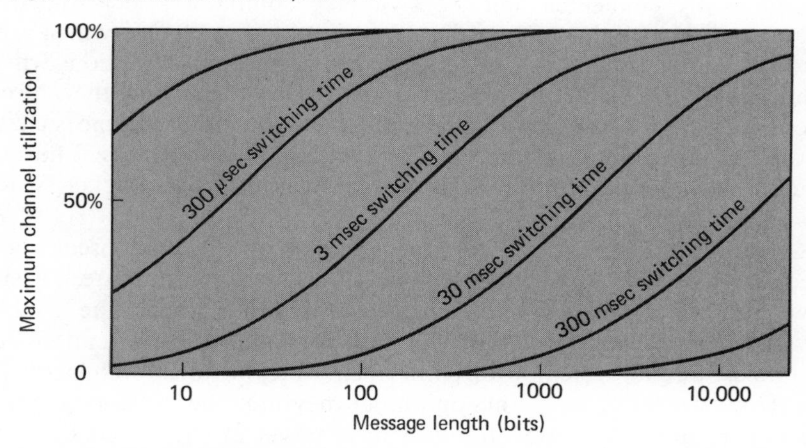

FOR TRANSMISSION AT 1,544,000 B/S:

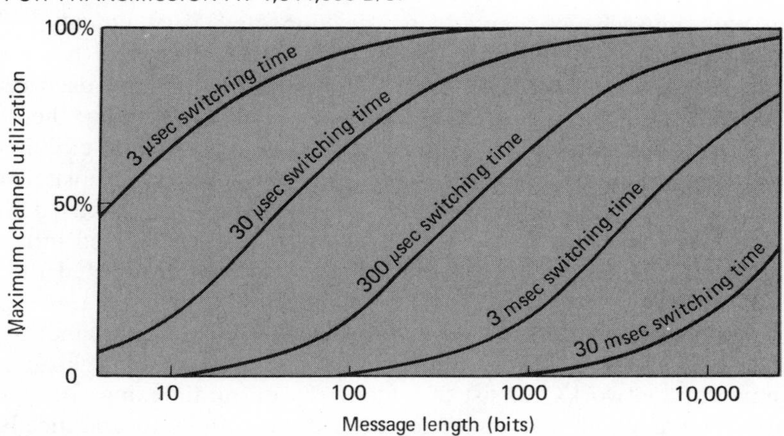

Figure 10.5 For efficient use of switched transmission facilities, fast switching is needed.

speeds and message lengths, when item-switching rather than session-switching is used. It is clear that a switched network for interactive data transmission needs switching speeds several orders of magnitude faster than the telephone network. Such speeds require solid-state electronic switching under computer control.

CIRCUIT-SWITCHING VERSUS PACKET-SWITCHING

There are two fundamentally different types of ways in which high-speed burst data switching can be achieved—circuit-switching and packet-switching.

Packet-switching refers to a network in which the data items are *stored* at the switching nodes. The nodes are minicomputers which receive the data burst, store it in their memory, and then forward it onward to the next part of the network. There is no physical switching of lines; the lines are permanently connected and the switching computer selects the line over which it sends which data. The data are transmitted in blocks, which are called "packets." Each packet contains a header giving routing instructions which each switching computer examines, rather like mail offices examining the addresses on envelopes. The packets cannot exceed a certain size. If a long message is to be sent it is broken into multiple packets each with their own header. The switching computers pass the packets onward as quickly as they can. Packet switching is discussed in Chapter 29.

Circuit-switching, unlike packet-switching, makes and breaks a circuit connection. The data flow through the connected circuit and are not stored at the switch location. Thus whereas packet-switching is rather like a mail delivery system with packets each carrying their own addresses, circuit-switching is more like a railroad system with points being moved to interconnect the train tracks. However all of the points on the journey must be switched correctly before the data are transmitted because data travel at the speed of electricity, not at the speed of trains. To handle high-speed burst traffic efficiently the physical path must be connected and disconnected very quickly. The connection remains only for the brief period of time that the data burst needs it. The terms *fast-connect switching* and *fast-connect network* are used relating to networks with this form of switching. As with packet-switching, the switching nodes are small computers. The path connected is in solid-state circuitry.

The circuit connection which is formed in a fast-connect switch may not be a physical electrical path. Time-division switching may be used, in which bits or characters of data from different messages are interleaved in a synchronous fashion as they flow through the switch. The path connected through the switch is then no more continuous than the beam of light from a movie projector, but to the user it appears continuous because the user handles a much lower data rate than the switch circuits.

Some data networks consist of a hierarchy of multiplexing points in which data streams are combined and uncombined according to the needs of the users. The interleaving of the data streams is changed dynamically. The streams can use interleaved bits, characters, blocks, or messages. Multiplexing with interleaved bits or characters is similar to time-division circuit-switching. Multiplexing with interleaved blocks or messages is more similar to packet-switching.

Data networking alternatives are discussed in more detail in Part III of this book.

REFERENCES

1. These dialogues are illustrated in detail in James Martin, *Design of Man-Computer Dialogues,* Prentice-Hall, Inc., Englewood Cliffs, N.J., 1972.

Marconi's demonstration of the wireless on Salisbury Plain, England to skeptical officials of the Armed Services and Post Office. The Italian government had refused his offer of a demonstration and Marconi had come to England where he established the company that first developed the radio. (*Courtesy The Marconi Company, Ltd., England.*)

11 DATA RADIO

Cables are not the only means of providing interactive services. Radio can also be used. The main uses of broadcasting today are for speech, music and television. *Data* can be broadcast equally well and a large amount of information can be transmitted in a short time. The entire text of the Bible can be transmitted over one television channel in four seconds.

There are three possible advantages to sending data by radio rather than by wire or coaxial cable. First, it can be inexpensive in that no cabling has to be installed. Second, the machine which receives the data, and possibly transmits, can be moveable. It could be a terminal in a vehicle, or a portable receiver. Third, large numbers of devices could receive broadcast data.

The main disadvantage of sending data by radio is that the radio spectrum is limited and is a uniquely valuable resource which must be allocated with care to its many would-be users. Efficient data encoding, however, can make a little bit of spectrum go a long way.

RECEIVE-ONLY TERMINALS

Data terminals for use with radio can be either receive-only, or can transmit and receive.

Receive-only sets can be cheaper than sets with transmitters, and generally cause less problems with spectrum allocation. There is one powerful transmitter in an area rather than many small ones. A device which transmits as well as receives, on the other hand, is capable of more versatile computer usage. It is now possible to build inexpensive computer terminals which can be held in the hand and which connect to computer networks via radio links. The potential uses of such devices are endless.

In this chapter we first discuss receive-only radio terminals, then two-way data radio.

BROADCAST DATA

A sound radio channel can be modulated to carry data and could operate inexpensive printers in the home or businesses, perhaps carry such information as Reuters news, the stock market ticker tape, sports reports, and weather forecasts. A television channel can carry far more information than a sound radio channel. With equipment of reasonable cost it can be made to carry several million bits per second. A terminal might receive blocks of a few hundred or a few thousand bits at a time, depending upon the application. The channel could therefore *either* transmit different blocks to many different users, *or* transmit a large number of blocks to every user, offering them a choice.

Of particular interest is the fact that a television channel can broadcast data *at the same time* as broadcasting television. A system developed by Hazeltine Research superimposes the data on the television lines without significantly degrading the picture quality. If one bit is imposed on each line, that gives a data rate of 15,750 bits per second. Other systems transmit data in the gap between the television frames. A television signal transmits the even lines on the screen, pauses while the scanning of the screen flies back to the top lefthand corner, then transmits the odd lines, pauses again, and repeats the process. The pauses are referred to as the *vertical blanking intervals*. Each interval lasts about 1.6 milliseconds. There are 60 intervals per second in North American television, and 50 in European (these numbers being equal to the electricity supply frequency). In Britain a standard exists for the transmission of data in the vertical banking intervals. 720 bits are transmitted in each interval.

TELETEXT

The transmission of data (text) superimposed upon a television signal is known as a *teletext*. The RCA Corporation designed an experimental system in the mid-1960s to transmit text in an analog form in the vertical blanking interval (Fig. 11.1). With today's digital circuits it is much better sent in a digital form.

The Public Broadcasting System in the United States uses the vertical blanking interval to transmit captions which can be superimposed upon the television picture, using a decoder unit. The captions could be used for subtitling foreign television or films, but the main use envisioned is to make television intelligible to persons who are deaf.

CEEFAX

In Britain a wide variety of data has been transmitted for display on television using *the Ceefax service*, jointly developed by the BBC and Britain's Independent Broadcasting Authority. Ceefax began operating in 1975 on a nationwide experimental basis, to a limited audience.

Figure 11.1 RCA experimental Home Facsimile System. RCA developed an experimental television system that made it possible to broadcast printed copy and standard programs to the home simultaneously and over the same channels by means of an electromagnetic "hitchhiking" technique. An electrostatic printer associated with the TV receiver picks off the print signal from the TV antenna without disturbing the TV program and converts it to printed form. The information to be printed is sent in gaps between frames as with the CEEFAX signals in Fig. 11.2.

The Ceefax service transmits pages of text designed for display on the television screen (Fig. 11.2). The data are organized into pages, each page being edited by the Ceefax staff to display information effectively on the screen. In the British teletext standard, which Ceefax uses, the pages consist of 24 rows of 40 characters per row, giving 960 characters per screen. Many characters may be blank to give attractive screen formats and color can be used. 8-bit rg298 CEEFAX 235 Thu 26 Aug 13:16/48

BBC CeeFax

NEWS HEADLINES 201
NEWS IN DETAIL 202-214
NEWSFLASH 250
WEATHER 215

FINANCE HEADLINES .220
FT INDEX 225

SPORT HEADLINES ...230

TRAVEL NEWS ...216-219
CONSUMER NEWS 241
SHOPPING GUIDE 242

PROGRAMME GUIDE
BBC1.. 251 BBC2.. 252
ITV... 253 Radio. 254

FULL INDEX
PAGE 298

EDINBURGH
FESTIVAL
NEWS from 290

RECIPE 243

SUBTITLE 270

NEW
TEST 263
PAGES 264

Table of Contents

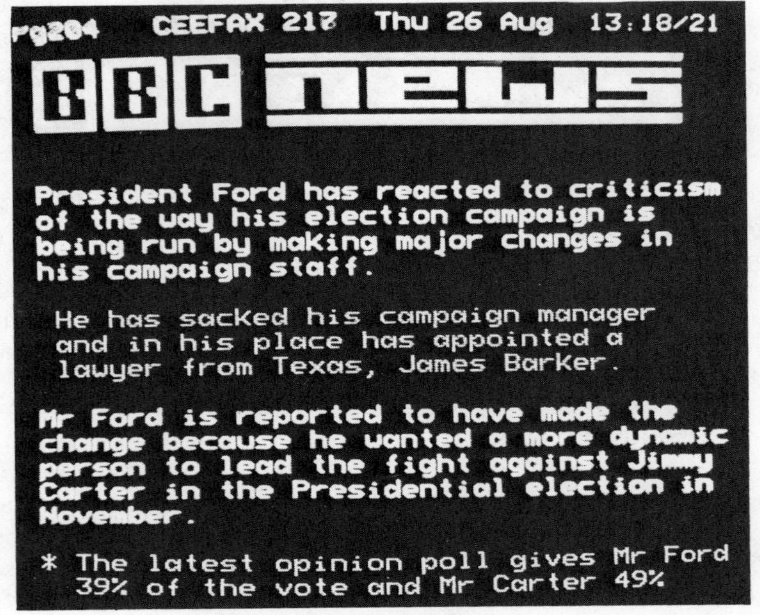

Part of the News.

Figure 11.2 CEEFAX screens. A BBC system for enabling television viewers to display frames of *broadcast* data on their screens. Britain's Independent Television Authority operates a similar system. Screens on television sets in a compatible format can accessed using a local telephone call to the British Post Office's Viewdata system, operating on a pilot basis.

characters are used, coded in 150-7 format (like the U.S. ASCII code). One row of text is transmitted in each vertical blanking interval, and consists of 360 bits (sync pattern, address, 40 characters, and error detecting code). 100 such rows can be transmitted per second.

The British television standard specifies 625 scanning lines per picture. 50 of these lines do not appear on the screen but are produced during the vertical scanning intervals—25 lines per interval. Of the 25 lines, only 2 are used in the Ceefax service. This gives a bit rate of 36,000 bits per second on each television channel. One page is transmitted every 0.24 second.

To display a page, the set requires a device attached to it which repetitively displays the characters on the screen. The television character generator can display graphics and colors in the display.

The user has a small keyboard with which he can key in the number of the page he wants to see. The standard allows for up to 800 separately numbered pages to be transmitted per television channel. Each page is preceded by its address. The Ceefax device examines the address of each page transmitted and when it receives the page requested, it loads it into the buffer and displays it.

The contents of a given page need not remain the same. It could be constantly changing like the stock market ticker tape or the news flashed in Times Square. The user could select one page, programmed to change, and watch screen after screen of news, advertising, or stock market reports.

PSEUDO-INTERACTIVE CHANNELS

One-way transmission, for example data broadcast over the air waves, can be made to appear like interactive transmission if a large quantity of data is sent.

The user reads one screen and keys in a response. The response, in effect, is the address of another screen which is then displayed. The response may be very short, for example, "1" or "2" meaning "yes" or "no." The receiving device adds this response to a constant associated with the current screen to obtain the address of the next screen. In this way a user could carry out a dialog with his receiver which could be used for applications such as medical diagnosis, computer-assisted instruction, or interactive advertising.

The response time on such a system depends on how frequently the same screen is retransmitted. The Ceefax service, if current intentions are met, will transmit a "magazine" of 100 pages on each television channel. A page would be retransmitted every 24 seconds (possibly containing different information). The response time would therefore vary from 0 to 24 seconds, depending upon when the page was requested.

The Ceefax service employs only two lines in each vertical blanking interval—that is 100 lines per second. If the entire television channel were used, either in North America, Britain or elsewhere, more than 15 thousand lines per second could be employed. With more efficient page encoding, a thousand pages per second could be transmitted.

If such a system were designed for an average response time of 5 seconds, each page being transmitted every 10 seconds, the system could make a total of 10,000 pages available on one television channel. This would be enough to allow highly elaborate dialogues to take place at a conventional television set equipped with an adapter.

The adapter which the consumer needs to pick up a service like Ceefax will probably be built into future television sets in some countries. It is illustrated in Fig. 11.3. Several experimental teletext services of different types are in use in various countries, some employing cable television. Many of the early transmissions were oriented to businesses rather than home subscribers. Some companies plan to transmit racing results and sports news to bars and betting parlours.

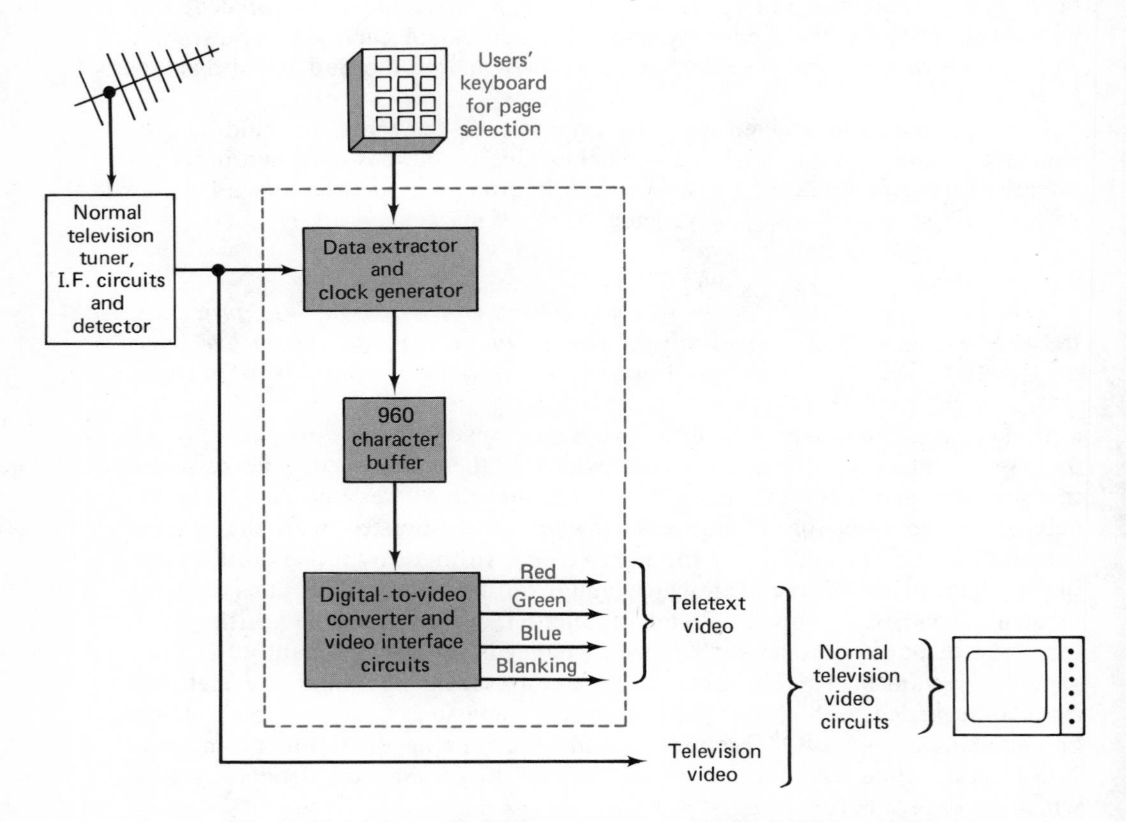

Figure 11.3 A small addition to a conventional television set converter permits it to receive teletext signals such as those in Fig. 11.2.

How much will the adapter cost? It has been estimated that the Ceefax adapter could be marketed for about $300 [1]. If a full television channel were used instead of merely the vertical blanking intervals, the cost should not be much different. The adapter would be built in LSI circuitry which is rapidly dropping in cost. With consumer acceptance the cost could drop like that of pocket calculators.

The television studio equipment used by the BBC for the Ceefax service was not expensive. It consists of a minicomputer with a disk file which stores the pages. Studio staff use visual display units attached to the computer to compose, edit, and update the pages. The computer is connected to the interface equipment which inserts the rows of data into the vertical blanking intervals in the television signal ready for transmission.

TWO-WAY DATA RADIO

Two-way radio devices can be imagined as being like radio telephones except that they are used for dialogue with a computer, transmitting and receiving data, not voice. Radio telephones are discussed in the next chapter, and one of the problems with them is shortage of spectrum space. The UHF frequencies which such devices use are grossly overcrowded. Interactive data transmission uses a channel only in brief bursts, rather than continuously like voice transmission. This was illustrated in Fig. 10.1. Consequently interactive radio terminals can be devised which use only a fraction of the radio spectrum space of radio telephones.

A technique that has great potential for radio terminals is called *packet radio*. Small hand-held radio units can be built, or small radio units can be attached to computer terminals to transmit and receive brief bursts of data. Packet radio was conceived by L. G. Roberts [2] and other workers on packet-switching techniques. Although it is attractive as an input/output means for data networks such as ARPANET and TELENET, it could also be of general use in computing and need not necessarily be associated with packet-switching networks. The reader should think of it as an inexpensive means of attaching any form of terminal to remote computers without using the local telephone loops.

LOW DUTY CYCLE

A packet radio transmitter is different from a conventional radio transmitter in that it does not transmit continuously; it transmits occasional very brief bursts of data. Because of this a relatively small power source can be used—a source capable of occasional brief bursts of high power but not continuous transmission. A small hand-held transmitter is therefore practicable.

It is indeed desirable that the power source should *not* be capable of powering continuous transmission because otherwise a transmitter accidentally stuck ON would play havoc with the other transmissions.

One of the first packet radio systems used the early NASA satellite ATS 1 via relays. This satellite was regarded as "dead" in that it did not have adequate power left to relay telephone signals. It did, however, work satisfactorily for packet radio because it switched its transmission power off after transmitting each packet. Operating in this mode it conserved enough energy to transmit the packets so that they were received correctly. It is likely that other "dead" satellites will remain valuable for packet radio.

POTENTIAL USES

There are many potential uses of radio transmitting brief bursts of data. Packet radio *could* come into widespread usage within five years or so if appropriate radio frequencies were allocated to it. A portion of the UHG band around 900 MHz has recently been set aside for land mobile radio applications, and a portion of this could be reserved for data radio. One of the attractive features of packet radio is that it does not require a very large radio allocation. However the regulatory authorities can be slow to react to new demands for the UHF spectrum. There will doubtless be a number of packet radio systems in operation before long but it may be slow to reach widespread usage.

Packet radio, given the spectrum allocation it needs, has the potential of revolutionizing some aspects of interactive computing. The following are potential uses of packet radio:

1. Radio units built into terminal controllers to permit them to continue operating when failures occur on their telephone connections.
2. A large population of portable hand-held terminals which fit into the user's pocket—perhaps about the size of an HP65 pocket computer.
3. Terminals and computers linked by radio to communication satellite antennas, thus providing a nationwide or worldwide network.
4. Burglar and fire alarms capable of radioing immediate details of violations to police or fire headquarters.
5. Burglar alarms on vehicles.
6. Utility meters transmitting their reading periodically.
7. Weather, seismic, and other monitors parachuted into forest areas, or monitoring potential earthquake areas.
8. Radio devices on every vehicle which identify the vehicle for automatically paying tolls and parking fees, for opening garage doors, for security control, and for enabling police to detect traffic violations or stolen cars.
9. Terminals in vehicles for sending messages, making inquiries, or fleet scheduling.
10. Terminals on boats for navigation and other purposes.
11. Portable terminals for children to act as a super-toy and give them access to libraries, computer-assisted instruction, and entertainment media.
12. Graphics terminals connected to remote computers and requiring bursts of transmission at a speed too great for connection by telephone lines.

13. Computer terminals in developing nations and other areas where the local telephone lines are inadequate.
14. Portable terminals for military field operations in areas where other forms of communication are poor.
15. Applications such as calling taxis, calling police, etc.

HAND-HELD TERMINAL

The hand-held terminal, described by Roberts [2], might be small enough to fit in one's pocket. Today's technology could provide a highly reliable, secure, communications device which would enable individuals to send or receive messages. They could communicate with other individuals, computers, data libraries, and possibly machines on the other side of the world via satellite relays.

Such a device could have an alphanumeric keyboard; some pocket calculators have enough space on their keyboard for 40 keys. To be generally capable of man-computer dialogues they would need a screen which could display short text messages. A liquid crystal screen could be used. The screen could either be designed to display alphanumeric characters or else could be a rectangular dot matrix.

Roberts visualized a dot matrix with 80 dots per inch resolution and 2.8″ × 1″ in size, i.e., 224 × 80 dots. This could display 8 lines of 32 characters, or small diagrams. The messages to the screen would be sent in character blocks, each block being a 7 × 10 dot matrix. If, instead, the machine handled 7-bit characters an alphanumeric message would require one tenth as many bits, but the machine would need character generation logic. There is a trade-off between bandwidth and logic requirements.

LINK CONTROL

Terminals attached to a communication line have a link control discipline. Such a discipline is needed also with radio terminals to stop the transmissions from different machines from interfering with one another. The radio link discipline used with existing packet radio systems is effective and surprisingly simple. It originated with the University of Hawaii systems [3] and is described in Chapter 32.

The possible social implications of portable data radio units are discussed at the end of the next chapter, after a description of radio telephones and other mobile radio units.

REFERENCES

1. T. Johnson, *A New Read-Only Terminal: Your Television Set,* Data Communications, May 1975.
2. L. Roberts, *Extension of Packet Switching to a Hand Held Personal Terminal,* AFIPS Conference Proceedings, SJCC, 1972.
3. N. Abramson, *Another Alternative for Computer Communications,* AFIPS Conference Proceedings, FJCC, 1970.

The earliest form of mobile radio, 1901. (*Courtesy The Marconi Company, Ltd., England.*)

12 MOBILE RADIO TRANSCEIVERS

For many years science fiction has been full of small, portable, radio-operated devices like Dick Tracy wristwatch intercoms, and paging devices that find James Bond no matter what he is up to. With the transistor, portable radio receivers suddenly became a reality—often a menace. Now, with mass-produced LSI circuitry, more elaborate portable devices are possible. A miniaturized telephone keyboard *could* be worn on the wrist. Once again, in the age of the computer, all manner of applications are possible for portable machines.

One of the fastest growth rates in the telecommunications industry is that of radio telephones, Citizens Band radio, and other mobile radio communication devices. New techniques now offer the potential of greatly expanding the sale and use of mobile radio equipment. The rising price of gasoline increases the benefits of radio use by delivery trucks, salesmen, maintenance engineers, and executives. A typical organization can increase the utilization of its delivery fleet by 50% with radio control. "Dial-a-bus" and other radio controlled schemes have become available for suburban public transport.

THE RADIO SPECTRUM

There is one snag, however. The radio spectrum is severely overcrowded. There is a critical shortage in frequencies allocated to certain two-way radio services, including police, taxicabs, fire engines, ambulances, and business users. The boom in Citizens Band radio is constrained by channel shortage. In London the number of unfulfilled applications for mobile telephones exceeds the total number in use. The economic and societal potentials of mobile radio cannot be achieved unless the radio spectrum is employed differently from in the past.

Improvements in spectrum usage are being made in ways such as the following:

1. Large blocks of radio frequencies are allocated to services which could be handled by cable rather than radio. This includes some of the UHF television frequencies. Major portions of the UHF spectrum could be made available for mobile radio applications. In the mid-1970s, the FCC did indeed reallocate 125 MHz of spectrum to mobile usage thereby making possible the widespread use of future mobile radio devices.
2. Frequencies need not be allocated on a fixed basis to users. Instead, users should be automatically allocated any channel that is free from a group of channels, when they need it—i.e. dynamic rather than fixed channel allocation. Modern equipment can do this.
3. The *same* frequency can be used by different callers in different locations provided that the transmitters are designed and located such that they do not interface with one another. The zones over which frequencies can be reused will be made relatively small in future systems.
4. New frequencies, higher up the spectrum, could be brought into use for mobile radio, given appropriate transmission design.
5. Transmission of data, rather than voice, should be used where possible because of the much greater use of channel capacity that is possible with data.

Using techniques such as these, the radio spectrum could be made to support hundreds of times as many mobile users as today. The U.S. President's Task Force on Communications Policy in 1969 illustrated the growth in mobile radio transmitters as shown in Fig. 12.1 [1]. The chart indicated that the mobile radio industry would be severely constrained if existing methods of frequency managements and utilization continued. With better techniques for using the spectrum, it would be possible to have as many mobile transmitters as there are vehicles, or even as many transmitters as there are people in the United States. Since that report, some major improvements in spectrum management have developed and the growth is likely to lie somewhere between the curves of Fig. 12.1. The new techniques could make mobile radio a massive growth industry. Some market studies [2] have estimated that its annual sales and service revenues may grow to $10 billion over the next 15 years.

CITIZENS BAND RADIO

In the mid-1970s the sales of citizens band radio, CB, took off as rapidly as television had done around 1950. This sudden surge in a communications medium which had existed since 1958 took everyone by surprise, not least of all the FCC. It is an example of the explosive growth which can occur when a telecommunications facility suddenly becomes fashionable. The FCC has a CB "family plan" permitting all members of a family residing at the same address to broadcast with a single license. In 1974 very few U.S. families had heard of CB. By the end of 1976 one in every 16 families used it and new license applications were flooding in at a rate of half a million each month.

CB is a service intended for short-distance personal or business radiocommunications, radio signalling, and the control of remote objects or devices.

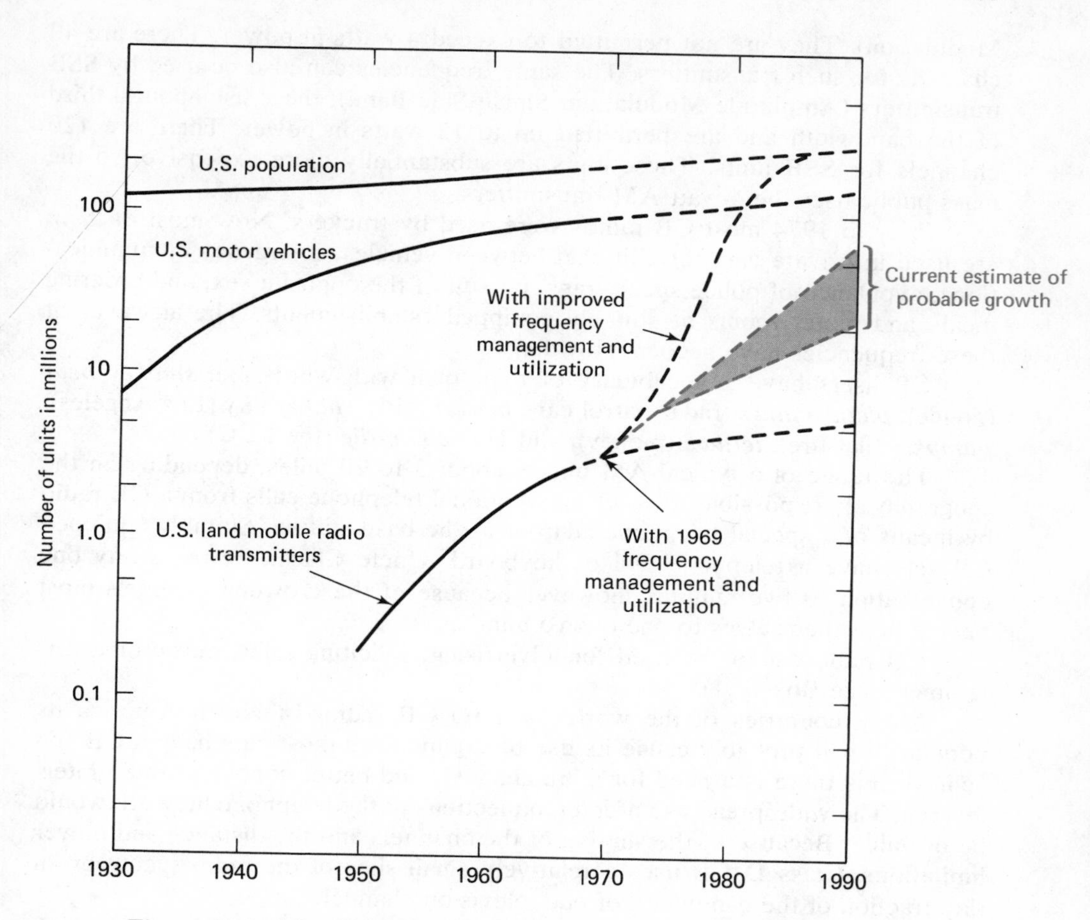

Figure 12.1 The U.S. President's Task Force on Communications Policy, 1969, contained this estimate of the future numbers of land mobile radio transmitters, with and without better management of the radio spectrum. Use of the spectrum has improved somewhat since then, and the shaded area shows the growth that now seems probable.

There are three classes of station:

Class A: A station operating in the 460–470 MHz band with a transmitter output power of not more than 50 watts.

Class C: A station operating in the 26.96–27.23 MHz band, or at 27.255 MHz for the control of remote objects or devices, or in the 72–76 MHz band for the radio control of models used for hobby purposes only.

Class D: A station operating in the 26.96–27.41 MHz band, or at 27.255, used for radio telephony only.

The latter class is that used by the public in large quantities for car and home radiotelephones, and walkie-talkies. Most transmitters are AM (Amplitude

Modulation). They are not permitted to exceed 4 watts in power. There are 40 channels for such transmitters. The same frequencies can also be used by SSB transmitters (Amplitude Modulation, Single Side Band); these use about a third of the bandwidth and are permitted up to 12 watts in power. There are 120 channels for SSB units. These units are substantially more expensive, so the mass public uses the 4-watt AM transmitters.

Prior to 1974 most CB radios were used by truckers. Now most of them are used in private cars for chit-chat between vehicles, emergency communications, avoidance of police speed traps, pursuit of the opposite sex, and ordering meals and motel rooms at suitably equipped establishments. The airwaves at these frequencies have become crowded.

CB users have a vocabulary of their own with words like *smokey bear* (police), *picture taker* (radar patrol car), *beaver* (girl), *shakey city* (Los Angeles), *pumpkin* (flat tire), *lettuce* (money), and *Uncle Charlie* (the FCC).

The range of a typical AM unit is about 3 to 20 miles, depending on the geography. It is possible to place conventional telephone calls from a CB radio by means of a special telephone adaptor at the base station. Suitably equipped CB sets have a telephone dial or keyboard. Uncle Charlie restricts any one conversation to five minutes, however because of the crowded channels most people limit themselves to one or two minutes.

CB radio cannot be used for advertising, soliciting sales, music or entertainment (see Box 12.1).

Many countries of the world have no CB radio. In North America its popularity will probably cause its use to expand until most cars have a CB. To achieve this there is a need for more channels and better control of radio interference. The widespread use of interconnections to the telephone network would be of value. Because of the sharing of the channels and the distance and power limitations, Class D CB uses a relatively small slice of the radio spectrum—a tiny fraction of the bandwidth of one television channel.

BOX 12.1 Restrictions on the use of CB radio sets.

a. A citizens radio station shall not be used:

(1) For any purpose, or in connection with any activity, which is contrary to Federal, State, or local law.

(2) For the transmission of communications containing obscene, indecent, profane words, language, or meaning.

(3) To communicate with an Amateur Radio Service station, an unlicensed station, or foreign stations.

Box 12.1 *Continued*

(4) To convey program material for retransmission, live or delayed, on a broadcast facility. Note: A Class A or Class D station may be used in connection with administrative, engineering, or maintenance activities of a broadcasting station: a Class A or Class C station may be used for control functions by radio which do not involve the transmission of program material; and a Class A or Class D station may be used in the gathering of news items or preparation of programs: Provided, that the actual or recorded transmissions of the Citizens radio station are not broadcast at any time in whole or in part.

(5) To intentionally interfere with the communications of another station.

(6) For the direct transmission of any material to the public through a public address system or similar means.

(7) For the transmission of music, whistling, sound effects, or any material for amusement or entertainment purposes, or solely to attract attention.

(8) To transmit the word "MAYDAY" or other international distress signals, except when the station is located in a ship, aircraft, or other vehicle which is threatened by grave and imminent danger and requests immediate assistance.

(9) For advertising or soliciting the sale of any goods or services.

(10) For transmitting messages in other than plain language. Abbreviations including nationally or internationally recognized operating signals, may be used only if a list of all such abbreviations and their meaning is kept in the station records and made available to any Commission representative on demand.

(11) To carry on communications for hire, whether the remuneration or benefit received is direct or indirect.

b. A Class D station may not be used to communicate with, or attempt to communicate with, any unit of the same or another station over a distance of more than 150 miles.

c. A licensee of a Citizens radio station who is engaged in the business of selling Citizens radio transmitting equipment shall not allow a customer to operate under his station license. In addition, all communications by the licensee for the purpose of demonstrating such equipment shall consist only of brief messages addressed to other units of the same station.

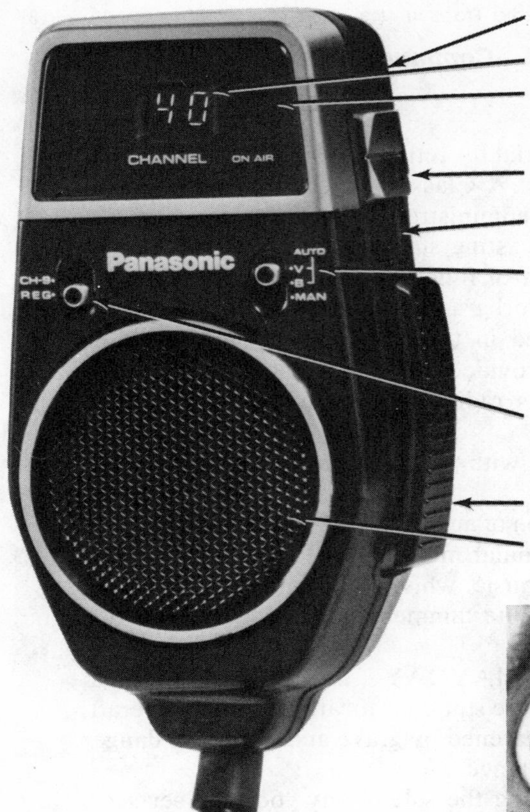

Squelch control, silences background noise.

LED channel number display.

ON-AIR indicator lights up when the set is on the air.

Manual channel selector.

Power switch and volume control.

Channel scan selector. Set to AUTO, it automatically finds the first busy channel, or the first vacant one. Set to MANUAL, the user can scan all 40 channels, one at a time.

Channel-9 EMERGENCY switch. Gives instant use of the emergency channel.

Push-to-talk button.

Microphone and loudspeaker.

The Panasonic *Big Mike* CB puts all the controls in the user's hand set.

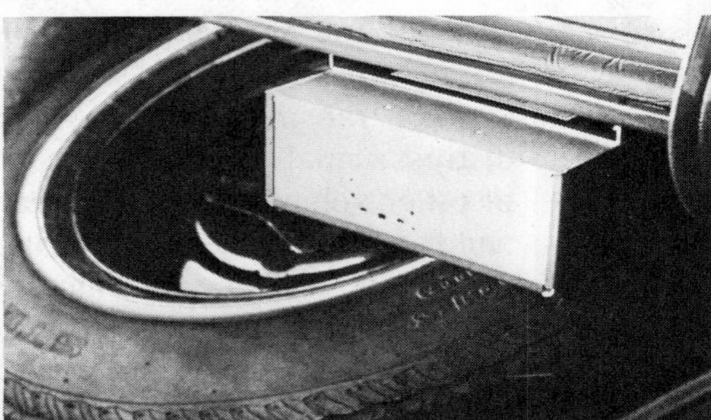

The transceiver and expensive circuitry are hidden in the trunk.

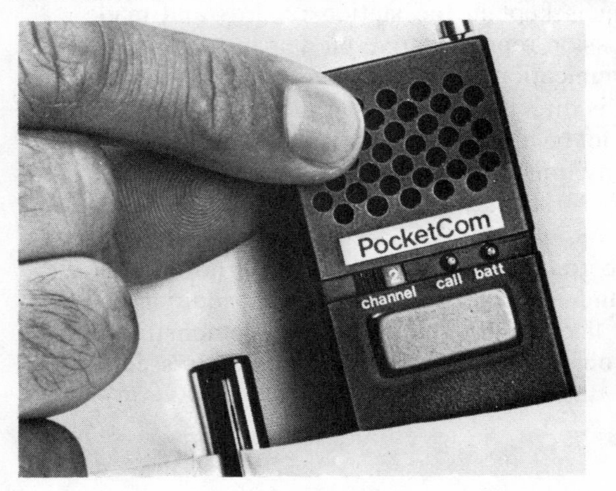

A tiny pocket CB with a range of 5 miles. Its power is 100 milliwatts and it contains 120 transistors to give a high amplification of weak signals with automatic volume control. *(Photo courtesy J.S.&A.).*

Figure 12.2 CB radio. A telecommunications market which exploded in the U.S.A.

OTHER CATEGORIES OF MOBILE RADIO SYSTEMS

There are three other main categories of mobile radio system in common use: dispatching systems, telephone systems, and paging systems. A fourth category which could be important in the future is interactive data systems such as the packet radio systems discussed in the previous chapter.

Dispatching systems are used mainly for communications between fleets of vehicles. The public are familiar with the radios that taxi drivers use. Many vehicles share in radio channel. An operator controls the fleet and can hear the messages from all drivers. In many systems, a driver can hear the operator's messages to all other drivers, although more elaborate addressing schemes can prevent this. Dispatching systems are used for controlling delivery trucks, police vehicles, fire engines, ambulances, military vehicles, and so on. Portable transceivers can be used in a similar fashion, and are used by police and security guards.

In most dispatching systems, the users talk only to a control operator. There are some radio nets in which the users can also talk to each other on an open channel. These are used in military operations. Because the users share an audible channel, the conversations must be disciplined by using words like "over" and "out" at the end of speech segments.

Radio telephones provide a switched two-way channel between two conversing parties, like a conventional telephone. The radio telephone channel is connected to the facilities of the public telephone network. Calls can therefore be placed to any telephone subscriber from a telephone in a car, and if the car is in a suitable location it can be called from a conventional telephone. Because of the shortage of radio channels, the few users who have radio telephones receive a poor grade of service. Much of the time when they want to place a call all channels are busy. New developments will attempt to make channel busy signals almost as infrequent as they are with conventional telephones.

Radio paging systems provide one-way channels to wandering users. The user has a small unit clipped to his belt or in his shirt pocket. Many such units share the same radio channel. Each unit reacts only to signals which are addressed to it, ignoring those to other units. The unit makes a beep to attract the attention of its wearer. Some units receive a voice message following the beep; some can only receive beeps. A paging user receiving only a beep goes to a nearby telephone. He may dial a paging operator who has a message for him or who connects him to another caller. Some private branch telephone exchanges handle paging so that the person receiving a beep dials his own number with a paging code and is automatically connected to the extension which paged him. Paging receivers can be small; one is designed to be worn on the wrist (like Dick Tracy). Some paging systems are connected to the public telephone network so that a person may be paged from any telephone.

Paging systems are used for locating executives, sending duty calls to volunteer firemen, and sending messages to reps, newspaper reporters, and other

Figure 12.3 This RCA microelectronic two-way radio can be used as a car radio or a portable via a repeater station in the car. The unit in the car also charges the batteries of the hand set.

A radio telephone unit for emergency medical services, transmitting voice and electrocardiogram signals. The vehicle in the background contains a repeater. *(Photo courtesy Motorola.)*

Figure 12.3 (continued).

Figure 12.4 A radio paging unit. A tone or a voice message is transmitted to the unit with FM radio. With appropriate transmitters and switches, individuals can be contacted over a wide geographical area. *(Photos courtesy Motorola.)*

Conventional land mobile radio systems:

CONVENTIONAL RADIO DISPATCH SYSTEM

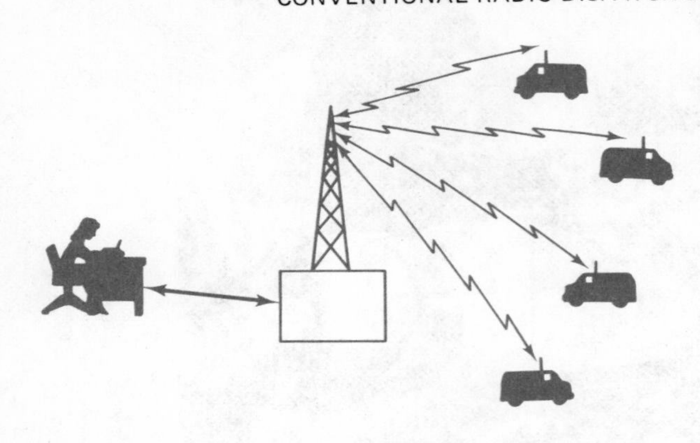

- Single channel
- Many transceivers share the channel
- Transceiver operators talk to a central control
- User owned and operated
- Used to control vehicles and other resource
- Typical applications: taxis, firemen, ambulances, police, delivery fleets

CONVENTIONAL RADIO TELEPHONE SYSTEM

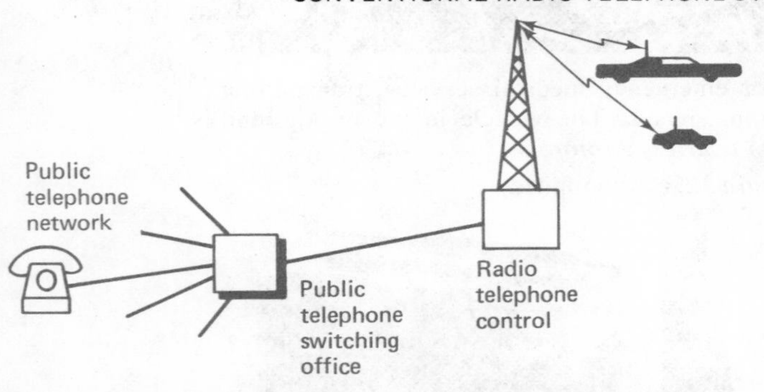

- Telephones in vehicles which can dial calls and receive calls via the public telephone network
- Multiple channels
- Random allocation of channels to users
- Users queue for a free channel sometimes with a long wait
- Can have automatic control of channel allocation and switching

CONVENTIONAL RADIO PAGING SYSTEM

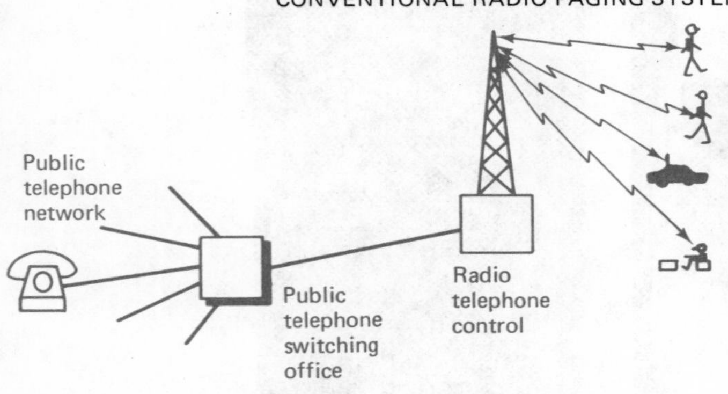

- Single channel
- Many receivers
- Normally one-way transmission to receivers
- Some systems have *voice* paging units
- Some systems have *bleep* paging units
- Some paging systems are accessible via public telephones or via a PBX
- Typical applications: locating staff, calls maintenance engineers

Figure 12.5

Future land mobile radio systems:

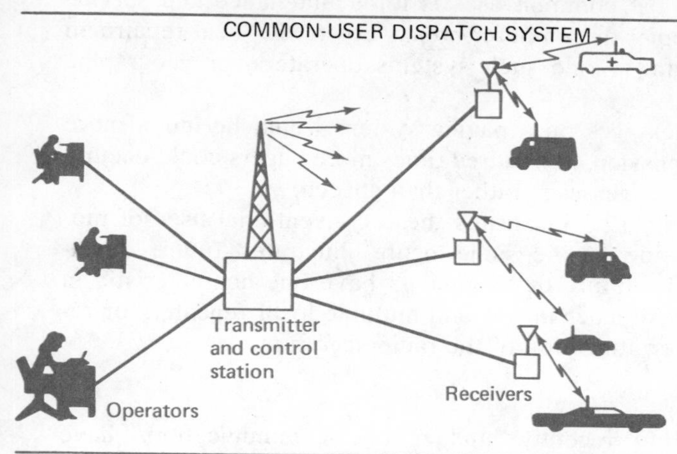

- Multiple transceivers share multiple channels
- Equipment automatically selects a free channel
- Requests for channels are automatically queued
- Calls to individual transceivers or fleet calls can be made
- The system can leave an indication of messages waiting
- Multiple users share a radio common carrier service

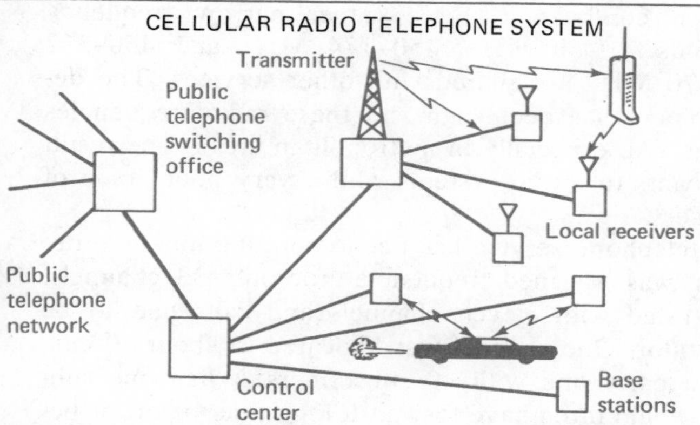

- Many channels
- Each frequency used for multiple, simultaneous, geographically separate calls
- The area covered is divided into cells as in fig. 12.8
- Complex computer-controlled allocation of frequencies to users who may move between cells during calls
- Has the potential of eventually providing mobile telephone service as good as conventional telephone service

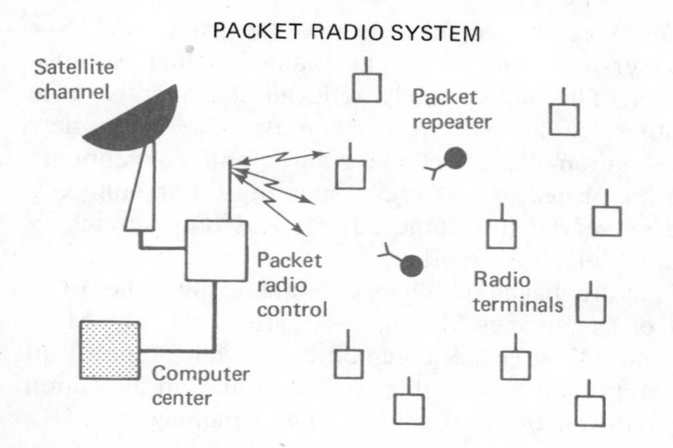

- Designed for interactive computer terminals
- Many terminals share one channel
- Short digital burst from terminals time division multiplexed on channel
- Aloha channel control
- Economic spectrum usage
- Many potential computer applications

travelling staff. One of the most common uses is for maintenance and service personnel. When an office copier or computer breaks down, the local repairman is contacted quickly by paging. Some such systems operate over geographic areas of 50 or 100 miles.

The sending of data messages on a paging system would be much more economical than voice transmission. A hundred times more pagers could occupy a channel if they received data messages rather than speech.

The left-hand side of Fig. 12.5 illustrates these conventional uses of mobile radio. The right-hand side of the same figure illustrates future developments of such systems. The future developments have one characteristic in common: more complex control mechanisms and multiple local repeaters or receivers designed to give better utilization of the radio spectrum.

FREQUENCIES ALLOCATED

Until recently land mobile communications have been confined to three rather narrow frequency bands, 30–50 MHz, 150–174 MHz and 450–512 MHz, and all but the 450–470 MHz are shared with other services. The demand for mobile radio equipment has been high and these radio frequencies have become extremely congested, especially in metropolitan areas. The result is a vast waiting list of applicants for radio systems, and a very poor grade of service for radio telephone users.

The Bell System radio telephone service has been operating for three decades, but by the mid-1970s, was assigned frequencies for only 33 channels. Until the late 1960s, it operated with eleven channels and calls had to be placed through a special operator. One channel was allocated to about 40 customers. That is like having a party line with 40 subscribers. When one subscriber wanted a channel, he would often have to wait a long time before it became available to him.

In the late 1960s, an improved technique came into use (called IMTS—Improved Mobile Telephone Service). Any of the 11 channels could be automatically allocated to any user. This substantially reduced the average time that a user had to wait. Another 12 channels were made available in the new 450 MHz band. IMTS used automatic telephone dialing, and conventional two-way telephone operation replaced push-to-talk operation. But still only about 1000 customers could be served in a large city. It was only scratching the surface of the potential radio telephone market.

In the mid-1970s, the picture began to change dramatically. The FCC made available a large block of frequencies high in the UHF band. 115 MHz were allocated between 806 and 947 MHz. A common carrier allocation of 40 MHz was made (it was 75 MHz at first, but then reduced) and an allocation for private use of 30 MHz (reduced from 40 MHz). The remaining 45 MHz were reserved for future allocation.

Common carriers such as AT&T could have more than 20 times their previous spectrum available to them. Equally important, the new frequencies will be used with techniques which permit a far higher level of sharing. The same frequency can be used multiple times in the same city. The combination of these two factors makes it possible to multiply the numbers of radio telephone users in a city by several hundred. AT&T estimates that large cities which today can have only a few hundred mobile telephone subscribers can, with the new techniques, expand to almost half a million.

To further enhance the spectrum usage, the FCC abandoned its block allocation system which gave separate allocations to private subservices such as fire, police, forestry, etc., in favor of shared systems.

CELLULAR SYSTEMS

The key to making a limited band of frequencies serve a large number of users is to reuse the same frequencies over and over again. The radio transmitters should be designed so that they cover only a small area, and there should be many such transmitters. The same frequencies can then be used in nearby, but not adjacent areas. This is the basis of the new mobile radio systems referred to as *cellular*.

The cellular principle is illustrated in Fig. 12.6. The cells are regular hexagons, a few miles in width. A group of frequencies is allocated to each cell. Different frequencies are allocated to adjoining cells. The digits in Fig. 12.6 each represent a unique group of frequencies. Each group is repeated in a cell nearby.

The cells are grouped into blocks. In Fig. 12.6, each block has seven cells. The pattern of frequencies is repeated in each block. Within one block no frequency is allocated twice, but each frequency is reused in each neighboring block.

The transmitters in a cellular system are tailored to the size and shape of the cell. They are low in power so that their signal does not reach that cell in a nearby block which reuses the same frequencies. The transmitters are designed to cause as little interference as possible. They may be lower in cost than today's high-power transmitters, but there will be many of them. This is the basis of an AT&T scheme called *Hicap* (High capacity) which will be implemented first in Chicago in the late 1970s.

A mobile user placing a telephone call may be in any cell. The system must be able to determine the location of the mobile set and allocate any one of that cell's available frequencies to the set. The user, talking on the telephone, may be in a vehicle travelling at high speed from one cell to another. As he crosses the boundary, the system must detect this and switch his conversation to a new frequency. The changeover must be made smoothly without disrupting the telephone call or causing too much change in the quality of the channel.

A form of cellular system has been used for some time on certain trains,

BOX 12.2 The Metroliner telephones.

The high-speed trains (the Metroliners) between New York and Washington use telephones which operate in a cellular fashion [3]. Calls can be placed either to or from persons on the trains. Any kind of telephone call can be placed, including credit card, collect (reverse charges), and data calls. When a person on the train is called, the caller must know the number of the train, which is listed in the train schedule. He does not have to know where the train is. The called party will be paged by a train attendant and asked to go to the telephone booth in his car.

The 225-mile track length is divided into nine zones of approximately equal length, and each zone has its own fixed radio transmitter (and receiver). As the train hurtles at 100 miles per hour from one zone to the next, calls in progress must be automatically switched from one fixed transmitter to the next. The switch happens without the customer being aware of it. As the train races into the Baltimore tunnels, the calls must again be switched to special transmitting and receiving equipment on the roof of the tunnel.

All calls go through the Bell System central office in Philadelphia where a special terminal remotely controls fixed transmitters. The terminal examines the noise-to-signal ratio from all the different transmitters and automatically transfers calls from one transmitter to another as the train rushes onward. When any available transmitter maintains a better noise-to-signal ratio than the one in use for several seconds, it is automatically switched in. The new transmitter is turned on 200 milliseconds before the old one is turned off. The user does not normally detect the change. The transmitter selected is not always the one closest to the train because of differences in the terrain or local conditions.

Frequencies in the 400-MHz (low UHF frequencies) range are used. A single frequency can be used more than once in areas geographically separate, as will be necessary with other mobile radio services. A telephone conversation may take place on one frequency near Washington while an entirely different conversation uses the same frequency near New York. The range of one frequency is about 100 miles, and two uses of the same frequency must be separated by at least 75 miles as shown in the figure.

The automatic switching between antenna zones as the train travels and the reutilization of frequencies are demonstrations of two essential features of future mobile radio systems.

BOX 12.2 *Continued*

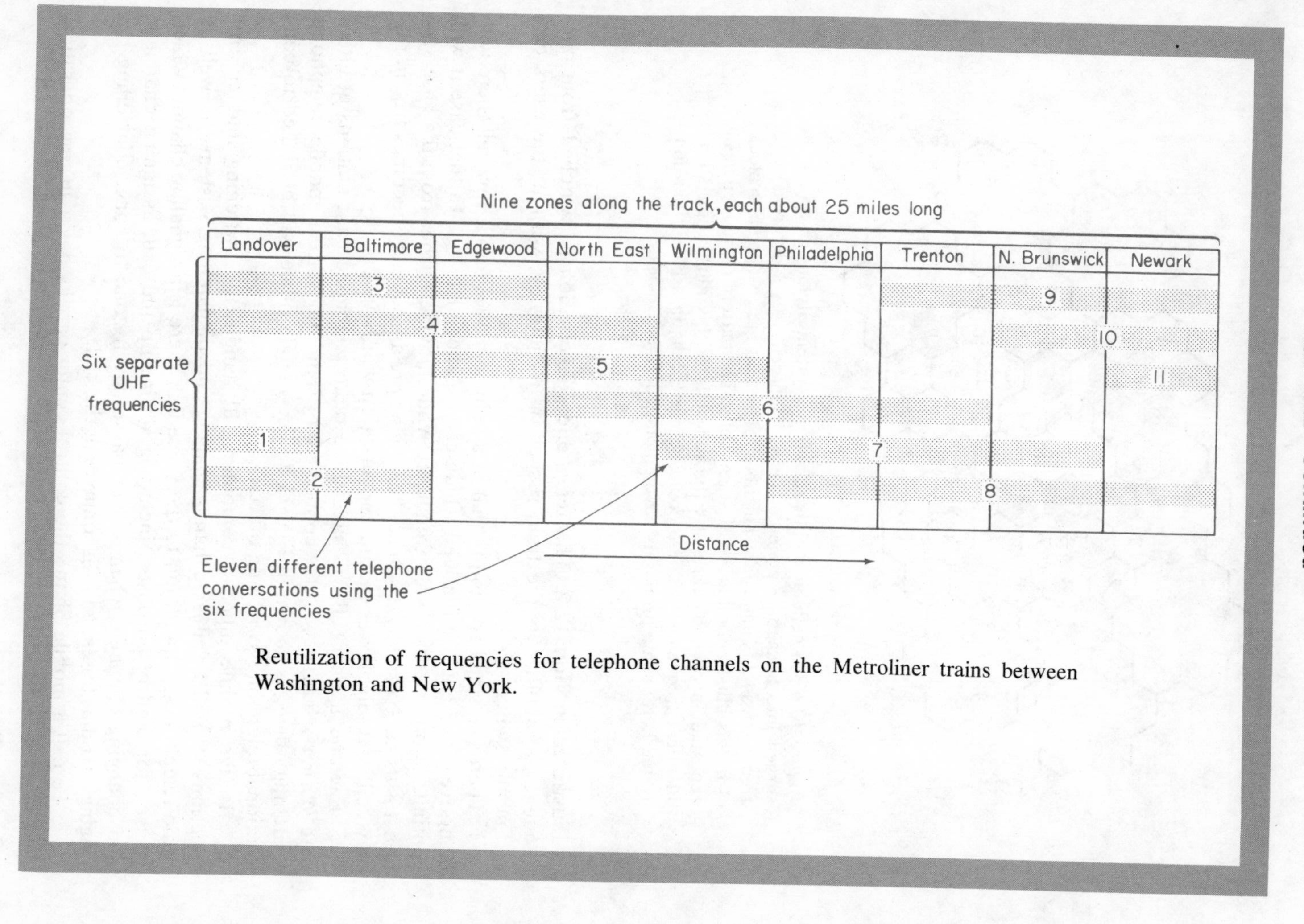

Reutilization of frequencies for telephone channels on the Metroliner trains between Washington and New York.

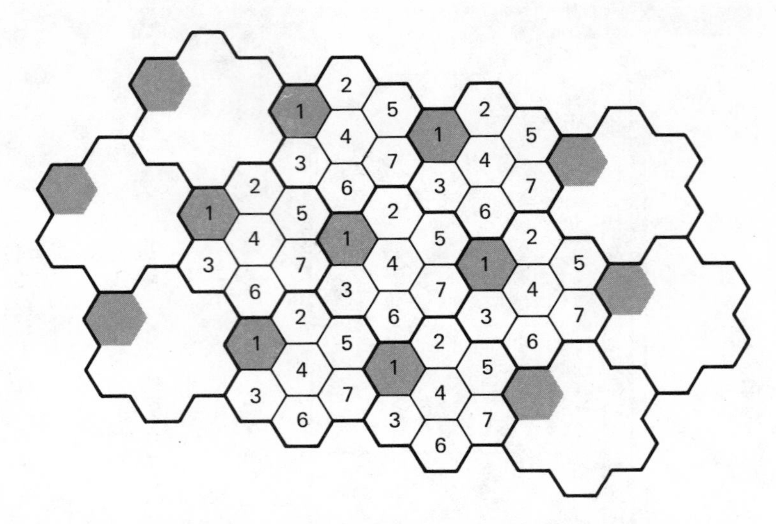

Figure 12.6 A cellular mobile telephone system divides an area to be served into hexagonal zones or cells. Each cell uses a different set of frequencies to its immediate neighbors. The cells are grouped into blocks —in this illustration blocks of seven cells. The pattern of frequencies is repeated in each block. Thus the red cells in this illustration all use identical frequencies. There could be more cells in a block than illustrated here; some proposals have employed 12-cell blocks.

as illustrated in Box 12.2. The telephone conversation is switched from one frequency to another as the train rushes on its journey, without the users detecting the switch.

There is a small possibility that as a vehicle moves into a new cell there will be no free frequency in that cell which can be allocated to it. In that case it will continue to operate on its old frequency while moving further from the base station it is using. Normally it will not be long before a frequency becomes free in the new cell, and telephones in mid-conversation will have priority.

Each hexagonal cell will have three radio telephone base stations at alternate corners. These will be connected by land lines to a mobile telephone switching office, as shown in Fig. 12.7. The mobile office will be linked to conventional telephone switching offices.

In the mobile office a computer will handle the switching and control functions. It will be able to handle paging and dispatching systems as well as radio telephones. In addition to the voice link there is a signaling channel which is used for control purposes, including setting up the call, assigning channels, and changing channels when a user passes from zone to zone, controlling the ringing and dial tones, and disconnecting the call.

To call a mobile phone the control center carries out a paging operation.

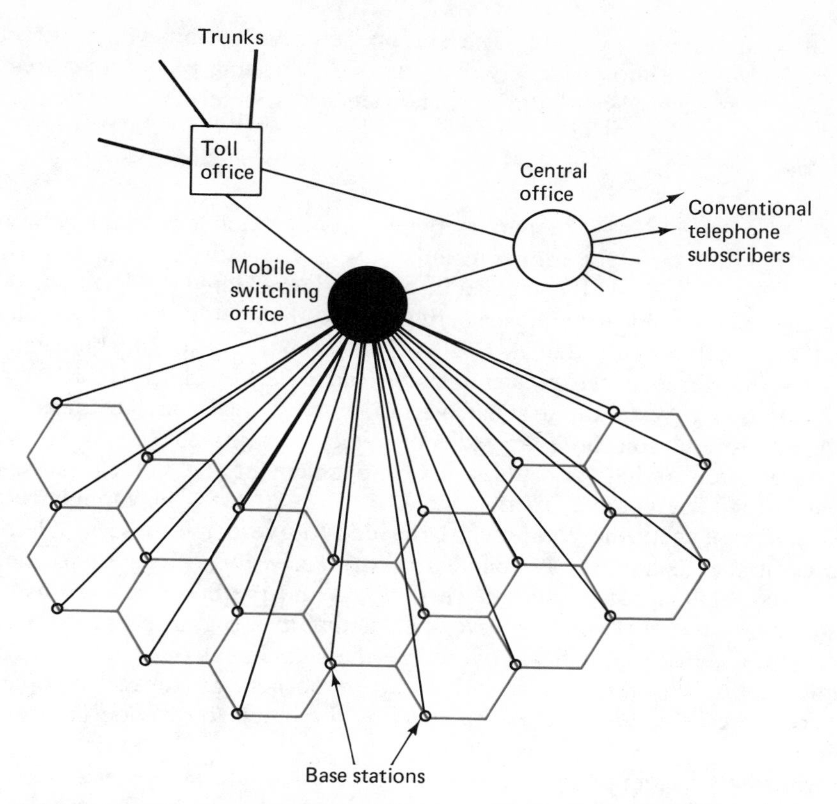

Figure 12.7 A mobile radio base station will exist at the three alternate corners of each hexagonal cell. Land lines will connect these to a mobile telephone switching office which is responsible for cell control. The mobile office will be linked to connecting telephone switching offices.

A coded signal containing a mobile telephone number is transmitted throughout the city and is received by all phones which are switched on. Each phone compares the number transmitted with its own number. The phone that is called responds by selecting a free data channel for response. It may scan all channels and select the one with the highest signal strength. When it receives the response, the mobile switching office allocates a voice channel to the unit, and transmits a signal which causes its bell to ring. There are many different ways in which the data channel could be organized. One is to use the ALOHA protocol discussed in Chapter 32. The control mechanisms needed in the telephone are quite complex but not necessarily very expensive if mass produced in LSI circuitry.

In order to establish the link it is necessary to locate the mobile unit. Three directional antennas are located at each base station and point into the three cells which intersect there. The station determines from which direction the signal is received most strongly. The mobile switching office compares the information from several base stations, and hence can determine which cell the telephone is in.

RECEPTION CONDITIONS

The path between mobile telephones and their base station presents severe transmission characteristics. The signal in a city is reflected by buildings and reaches the receiver by several scattering paths. The paths can have slightly different propagation time delays and so interference can occur between the paths. This interference causes signal fades and as a vehicle moves down the street it encounters very rapid signal variations with peaks and troughs a few inches apart. In addition there are fades of longer duration caused by gross terrain features such as hills and cliffs. The movement of the vehicle causes an apparant change in frequency of the signal of up to 80 Hz—the Doppler effect. Careful design of the radio equipment is needed to overcome these problems.

One of the techniques is called *diversity reception.* More than one antenna is used. The signals received are sampled and the best ones are used, or else the signals are combined to give the maximum signal-to-noise ratio. Two antennas are used on a vehicle to overcome the signal peaks and troughs which are inches apart. The three base stations at the corners of the cell receive the signals from the vehicle, and these signals are combined to provide the best reception.

FM radio receivers can be designed to discriminate between a weak and strong signal which would otherwise interfere with one another. They can capture the strong signal and reject the weak one. This is of value in cellular systems because a signal in one cell might be interfered with by a signal which occasionally strays in from a cell some miles away using the same frequency.

In general, AT&T and others comment that the techniques they have developed can make the quality, operation, and grade of service of mobile telephones similar to that of conventional telephones.

SUBSCRIBER GROWTH

Subscriber density is likely to increase over a wide range in future cellular telephone systems. The system can change to accommodate this growth in two ways. First, the cells can be made smaller, as illustrated in Fig. 12.8. The dense center of a city may have many small cells, larger cells are used in the outskirts of the city and still larger cells in the suburbs. Second, more frequencies can be assigned to each cell if enough spectrum space is available.

Fig. 12.9 shows an AT&T estimate of the growth of subscribers in Phila-

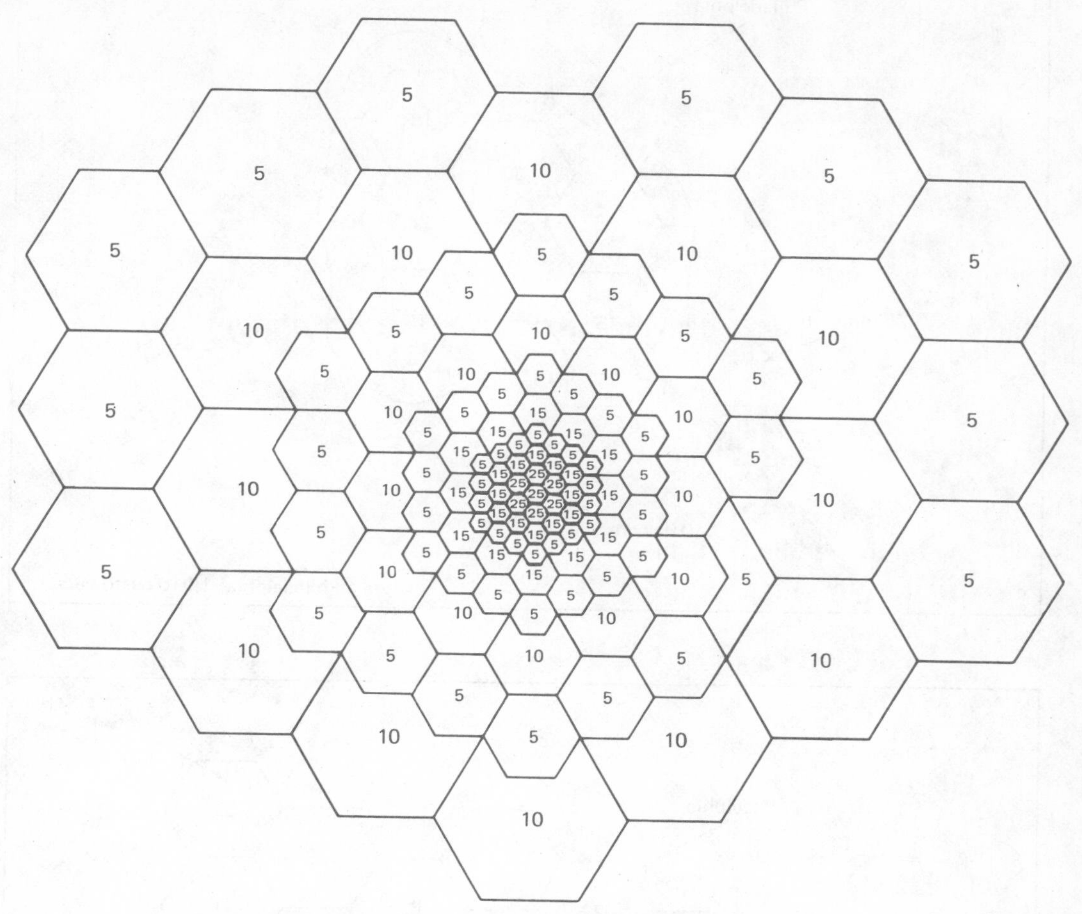

Figure 12.8 Urban areas vary greatly in subscriber density. The variation is accommodated first by using differing cell sizes as shown here, and second by allocating different numbers of channels (frequencies) to the cells. The figures in this illustration represent the numbers of channels per cell.

delphia [4]. Today, there is one mobile radio transmitter and receiver covering the Philadelphia area and its population of two million. This provides telephone service to several hundred subscribers. A long waiting list of would-be subscribers cannot obtain service. The initial cells might be about 8 miles across. As the subscriber density grows 4 mile cells and 2 mile cells might be used. Subscriber growth to 101,000 is illustrated, and could go higher.

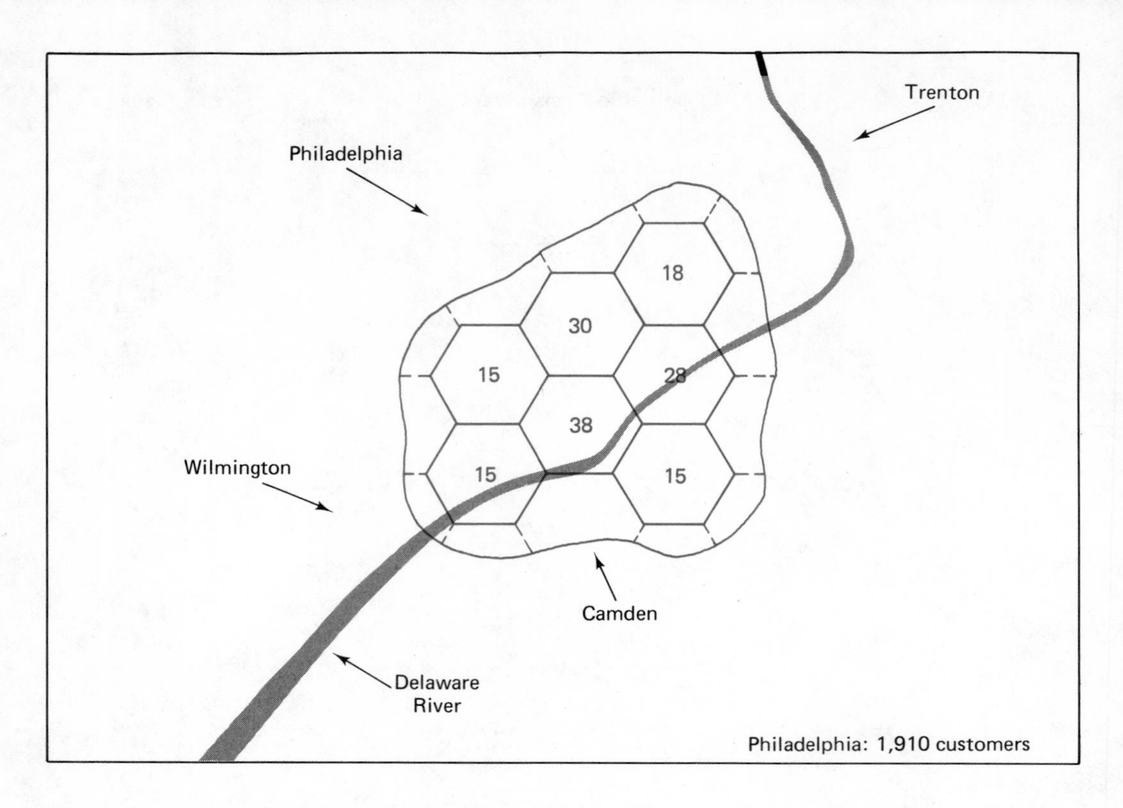
Trenton
Philadelphia
18
30
15
28
38
15
15
Wilmington
Camden
Delaware River
Philadelphia: 1,910 customers

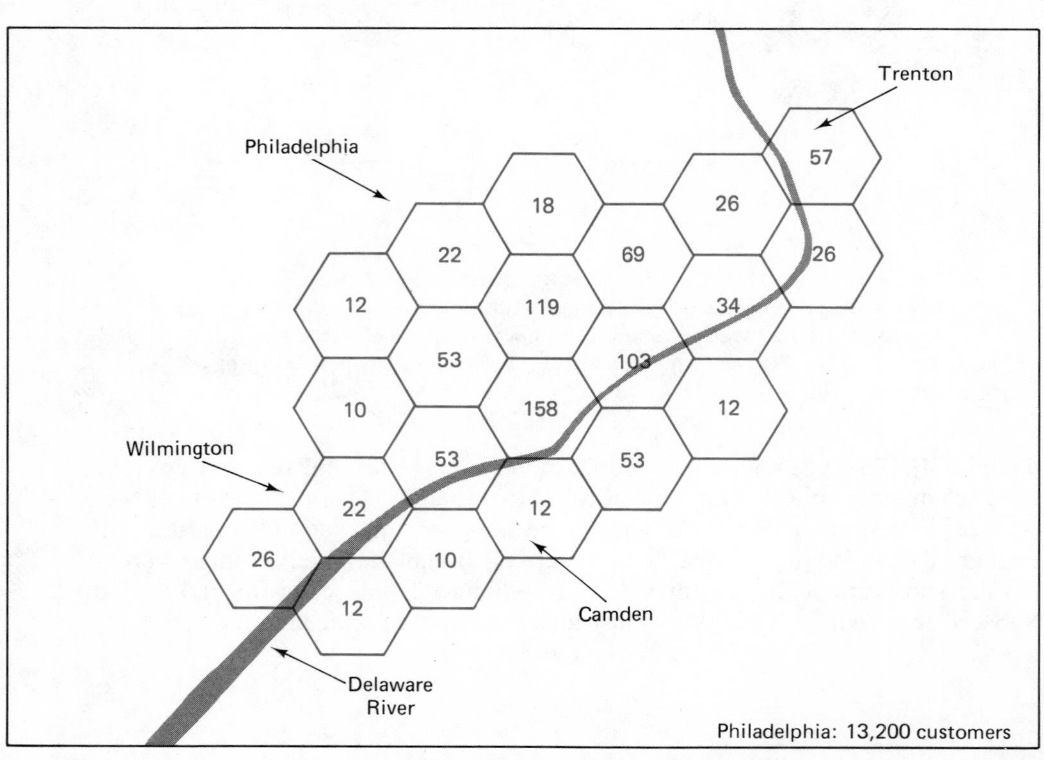
Trenton
Philadelphia
57
18
26
22
69
26
12
119
34
53
103
10
158
12
Wilmington
53
53
22
12
26
10
12
Camden
Delaware River
Philadelphia: 13,200 customers

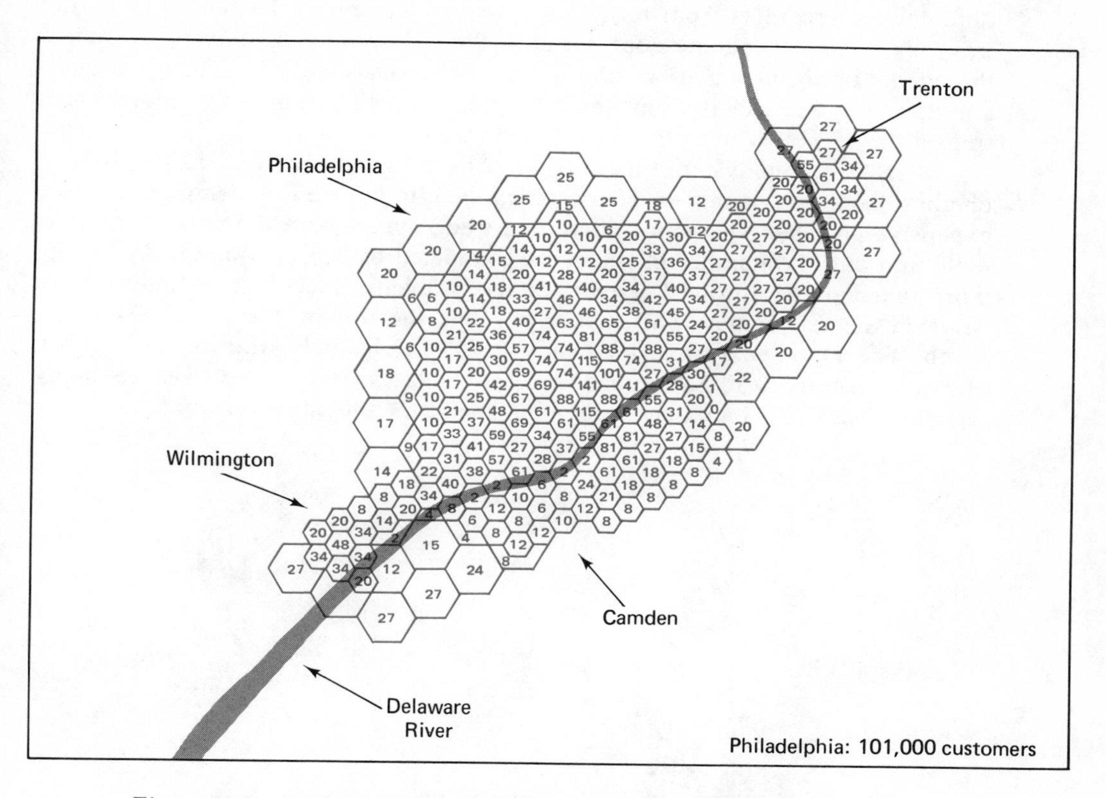

Figure 12.9 As the subscriber density for mobile telephones increases, the cell pattern can change to accommodate it. These charts show subscriber growth in Philadelphia (population 2 million). The numbers in the cells represent the numbers of calls in the peak hour. The radiotelephone common carrier first increases its number of channels per cell, then decreases the size of the cells. (*Courtesy AT&T.*)

HAND TELEPHONES

The cellular technique makes possible portable telephones not much larger than pocket calculators, as well as vehicle telephones. Fig. 12.10 shows a light portable telephone built by Motorola. Motorola has a control system for portable telephones called the Dynatec system. Hand-held telephones have certain different requirements from telephones in vehicles. First, the telephone is not traveling at high speed and does not need the elaborate techniques to overcome the effects of signal variation, and to inconspicuously switch channels in mid-conversation. Second, the unit must be small and light, and so cannot have a transmitter as powerful as that in a vehicle telephone.

The Motorola system therefore uses many small receivers in the area of one base station transmitter. This enables them to have a less complex and less expensive portable telephone. There is a trade-off between the cost of the portable unit and the number of receivers connected to a base station. By having more receivers Motorola also reduces the problem of co-channel interference between the telephones. This makes possible the use of a smaller FM bandwidth. The Dynatec system uses only 25 KHz channels whereas the AT&T Hicap system uses 50 KHz channels. Motorola can thus have twice as many telephones as AT&T in a cell of a given spectrum allocation.

Figure 12.10 Unfair to foxes? The Motorola Dynatec portable telephone. *(Photos courtesy Motorola.)*

In the Motorola view, portable telephones, carried in briefcases or hanging from a belt, need a different system to vehicle telephones, and the two systems can coexist in the same city.

SPECIAL CELLS

The cells of a mobile radio system will not all be hexagonal areas like those in Fig. 12.9. In some cases, the portable units will operate inside a building such as a school or factory and receivers will be designed for that building. Transmitters and receivers of a highly directional nature may be placed in a ribbon along long-distance highways.

Many different designs for low-power distributed systems are now being explored. Highly directional antennas are being designed to beam signals down city streets. The shielding properties of large buildings could be taken advantage of. Long twin-lead antennas, which extend along highways or city streets, are being investigated. These antennas are being used for transmission in subways and tunnels—for example, by the New York City Transit Authority. Wire antennas are also being used within buildings for paging systems. Such antennas can confine their radiation to within a few hundred feet. Small walkie-talkie transmitters are being used to enable commentators to follow golf matches and other sporting events. Different means of signal distribution will meet different situations, and the radiation patterns must be carefully tailored to give efficient use of the available frequencies.

MOBILE TELEPHONE NUMBERS

An appropriate telephone-numbering scheme would be needed if subscribers carry their transceiver from one city to another. In a scheme discussed by F. R. Eldridge [5], each private transceiver would be permanently associated with its owner. The number of the individual would be dialed, or perhaps even his name would be keyed into the call-originating transceiver—"John Smith 42" where there are many John Smiths. When his prime location is different from that of the call originator, it would be necessary to key also a number giving his prime location, just as an area code is dialed on today's telephone network.

Each subscriber would be listed in the machine serving his prime location. When that person is "dialed," the call would be transmitted to the computer at his prime location and the computer would look up his whereabouts. It would have a record that tells whether he is within the area covered by that computer. If he is within that area, the record would say which transmitter zone he is in at present. The call would then be switched to this transmitter and the appropriate transmission frequency selected.

If the called person is not in the area covered by his prime-location computer, the record examined by that computer would say where he is, if that fact is known. The call would be routed by land line to the computer serving the area in question. This computer would then look up which transmitter zone he is in and switch an appropriate circuit path to that transmitter.

In this way a call dialed to a person could reach him at any location served by the system. If he is in his car, the radiotelephone could ring. If he is on the Metroliner train to Washington and is wearing a pocket transceiver, the call would come over the train's radiotelephone and be transmitted to its passengers over antenna wires within the train. When he is on an airliner to Hawaii, the call in the future might go via aeronautical radiotelephone if one of the channels to that plane is free. If they are all busy, a recorded voice from the computer would tell the caller to wait or, alternatively, it could offer to call him back.

The whereabouts of the called person may not be known to the network. Also, he may have switched off his transceiver because he does not want to be disturbed. Then the stored-computer voice would inform the caller that such was the case.

KEEPING TRACK OF MOBILE TRANSCEIVERS

How does the computer network know the location of the called parties?

There must be a mechanism for keeping track of the portable transceivers. When they move from one transmitter zone to another, this fact must be recorded. Possibly the easiest way to do so is to design each transceiver so that it transmits its identification number at intervals. This number will be picked up at the nearest fixed transmitting station, which will add its own code and send the signal over land line to the area computer. The address transmission from the portable transceiver may be received by more than one fixed station. The area computer would record which base station received the best signal from each portable transceiver.

When a subscriber moves from the zone of one transmitter to the zone of another, the subscriber record in the area computer would be appropriately modified. When he moves from the region covered by one area computer to that of another, the computer would transmit messages to the computer in his prime location, saying which area computer could now contact him. (Fig. 12.11.)

DATA DEVICES

Many of the uses to which a mobile telephone is put could be handled equally well by the two-way transmission of data. In some cases data machines are more satisfactory because printed messages can be left—for example address lists in a delivery truck schedule. On mobile radio links it makes sense to encourage data transmission rather than speech when possible because of the enormous spectrum saving that can result. The words occupying one second of speech can be encoded

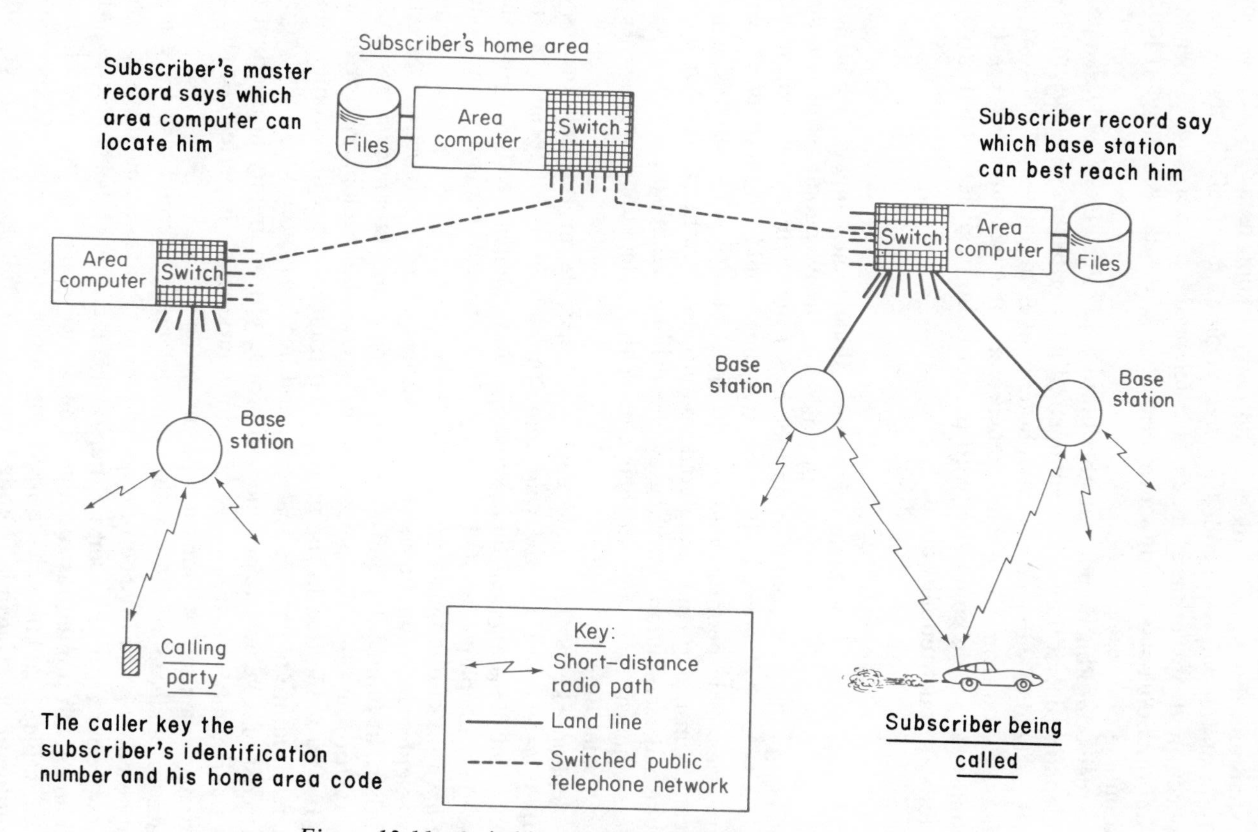

Figure 12.11 Switching between mobile radio transceivers.

into 100 bits of data. 50 KHz is needed for a telephone channel in the AT&T cellular system and this same channel could carry 10,000 data bits per second or more. Information can be encoded a hundred times more efficiently as data than as speech, and furthermore, much of a telephone conversation is wasted from the information point of view in chit-chat, social talk, wooliness, redundancy and human courtesy.

If the cells in Philadelphia at the bottom of Fig. 12.9 could support 101,000 telephone users, they could support 10 million packet radio users. Every channel, instead of carrying one telephone conversation, would carry interleaved bursts of data from many radio transceivers. It is technically possible to provide almost the whole population with packet radio devices, without greatly changing today's spectrum allocation.

USES OF VEHICLE TRANSCEIVERS

Vehicle telephones alone would make mobile radio usage worthwhile for many people. Businessmen could set up meetings from their car on their way to work. Persons traveling could be in contact with their secretaries. Salespeople could talk to their clients while driving to see them; if caught in traffic jams, they could phone ahead rather than sit helpless. They could make restaurant bookings while driving. Highly active men presently waste much of their time in their car. With a telephone, they could employ some of that time.

A telephone keyboard would enable salespeople to transmit orders as soon as they were taken. On the way home they could use it to enquire how their stocks had fared and tell their spouse to have dinner ready. If lost, they could use their telephone to find the way. In fact, one channel on highways might be used for continuously giving directions.

Car telephones could be used for a variety of safety functions as mentioned earlier, including fog and accident warning and breakdown assistance. When a crash occurs the first person on the scene could telephone for an ambulance. It is possible that in the future radio transceivers will become a mandatory safety feature of cars, as seat belts and headrests are today.

The cost of a vehicle telephone will probably drop greatly in the next ten years as their numbers increase, as LSI mass production is perfected, and cellular systems lessen the transmission power needed.

Data terminals in vehicles, rather than telephones, could also have many uses and be economical in spectrum consumption. Police, ambulances, delivery fleets, newspaper reporters, service personnel, etc., could be contacted when required and given printed instructions. The device might be used for obtaining assistance in parking. The vehicle need not stop at toll booths. Its transceiver could automatically transmit a coded packet. A toll booth receiver would record the vehicle identification and transmit messages for deducting the toll from the owner's bank account. Automatic toll collection may be used in some countries for regulating the flow of traffic into congested city centers. Parking fees may be similarly collected, and parking violations made expensive.

PERSONAL RECEIVERS

Two thirds of a million Americans already carry personal radio receivers in the form of paging devices. There are many applications of radio paging and its use is growing fast. It is used for calling foremen or expeditors in factories, locating buyers or sales staff in large stores, calling service personnel, calling auxiliary firemen and telling them the location of a fire, requesting roving executives to come to the phone, and so on.

Some paging systems are strictly local, for example within a factory. Others cover a large area. Eventually it will be possible with the aid of switching computers to locate a person almost anywhere in the country, provided that his transceiver is switched on and working. This situation need not necessarily constitute an invasion of privacy because he would always be free to switch off his transceiver or leave it at home. His home area computer might store the identifications of persons who tried to reach him when his transceiver was off the air.

In industry or government, key persons may be made permanently accessible by means of a transceiver which they are not allowed to switch off during working hours.

As with other forms of telecommunications, paging may spread from business usage to consumer usage.

Electronic miniaturization is such now that a wrist receiver could be built which gives the time, date, telephone numbers which paged the owner and which he should call back, and possibly other brief information (Dow Jones, weather information, traffic report, racing results, etc.). The device would be selecting data from a one-way information stream like the Ceefax system now operating in England. Indeed future teletext systems, designed to display information on television sets, could be organized so that a few of the pieces of data displayed are brief and can be received on portable data receivers with a small keyboard for selecting displays (possibly five keys labelled 1, 2, 4, 8, and *, although the public might prefer a bulkier 10-key keyboard).

There are many potential business and societal applications of pseudo-interactive receivers. They could be used by nurses in hospitals, bookies for obtaining odds, traveling salesmen, ships which need navigation and weather data, or any individuals on the move who need access to variable information.

PERSONAL TWO-WAY DEVICES

When the capability to send back packets of data is added to a personal unit, it has far more applications, and becomes a general-purpose portable computer terminal. It could be used by patrolling security guards, visiting doctors, persons who want to work while on board their boat, maintenance staff could use it for diagnostic functions, contractors or users of earth-moving equipment could contact their base computer. Pocket calculators could be designed so that they could be loaded with programs on request from remote program libraries.

A portable radio in public use would be used for listening to broadcasts, as today, as well as for telephoning, obtaining time and date, and communicating with computers. The commuter on his tedious train ride could telephone home or arrange meetings in his office like the businessman with a car transceiver. He could fill in his time by dialing computers and using telephone keyboard overlays as described in Chapter 7. One can imagine carrying a wallet full of keyboard overlays the size of credit cards. He might learn French on the train from a computer, inquire after sports scores, or balance his checking and credit accounts. The radio facilities in the train would be a simple extension of what existed in the New York–Washington trains. The call booths in the train would be replaced by a less expensive twin-wire antenna running the length of the train. In the earlier days of this technology, travelers without a pocket telephone might borrow one from the train attendant, with charges being automatically radioed for deduction from bank or credit accounts, as with everything else in this telecommunications-dominated society. The same facilities would be even more welcome on tedious airplane trips if appropriate channels were made available.

The portable transceiver would give people a means of immediately contacting the police. This might prove very popular in a society in which the growth rate of crime rivals the growth rate of electronics. The police computers now spreading across the country would instantly pinpoint the emergency call and dispatch a patrol car. The fire and ambulance services could be contacted equally quickly. In a system with relatively small antenna segments, the approximate location of the transceiver sending an emergency call could immediately be found and a system could be devised so that precise positioning by triangulation is possible. Automatic positioning has been advocated for use in computer control of delivery trucks, taxis, "dial-a-bus" vehicles, and so on. Without automatic positioning, the emergency caller would have to give his whereabouts. Such a scheme would probably be a deterrent to crime on the streets. One imagines the running criminal being tracked by the calls from persons near the scene of the crime as the patrol car closes in on him.

The portable transceivers may become an important part of the financial scene—a means of obtaining credit, for example. An individual might be securely identifiable by a combination of the transceiver he carries, which transmits a coded identification, and some other factor such as a memorized number or his "voice print," which is checked by a computer when he speaks his name. The system could be devised so that a stolen transceiver would be valueless to thieves, unlike today's credit cards. Nevertheless, each person would guard his personal transceiver as he does his wallet and would report any loss immediately. Many applications of personal transceivers depend on whether the security problems can be solved, but the technology to do so now appears available.

The transceiver could also be used as a personal information source. The user could have access to his own records stored in computer memory. His secretary or wife could store entries on his shopping list or diary.

If large numbers of portable transceivers come into use, new types of applications will become available. Those new functions would in turn increase the market demand. As with radio, television, and the telephone in their day, a snowballing effect would set in until most of the public owned one. However, as with interactive CATV the initial takeoff might be slow. On the other hand the sudden popularity of CB radio might extend into other types of devices. Again the reader may glance at Fig. 2.1 and reflect upon the possible growth curve for portable transceivers.

Like all new technology, radio transceivers have potential for ill as well as for good. Bugging devices will be further improved. The police could arrange for automatic surveillance of persons possessing transceivers. Nevertheless, it seems likely that the social benefits will override the disadvantages.

The world of the ubiquitous portable transceiver may be two decades away, but sooner or later it may be as important to our society as the telephone is today.

REFERENCES

1. The President's Task Force on Communications Policy, Eugene V. Rostow. Staff Paper No. 1, Appendix F. "Concepts for Improving Land Mobile Radio Communications" by F. R. Eldridge, Washington, D.C., 1969.

2. *Business Week:* June 30, 1975.

3. C. E. Paul, "Telephones aboard the Metroliner," *Bell Laboratories Record,* March 1969.

4. "High Capacity Mobile Communication System," *Bell Laboratories Record,* March 1972

5. Appendix F of the President's Task Force on Communications Policy, Staff Paper No. 1: "Concepts for Improving Land Mobile Radio Communications" by F. R. Eldridge, Washington, D.C., 1969

H.M.S. Agamemnon laying the first transatlantic telegraph cable in 1858. The cable had a spectacular but short life. After some of the toughest financial wrangling of the nineteenth century and more than a year of heartbreaking failures on the high seas, it was eventually laid successfully. The signals that trickled through it were so minute that only the most sensitive suspended-mirror galvanometer could detect them. The first message of ninety words from Queen Victoria took 16 1/2 hours to transmit. Press headlines were sensational beyond precedent. However, some days after the Queen's message, the insulation of the cable failed and it never worked again. It was eight years before another cable was laid. One American newspaper called the cable a hoax, and an English writer "proved" that it had never been laid at all.

13 SATELLITE ANTENNAS ON THE ROOFTOPS

Perhaps the most exciting and most rapidly evolving of all the developments in transmission is the communications satellite.

On April 6, 1965, the world's first commercial satellite, Early Bird, rocketed into the evening sky at Cape Kennedy. The success of the transmission experiments that followed this was spectacular. Before long earth stations were being built around the world, and new and more powerful satellites were on the drawing board.

TRANSPONDERS

A communication satellite (Fig. 13.1) is, in essence, a microwave relay in the sky. It receives microwave signals in a given frequency band and retransmits them at a different frequency. It must use a different frequency for retransmission otherwise the powerful transmitted signal would interfere with the weak incoming signal. The power of satellites lies in the fact that they can handle a large amount of traffic and send it over most of the earth.

The equipment which receives a signal, amplifies it, changes its frequency, and retransmits it, is called a *transponder*. Most satellites have more than one transponder. The bandwidth handled by a transponder has differed from one satellite design to another, but most contemporary satellites (e.g., INTELSAT IV, ANIK, Western Union's WESTAR, and RCA's SATCOM) have transponders with a bandwidth of 36 MHz. How this bandwidth is utilized depends upon the earth station equipment. The WESTAR satellites, which are typical, have transponders which may be used to carry any of the following:

1. One color television channel with program sound
2. 1200 voice channels
3. A data rate of 50 Mb/s

Figure 13.1 Communications satellites.

Western Union's WESTAR satellite, the first USA domestic satellite. Canada's ANIK satellite launched two years earlier is similar. Twelve transponders of 36 MHz. (*Photo courtesy Western Union.*)

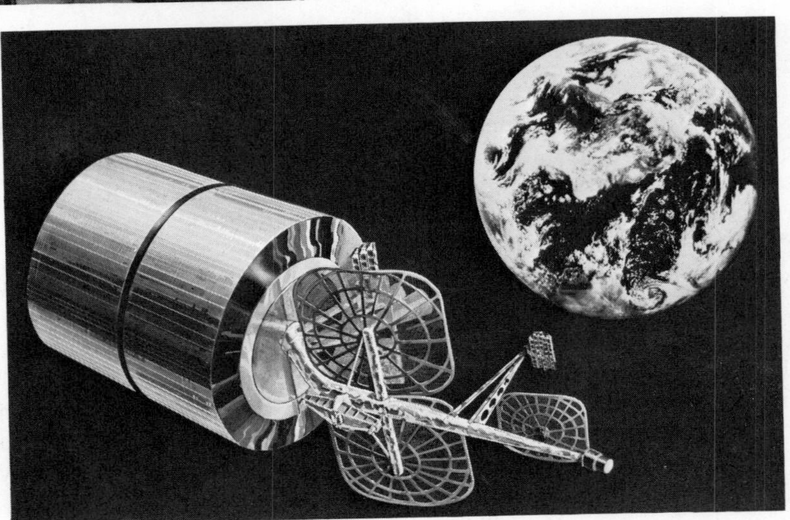

INTELSAT IV-A, also twelve transponders of 36 MHz, designed for intercontinental transmission. (*Photo courtesy COMSAT.*)

The heaviest communications satellite, the US military TACSAT I, for tactical communications with mobile military field units, aircraft and ships. Remotely deployed field units can communicate using antennas only 1 foot in diameter. The 5-element helical array is for UHF transmission. (*Photo courtesy Hughes Aircraft.*)

An experimental satellite for aeronautical communications, planned by COMSAT General, ESRO (the European Space Research Organization) and the Government of Canada. (*Photo courtesy COMSAT General.*)

4. The center 24 MHz of each band may relay either
   a. 16 channels of 1.544 Mb/s or
   b. 400 channels of 64,000 b/s or
   c. 600 channels of 40,000 b/s

The WESTAR satellites each have 12 such transponders, two of which are spares used to back up the other ten in case of failure. RCA's SATCOM has 24 such transponders.

WORLD COVERAGE

Communication satellites are stationed at a special position in space, in a ring 23,000 miles over the equator. In this unique orbit they revolve around the earth in exactly the time that the earth takes to rotate, and hence appear to hang stationary in the sky. We can thus think of a communications satellite as a stationary microwave relay 23,000 miles above the equator. Because they are so high they can transmit to much of the earth. Three satellites can cover all the inhabited regions of the earth with the exception of a few dwellings close to the poles.

The cost of satellite channels is dropping remarkably fast and a main thrust of the technology is to find ways to make the receiving equipment on earth cheap and mass-producible, and to devise a way of giving corporate and government locations their own transmitters.

THE CHANGING PERCEPTION OF SATELLITES

The perception of the value of communication satellites has changed since man's first satellites were launched. At first satellites were perceived largely as a means to reach isolated places. Most of the world's population is not served by the telephone and television networks that so greatly influence Western society. The cost of lacing Africa and South America with Bell System engineering would be unthinkable. Satellites were perceived as a counter technology, and earth stations began to appear in the remotest parts of the world. Countries with only the most primitive telecommunications put satellites on their postage stamps.

As satellites dropped from their initial exorbitant cost it was realized that they could compete with the world's suboceanic cables; satellites then had a part to play in the industrial nations, linking the continental land masses. The owners of the suboceanic cables took political steps to protect their investment at the expense of satellites, but soon more transocean telephone calls were made by satellite than by cable. Television relayed across the ocean by satellite became common, as the cable of the 1960s did not have the capacity to send live television.

COMSAT (the Communications Satellite Corporation) launched four generations of satellites in six years. EARLY BIRD, the first *commercial* satellite to retransmit signals from a fixed position in space, was followed by INTELSAT II in 1967, INTELSAT III in 1968, and INTELSAT IV in 1971.

When the first INTELSAT birds brought competition to subocean telephone cables, the domestic telephone networks seemed immune from the threat. The cost per telephone channel of the early satellites was high, and the United States Communication Satellite Act said that only COMSAT could operate satellites and that they could be used only for international transmission.

As often technology changed more rapidly than the law, and the first North American domestic satellite was launched in 1972 by Canada. It was originally perceived as a means to communicate with Canadians in the frozen North and was called "ANIK" which means "brother" in Eskimo language. However, it was soon realized that the ANIK satellites would provide cheaper long distance telephone or television circuits than those of the established common carriers. Antennas were set up in the United States to use the ANIK satellites, and for their first two years in orbit these satellites earned a return on capital investment that was virtually unprecedented in the telecommunications industry.

A flurry of legislation resulted in 1972 in the United States Federal Communication Commission's *Open Skies Policy* encouraging private industry to submit proposals for launching and operating communication satellites. The first United States common carrier to take advantage of the Open Skies Policy was Western Union which launched two WESTAR satellites in 1974—the first U.S. domestic satellites.

A price war ensued for long distance leased communication channels in the U.S.A. A leased telephone circuit from coast to coast via WESTAR was less than half the cost of similar channel from the terrestrial common carriers. It seemed clear that the price could drop further with more advanced equipment.

It became clear that there were major *economies of scale* in satellites. A big satellite could give a lower cost per channel than a small one. To take advantage of the economies of scale, satellites should be employed where the traffic volume was heaviest. Nowhere was it heavier than in United States domestic telecommunications, and so it began to appear, contrary to the earlier view, that there was more profit in domestic satellites than international satellites.

Nowhere was this perceived more clearly than in Bell Laboratories, the birthplace of the first commercial communication satellite, TELSTAR. A Bell Laboratories study showed that a few powerful satellites of advanced design could handle far more traffic than the entire AT&T long-distance network. The cost of these satellites would have been a fraction of the cost of equivalent terrestrial facilities. However, government regulations prevented AT&T from developing the satellites which it, more than anyone else, could make good use of. The field was left open for competition. A number of corporations, some very small beside AT&T, announced that they would operate satellites, and

AT&T continued to spend many billions of dollars per year on expanding its terrestrial facilities.

For WESTAR users the perception of satellites had now become that of communication pipelines linking five earth stations in one country. A further perceptual change was to follow.

BROADCASTING

While corporations and computer users perceived the satellite as providing two-way channels between the relatively few earth stations, broadcasters or would-be broadcasters perceived it as a potentially ideal way to distribute one-way signals. Television or music sent up to the satellite could be received over a vast area. If a portion of the satellite capacity were used for sound channels for education or news, a very large number of channels could be broadcast. The transmitting earth stations would be large and expensive, but the receiving antennas could be small and numerous. The Musak Corporation envisioned small receiving antennas on the roofs of their subscribers' buildings. Satellites offer the possibility of broadcasting television to vast areas of the world that have no television today. If more powerful satellites were launched, television could be broadcast directly to the hundreds of millions of homes in industrial countries. The Japanese broadcast satellite will beam programs directly to Japanese homes which can use relatively inexpensive home receivers. With satellites of lesser power, television is today being distributed to hundreds of regional stations for rebroadcasting over today's cable television links.

Television used well can be an extremely powerful medium for education. The majority of television, however, has been used very poorly for this purpose. America has thousands of classrooms or lecture rooms in which the television sets have been removed or are unused. Nevertheless the best examples are very good. In some classes teachers have used television to powerful effect to augment their teaching. Britain's University of the Air has programs deserving worldwide multilanguage availability. And what better way for most people to learn history than from programs such as Alistair Cooke's "America"? The dream of superb educational facilities via satellite will probably be achieved one day, but much more than advanced technology is needed to bring it about.

Broadcasting is usually thought of as having one transmitter and many receivers. However when a satellite is used for two-way signals, a form of broadcasting is taking place in which there are many transmitters. Each earth station is, in effect, a broadcasting transmitter because its signal reaches all other earth stations, whether they want it or not. Each earth station, like a radio set, tunes in to only that signal it wants to receive.

Because of this broadcasting nature of satellites it is limiting to think of a satellite as a "cable in the sky." It is much more than that. A signal sent up to the satellite comes down everywhere over a very wide area. To maximize the usefulness of the satellite for telecommunications any user in that area should be able to request a small portion of the vast satellite capacity at any time, and have it allocated to him if at that moment there is any capacity free. Just as with a telephone network, to make it really useful any user should be able to call any other user when he wants; however it is not desirable to put a telephone exchange in the satellite, at least not yet. The equipment on the satellite needs to be simple and reliable because equipment failures cannot be repaired, and needs to be light and consume as little power as possible. To achieve the desirable *multiple-access* capability, ingenious ways have been devised of allocating satellite capacity to geographically scattered users, permitting them to intercommunicate.

SATELLITES FOR DATA

For the first decade of communication satellite operation most of the capacity of the satellites was used for telephone traffic and television. The technology has evolved, however, so that, in a sense, satellites are much more powerful for the transmission of *data* between computers and computer users, or between telegraph machines. It now appears desirable technically to carry telephone traffic, and possibly television, in a *digital* form, as we will discuss later. When the telephone voice is digitized in a simple manner it again becomes 64,000 bits per second. When digitized in a more compressed fashion, fewer bits can be used. Digitized television requires a bit rate between 40 million and 92.5 million bits per second depending upon the technique used and can be sent via one transponder. A substantial quantity of *data* can be sent using these bit rates. Such bit rates make telephone voice appear expensive by comparison with *data* transmission as a means of transmitting information.

The potential power of satellites for the computer industry can be illustrated by means of a simple calculation. When a person uses a computer terminal he does not transmit continuously at the full speed of the line the terminal is connected. He transmits a number of characters, reads the response, thinks about it, and keys in more characters. The dialogue between the terminal user and the computer results in bursts of data, often short bursts, passing across the line with pauses between them. Much more data could be sent in the pauses than is actually transmitted in most cases. As we illustrated in Chapter 10 most man-terminal dialogues result in not more than 10 bits per second passing backwards and forwards *on average,* although in *certain* seconds a much larger number of bits is sent (see Table 10.1).

Today's domestic satellites carry a number of transponders. RCA SATCOM users can transmit 60 million bits per second via one transponder. The satellite has 24 such transponders. Such a satellite could thus have a total data throughput of 24 × 60 million bits per second. If this capacity were employed entirely for interactive terminal users, it would not be possible to achieve 100% efficiency. A conservative assumption is that 15% efficiency could be achieved (much lower than today's equivalent on well-organized terrestrial lines). 15% efficiency would give a usable capacity of *216 million bits per second.*

The combined population of the United States and Canada is about 240 million. Let us suppose, for the sake of this illustration, that *every* person makes substantial use of computer terminals. The *average* working person uses them one hour per day, and the average non-working person half an hour per day. This gives a total of about 160 million hours of terminal usage in total per day. Let us suppose that in the peak hour of the day the usage is three times the daily average. The total data rate in the peak hour is then

$$\frac{160 \text{ million} \times 3 \times 10}{24} = \textit{200 million bits per second.}$$

In other words one satellite using today's state of the art could have enough transmission capacity to provide every man, woman, and child in the United States and Canada with a computer terminal. In addition, as we have done the calculation for the peak hour, twice as much nonreal-time data could be sent over the same satellite.

This calculation assumes, as does other such discussion of the power of satellites, that it is possible to organize terrestrial facilities for the satellite channels in an appropriate manner. In the above case, how does one enable an extremely large number of users to share the same channel without interfering with one another excessively? As we will see later there *are* types of organization that could work well. However when a large number of users share the large capacity available, the design becomes dominated, not by the satellite relay itself as on a point-to-point link, but by the architecture of the ground facilities that permit multiple access to the satellite.

In spite of the satellite's power for data transmission it would not be a sound business operation to launch a satellite solely for use with computers. Of all of the traffic that might be sent by satellite, a relatively small proportion of it is computer traffic. Whatever the mix in the future, most of *today's* traffic is "Plain Old Telephone Service." To maximize its potential profit a satellite should be capable of carrying many different types of signals—real-time and nonreal-time, voice, data, facsimile, and video. For all of these signals it should be regarded as a broadcasting medium accessible from anywhere beneath it, not as a set of cables in the sky.

SUMMARY

To summarize, the perception of what a communication satellite is has changed, and different perceptions have been:

1. A means to reach isolated places on earth.
2. An alternative to subocean cables.

3. Long-distance domestic telephone and television links.
4. Television and music broadcasting facilities.
5. A data facility capable of interlinking computer terminals everywhere.
6. A multiple-access facility capable of carrying all types of signals on a demand basis.

THE DROPPING COST OF SATELLITE CHANNELS

The changing perception of satellite potential has been related to the change in satellite cost. This has been dramatic.

The first four generations of INTELSAT satellites carried increasing numbers of channels and had progressively longer design lifes as shown in Fig. 13.2. Consequently the cost per voice channel per year dropped dramatically. The process will continue with INTELSAT V.

The bottom line of Fig. 13.2 shows the drop in cost per satellite voice channel per year. Figure 13.3 plots the trend. The figure shows the *investment* cost of the satellite and its launch. The cost to a subscriber will be much higher because it must include the earth station and links to it, and must take into consideration the fact that the average channel utilization may be low.

The extraordinary cost reduction shown in Fig. 13.3 will probably continue, but not at such a spectacular rate. Massive reductions in the cost per voice channel could result if satellites with a much larger capacity than today's were launched. In the 1980s the space shuttle and associated equipment may lower the launch costs substantially.

The satellites and their launch costs are referred to as the *space segment* of satellite communications. The comment is sometimes made among system planners that the space segment costs are dropping to such a low level that overall system costs will be dominated by the organization of the ground facilities.

The cost of an earth station, however, has dropped more spectacularly than that of a satellite. The first COMSAT earth stations cost more than $10 million. (The first Bell System earth stations for TELSTAR cost several times that). Earth stations have dropped in cost until now a powerful transmit/receive facility can be purchased for less than $100,000. Receive-only facilities are a fraction of this cost. At the same time the traffic that can be handled by an earth station is increasing as satellite capacity increases. Combining these two trends we find that the investment cost per channel per earth station is dropping as shown in Fig. 13.4.

The *total* earth segment costs are *not* dropping because to provide increased accessibility to the satellites many earth stations are being built. Prior to 1973 the United States had only a handful of earth stations. Now many are being installed. Some corporations are setting up their own satellite antennas, and we can look forward to an era of many small private earth stations. The

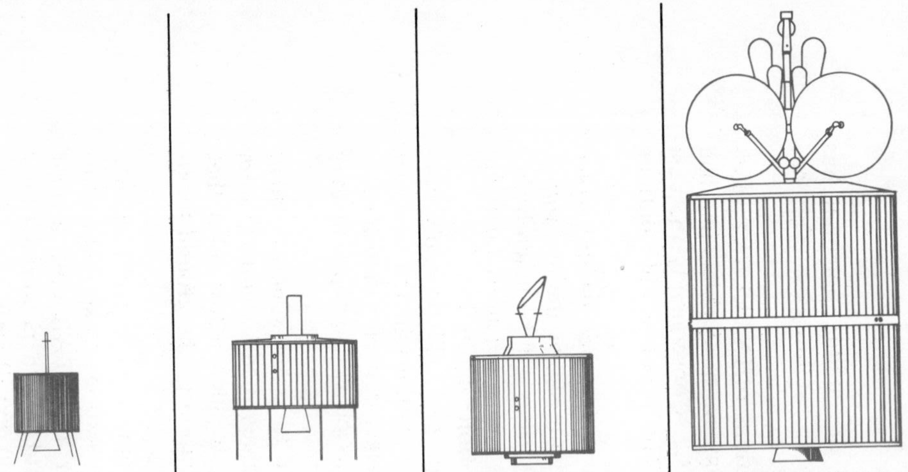

					Being designed
Name	Intelsat I (Early Bird)	Intelsat II	Intelsat III	Intelsat IV	Intelsat V
Year of launch	1965	1967	1968	1971	1979
Diameter	28 inches	56 inches	56 inches	93 inches	600 inch sails
Height	23 inches	26 inches	78 inches	111 inches	264 inches
Weight in orbit	85 lbs	192 lbs	322 lbs	1547 lbs	3200 lbs
Number of antennas	1	1	1	3	6
Primary power (watts)	40	75	120	400	1000
No. of transponders	2	1	2	12	27
Bandwidth of transponder	25 MHz	130 MHz	225 MHz	36 MHz	
Cost of satellite	$3.6 million	$3.5 million	$4.5 million	$14 million	≈ 25 million
Cost of launch	$4.6 million	$4.6 million	$6 million	$20 million	≈ 23 million
Design lifetime	1.5 years	3 years	5 years	7 years	10 years
Total cost per year	$5.47 million	$2.70 million	$1.90 million	$4.85 million	$4.8 million
Maximum No. of voice circuits	240	240	1200	6000	≈ 24,000
Cost/voice circuit/year	$23,000	$11,000	$1600	$810	≈ $200

Figure 13.2 The INTELSAT satellites.

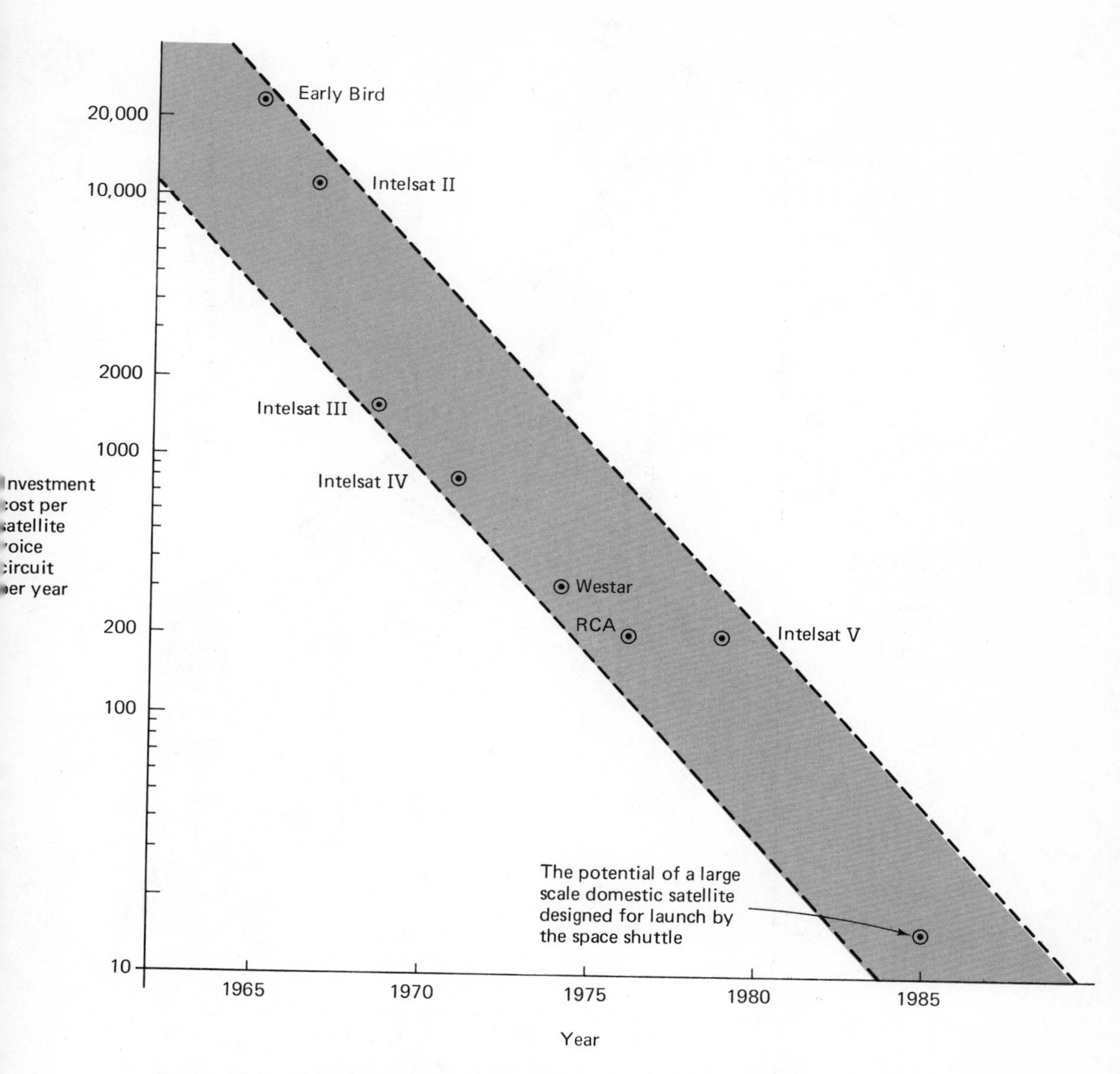

Figure 13.3 The falling investment cost of satellite voice circuits. The trend could continue if economies of scale are permitted. However, they may not be permitted for telephone traffic.

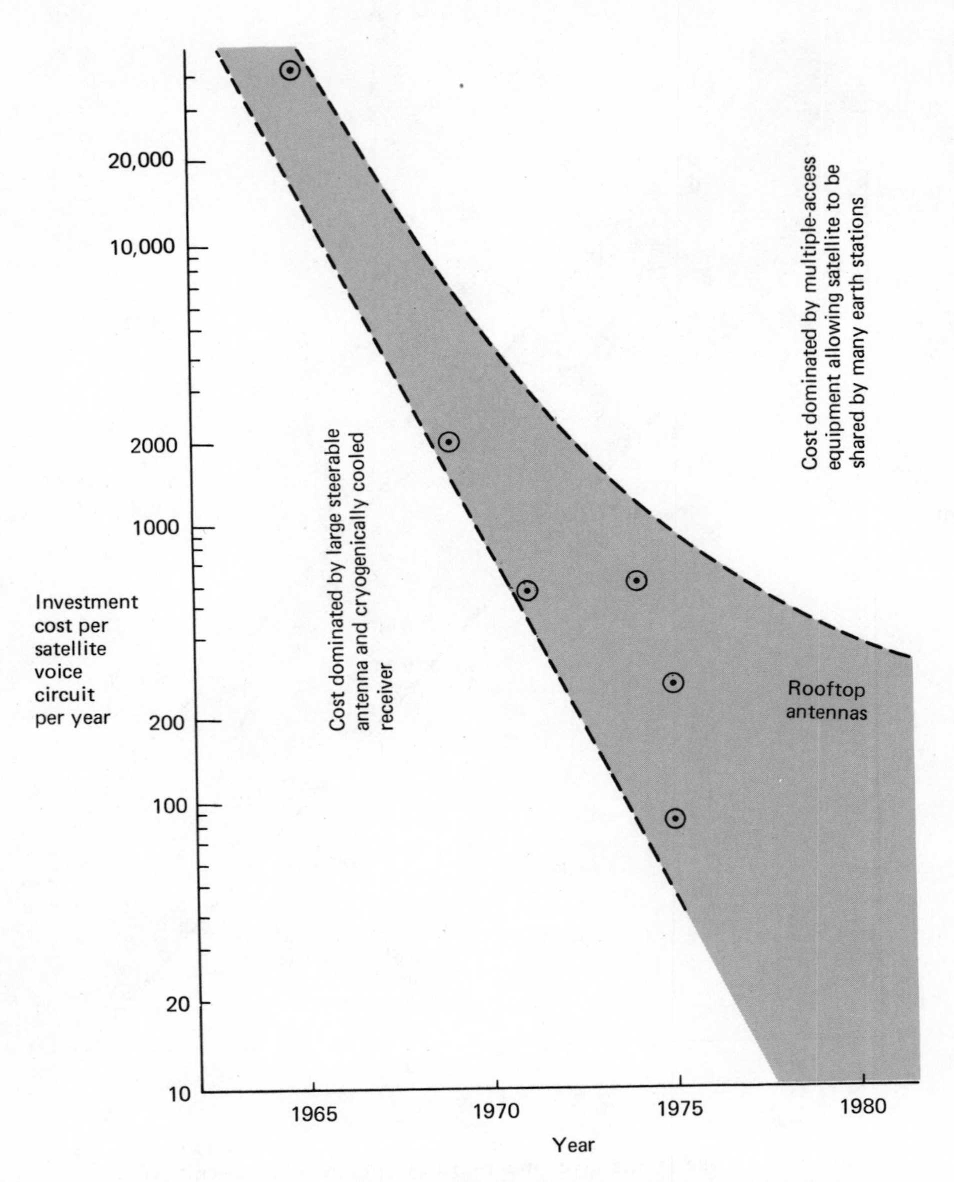

Figure 13.4 The falling cost of satellite earth stations. The cost per channel with private satellite antennas will vary greatly with the number of channels handled.

earth station cost in the 1960s was dominated by its large steerable antenna, 30 meters in diameter, with automatic tracking facilities and the hypersensitive cryogenically cooled receivers. Small corporate earth stations will use relatively low-cost nontracking antennas and uncooled cheaper receivers; their cost will be dominated by equipment which enables them to share the satellite with many other earth stations. A low earth-station cost does not necessarily mean a low cost per channel because it may be used at a location which employs relatively few channels.

There is a trade-off between the cost of the satellite and the cost of its earth station. If the satellite has a large antenna and considerable power, smaller stations can be used. If the satellite makes more efficient use of its frequency allocation, the cost per channel will be lower. There is a limit to satellite efficiency and so the main effect of increasing satellite cost will be to reduce earth antenna size and cost.

As the earth facilities drop in cost, more antennas will be constructed and more traffic will be sent, making it economical to use more powerful satellites, which will make the earth facilities drop in cost further. If satellites use large numbers of small earth stations, however, the overall system architecture which permits the earth stations to share the satellite is extremely important and eventually will dominate the cost of satellite systems.

DELAY

A disadvantage of satellite transmission is that a delay occurs because the signal has to travel far into space and back. The signal propagation time is about 270 milliseconds, and varies slightly with the earth station locations. A telephone user waits for the reply of the person he is talking to for an extra 540 milliseconds if the call goes via satellite in both directions.

The bad effects of this delay have been much exaggerated by organizations which operate long distance terrestrial links. The claim is frequently heard that the delay is psychologically harmful in telephone conversations and renders satellite links useless for interactive data transmission. Some common carriers (without prospects of owning any satellites) have claimed that the delay is unacceptable to telephone users. Arthur C. Clarke, normally the most optimistic of writers about technology, once suggested that satellite users would end each stretch of conversation with the word "Over" [1]. In practice a telephone user certainly notices the delay but very quickly becomes used to it if he makes many satellite calls. It is much less annoying than having a noisy local loop. Assessment of psychological effect should not be based on the first call a person makes (an error made in some of the published studies). In a report to the Federal Communications Commission on the experimental use of a satellite link in its corporate telephone network, IBM commented on the delay as follows [2]:

> For most cases with one talker unaware of the interposition of the satellite path (the originating talker is aware and primed in his expectations) the conversation suffered only initial awkwardness with a rapid adjustment made in terms of accommodating the delay effect by more care in interruption. It has been observed by several participants that a widespread use of satellite connectivity would gradually increase the "politeness" of telephone conversations.

On many of today's transatlantic telephone calls, the satellite is used for one direction of transmission and subocean cable is used for the other direction. The result is a total delay slightly greater than a quarter of a second. It also sometimes results in the disturbing effect of person A being able to hear person B very distinctly, as though he were in the next room, but person B being able to hear person A only poorly, and hence feeling that there is a need to shout. The more clearly heard person shouts; the more poorly heard person reacts to the shouting by talking softly; the shouter reacts to the soft voice by shouting louder; and so on.

On transoceanic circuits, two-way delay seems less harmful to a person who is used to it than the effects of TASI. TASI (Time-Assigned Speech Interpolation) snatches the channel away from a person when he pauses in his speech and may allocate it to another speaker. The procedure increases the overall utilization of the circuit. When the circuit is heavily loaded, a speaker may not be reassigned a channel quickly enough when he speaks again, so his first spoken syllables are sometimes deleted. Intercontinental callers sometimes confuse the effects of TASI with the effects of satellite delay.

For telephone users it is particularly important to remove the *echo* on a satellite channel. If a talker hears his own voice echoed back to him with a delay of 540 milliseconds, this proves very disturbing. The echo is removed by means of an echo suppressor which inserts an impedance into the reverse path when a person talks, and removes the impedance when he stops talking.

While a telephone user can learn to ignore one or two 270 millisecond delays in a conversational response, four such delays (1080 milliseconds) may strain his tolerance. It is therefore desirable that the switching of calls should be organized so that no connection contains two or more round trips by satellite. Where satellites supplement the terrestrial toll telephone network the switching can usually be organized to limit the delay to 270 milliseconds.

In interactive data transmission via satellite a terminal user will experience a constant increase in response time of about 540 milliseconds. A system designer has to take this into consideration in designing the overall system response time. In many interactive systems it is desirable that the mean response time should not be greater than two seconds. This is achieved satisfactorily on many interactive systems using satellites today. However, appropriate line control procedures have to be used on satellite channels. Some of the most common equipment for data transmission over telephone lines performs very poorly if used on a satellite channel because it uses a control procedure that is inappropriate if the channel has a long delay. Two control procedures that should not be used on satellite channels are polling (in which a control station asks each station on the line in turn if it has anything to transmit) and stop-and-wait error control (in which the transmitting station stops after each message sent and does not transmit another message until it hears that the previous one was received correctly). Alternative forms of line control perform efficiently on satellite channels.

USER EARTH STATIONS

Perhaps the most important thrust in satellite technology is to bring earth stations to user locations—as many user locations as possible. Eventually many office buildings will have a satellite antenna on the roof, and factories will have an earth station beside the parking lot. Cable television systems are being fed directly from satellites. In Japan, a broadcasting satellite has been designed to permit individual homes to pick up television with relatively inexpensive earth stations.

There are several current indications of the trend toward earth stations inexpensive enough for mass use. The Musak Corporation has demonstrated the reception of four channels of hi-fi music, transmitted via one WESTAR transponder, using a 1-meter diameter receiving dish (Fig. 13.5). The quality is excellent and Musak applied for permission to provide their customers with satellite music distribution, the earth stations costing less than $1000. The larger the satellite antenna, and the greater the on-board power of the satellite, the cheaper the earth station can be. The largest antenna in use today is on NASA's experimental ATS-6, launched in 1974. This is a 10-meter antenna which opens in space like an umbrella (Fig. 13.6). ATS-6 spent a year of its orbital life over India relaying television to the Indian subcontinent. The receiving antennas were made locally using chicken wire, and cost less than $100 each. Eventually the Sears Roebuck catalogue may list satellite antennas.

Antennas which can receive music or television can equally well receive mail or other printed matter. The most expensive parts of a mail service are the sorting and delivery to the customer. These could be automated if mail is sent electronically. The mail could be sorted by computer, transmitted to a satellite and received by small dishes on the rooftops (which might also receive music, computer data and other forms of messages).

A transmitting earth station is more expensive than a receive-only station. It needs more power and a larger antenna. Most corporate and government uses of satellites need the capability to transmit as well as to receive. Some corporations already have their own satellite earth stations. Satellite technology in the near future will lower the cost of private earth stations and increase their versatility.

Figure 13.5 Satellite earth stations.

The first earth stations for commercial use were massive and cost more than $10 million. They had dishes almost 100 feet in diameter, electronics cooled by liquid helium, and automatic satellite tracking mechanisms. (*Photo courtesy British Post Office.*)

This earth station is 4 feet in diameter. It is used on board ships for worldwide voice and data communications using the Marisat satellite. (Its protective dome is removed in this photograph.) (*Photo courtesy COMSAT General.*)

A transportable earth station used by *Teleprompter* for demonstrating the feeding of CATV cables using the WESTAR and ANIK satellites. Many cables systems now receive programs via satellite. (*Photo courtesy Scientific Atlanta.*)

Canada's CTS (Communications Technology Satellite) makes possible excellent quality television reception with this 32-inch earth station. (*Photo courtesy Canadian Department of Communications.*)

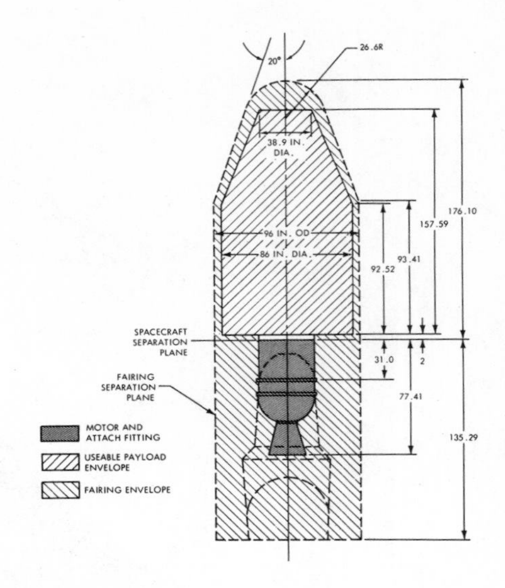

Figure 13.6 Today's communications satellites have been launched with relatively small rockets like this Delta. In the 1980s new vehicles including the Space Shuttle below, will make possible much greater satellite weight.

The Space Shuttle orbiter is a reusable space vehicle about the size of a DC-9 jet liner. Its cargo bay is about 60 feet long and 15 feet wide. After missions in low earth orbit, with a crew of two, it will return to earth and land like an airplane. When it is used to launch a communications satellite, a perigee kick motor will be fired to send the satellite to the location of the geosynchronous orbit.

MICROWAVE INTERFERENCE

With current satellites there is a serious problem which prevents the widespread deployment of private earth stations—microwave interference.

It is customary to refer to the frequencies used by satellites with two figures such as 4/6 GHz. The first figure is the frequency of the down-link—the frequency of which signals are received from the satellite (4 GHz). The second figure is the frequency of the up-link (6 GHz).

Current commercial satellites operate at 4/6 GHz, and unfortunately these are the frequencies most commonly used for terrestrial microwave links. The major cities of the world are becoming highly congested with traffic at this frequency. Figure 13.7 shows microwave traffic in the New York City area.

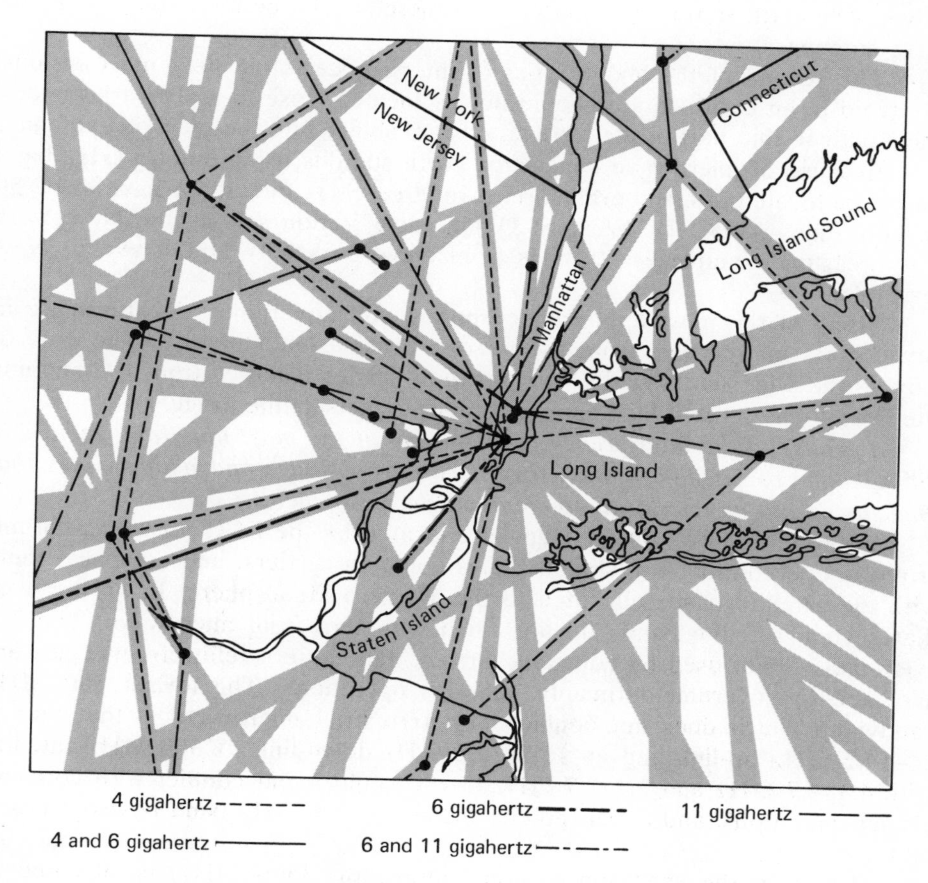

Figure 13.7 Criss-crossing microwave beams in New York. The microwave congestion is so great that 4–6 GHz earth stations have to be many miles from city centers.

Four types of interference are theoretically possible:

1. Transmission from earth station interferes with terrestrial link receiver.
2. Terrestrial link transmission interferes with reception from satellite.
3. Transmission from satellite interferes with terrestrial link receiver.
4. Terrestrial link transmission is received by satellite.

The first is by far the most serious. An earth station must transmit a powerful signal to compensate for the vast distance of the satellite. The antenna transmits a highly directional beam towards the satellite, but nevertheless some of the signal spills in other directions and may interfere with a microwave receiver. The earth station transmitter must therefore not be too close to a microwave antenna.

The second of the above types of interference is the next most serious. To avoid it, an earth station should not be located close to a terrestrial microwave path so that part of the terrestrial beam shines into the receiving antenna.

Because of microwave congestion earth stations using the 4/6 GHz band cannot be located in many urban areas. In large cities they often have to be 50 or more miles away. The Western Union earth station serving the New York City area, for example, is 50 miles from New York City at Glenwood, New Jersey.

Many of the most important corporate and government locations are in city centers many miles from the nearest possible antenna site. The cost of moving the wideband signals over common carrier facilities from the antenna site is very high—often high enough to discourage satellite usage.

Because of the interference problem one of the most important factors in the selection of satellite frequencies for corporate and broadcasting use is that they should not be terrestrial common carrier frequencies.

Table 13.1 shows the frequencies allocated by the FCC for terrestrial microwave operation by the United States common carriers, and compares them with the satellite bands allocated to the Western Hemisphere. It will be seen that several satellite bands do not clash with terrestrial microwave. The 7/8 GHz band is not used by common carriers, but these satellite frequencies are allocated to government (mainly military) operations. The lowest 500 MHz bandwidth which does not conflict with terrestrial microwave is that with a 14–14.5 GHz up-link and an 11.7–12.2 GHz down-link A 300 MHz satellite with 4.4–4.7 GHz and 3.4–3.7 GHz down would avoid conflict with common carrier operations, and a transponder in the 2.5–2.6 GHz band is also a possibility.

Higher in the spectrum a large bandwidth (3500 MHz) is allocated at 20/30 GHz. This will become a very important satellite band in the future but more development work is needed in using these frequencies. Transmissions at these frequencies suffer severe degradation from heavy rainstorms, as we discuss later in the book.

Table 13.1 A comparison of the frequencies allocated by the FCC for terrestrial microwave transmission in the United States, with those available to the United States (Region 2) for communication satellites.

Terrestrial Common Carrier Band (GHz)	International Satellite Frequency Bands Down-link (GHz)	Up-Link (GHz)	Satellite Bandwidth (MHz)	
2.11–2.13			20	
2.16–2.18			20	
	2.5–2.535		35	
		2.655–2.69	35	
	3.4–3.7		300	Frequencies used by commercial satellites and most terrestrial microwave links in the 1970's. (3.4–8.4 GHz rows)
3.7–4.2		3.7–4.2	500	
		4.4–4.7	300	
5.925–6.425		5.925–6.425	500	
	7.25–7.75		500	
		7.9–8.4	500	
10.7–11.7	10.95–11.2 11.45–11.7		500	
	11.7–12.2		500	Frequencies of the next generation of satellites. (11.7–12.2 and 14–14.5 GHz rows)
		14–14.5	500	
	17.7–21.2		3500	
		27.5–31	3500	

12/14 GHZ SATELLITES

The next major thrust in commercial satellites will be at 12/14 GHz. The 12/14 GHz band has a mixture of advantages and disadvantages over the traditional 4/6 GHz satellites.

The advantages of 12/14 GHz are as follows:

1. The band is generally not used for terrestrial common carrier links, so 12/14 GHz earth antennas can operate in city centers on the rooftops of buildings. Many major corporate locations could have their own antenna. In a crowded center there may be one earth station serving many local users, and linked to them by short line-of-sight microwave or millimeterwave radio hops.
2. The beam width from an earth antenna of a given size is less than half of that for a 4/6 GHz satellite. Therefore, about twice as many satellites could be used without interference, thus lessening potential congestion in the equatorial orbit.
3. An antenna of a given weight on the satellite can be made more directional, the beam to earth being narrower. The signal received on earth is stronger than with a more diffuse beam, thereby permitting the earth station cost to be lower. The gain of a satellite antenna of given size is 5.33 times greater on the up-link and 9.15 times greater on the down-link than at 4/6 GHz—i.e., 48.77 times greater in total.
4. The satellite could have multiple spot beams to earth like searchlight beams, the separate beams reusing the same frequencies. The satellite could therefore have more transponders without exceeding its allocated 500 MHz bandwidth.

5. At 4/6 GHz there are restrictions on the power of satellites in order to limit interference with terrestrial microwave. At 12/14 GHz there are no such restrictions and so a very powerful broadcast or commercial satellite could be launched. This could result in very low cost earth stations.

The disadvantage of the higher pair of frequencies is that rain, fog, and cloud have a worse effect on the signal. They attenuate the signal more and also introduce more noise into the signal. The improved signal gain is normally greater than the increased loss caused by rain, fog, or cloud. However under very extreme weather conditions the signal-to-noise ratio is substantially worse than at 4/6 GHz. Weather conditions this bad occur only very infrequently in most nontropical locations and are almost always short in duration. Such storms are usually not more than 5–10 miles across.

There are therefore options in the design of the higher-frequency systems. Relatively expensive earth stations can be designed which survive the worse the weather can do. Alternatively, fair-weather earth stations can be built which are substantially cheaper but operate with degraded performance when the weather is foul. The degraded performance can be engineered to take a variety of different forms. Voice channels may become noisier. Video channels may have snow on the screen but no voice degradation. The numbers of channels may be cut. Some signals may be rerouted via toll telephone circuits. Nonurgent or nonreal-time transmission may be temporarily suspended in favor of real-time signals.

Alternatively two earth stations 10–20 miles apart can be built and the better signal of the two used.

MULTIPLE ACCESS

Designing the transmission links so that corporate and other locations can have their own satellite antennas is a key to the growth of this powerful technology. Very much related to it is the use of control mechanisms that allow many users to share the satellite channels. The sharing of communication links is essential in almost all telecommunications trunking. With satellites there is a difference in that the users who transmit using one transponder may be geographically scattered over a wide area. If a thousand corporate locations all transmit over one future satellite, how will the signals be organized so that they do not interfere with one another? This is a problem of control somewhat similar to that with mobile radio transceivers. There are several different types of control mechanisms which can be used. The system which allows geographically scattered users to share a satellite is called a *multiple-access* system.

Telephone users make their calls at random. So do computer terminal users. A problem in telecommunications is that of assigning channels in a sufficiently flexible way that users can have a channel whenever they have the whim to communicate. If channels are permanently assigned to user locations, they will not handle the highly fluctuating demand as efficiently as if they are allocated dynamically when and where the demand arises. This variable allocation of channels on a demand basis can be done using computers. It is referred to as *demand-assignment multiple-access* (DAMA).

A unique satellite problem is: How do you allocate subchannels to users when the users are scattered across the earth and their demands vary constantly? An efficient solution to this *demand-assignment multiple-access* problem is necessary if users are to have their own antennas, or if satellites are to serve remote locations with little regular traffic. Multiple-access equipment with either fixed channel assignment or demand assignment, will be an important component of corporate satellite earth stations.

Satellites, unlike terrestrial trunks, will provide complete interconnectability between earth stations without switching offices. Figure 13.8 illustrates this. Any location on the network, using its multiple-access equipment, can communicate with any other location.

The mechanisms of multiple access are discussed in Chapter 26.

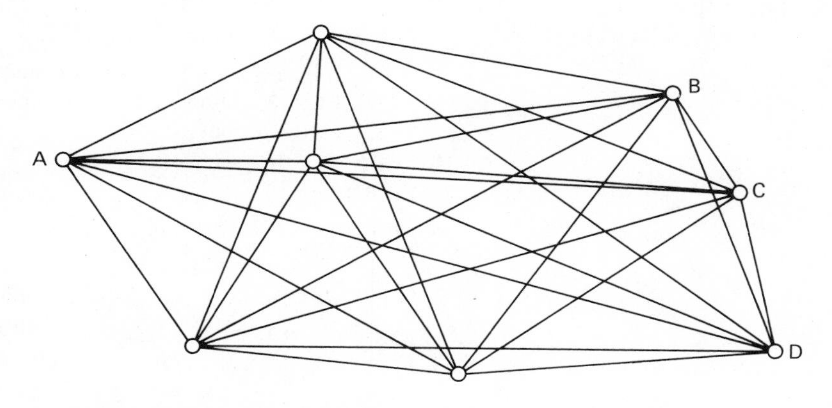

Figure 13.8 A multiple-access satellite system provides a complete set of interconnections between the earth station locations. This would require $\frac{1}{2}N(N-1)$ paths if terrestrial links were used without switching.

DIGITAL TRANSMISSION

For their first decade, most satellites used analog transmission. It has now been demonstrated that more telephone channels can be derived from existing satellites if digital, rather than analog, transmission is used. Furthermore, digital transmission yields itself to the digital multiplexing techniques that will be used for future demand-assigned multiple-access equipment. Digital circuits under computer control intermix the signals in accordance with the varying demands of the users.

In order to interleave the transmissions from many different earth stations each station sends high speed bursts of data. The bursts are very precisely timed and are sent using *a burst modem*. Today's burst modems operate over existing satellites at speeds up to 60 million bits per second. In Fig. 13.8, station B may send a burst, then station C, then station D. The bursts may contain traffic for any other station. Station A receives bursts and extracts only the traffic which is addressed to A. Security procedures can prevent A decoding the traffic addressed to other stations.

This is a highly flexible mechanism because all types of signals can be digitally encoded. B may be sending facsimile documents to A. C may be using telephone channels to D. D may have set up a video conference with B. A may be carrying out computer dialogues with B, C, and D, and so on.

It is important to note that in all of these diverse interconnections there is no use of *switching offices*. Switching of the type needed in terrestrial circuits is not used in Fig. 13.8. It is replaced by the demand-assigned multiple-access control in the earth stations. In reality not every location in such a system would have its own earth station equipment and so local switching would be used at the earth station sites. The same locations are drawn again in Fig. 13.9

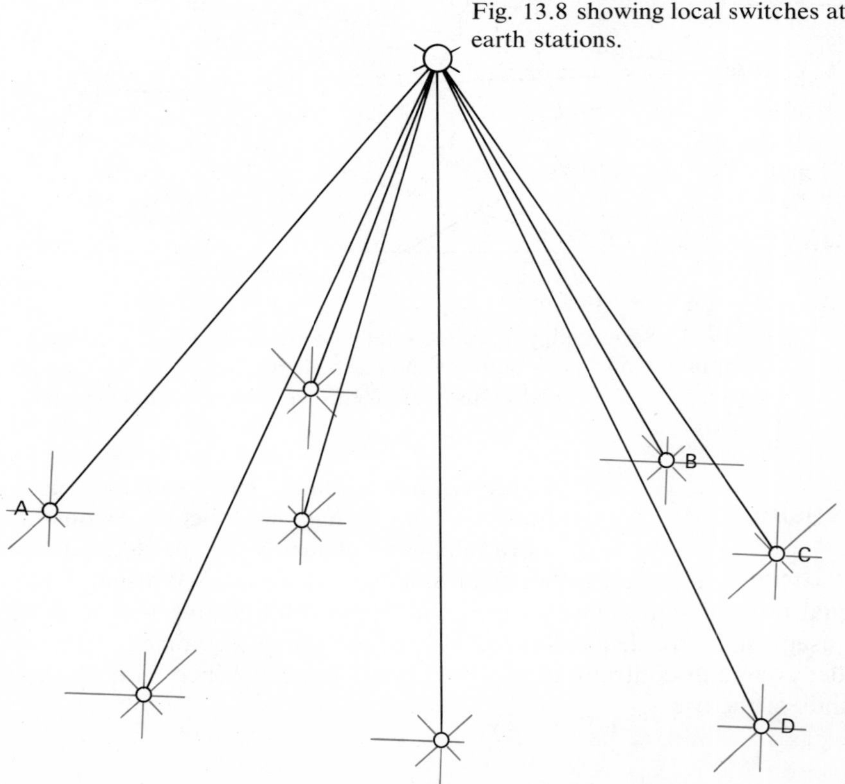

Figure 13.9 The same locations as in Fig. 13.8 showing local switches at the earth stations.

showing the local switched connections. The local earth station switching facilities may be, in essence, computerized PABXs.

This new technology has a potential for greatly changing the way large corporations use telecommunications. It will permit the transmission of signals of too high a bandwidth for the local telephone loops, e.g., video conferencing, interactive facsimile display, data base transmission, brief data bursts at multi-megabit rates, or remote file dumping for security purposes. It has the potential of lowering the overall cost of corporate networks. And above all it has the potential of bringing data transmission facilities of much lower cost than today.

UNIQUE PROPERTIES

It should be noted that satellites are not simple replacements for terrestrial lines. They have several uniquely different properties. These are summarized in Box 13.1.

BOX 13.1 Unique properties of satellite links.

A satellite channel is often used simply as a substitute for a point-to-point terrestrial channel. However, it has certain properties which are quite different to conventional telecommunications. It should not be regarded as merely a cable in the sky. New types of communications architecture are needed to take advantage of satellite properties and avoid the potential disadvantages. This is especially so in the design of interactive computer systems.

A satellite channel is unique in the following respects:

1. There is a 270 millisecond propagation delay.
2. Transmission cost is independent of distance. A link from Washington to Baltimore costs the same as a link from Washington to Vancouver. A computer center can be placed anywhere within range of a satellite without affecting transmission costs. It is becoming economical to centralize certain computing operations. In an international organization worldwide links can be similar in cost to national links if the regulatory authorities so permit.
3. Very high bandwidths or bit rates are available to the users if they can have an antenna at their premises, or radio link to an antenna, thereby avoiding local loops.
4. A signal sent to a satellite is transmitted to all receivers within range of the satellite antenna. The satellite *broadcasts* information unlike a terrestrial link.

(Continued)

BOX 13.1 *Continued*

5. Because of the broadcast property, dynamic assignment of channels is necessary between geographically dispersed users. This can give economies, especially with data transmission, on a scale not possible with terrestrial links but needs new forms of transmission control.
6. Because of the broadcast property, security procedures must be taken seriously.
7. Most transmissions are better sent in digital form. Digital techniques can therefore be used to manipulate and interleave the signals in a variety of ways. The high bit rates make possible new uses of telecommunications not economical on terrestrial links.
8. A transmitting station can receive its own transmission and hence monitor whether the satellite has transmitted it correctly. This fact can be utilized in certain forms of transmission control.

The first use of a 12/14 GHZ satellite for digital multiple-access operation may be in 1979. SBS (Satellite Business Systems), a partnership sponsored by IBM, Comsat General, and the Aetna Insurance Company, has filed for permission to launch two such satellites in 1979 to provide business and government users with wideband, switched, digitized communications networks for voice, data, and image traffic. The users would have small earth stations at or near their premises, sharing the capacity of the transponders by interleaving high-speed bursts of digital transmission.

This technology is explained further in Chapters 26 and 31.

REFERENCES

1. Arthur C. Clarke, *Voices from the Sky,* page 147, Harper and Row, New York, 1965.

2. Report to the Federal Communications Commission on a Satellite Link Test conducted by IBM Corporation at Poughkeepsie, New York. IBM System Development Division, Poughkeepsie, 1974.

1845. Samuel F.B. Morse sending his first public telegraph message: "What hath God wrought."

14 HIGH VELOCITY MONEY

One of the most intriguing new applications of data networks is *electronic fund transfer, EFT*.

Over the next ten years the nature of the payments mechanism in some countries will swing from being predominantly paper-oriented to being in part electronic, with vast quantities of financial transactions traveling over data networks. Some of the world's largest data networks will be involved in this application. Some large banks are now planning private networks with tens of thousands of terminals. There are more than 1400 banks in the United States and eventually they will be interlinked into nationwide networks for transferring money. Many institutions other than banks handle money, hold deposits, and offer credit. The financial data networks affect all such institutions and present sudden new opportunities that will generate fierce competition in the money-handling business. Eventually in the United States many billions of messages per year will be passing over the financial networks.

MONEY IS INFORMATION

Before societies used money, trade used to be carried out by means of barter. Later man devised systems of exchange in which certain commodities became standards of value against which all others were measured. In early societies these standards had intrinsic usable value, such as corn, cattle, or wives. Later, gold became a standard. Gold was rare, divisible, unattacked by rust and lichen, and beautiful enough to inspire poets. For millennia gold has been the world's standard of value, fought over, traded, ornamented, stolen, and worshipped. Only in recent years has man gathered the effrontery to question the necessity of this "symbol of pre-eminance ordered by the celestial will."

Paper money was invented in 1694, and was at first regarded by many as a sinister banker's trick. Until this century paper money only seemed respectable if it was backed by an equivalent amount of gold in some banker's vault. Today the currency note is no longer convertible into gold. Currency bills used to say that a central bank "promised to pay the bearer on demand" the value represented by the bill. Today they merely say "In God We Trust."

In relatively recent years checks came into common use by the man in the street, replacing the need to pay with currency. In the late 1960s check and currency usage began to give way to credit cards, removing the payments still further from the backing by gold or commodities. Today, what may be the ultimate payments mechanism is gaining momentum. The transactions are neither gold, nor currency, nor paperwork, but instead are electronic bits flowing between computers—electronic fund transfer.

EFT recognizes that money is merely a form of information. The dollar bills which pass from pocket to pocket have become merely a demonstration of man's ability to pay. If money is merely information, then that information can reside in computer storages and payments can consist of data transfers between one computer and another. EFT enthusiasts began to talk of a cashless, checkless society.

In reality society will be neither cashless nor checkless for the foreseeable future. Rather, what has worried bankers is that the paperwork associated with checks and credit cards is growing by leaps and bounds. One hundred million checks a day are written in the United States, and without automated fund transfer this number would double in the next ten years. While checks are expensive because of the paperwork costs, credit card transactions are even more so—their cost is approximately 50 cents per transaction and rising. Electronic fund transfer offers a way to slow and later reverse, the growth of paperwork.

The replacement of gold by paper money and of paper money by checks were each revolutionary in their day. Now we must become used to financial transfers occurring in the form of electronic pulses on a data link. The paperwork associated with the transaction will now merely inform us about the transaction, rather than represent the transaction itself. Nor will it have to be punched into cards and fed into a receiving computer. The eventual consequences of the simple idea of automatic credit transfer will be enormous. Vast random-access computer files in banks will hold full details of all accounts. As a transaction is entered into the system, transmitted data will cause the appropriate amount to be deducted from an account in one computer and added to an account in another. Eventually, the financial community will become one vast network of electronic files with data links carrying information between them.

Thomas J. Watson, Jr., then President of IBM, foresaw the revolution in banking as follows in 1965:

> In our lifetime we may see electronic transactions virtually eliminate the need for cash. Giant computers in banks, with massive memories, will contain individual customer accounts. To draw from or add to his balance, the customer in a store, office, or filling station will do two things: insert an identification into the terminal located there; punch out the transaction figures on the terminal's keyboard. Instantaneously, the amount he punches out will move out of his account and enter another.
>
> Consider this same process repeated thousands, hundreds of thousands, millions of times each day; billions upon billions of dollars changing hands without the use of one pen, one piece of paper, one check, or one green dollar bill. Finally, consider the extension of such a network of terminals and memories—an extension across city and state lines, spanning our whole country. [1]

Perhaps payments of the future, instead of carrying the message "In God We Trust" should say "In Mother Bell We Trust."

FOUR TYPES OF EFT

There are four main types of electronic fund transfer representing successive steps towards an EFT society. The first involves transfers of money between banks, to carry out clearing operations. Second there are transfers between the computers of other organizations and the bank computers. A corporation may pay its salaries, for example, by giving a tape or transmitting salary information to a bank clearing center which distributes the money to the appropriate accounts. Third, the general public use terminals to obtain banking services. These terminals include cash-dispensing machines in the streets. There are a variety of such terminals with different functions, and bankers refer to them as CBCTs (Customer Bank Communication Terminals). They threaten to play havoc with the traditional structure of banking, at least in the United States.

To operate the CBCTs, customers are equipped with machine-readable bank cards. These cards, which look like credit cards, make possible the fourth and ultimate phase of electronic fund transfer, in which consumers pay for goods and services in restaurants and stores by using their bank cards or similar cards provided by American Express, large retail chains, petroleum companies and other organizations. Today's credit card devices (which create paperwork) are replaced by inexpensive terminals which accept the new machine-readable cards. Thousands of such machines are already in use.

EFTS (Electronic Fund Transfer Systems) thus describes a wide variety of different computer systems, but in general has become synonymous with advanced new technical directions in banking.

The present payments mechanism is highly labor intensive. Credit cards have increased, not decreased, the quantity of paperwork and manual operations. Labor costs are rising and it is becoming more difficult to obtain workers for dull, boring, but high-accuracy tasks. It has been estimated that the overall cost of using credit cards exceeds 50 cents per transaction in the United States, and that the cost of equivalent EFT transactions could be dropped to 7 cents.

Furthermore EFT can make cash available to bankers faster, and time is money, especially with today's high interest rates. The volume of checks alone in the United States is about $20 trillion per year. If the money from these could be available to banks one day earlier on average, because of faster processing and clearing, that represents a float of $54 billion per year. It is worth installing some expensive automation schemes to capture a portion of this float.

AUTOMATED CLEARING HOUSES

A bank clearing house takes checks drawn against many banks, and allocates the funds appropriately. The automated clearing house (ACH) movement in the banking industry is an attempt to create an electronic infrastructure which can reduce the labor in check clearing. This will both lower the cost of check clearing and speed it up. It will also enable banks to offer new services to their customers. For example computers in some corporations deliver the payroll in electronic form to a clearing house, and from there the money is moved into the banks of the employees. There is then no need to print and read payroll checks. Initially these electronic payrolls were delivered on magnetic tape. It would be quicker to transmit them by telecommunications.

In the United States the number of automated clearing houses is growing and may eventually reach 35 or so. Between these centers a telecommunications network will operate transmitting many millions of transactions per day by the early 1980s. The National Automated Clearing House Association (NACHA) coordinates the development of the clearing house facilities which must remain neutral to the competitive banking industry. The automated clearing houses and their network will have a vital role to play as EFT systems spread to the consumer level.

PREAUTHORIZATIONS

Many of the payments that are made by check are repetitive payments, the same sum being paid at regular intervals, or at least a sum which can be calculated well in advance. Such payments include rents, mortgages, local taxes, society dues, interest payments, social security payments, salaries, installment credit payments, and so on. Much work can be saved if these payments are made by *preauthorization* (the term *standing order* is used in British parlance). An instruction to make the payment repetitively is given to a bank computer, and the payment is made without further paperwork.

The U.S. government handles some military payroll and many social security payments in this way. Some labor unions have discussed having workers paid *daily* by electronic means. The preauthorized payments may be handled by an automated clearing house or by a suitably prepared bank. Many bank customers would welcome a bank service which pays their rent, mortgages, society dues, and so on, without involving them in further paperwork.

The situation is slightly more complicated if the payments vary each time they are made. Dividends, like wages, can be paid automatically into customer accounts and most customers, once they are used to it, welcome rather than resist this form of computer-to-computer payment. In a similar way, telephone companies and other utilities could send their bills directly to the bank clearing system. This, however, is a much more drastic step because money is being taken from the accounts of individuals rather than added to them. Many consumers feel that they should have the option of not paying their telephone bills! Nevertheless the majority would probably welcome an automatic bill-paying service.

If preauthorized payments of these types were fully used, the total volume in the United States would exceed 5 billion transactions per year. The only paperwork would be periodic statements informing the bank users what transactions had been made. The paperwork would be statements about the transactions, and *would not be the transactions themselves*. There would be no writing or processing of checks, or rooms of keyboard operators laboriously and sometimes erroneously entering details of transactions.

AUTOMATED CREDIT

If money is *deducted* from consumer accounts by electronic fund transfer, some of these accounts are likely to go into the red periodically. The consumer does not have quite the same control as when he can add up every payment he makes with his checkbook (though few today do so).

An essential aspect of EFT, therefore, is the ability for customers to have negative balances in their banks. The magnitude of the permissible negative balance would be set by a bank officer (or possibly set automatically). The customer would be automatically charged interest on his negative balance. There are various forms of automatic negative balance in operation today. Some of them seem designed to create customer resistance by charging exorbitant interest rates on an automatic loan which does not drop when new funds come into the account. A floating negative balance with a reasonable interest rate is required.

Automated credit offers a constant temptation to overdraw. This might have great appeal to bankers who would make money on the interest charged. Customers, on the other hand, might resent losing their float or losing their ease of refusing to pay. A wide variety of incentives have been devised for making

EFT appealing to customers including discounts, cash-dispensing terminals in corporate offices, EFT terminals in corporate cafeterias, lower bank charges, and general ease of obtaining money.

INTERNATIONAL FUNDS TRANSFER

Systems for transferring funds between banks electronically are coming into operation not only on a national scale, but also internationally. The first major international network is the SWIFT system.

SWIFT, Society for Worldwide Interbank Financial Transactions, is a nonprofit-making organization set up and wholly owned by banks in Europe, Canada, and the United States. SWIFT implemented and operates the network shown in Fig. 14.1, the purpose of which is to send money, messages, and bank statements, at high speed between banks. The participating banks financed the system and a tariff structure charges for its use on a per-message basis plus a fixed connection charge and an annual charge based upon traffic volumes. The banks range from very small to banks with 2000 branches.

As in other telecommunication systems, SWIFT imposed standards for procedures and message formats on its users. These enable the banks to send and receive messages between countries in computer-readable form. Figure 14.2 shows typical SWIFT messages.

The SWIFT system is a message-switching network which originally has two switching centers as shown in Fig. 14.1. It can expand without functional redesign to have multiple centers. It uses voice-grade circuits and most traffic is delivered in less than one minute. All traffic is stored at the switching centers for ten days and during that period can be retrieved if necessary. Transactions can be entered into the system regardless of whether the recipient bank's terminals are busy or not. The originator of an urgent message will automatically be informed by the system if there is a delay in delivering the message.

It is estimated that by the late 1970s the SWIFT traffic will be about one third of a million messages per day. The initial switches were each designed to handle 23 transactions per second. The system can accept either single messages or bulk traffic from computers or magnetic tape. Transactions can have priorities allocated to them.

BANK CARDS

In the mid-1970s a new wave of banking automation swept across America triggered by the advent of machine-readable bank cards. These plastic cards have the size and appearance of a credit card, but unlike conventional credit cards carry invisible data which can be read by a terminal. On most such cards the data are encoded on two magnetic stripes. A third magnetic stripe is being added on which data could be *written* by the terminal as well as read. There are some exceptions to the mag-

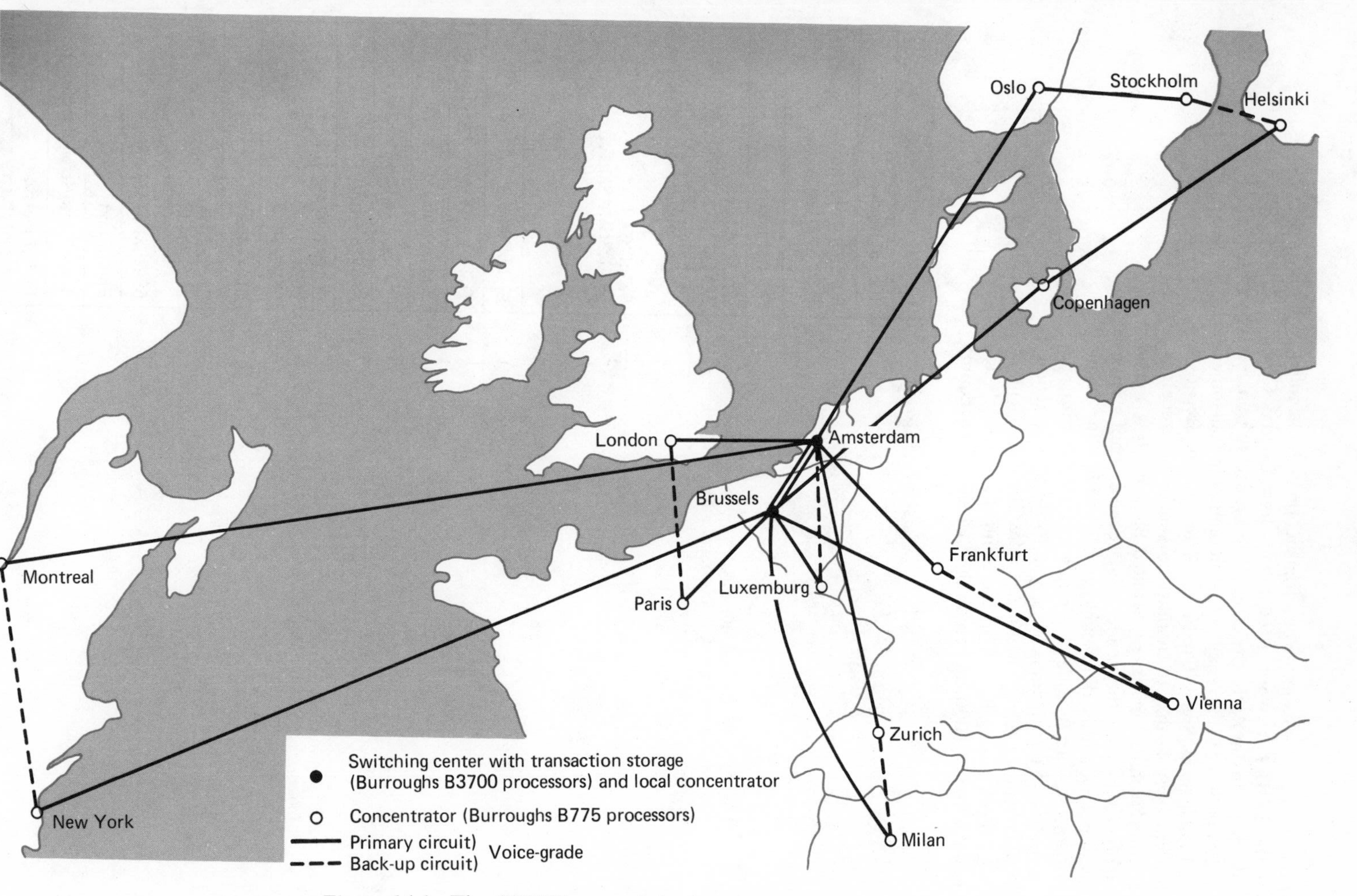

Figure 14.1 The SWIFT network for international fund transfer.

Figure 14.2 An example of a transaction being handled by the SWIFT network. A customer, John Loeb & Co., ask their bank in Paris, the Banque de France, to transfer $750,000 in U.S. currency to the account of the customer J. Blanagan in Swiss Credit Bank in Zurich. Because the currency is that of neither the sender nor receiver country, a third bank, the Chemical Bank in New York, is involved. Both the sender and receiver banks have accounts with this third bank which handles the reimbursement. The Banque de France first sends a message to the Swiss Credit Bank with details of the transaction. It then sends a related message to the Chemical Bank in New York asking it to debit the Banque de France's account with $750,000, and credit the Swiss bank's account.

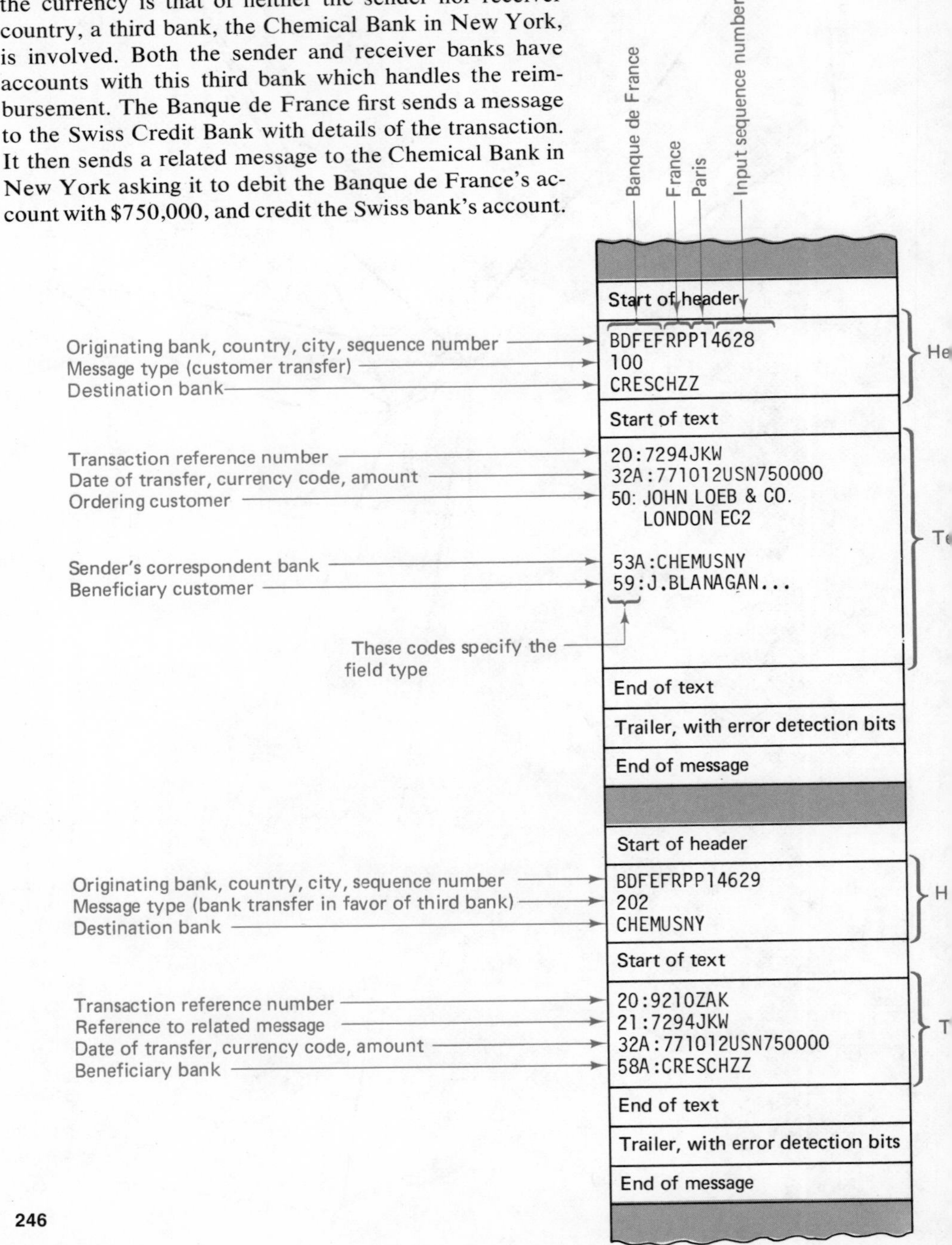

netic stripe technology, notably New York's First National City Bank's cards which contain a stripe read using ultraviolet light.

CUSTOMER ACTIVATED TERMINALS

Using a bank card at an appropriate terminal, a bank customer can inquire about the status of his accounts. He can deposit or withdraw cash, borrow money if it is not in his account, or transfer money between different types of accounts. Such a terminal is shown in Fig. 14.3. In fact he can do virtually everything that he would previously have done by going to a branch of the bank, standing in a queue, and talking to a teller. The interesting question arises: if all of a bank's customers used bank cards and terminals, would the bank need tellers? Perhaps it would only need officers, who deal with situations needing human interaction and decisions. A bank could close some of its branches in expensive city streets and yet give its customers more convenient service because the automated teller terminals are becoming located in stores, shopping plazas, airports, factory cafeterias, and office buildings. Furthermore, the customers could obtain cash or other banking services when the bank was closed.

Some customers have an initial hostility to banking by machine, but once used to its convenience few want to go back to queuing in marble-pillared branches.

The prospect of doing away with the bank teller is revolutionary enough, but another implication of automated teller terminals threatens to play havoc with the entire structure of banking. Banking in the United States has traditionally been regulated by state and federal laws saying where a bank may have its branches. The McFadden Act of 1927 prohibits interstate branching and makes national banks conform to restrictive state laws. No bank can have branches in more than one state; some can operate only within a city; some can have no branches other than at the head office location. Persons living near state boundaries, for example near New York City, are sometimes flooded with advertising from banks they cannot use. In 1974 the Comptroller of the Currency, who regulates banking activity, made the ruling that a remote terminal which customers use does not constitute a bank "branch." Following this ruling, banks rapidly started to spread their tentacles into geographical areas from which they had earlier been excluded. The controversial ruling was then challenged in the courts and partially reversed but nevertheless it seems certain that the structure of American banking will change fundamentally.

New York's First National City Bank has almost 300 branches in New York State and cannot have branches elsewhere. However it has several thousand bank-card terminals in the New York area with some across the state boundary in New Jersey. A landmark court case ruled the bank terminals in the Hinky Dinky supermarkets in Nebraska were legal, and did not constitute

An unmanned office of Citibank in New York. Legally this is not a *branch* of the bank.

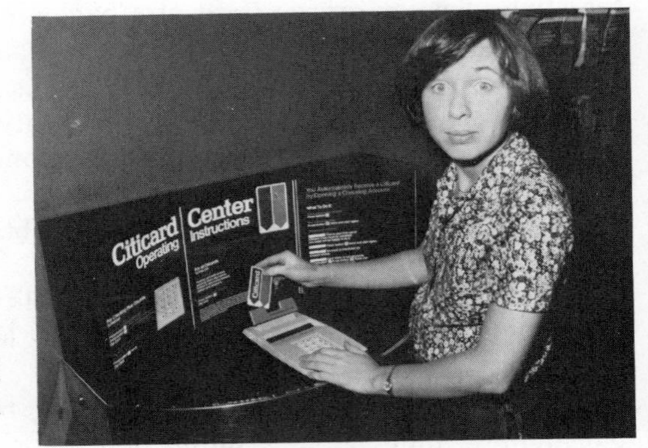

A variety of machines enable a customer to do almost anything she could do by communicating with a human bank teller, face-to-face.

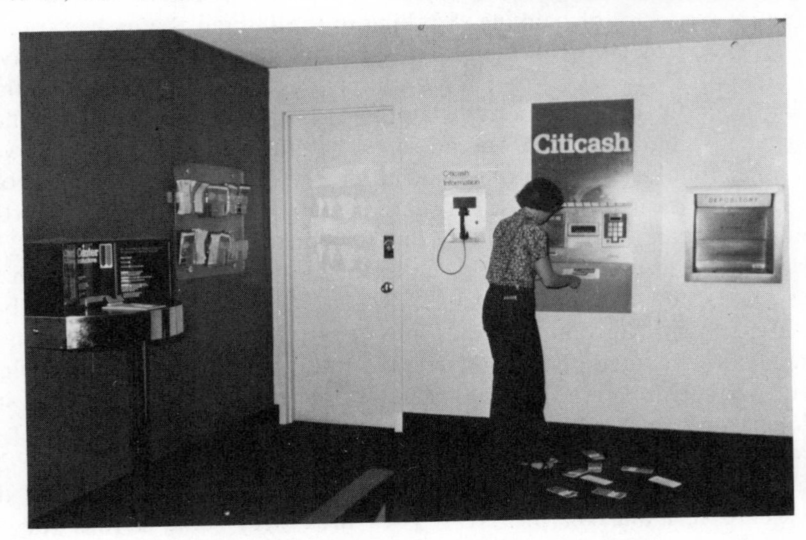

Figure 14.3 Banking by telecommunications.

branches. In a Nebraska supermarket, an individual can elect to pay for $25 of groceries by switching the $25 from a savings account to his account with the supermarket. The transaction is completed without the use of currency or checks.

Automated teller terminals can be operated by organizations other than banks. Several savings and loan associations operate them. Consumer finance companies, large chain stores, gasoline companies, credit unions, and other organizations, collectively extend more consumer credit than the banks. Many of these operate credit checking terminals and some have applied for permission to hold customer balances in which case their terminals could have most of the functions of a banking terminal. The banks, in other words, could face electronic competition from a variety of other organizations.

The 1974 ruling that a banking terminal was not a branch opened up the possibility of large banks developing nationwide terminal networks. Later regulation may prohibit that because the competition could become too severe for small local banks. However, the bank customers who have become used to electronic banking certainly want to have its facilities nationwide. A New Yorker wants to be able to use his New York bank card across the river in New Jersey and across the country in California. He not only wants to be able to use the card in stores and restaurants, but also wants to be able to obtain cash from cash-dispensing terminals.

Nationwide networks for checking a customer's credit already exist. Nationwide networks for transferring consumer funds and accepting the bank cards will be built. Hooking together many localized systems of banking terminals is a problem not unlike hooking together many localized telephone companies. There must be national standards for the terminals that are used. A nationwide network may develop that serves many different banks, as AT&T Long Lines serves many different telephone companies. A small-town bank, like a small-town telephone company, will be able to say to its customers, "We can hook you into the world."

TERMINAL SHARING

Although the simplest form of bank-card terminal can be cheap, a full-function automated teller terminal is expensive. It is desirable that banks should be able to share the services of these terminals, and the banking regulations appear to encourage such sharing. A big city store or restaurant would not want to install dozens of different terminals connected to different banks.

The International Standards Organization, ISO, has established standards for the cards that are used, and the coding that will be employed on their magnetic stripes.

When terminals are shared between banks, they relate to automated clearing house operations. Most customer-activated terminals are today installed by specific banks. In the view of some authorities (e.g., Reference 1) as electronic fund transfer matures the terminals and supporting networks will become more and more neutral, and competition will be based on the design of services not on control of the electronic delivery system.

Many small banks regard the spread of automated-teller terminals with great apprehension. They have neither the finances nor the expertise to establish such systems themselves, but fear that the terminals of larger banks will invade their territory and erode their customer bases. Organizations of small banks in the United States have been fighting the new regulations which permit the spread of customer-activated terminals. It is desirable that the regulations evolve in a manner which permits small banks to remain competitive and yet allows the desirable automation to spread as fast as possible.

In the long run the 1400 small banks in the United States cannot be *prime producers* of complex machine-based services, any more than the 1700 small telephone companies can be the prime producers of telephone equipment. Instead they will become service organizations which often provide a pleasanter service than the larger banks, and which enable their customers to use the nationwide EFT facilities. The small bankers will have an in-depth knowledge of their customer's needs and will be able to advise them how to use the increasingly complex banking services. They will act as franchisees of the big electronic systems and bank cards, and as with other franchise businesses this can be both highly lucrative and competitive.

BANKING FROM HOME

The simplest customer-activated terminal is the telephone. Some banks offer services which enable customers to make payments by telephone. There have been experiments in which some banks have attempted to automate banking from the home by using a touchtone telephone and voice answerback systems.

One of the costliest operations in the use of computers is the preparation or keyboard entry of data. To reduce costs it is desirable to persuade customers to enter transactions themselves in an electronic form. This can be done at customer-activated terminals in banks or in the street. It could also very conveniently be done from the home on a 7-day, almost-24-hour basis. Paying one's bills at home with a touchtone telephone could be very convenient and appealing to many customers. Some banking authorities believe that banking from the home will become commonplace [2].

Catalogue or mail order shopping could be done from the home telephone, using EFT. This is done today without EFT by one large store in Canada, using a voice answerback computer system. In the more distant future television, or still-frame CATV, selling may enable viewers to purchase goods and pay for them using a keyboard.

POINT-OF-SALE TERMINALS

A major amount of human drudgery will be saved when the payments made by consumers in stores and restaurants are entered directly into the banking systems instead of being made by credit cards or checks. The rapidly spreading bank cards are the means for such transactions, and consumers have demonstrated in a few early systems that they like the convenience of paying with bank cards.

The term applied to extension of financial services to stores and restaurants is *point-of-sale* (POS). Eventually there are likely to be 50 billion electronic point-of-sale transactions made per year in the United States. Much of the initial use of point-of-sale terminals is not for fund transfer, but for checking the consumer's bank balance so that checks can be cashed.

The networks needed for point-of-sale fund transfer are, as elsewhere in telecommunications, highly volume-sensitive. The initial systems may therefore be difficult to cost-justify, but once such systems exceed a certain volume they will become highly economic and will probably spread very rapidly, as did the use of credit cards.

Figure 14.4 The small terminal by the cash register is a TRW point-of-sale machine for checking and updating credit records. A similar machine could be used for transferring the funds when the customer makes a payment. Such machines will need magnetic-stripe bank cards.

One essential to the future of point-of-sale EFT is the availability of mass-produced inexpensive terminals. Such terminals ought to have some form of back-up, such as a tape cassette, for when the telephone link or computer system fails. Some terminals now in use are simple and low in cost. It is interesting to reflect that a bill-paying terminal for the home could be very inexpensive. Such a terminal could identify itself to the bank computer and have no need for a bank-card reader—merely a keyboard which sends signals like touchtone telephone beeps. The cost, manufactured in quantity, could be as low as that of pocket calculators.

Because of the volume-sensitivity, point-of-sale linkage to banks is coming from the very large banks, cooperative bank-card groups or from service bureau organizations. In many cases, it is not the commercial banks that are providing point-of-sale terminals. The banks face potential competition in this area from credit-card companies, credit unions, saving and loan associations, finance companies, and particularly from the large stores themselves. The large stores see point-of-sale terminals as carrying out functions other than merely the cash transfer. They can provide better inventory control and sales analysis, tighter credit control, improved cash flow, shorter checkout time and hence less checkout staff.

A factor which will make the area highly competitive is the question of what organizations handle the sums of money, gigantic in total, that consumers deposit and which are extended as consumer credit. The battle for these funds, each running into hundreds of billions of dollars, will be fierce and is very much related to what credit cards or banks consumers use.

Figure 14.5 shows some examples of how individuals relate to EFT. There are many alternatives to the details shown in Fig. 14.5.

CRIME

Eventually a high proportion of society's payments will be made with machine-readable cards. If appropriate security procedures are built into the systems, a criminal would be able to gain nothing by stealing a bank card. They could be made much safer than today's credit cards (and in some systems have been). One of the subsidiary benefits of an EFT society could be that street robberies are greatly reduced because pedestrians no longer carry much money or credit cards.

Major costs are loaded onto today's economy from robbery, theft, and fraud. It is reflected in the price of insurance, the reluctance of big-city police to investigate minor robberies, the fear of walking in city streets after dark, the inability to obtain some types of insurance. Insurance is impossible to obtain on many welfare check payments due to the risk, for example. Electronic payments systems with tight security controls could do much to lessen this burden of crime.

The EFT networks themselves offer new opportunities for ingenious computer crime, and tight security controls need to be built into the systems. The only way to make the transmissions safe from wire-tapping is to use cryptography. Some EFT terminals use cryptography having an extremely high level of safety. On one network in the author's experience which did *not* use cryptography funds were in fact stolen by an ingenious programmer. Like most such crimes no mention was made of it outside the organization in which it occurred. The technology does exist for making EFT networks sufficiently secure, though on some systems it may not be used adequately.

Successful crime in a computerized society will probably require, like other activities, longer training and a higher IQ.

NEW CONTROL REQUIREMENTS

There may be some subtle problems associated with the fact that money can move at the speed of electricity, and these problems will require new control mechanisms.

First, it may be a cause for concern that international transfers can take place at an unlimited rate. One can imagine a stampede to move funds out of a particular country, or sudden planned currency moves by power groups, adding to world financial instability.

An extensive use of EFT will reduce the "float" money within a country, like taking up the slack in an anchor rope. It will make the float available for people or corporations to spend. Economists use the phrase "velocity of money" to refer to the rate at which money turns over. Today one dollar issued by the government might be spent 17 times in the course of the year. With EFT, the velocity would be higher, perhaps much higher; the same dollar issued would be spent many more times. Unless compensating controls were devised this increase in the velocity of money would probably have an inflationary effect. As with a vehicle, the faster the velocity the tighter the controls must be. Today's mechanisms would provide inadequate control.

Furthermore, the nature of monetary controls will be complicated by the spread of EFT to institutions other than banks. The Federal Reserve authorities today control what is referred to as M_1—currency in circulation plus the demand deposits in banks. With the electronic systems now developing, that will become a steadily smaller proportion of the money that is actually available for spending. Savings banks, savings and loan associations, and credit unions are talking of making their deposits electronically available, and have been told that they will have direct access to the electronic clearing houses rather than going via regulated banks as they do with checks today. These institutions are likely to resist any attempt to regulate them like commercial banks. Even some large chain stores are considering accepting customers' deposits as well as making credit available to customers.

Individuals, as EFT spreads, will have multiple lines of credit available to them via their EFT cards. They will have credit available from stores, from travel and expense organizations like American Express, from negative balance

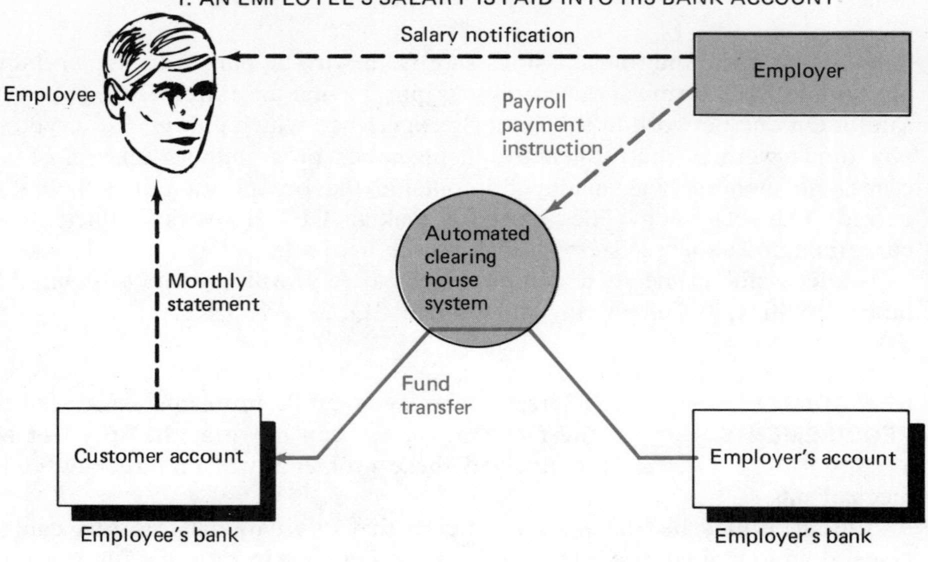

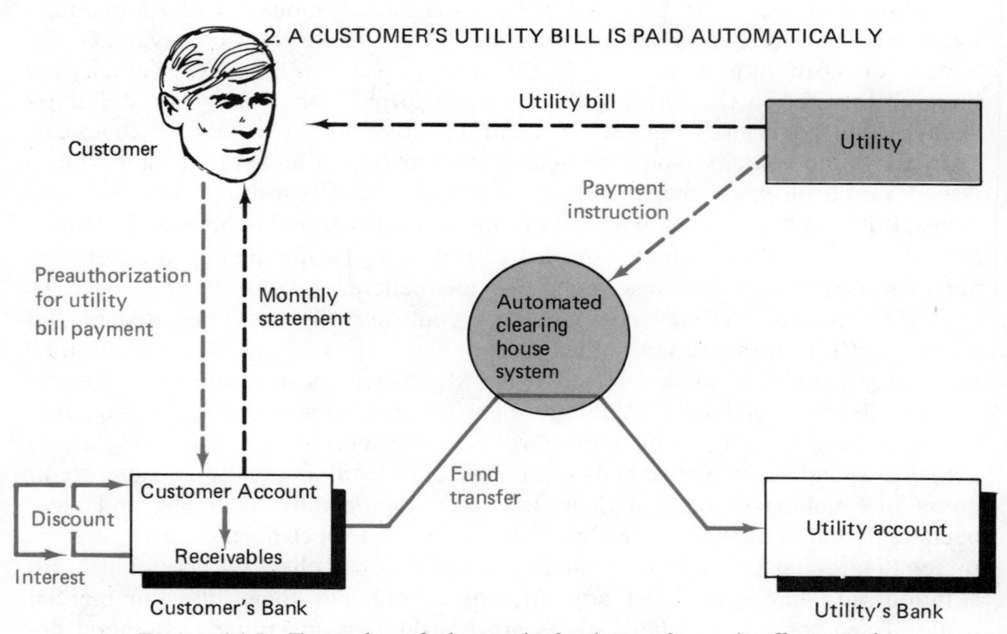

Figure 14.5 Examples of electronic fund transfer as it affects an individual. The red lines represent payment instructions or fund transfers. The black dashed lines represent information external to the system for the benefit of the individual. There are many variations on the operations shown here.

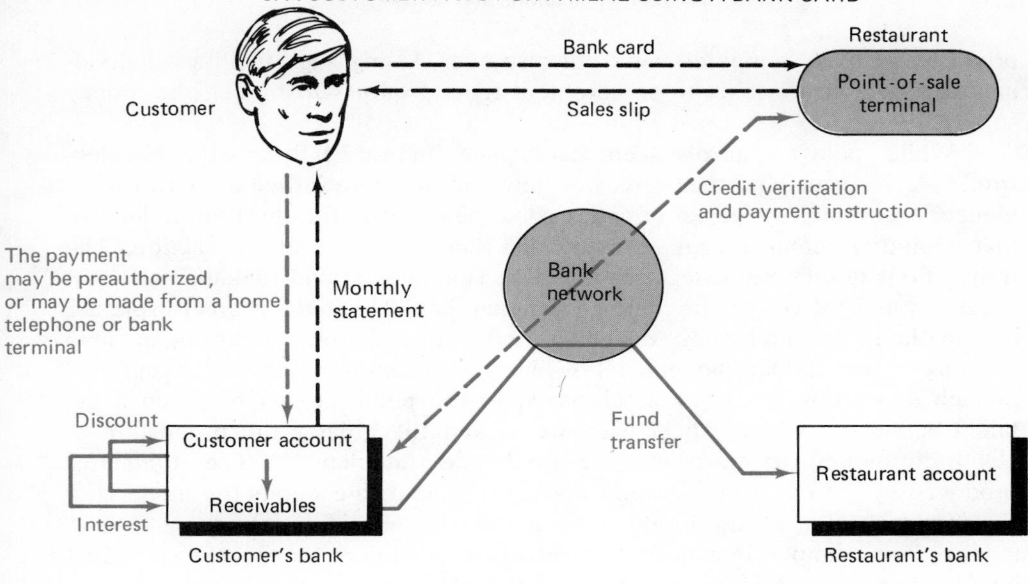
3. A CUSTOMER PAYS FOR A MEAL USING A BANK CARD
Restaurant
Bank card
Point-of-sale terminal
Customer
Sales slip
Credit verification and payment instruction
The payment may be preauthorized, or may be made from a home telephone or bank terminal
Monthly statement
Bank network
Discount
Customer account
Fund transfer
Restaurant account
Receivables
Interest
Customer's bank
Restaurant's bank

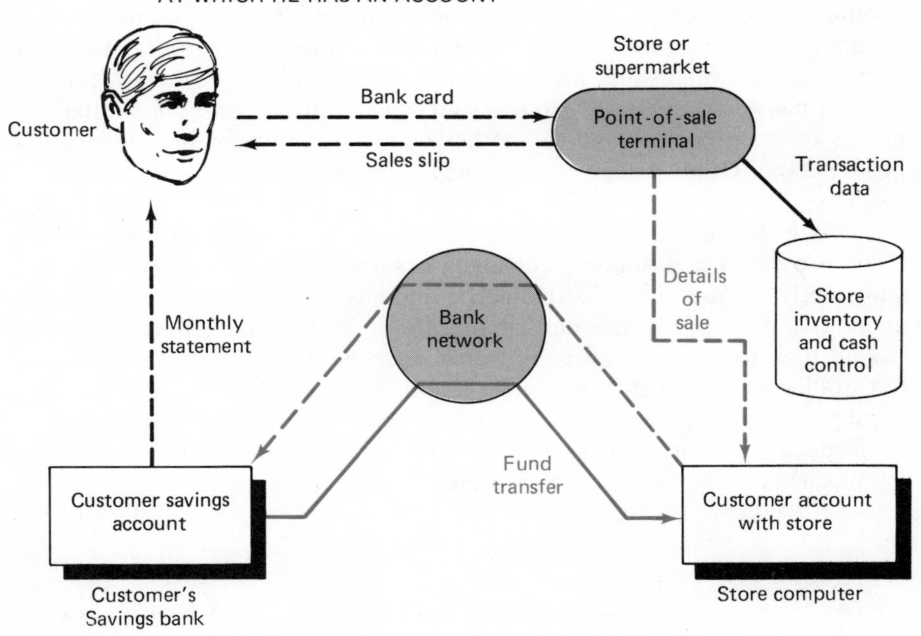
4. A CUSTOMER BUYS GOODS IN A STORE OR SUPERMARKET AT WHICH HE HAS AN ACCOUNT
Store or supermarket
Bank card
Customer
Point-of-sale terminal
Sales slip
Transaction data
Details of sale
Store inventory and cash control
Monthly statement
Bank network
Fund transfer
Customer savings account
Customer account with store
Customer's Savings bank
Store computer

Figure 14.5 cont.

privileges in bank computers, and other sources. Using such facilities, individuals will be able to offset to some extent federal efforts to tighten the money supply.

While today's controls seem inadequate for an EFT society, the electronic mechanisms lend themselves to new and far more intricate control techniques. A comprehensive set of codes could be appended to the transactions so that computers could recognize many different categories of transaction. The transactions in different categories could be summarized and the summaries digested by federal computers. The government could be given a precise picture of how the electronic money was being used, and regulators would obtain rapid warning of any sudden movements or changes in trends which are exceptional enough to warrant attention. Legislators, or the public, could be given a detailed economic balance sheet each week, and the greater the proportion of electronic money the more accurate would the statement be. The summaries produced by the electronic systems could form input to econometric models.

Using EFT it would, in effect, be possible to have more than one type of money. For example, money for individual consumption, money for individual liquid savings, money for individual investment as in bonds and stocks, industrial capital, and government money. The government attempts to regulate the proportion of individual wealth that is used in consumption and saving or investment, but the regulating mechanisms are very crude. If the users of money were categorized it would be possible to regulate the flow from one category to another in an attempt to influence the economy. This might be done by tax mechanisms, the controlling of interest rates, possibly by more direct mechanisms such as price and wage controls. Money might be taxed, for example when it passes from the *investment* category to the *consumption* category. If it passes from one investment to another within the investment category it would not be taxed. Other categorizations might be used to effect the balance of payments.

Such regulators could be made much more sensitive than today, and could provide a fine-tuning mechanism assuming that their effect was modelled sufficiently accurately in econometric models. It is possible to visualize the economy, perhaps by the end of the 1980s being much more precisely controlled than today, with all bank transfers being electronically summarized and fed on-line to the highly intricate econometric models which form a basis for regulation and argument concerning the economy. Compare the sensitivity of such mechanisms with that of the Ford administration doing a once-only mailing of checks to consumers to give them a tax rebate.

PRIVACY

A concern with electronic fund transfer, as in other advanced uses of teleprocessing, is that individual privacy may be eroded. An individual's financial history might be laid bare to government authorities. Some doctors rebelled against using on-line services for billing patients, presumably because they revealed too much to the IRS.

The on-line terminal services, while accepted with delight by some, are regarded with distrust by others. There is still a fear, if not always consciously stated, that the remote machine will lose information, make mistakes, or print out one's financial details for other people to see. The computers can be made trustworthy, superbly accurate, and secure. The computer files can be made private, but it will take time for these facts to be understood by the public, and it may require new privacy legislation.

REFERENCES

1. Thomas J. Watson, Jr., "Man and Machines—The Dynamic Alliance," Proceeding of the ABA National Automation Conference, 1965.
2. R. H. Long and J. C. Poppen, "EFTS: A Look at the Future," Bank Administration, Feb. 1975.

A pony express rider in 1860 passing a telegraph line construction crew. The pony express went out of business shortly after the line was completed. (*Courtesy Western Union Telegraph Co.*)

15 ELECTRONIC MAIL AND MESSAGES

As we have commented, many telecommunication paths are now being built as digital links rather than analog links. The wire-pair lines that fill the cities and suburbs, and stretch along country highways, can carry more telephone calls if they are digital than if they are analog. Digital technology is spreading to the bigger telecommunications highways such as coaxial cable systems and the new high-capacity waveguide systems. Satellites, the most promising of new communication technologies, can also transmit more telephone calls if used in a digital fashion. A main trend throughout telecommunications will be the swing to digital techniques.

When telephone calls are carried in digital form, the relative cost of data transmission and telephone transmission swings in favor of data. Data are transmitted on terrestrial telephone lines at between 1200 and 9600 bits per second. Telephone calls are transmitted using 64,000 bits per second (an AT&T and CCITT standard). In space, 60 million bits per second of existing satellite modems represents a vast amount of data traffic.

Furthermore, telephone traffic has to be transmitted in *real time,* i.e., when a person speaks his speech must be transmitted almost immediately. To achieve a good grade of service, i.e., low probability of a caller encountering a *busy* signal, there have to be idle channels ready for immediate use. No real-time systems achieve 100% utilization of their facilities. Probability calculations determine how much idle capacity is needed to provide a given grade of service.

The transmission capacity must be designed for the peak telephone traffic. The traffic during the peak hour of the day is several times higher than the average traffic. The peak hour of the peak day is substantially higher than that of the average day. The unshaded part of Fig. 15.1 represents idle capacity. On

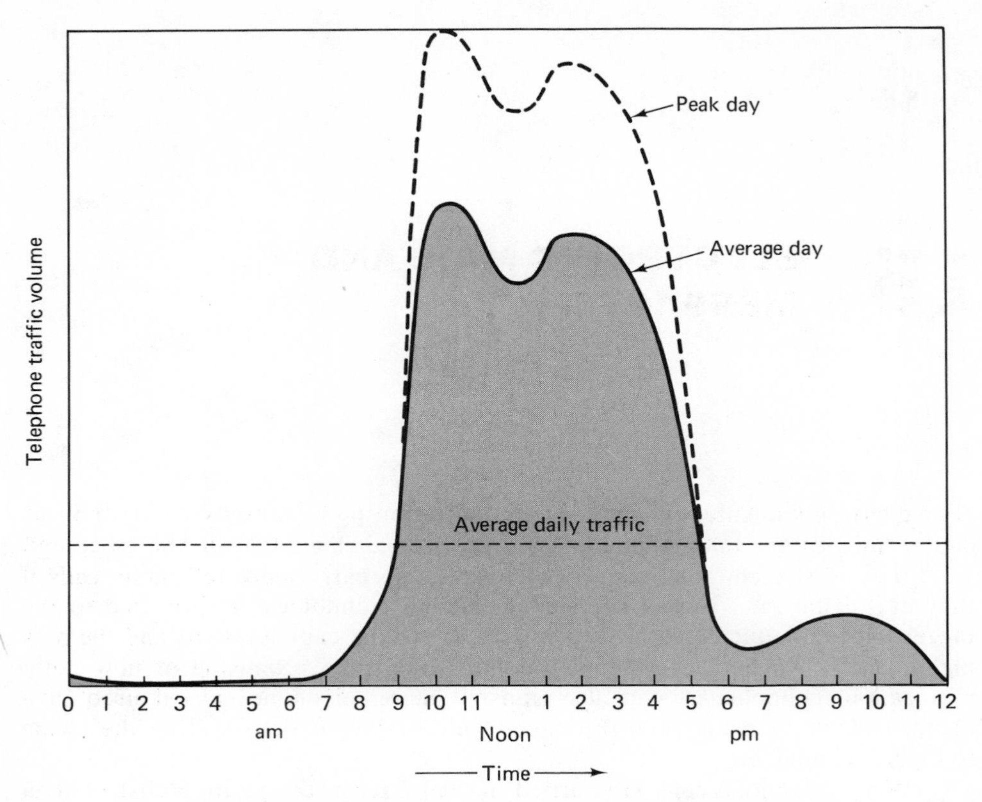

Figure 15.1 Telephone traffic. The unshaded part of the chart represents idle channel capacity.

a typical corporate network the channels are idle 80% of the total time. Such is the nature of telephone traffic.

Mail can be sent over telecommunication networks. We have already reached the time when it is cheaper in certain circumstances to send mail electronically than to send it by conventional methods. The cost of electronic mail will drop substantially whereas the costs of typing, addressing, delivering, receiving, opening, distributing, and filing paper mail are rising. Mail delivery is not "real-time." We are happy if it is delivered an hour, or a day, later. We write letters, leave messages, send telegrams, order catalogues, transmit batches of computer data, and request books from libraries. This information transmission has two important characteristics. First, it can wait until channels are not occupied with telephone or other real-time traffic. Second, it can be interrupted in the middle of transmission provided that the interruption is done in such a way that no information is lost.

Approximately 70% of all *first class* mail in the USA is originated by computer. Most of this, invoices, orders, receipts, payments, etc., is destined to be fed into another computer, often in another firm. It should be transmitted directly in alphanumeric form. Instead it is usually printed, bursted, fed in envelopes, sent to a mail room, stamped, sorted, delivered to Post Office, sorted again, delivered to the destination Post Office, sorted again, delivered to corporation, handled in the mail room, opened, and laboriously keyed into a medium which the receiving computer can read. All this, when a packet sent on a value-added network costs a small fraction of one cent.

The most efficient way to utilize communication channels is to organize them so that real-time and nonreal-time traffic can be intermixed, and so that real-time traffic has absolute priority over nonreal-time traffic. Nonreal-time traffic should never delay the real-time traffic for more than a small fraction of a second—50 milliseconds, say. Most telephone networks today do not carry interruptable nonreal-time traffic, and consequently 75% to 85% of their total daily capacity is unutilized. When digital equipment is used for multiplexing voice signals, a variation of that equipment can accommodate the intermixing of real-time and nonreal-time traffic. If the mechanism exists for doing it, a remarkably large capacity for mail and other nonreal-time traffic is available. It is worth looking closely at society's nonreal-time uses of information, to see how the unused transmission capacity could be employed.

DIGITIZED MESSAGES

It is possible to convert any type of message into a digital form for transmission. Different messages require different numbers of bits. Box 15.1 gives some examples of approximate message lengths when converted to bits, and compressed ready for transmission or storage. The figures are conservative in that a smaller number of bits could be used in some of the items if complex encoding techniques are employed. Telephone speech, for example, has been economically encoded into far fewer than 32,000 bits per second with a level of distortion acceptable for one-way messages. A vocoder is a device which synthesizes speech from encoded parameters describing the speech-production mechanism and the sounds it is making. Vocoder telephone speech, listed in Box 15.1, can be designed to make the spoken words intelligible but not necessarily make the speaker's voice recognizable. Using vocoder techniques, spoken messages can be transmitted in one tenth of the number of bits of speech using PCM or delta modulation.

Much more compaction can be achieved with still less natural sounding speech if the messages are composed of prerecorded words. A prerecorded vocabulary might have up to a thousand words, each addressed with 10 bits, thus permitting a 30-word message to be sent with 300 bits plus addressing and error-detection bits. After transmission, the 300 bits would trigger a spoken message from a voice response unit (many of which are in use today). A code book of words commonly used in business was employed for a similar purpose in the early days of telegraphy.

Varying degrees of compaction can be achieved with other types of message delivery. There is a trade-off between message length and encoding complexity.

A digital transmission system with the high bit-rate of today's PCM telephone links can be given the capability to transmit any of the message types in Box 15.1, the nonreal-time messages fitting in gaps between the real-time traffic. The signals will be interweaved in the bit stream with whatever is the most appropriate form of multiplexing.

BOX 15.1 Numbers of bits needed for different message types.

Message type	Bits
1. A high-quality color photograph	2 million
2. A newspaper-quality photograph	100,000
3. A color television frame	1 million
4. A picturephone frame	100,000
5. A brief telephone voice message (voicegram)	1 million
6. A vocoder telephone voice message	100,000
7. A voice message of codebook words	400
8. A document page in facsimile form	200,000
9. A document page in computer code	10,000
10. A typical inter-office memo	3000
11. A typical flip chart	1000
12. A typical computer input transaction	500
13. A typical electronic fund transfer	500
14. A typical telegram	400
15. A typical airline reservation	200
16. A coded request for library document	200
17. A fire or burglar alarm signal	40

PRIORITIES

It is necessary to have some form of priority structure in the system. At its simplest there could be two priorities: *real-time* and *nonreal-time*. There are, however, different degrees of urgency in the nonreal-time traffic, so several priority levels may be used to help ensure a fast delivery of messages which require it. The system organization may be designed to permit the following categories of end-to-end delivery time:

1. Almost immediate (as with telephone speech).
2. A few seconds (as with interactive use of computers).
3. Several minutes.
4. Several hours.
5. Delivery the following morning.

When more than one message is waiting for transmission at any point, the higher priority messages will be sent first. The fact that much of the traffic is not in the highest priority category will make it possible to achieve a substantially higher line utilization than on a network which guarantees real-time transmission for all messages. In addition the facilities will be well utilized at night, when on other systems they would be largely idle.

TRAFFIC GROWTH

If transmission systems come into existence for transmitting nonreal-time information at a low cost, there can be a major growth of such traffic, and the growth will probably incorporate traffic which is not sent electronically today.

Table 15.1 is a forecast from a NASA study showing demand trends for data record transmission. In addition to data records, there are other types of traffic of very high volume, such as mail.

Table 15.1 Demand trends for transmission of records *(from a study of satellite uses commissioned by NASA [1])*

		1950	*1960*	*1970*	*1980*	*1990*
Stolen vehicle information transfer	cases/yr × 10^3	160	320	820	1950	4600
Facsimile transmission of "mug shots," fingerprints, and court records	cases/yr × 10^6	2	4	7	13	25
Stolen property information transfer	cases/yr × 10^3	430	880	1700	3500	7000
Motor vehicle registration	items/yr × 10^6	49	74	110	164	245
Driver's license renewal	items/yr × 10^6	38	48	60	75	90
Remote library browsing	accesses/yr × 10^6	0	0	low	5	20
Remote title and abstract searches	searches/yr × 10^6	0	0	low	8	20
Interlibrary loans	books/yr × 10^6	...	...	low	40	100
Remote medical diagnosis	cases/yr × 10^6	0	0	20	60	200
Remote medical browsing	accesses/yr × 10^6	0	0	20	60	200
Electrocardiogram analysis	cases/yr × 10^6	0	low	20	60	200
Patent searches	searches/yr × 10^6	6	6	6.5	7	7
Checks and credit transactions	trans/yr × 10^9	11	25	56	135	340
Stock exchange quotations	trans/yr × 10^9	0	0	1	2	4
Stock transfers	trans/yr × 10^6	290	580	1200	2500	4900
Airline reservations	pass/yr × 10^6	19	62	193	500	1400
Auto rental reservations	reserv/yr × 10^6	0	low	10	20	40
Hotel/motel reservations	reserv/yr × 10^6	...	...	25	50	100
Entertainment reservations	reserv/yr × 10^6	...	...	100	140	200
National Crime Information Center	trans/yr × 10^6	0	0	6	20	70
National legal information center	trans/yr × 10^6	0	0	low	5	30

ELECTRONIC MAIL

The *total* cost of mail delivery is gigantic, especially in North America. Americans are not only the most communicative people by telephone; they also receive the most mail. More mail is sent in New York City than the whole of Russia.

Some types of mail could be sent and delivered by electronic means, and where the volumes are high this could be done at a fraction of the cost of manual delivery. To send a handwritten letter electronically it is fed into a facsimile machine, transmitted, and received by another facsimile machine which produces a copy. Most of today's facsimile machines transmit an analog signal over telephone lines. They can be designed to transmit a digital signal, and some digital facsimile machines are in use.

Box 15.2 breaks down the U.S. mail by type. The asterisks indicate which mail could be sent by electronic means, and hence potentially by digital channels. A single asterisk refers to mail which could be delivered electronically to the end user. It is assumed in the table that individual households can neither send nor receive electronic mail. They have neither the equipment nor the desire to change their mail-sending habits. At some time in the future electronic mail will reach into consumers' homes, but we will make the conservative assumption for this chapter that for the time being only businesses and government will use it. When government and businesses send mail to households, this mail could be delivered to the local post offices already sorted for delivery. All local post offices could have a receive-only satellite antenna on the roof (like the Musak antenna) and a high-speed facsimile printer. Advertising letters and promotional materials have not been included as potential electronic mail because they may contain glossy or high quality reproductions. Some advertising letters could be sent by facsimile machines. Newspapers and magazines have not been included although there has been much discussion of customized news sheets being electronically delivered to homes.

On this basis, 22.7% of all mail is potentially delivered to end users by telecommunications, and a further 22.8% is potentially deliverable to post offices. In 1980 this will be a total of about 50 billion pieces of mail per year. The 50 billion pieces of digitizable mail would require on average approximately 200,000 bits each to encode; some would need more than this; many would require less because alphanumeric/encoding rather than facsimile would be used. The annual total would be roughly 50 billion $\times$ 200,000 = 10^{15} bits.

CONTROL MECHANISM

It is possible (some would say almost certain) that the control mechanisms necessary to interleave telephone and nonreal-time traffic will not be built as part of the traditional common carrier systems. We are, however, embarking on the construction of new types of networks—value-added networks, satellite systems, government and defense networks, and private industry networks. The designers of these networks, seeking to maximize the utilization of their facilities, are likely to intermix real-time and nonreal-time traffic in a way traditional telephone systems do not.

Figure 15.2 Facsimile machines.

The Xerox Telecopier 200 receives and transmits over telephone lines without an attendant. It used plain paper, and transmits a page in 2 minutes.

The 3M VRG Remote Copier can send and receive simultaneously over public telephone lines. It answers the telephone, receives, and disconnects, automatically for unattended operation. It transmits a 8½ × 11″ document in 4 or 6 minutes, depending upon the resolution selected.

BOX 15.2 **The composition of the U.S. mail. The red bands indicate which mail is potentially deliverable by telecommunications.**

Type of Mail		Percentage	
Individual households to:			
Business		5.8	
Individual households		14.0	
Government		0.4	
TOTAL			20.2
Government to:			
Business*		1.8	
Individual households**		3.8	
Government*		0.6	
TOTAL			6.2
Business to business:			
To suppliers*		3.9	
Intracompany*		1.4	
To stockholders*		0.7	
To customers: order acknowledgement*		0.2	
bills*		6.7	
product distribution		1.3	
promotional materials		5.4	
Other*		6.2	
TOTAL BUSINESS TO BUSINESS	25.8		
Business to households:			
Letters:			
Bills*		10.1	
Transactions**		1.2	
Advertising		12.6	
Other**		4.5	
TOTAL LETTERS		28.5	
Postcards:			
Bills**		0.7	
Advertising**		2.1	
Other**		0.4	
Newspapers and magazines		13.6	
Parcels		1.3	
TOTAL BUSINESS TO HOUSEHOLDS	46.7		
Business to government:*		1.2	
TOTAL BUSINESS			73.6

*: Potentially deliverable by telecommunications to the end user (22.7%).
**: Potentially deliverable by telecommunications, sorted, to a post office (22.8%).

Particularly interesting are the satellite networks now being designed and discussed. A multiple-access satellite, designed to transmit telephone and other traffic between many earth stations, can operate most efficiently in a digital fashion. (Part III of the book describes the technology.) With today's equipment one voice call typically requires 32,000 bits per second. If a satellite system were designed to carry a peak traffic of 100,000 voice calls, the total capacity of the satellite would be $100{,}000 \times 32{,}000 \times 60 \times 24 \times 365 = 1.68 \times 10^{15}$ bits per year. If 80% of this capacity were unutilized by telephone traffic because of the uneven traffic distribution shown in Fig. 15.1, as would probably be the case, then 1.35×10^{15} bits per year would be available for nonreal-time traffic.

In other words such a satellite system could carry all of the above 1980 digitizable mail, the required channel capacity being an unused byproduct of a telephone satellite. However, if only 1% of such mail were transmitted it would pay for a large satellite system. The practicality of satellite mail is aided by the fact that three quarters of all U.S. mail originates in only 75 cities, and only 20.2% of all mail originates from individuals—the rest is from business and government.

ELECTRONIC FUND TRANSFER

Nonreal-time digital transmission capacity will be required in the future for the transfer of funds between banks and other institutions. The Federal Reserve Board has made it clear that electronic fund transfer is essential for America, if only to halt the growing burden of paperwork such as check processing.

About 30 billion checks per year are written in the United States representing \$20 trillion per year. An electronic fund transfer network could speed up the clearing time for checks by at least one day on average; probably more. This represents a float of

$$\frac{\$20 \text{ trillion}}{365} = \$54.8 \text{ billion, savable by electronic check transfer.}$$

At 8% interest this gives a saving of \$4.38 billion per year.

If one check requires 500 bits for transmission, the total capacity needed is 30 billion $\times$ 500 = 1.5×10^{13} bits per year—a little more than 1% of the spare capacity in a 100,000-voice-channel satellite.

The number of credit transactions is almost double the number of checks, and the payment delay with these is much longer. In the electronic-fund-transfer society, when a customer makes a payment in a store or restaurant with a machine-readable card, a transaction would travel to a bank and a response would be received from the bank computer. If *all* credit card transactions in the United States in 1980 were handled this way, 200 billion such messages would be needed, or about 10^{14} bits per year. In fact such transactions will still be a small proportion of the total by 1980 and most will be to local banks on local telephone loops, but the quantity of long distance credit transactions will be growing rapidly.

FASTER-THAN-MAIL TRAFFIC

Today, message delivery which is faster than mail costs substantially more than mail. As the cost of long distance telephone calls has dropped, the number of telegrams sent has steadily declined. In many cases today it is cheaper to telephone than to send a telegram.

With digital transmission links, however, the relative costs of transmitting telegrams and telephone calls change. A typical telephone call lasts four minutes, and with telephone company PCM requires about 15 million bits transmitted in both directions. A typical telegram requires 2000 bits. Furthermore, the telegram can wait to be fitted in gaps on the real-time traffic that would otherwise be unused.

Even with such a dramatic cost difference most corporate communications users would probably still use the telephone because of its convenience, friendliness, and the cheapness of the instrument. If computerized corporate telephone exchanges (PABXs) place controls on user's communication expenditures, then digital transmission may bring new life to telegraphy.

TELEPHONE MESSAGES

The total number of telephone calls in North America is far higher than the total number of written messages. AT&T alone plans a capital expenditure over the next ten years more than 20 times higher than the likely capital expenditure in the U.S. Postal Service. Telephone callers are often greeted with busy signals or no answer and many of these callers would leave a brief message if they could. 32 long distance calls out of 100 are not completed today [2]. They receive busy signals, no answers or equipment failures. Of the business calls which are completed, on only 35 per cent does the caller reach the called party. It is estimated that this wastes 200,000 man-years of callers' time, which at $10,000 per year is equivalent to $2 billion [2].

A one-way telephone message could be digitized and stored so that the person it is sent to can retrieve it at his convenience. It could be transmitted and stored in any of the three forms mentioned earlier:

1. True-to-life :~ 32,000 bits per second
2. Vocoder : ~ 2,400 bits per second
3. Words from a prerecorded vocabulary: ~ 10 bits per word

Such a service could be designed so that subscribers could leave either a coded telephone message from a list of such messages, dialed on a conventional telephone, or they could leave a brief spoken message. The system would ring the called party periodically until it could speak the message. The called party could use his telephone dial to ask for repetition of the message, give confirmation of its receipt, or dial a response. The system may be designed so that a user can dial his stored message queue from any telephone, key in a security code, and have the messages spoken to him.

If 10% of all business telephone callers left such a message when they failed to complete their call, this would be approximately 100 million messages per year. If 10% of all business callers who failed to contact the individual they telephoned also left such a message, that would be about a billion messages per year.

Telephone messages could be sent for which simple responses are required. The receiver would dial the response on his telephone after a local computer speaks to him and the computer would receive the response and deliver it. The voice answerback unit would inform the called party what form of response was expected.

Organizations could send bulk messages in this way, in which one telephone message is sent to many individuals. Unsolicited messages could be composed by computers for verbal delivery, possibly expecting a response. One can imagine such a system being programmed to carry out opinion polls or gather statistics from individuals.

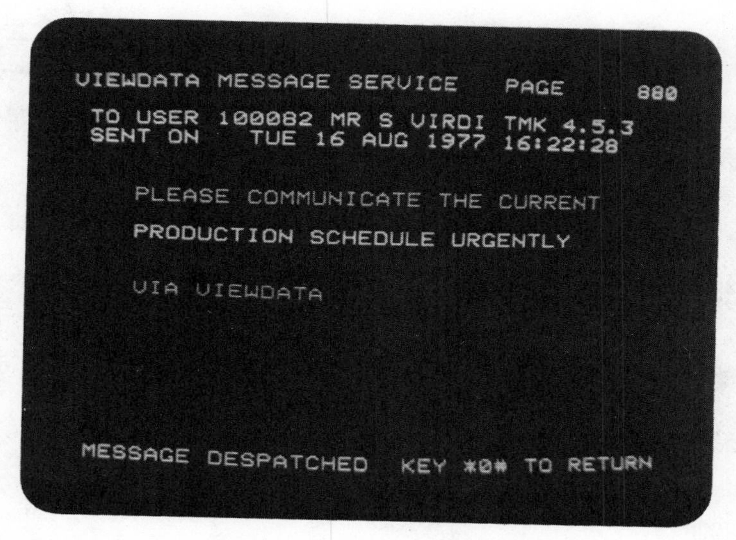

The British Post Office Viewdata service permits messages to be sent to and from domestic or business television sets. A light on the set would indicate that a message is waiting. See page 322. *(Photo courtesy British Post Office.)*

REFERENCE

1. Roger W. Hough, *Information Transfer in 1990,* AIAA 3rd Communications Satellite Systems Conference, Los Angeles 1970, presenting work supported by NASA Contract NAS2–5369.
2. Harry Newton, *Communications Lines,* Business Communications Review, Sept., 1976.

Telephone company building in Stockholm in 1897. The early telephone companies had humbler beginnings than the new common carriers of today. *(Courtesy L. M. Ericsson.)*

16 NEW BREEDS OF CARRIERS

In the 1970s the telecommunications industry of the United States changed from being a somewhat placid growth industry to an industry in upheaval. Changes in the regulatory climate discussed in Chapter 22 encouraged three new breeds of common carriers to come into existence:

1. The specialized common carriers
2. The satellite common carriers,

and 3. The value-added common carriers.

All three compete with the old-established carriers to provide users with different or cheaper communication facilities or with better service of various types. All three offer the classical American entrepreneur the opportunity to make a fortune. However all three are subject to regulatory risks rather than merely business risks. If there were a change in the regulatory climate, the giant established carriers could make it very difficult for the new carriers to exist.

Major legal battles and regulatory slowness have impeded the new carriers, and in some cases acted as a disincentive which has prevented new technology or operations from playing the part they should. Nevertheless the new carriers have a flexibility and inventiveness that permit them to move in a nimble fashion while the old large carriers are bogged down by vast capital investments, obsolete equipment, huge pension commitments, massive bureaucracy and inappropriate depreciation policy. New management taking control of one of the old carriers publicly described the job of adapting it to modern electronics as "like trying to turn around an elephant in a bathtub." Offset against this is the superb research achievement of the largest common carriers. Probably no other industrial laboratory in man's history has a track record of valuable innovations as spectacular as that of Bell Laboratories.

THE SPECIALIZED COMMON CARRIERS

In the early 1970s a new breed of common carriers developed which did not install telephone or telegraph units but provided specialized transmission facilities. The would-be "specialized" carriers seized upon the rigidity of the United States telephone system claiming that separate microwave and other transmission links were needed for specialized purposes, including business telephone and data transmission. In 1970, the FCC, summarizing the conclusion of a lengthy enquiry into computers and telecommunications stated that there was

> dissatisfaction on the part of the computer industry and by many data users who had been attempting to adapt their requirements to existing (communications) services.

In 1971, the FCC commissioners voted favorably on the concept of specialized carriers, saying:

> The entry of new carriers would have the effect of dispersing somewhat the burdens, risks and initiatives involved in supplying the rapidly growing market for new and specialized (communications) services.

This sudden injection of competition into an industry which for decades had little competition, has been dramatic in its effect. Many aspiring new carriers triggered new tariffs and service offerings (such as Bell's DDS) from the old carriers. It is possible that the most major effect of the new carriers will be edging the massive old carriers into competitive response.

There are major economies of scale in telecommunications and the large carriers have often quoted this fact to say that it is not economically beneficial for the country to have small carriers operating on similar routes. Why, then, can the specialized carriers charge lower rates? The first reason is that the old established carriers have tended to install high-quality but expensive equipment. The formula regulating their profits tends to encourage high capital expenditure. A typical microwave tower of a specialized carrier is much lower in cost than a typical Bell microwave tower.

The second reason is that there is a certain range of traffic volumes, which are handled by microwave, for which economics of scale do not apply. This can be seen in Fig. 20.1, as a kink in the microwave curve between 4000 and 10,000 circuits. Beyond a certain traffic volume a major increase in equipment is needed. This is a very temporary reason. It is true, and likely to remain true, that the cost per call on a trunking system handling 100,000 calls or more is substantially less than on a microwave route handling less than 10,000 calls.

A third reason, and perhaps the most important, is that new carriers are better able to introduce new technology such as satellites, small private earth stations, new switching techniques, CATV interconnections, and so on. The new carriers can also introduce new *services* competitive with the old.

It has been suggested by the Justice Department that AT&T Long Lines should be broken away from the Bell Telephone operating companies. If this happened, there could be open competition in long-haul transmission.

In view of the economies of scale and advantages inherent in an integrated technology the final argument in favor of the specialized carriers probably has to be a belief in the ultimate virtues of competition, the low overheads of small corporations, and the initiative of private entrepreneurs.

LOCAL LOOPS

While it is possible to compete with the giant telephone companies in long-haul transmission, it would be difficult to compete with them in local signal distribution—the copper wire pairs which go into homes and offices. Because of this the telephone companies will have less immediate competition in home telephone distribution, although there is great scope for the introduction of new home telephone instruments, conference devices, intercoms, telephone answering machines, and so on.

The specialized common carriers all have a problem in local distribution. Today the local distribution usually has to be done by having an interconnect agreement with the telephone company. This works satisfactorily but limits the link capacities to those of the telephone loops. It therefore precludes very high speed data transmission or video applications. To achieve these valuable applications the specialized carriers could forge links with CATV companies, or build new types of local distribution links including short high-capacity radio links at frequencies above 10 GHZ.

Most specialized carriers are concentrating on voice transmission and data transmission up to 9600 bps because this is where most of the business revenue exists, and telephone interconnect agreements are satisfactory.

An AT&T tariff specifies what transmission facilities it will provide to other common carriers. In particular, it specifies characteristics of local loops. There are three grades of wire pair line specified suitable for digital transmission:

Type W1: 2400 or 4800 bps

Type W2: 9600 bps

Type W3: 56,000 bps

The tariff also specifies voice grade and television grade interconnections.

MCI COMMUNICATIONS CORPORATION

The pioneer and pace-setter of the specialized carriers was MCI. In 1969, after six years of legal battling, (the telecommunications lawyers are making a fortune), the FCC gave MCI permission to build a microwave system between St. Louis and Chicago. This historic decision triggered a flood of 1900 new microwave station applications.

MCI was soon a $100 million group of corporations building a nationwide microwave network, and selling a wide variety of bandwidths to any customer who could use them. MCI's prices, like those of the other specialized carriers, were just sufficiently lower than those of the established carriers to attract customers. The MCI network is shown in Fig. 16.1.

AT&T bitterly accused MCI of "cream-skimming," that is providing service only to those parts of the country where there would be maximum profit, whereas Bell had to provide a similar service to its entire geographical area. The FCC ruled that in order to enable new corporations to come into business they were indeed allowed to operate in selected areas.

MCI originally described themselves as a "nontelephone" common carrier. It soon became clear, however, that their main business revenue was to come from corporate telephone service and their marketing drive became oriented largely towards this.

A particularly interesting service of MCI is *Execunet*. This employs switching units which connect the public telephone channels to the MCI trunks. They enable users to dial calls from telephone company telephones, for example in their homes. These public calls travel over the MCI trunks and may again be switched at the far end to a public telephone subscriber. The cost of such a call may be a fraction of a toll telephone call. This service introduces competition to the toll telephone service which AT&T is trying to stop by legislation. The FCC ruled against Execunet.

Increasingly, MCI is moving into network management services, in which it takes over and manages a corporate communications network, sharing with the company any savings that accrue.

DATRAN

Whereas MCI built an analog trunking system intending to sell primarily analog capacity, Datran (Data Transmission Company) built a digital trunking system intended to serve primarily data processing users. (Datran ceased operations after this book had been set into print.)

Datran uses a microwave route shown in Fig. 16.3 on which the radio signal is digitally modulated by a very high data rate. This bit stream is submultiplexed into streams of 1.344 million bits per second, compatible with the T1 carrier. Users can obtain channels derived from this at speeds up to 1.344 Mbps, thus:

Figure 16.1 MCI network.

The microwave network of MCI Communications Corporation. MCI offers leased channels of a wide range of bandwidths, and many innovative user services.

The nerve center of the MCI System. All customer circuits can be checked from here. Constant monitoring of those circuits takes place.

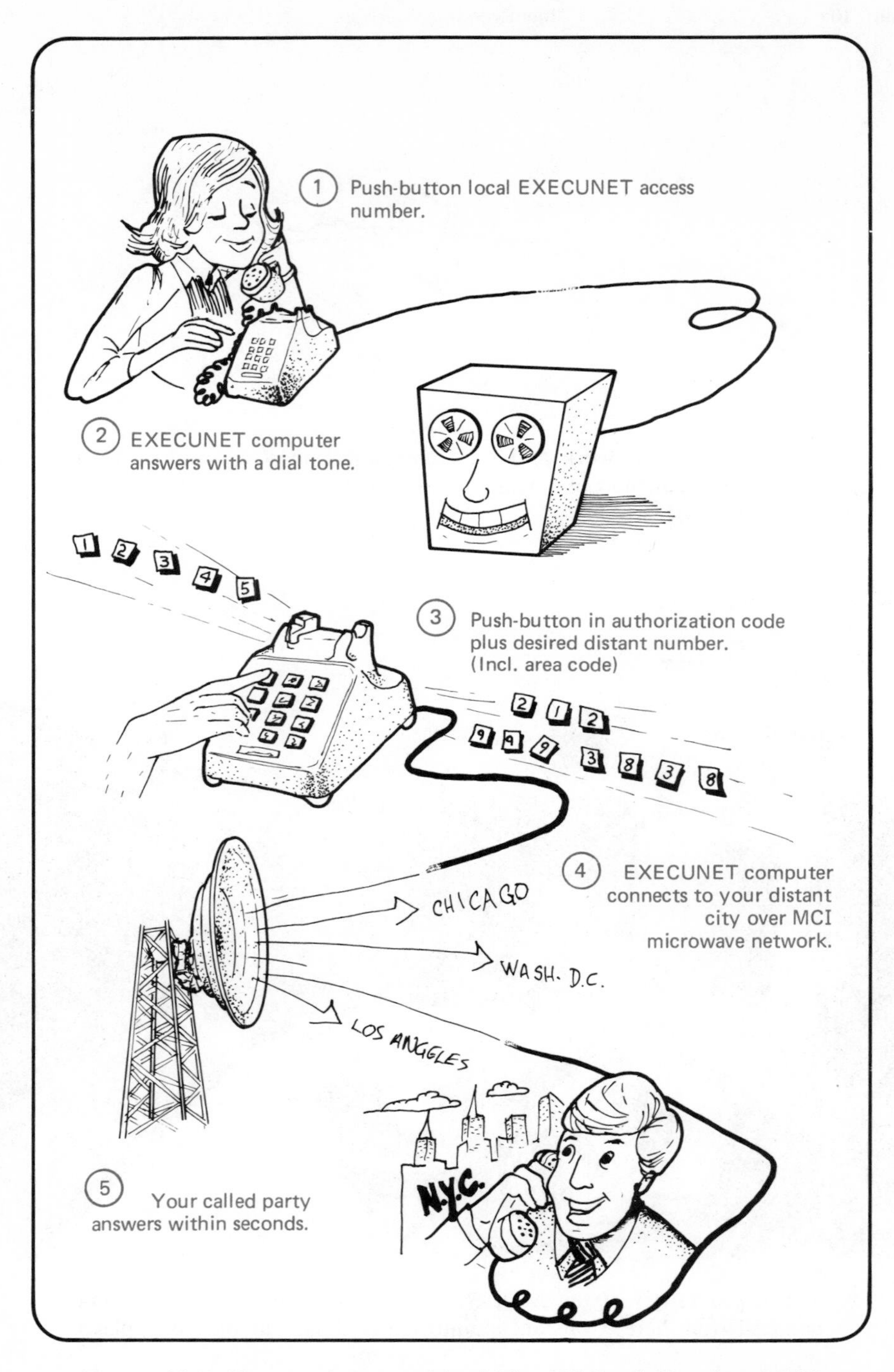

1 Push-button local EXECUNET access number.
2 EXECUNET computer answers with a dial tone.
3 Push-button in authorization code plus desired distant number. (Incl. area code)
4 EXECUNET computer connects to your distant city over MCI microwave network.
CHICAGO
WASH. D.C.
LOS ANGELES
N.Y.C.
5 Your called party answers within seconds.

Figure 16.2 Too much competition? The FCC ruled against MCI's Execunet service, illustrated in this MCI diagram. Execunet gave cheaper long-distance telephone calls from home or office telephones. During its brief existence it acquired many users and was very popular.

Figure 16.3 Datran operated a switched and point-to-point data network subdividing a 44 megabit microwave signal into channels of speeds usable in data processing. *(Datran subsequently went out of business and part of its network is now operated by Southern Pacific Communications.)*

asynchronous channels up to 1.2 Kbps
synchronous channels of: 2.4 Kbps
4.8 Kbps
9.6 Kbps
19.2 Kbps
56 Kbps
112 Kbps
224 Kbps
and 1344 Kbps

fraction of that on the telephone network. Because of the switching speed, such systems are referred to as a *fast-connect network*. The switching speed is reflected in the tariffs. The minimum time for which a user can be billed on the telephone network is 1 minute. The minimum billing time on this switched network is 1 second. Consequently if a user location has sporadic data messages to send, for example is using a remote computer interactively with an input and response every minute or so, then the charge to that user is much lower than with dialed or leased telephone lines.

When a single terminal user is employing a distant computer the network control unit, called Datalink, may redial the connection for each user input. The average connect time is 0.8 second. After the computer responds the link is disconnected. A minute or so later it may be redialed. The cost of such a connection is usually 1 cent—the minimum billing charge.

Most data transmission does consist of sporadic bursts of data as indicated in Fig. 10.2.

Because the channels are digital end-to-end, the users should not need modems (which must be used on analog telephone lines to convert the data transmitted in analog form). Southern Pacific like other specialized carriers, has to do its local distribution over telephone loops (wire pair cables). If the loop used is short enough to have no repeaters or load coils then it can be used digitally, and it needs a line driver, not a modem. A line driver transmits and receives suitably encoded digital signals on the wire pair. If, however, the user is a long way from the nearest network office he will have to be connected by means of a leased telephone trunk circuit. In this case a modem is used, capable of transmitting at 9600 bps. The modem or line driver is housed in an identical-looking unit which resides near the user's terminal or computer.

Probably the main attraction of this network is its lower cost especially for the brief 1 cent calls transmission. The channels have several other advantages over telephone channels, however:

1. Full duplex transmission. The channels transmit in both directions at the same time. Dialed telephone channels are half duplex, i.e., they transmit in one direction or the other but not both at once.
2. No modems on some end-to-end links.
3. An attractive easy-to-use control unit giving abbreviated dialing at 0, 1, 2, or 3 digits in addition to full 7-digit dialing (Fig. 16.4).
4. Fast connection. Average call set-up time 0.8 seconds.
5. *Guaranteed* good error rate. 99.95% of all call seconds will be error-free.
6. Higher speeds. Leased channels up to 1.344 Mbps are offered.
7. Automatic or manual dialing.

Figure 16.4 The customer key pad and test module of the Southern Pacific system.

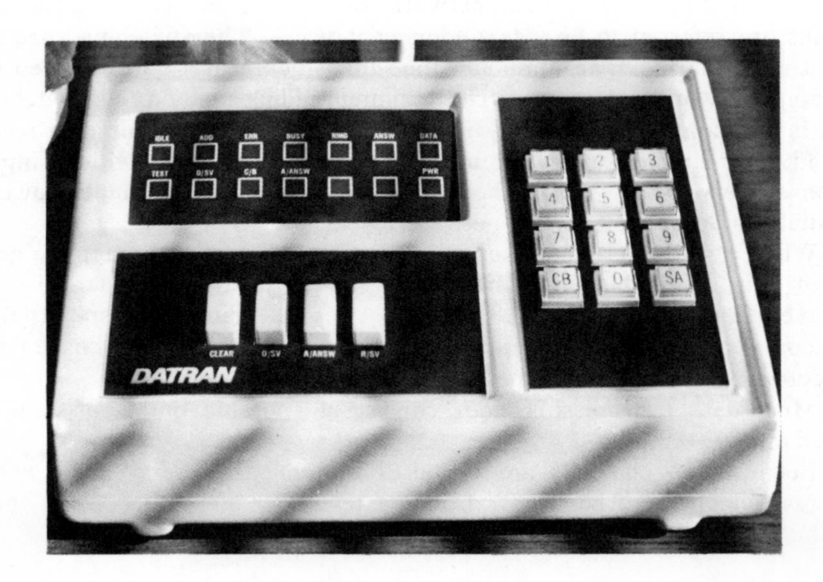

8. Built-in security features. Incoming calls from an unauthorized terminal are not accepted.
9. Automatic circuit testing during call set-up, and loopback testing of all services.
10. No minimum monthly usage charge (like WATS) but there is a maximum monthly usage charge. (See the datadial curve in Fig. 16.7.)
11. No blocking, in that calls which cannot be switched instantly are automatically queued.
12. Detailed billing reports.

In the mid-1970s, AT&T introduced DDS, their Dataphone Digital Service, which uses the digital network shown in Fig. 10.4. This provides users with digital channels which, at least initially, are not switched and operate at speeds of

2.4 Kbps
4.8 Kbps
9.6 Kbps and
56 Kbps

AT&T has the capability to introduce additional equipment which would make DDS more suitable for burst traffic, such as concentrators, packet switching, or fast (computerized) circuit switching. It is possible that they will do this on a nationwide scale before long. As the telephone companies convert their networks to PCM operation, and as the satellite operators use digital transmission, a tremendous digital transmission capacity will become available.

Initially, both Datran and DDS channels are available only in certain geographical locations. Often a potential user cannot obtain a local channel of such a network going into his premises. His alternative is to use a telephone connection to the nearest access point of the digital network. When this is done, however, some of the above advantages are lost. If he has a telephone connection at both ends of the otherwise digital link he needs a total of four modems whereas he would have only needed two if he had used a telephone line all the way. And he is restricted to telephone speeds.

It can be expected that the digital networks will grow and become more widespread so that increasing numbers of users will be able to connect to them directly.

OTHER SPECIALIZED COMMON CARRIERS

MCI and Datran were two of the earliest specialized carriers. There are now several others competing to form the core of this new segment of the industry, notably ITT. As is always the case when a new industry grows rapidly there will probably be many mergers and shakeouts.

The specialized carriers can be subdivided into those offering a *switched* service, and those offering only point-to-point links. There is more revenue as-

They can also be subdivided into those offering a *switched* service, again like Datran, and those offering only point-to-point links. There is more revenue associated with voice transmission than data transmission, so that majority of specialized carriers with terrestrial microwave facilities have designed them primarily for analog transmission. Data can be sent over them by using modems.

SATELLITE CARRIERS

A particularly interesting form of specialized carrier is that constructing a satellite network. Several new carriers in addition to existing carriers have been given permission to operate a satellite network in the United States.

Several new corporations obtained permission to become satellite carriers, but so far the domestic satellites have been launched by existing organizations such as Western Union, RCA, and Comsat. The Canadian ANIK satellite—the first domestic satellite—and later Canada's CTS satellite, were planned and launched by the Canadian government. SBS, Satellite Business Systems, a subsidiary of IBM and Comsat General and the Aetna Insurance Co., plans to launch satellites which permit earth stations on corporate premises in 1979.

Satellite carriers, like the other specialized carriers are restricted by the local distribution facilities. A high bandwidth signal could be sent via the transponder but cannot be delivered to the end user in most cases because the local wiring going to the end user is telephone copper wire cables organized so as to restrict the user to telephone bandwidths.

Satellite carriers can bypass the local loop bottleneck by taking the signal directly to some subscribers, giving them an antenna on their premises. As we discussed in Chapter 13, this will be much more practical with satellites operating at the new frequency band of 11–14 GHz which permits small earth stations that are not interfered with by terrestrial microwave. When 11–14 GHz satellites are in orbit there is probably a great future for the satellite carriers.

THE VALUE-ADDED COMMON CARRIERS

In 1975 the industry took another new turn. The first "value-added" common carriers (VACCs) became operational. A value-added carrier does not construct any telecommunication links. Instead it leases links from other carriers and creates a "value-added network" (VAN) with sophisticated computer control to provide new types of telecommunication services.

The first value-added common carrier to receive FCC approval was Packet Communications Inc. PCI intended to build a network providing packet-switched data communications. PCI later went out of business.

The second FCC approval for a value-added common carrier went to Graphnet Systems Inc., who intended to provide a network to interconnect facsimile machines, and to transmit from computers, telegraph machines, and data terminals, to facsimile machines. A Graphnet subscriber can deliver mail or

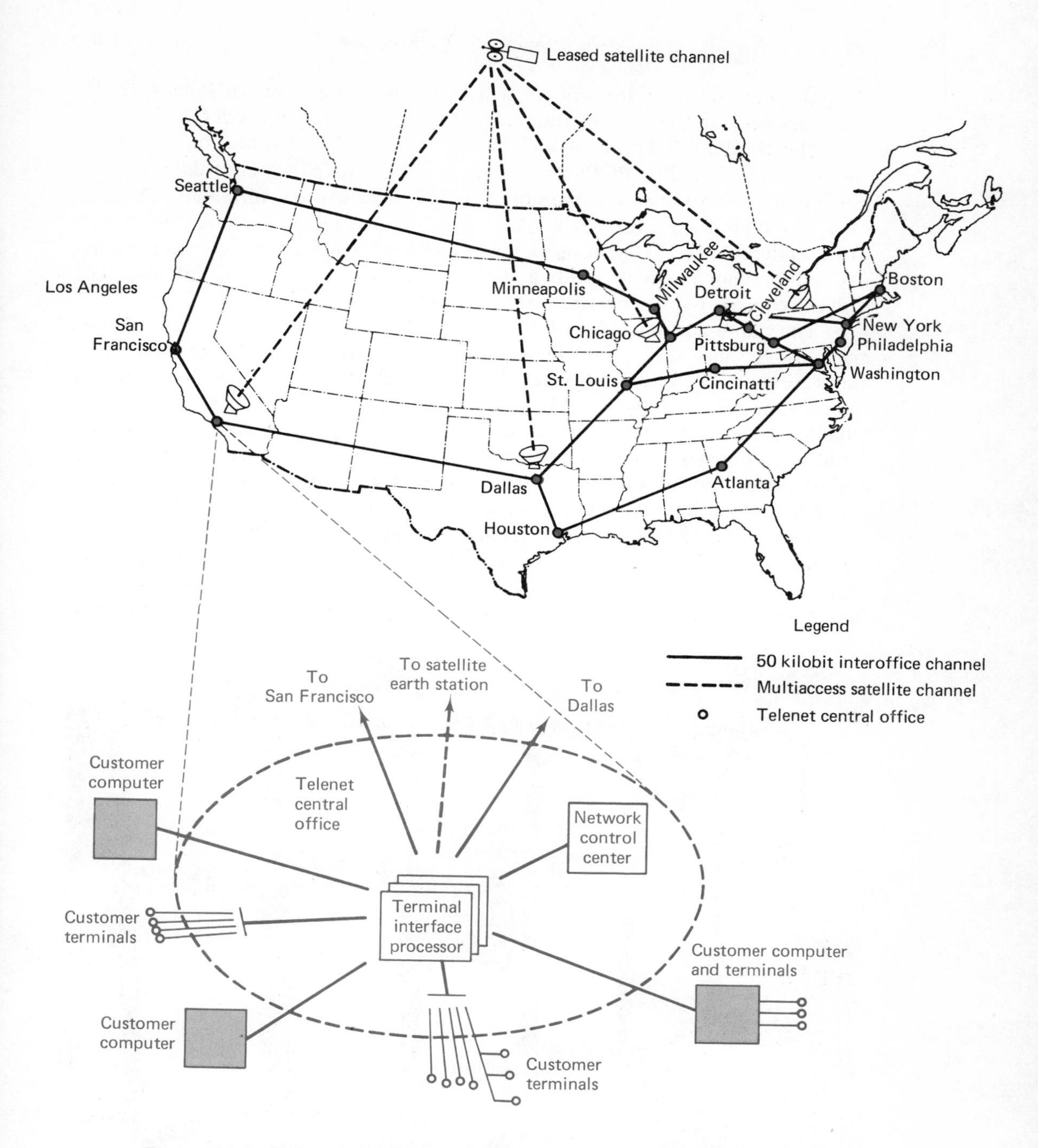

Figure 16.5 Telenet's rapidly changing packet-switching network—one of the first "value-added common carriers." *(Reproduced from the Telenet FCC filing, 1973.)*

computer data to other subscribers with facsimile equipment. In January 1975, Graphnet became the first operational value-added common carrier.

The third FCC approval was for the Telenet Communications Corporation network. This, like PCI's proposal, was for a packet-switched data network providing fast-response transmission among computers and terminals. The network is shown in Fig. 16.5. It was based on the previously existing ARPA network, built with Department of Defense (ARPA) funds to interconnect university computing locations and a few other research establishments. The techniques used by Telenet are discussed in Chapter 29.

The value-added networks take advantage of economies of scale. When many users share the same wideband communication channels they can communicate more cheaply. The wider the bandwidth they share, the faster the response time of the network. Also the communication path can be made more reliable, because the network can be designed to bypass failures. Users who could not otherwise communicate, because they have incompatible machines, can be connected. Graphnet will interconnect incompatible facsimile machines, and Telenet will interconnect incompatible computers.

The value-added common carriers do not *process* the messages they transmit and store. If processing takes place then the service is a computing or time-sharing network, not merely a common carrier. They may, however, enhance the messages by actions such as adding a corporate letterhead and logo to

Figure 16.6 Equipment in a typical Telenet central office.

messages sent in alphanumeric form from the telegraph machine or computer, and delivered in facsimile form. Similarly signatures may be added to certain documents if tight security controls are used. A facsimile delivery system may add a form to data created by a computer as though it were filling in that form.

Networks such as the ARPA network permit not only the communication facilities to be shared, but also the computing resources. Users far away can share a computer, a data bank, or other facilities. The networks are thus *resource-sharing* networks in two senses. Appropriate protocols and software have to be made available to make the sharing possible.

Many different types of "value-added" services are possible and we may see this breed of carrier develop in new directions. Any of the message types in Box 15.1 could become handled by such carriers. A carrier which transmits and stores facsimile messages could also transmit and store voice messages. A carrier which transmits packets of interactive computer data could equally relay telex conversations or provide a fast message delivery service. The more diverse types of traffic a value-added network handles the better, because of the economies of scale. Graphnet today delivers messages faster and cheaper than telegrams. If the traffic volumes were high enough it could handle mail at delivery costs lower than those of the postal service.

SERVICES

The value-added carriers are in the business of providing new *services* to society and industry which the traditional carriers do not provide. There are many possible opportunities for new types of service. Box 16.1 lists some examples.

The specialized common carriers can also offer new services such as setting up video conferencing studios, or transmitting bulk data at high speed if it is delivered to their office on a tape or disk. Using manual pick-up and delivery a file of say 10^9 bits could be sent from one computer center to another in an hour or so. A service that would be attractive to some corporations is the complete analysis of their communication needs, traffic monitoring, and the design of an optimum corporate network. "Turnkey" corporate network management could have great appeal if done in highly professional fashion with the latest equipment and tools. Several specialized carriers are moving into this business.

Much of the future of the specialized carriers may lie in their offering innovative *services* which the telephone companies do not supply.

HYBRID CARRIERS

It is likely that the specialized and value-added carriers will merge, to some degree, with the specialized carriers offering value-added services. To an increasing extent Datran, for example, is *leasing* digital transmission links from the telephone companies and adding its fast-connect switches to them to offer users a new pricing structure

BOX 16.1 Examples of services that might be handled by a value-added common carrier

Message delivery:	Telegrams Facsimile Mail Interactive computer data Batch computer data Interconnecting incompatible data machines Interconnecting incompatible facsimile machines One-way voice messages Monetary transfers Traffic when a bank card is used
Broadcasting:	Data broadcasting (like Britain's Ceefax but for different reception devices) Weather and marine forecast services News broadcasting by data, voice or video Financial information services Music delivery
Message enhancement:	Adding forms to computer data Adding corporate logos and letterheads Adding signatures under tight security control Adding form letters to a transmitted recipient's name Message editing Word processing functions
Message storage:	Document filing services Secure storage service for vital records or audit trails
Message retrieval:	Library services Information retrieval and search services Financial information services Data bank services Newspaper morgue searching Music library

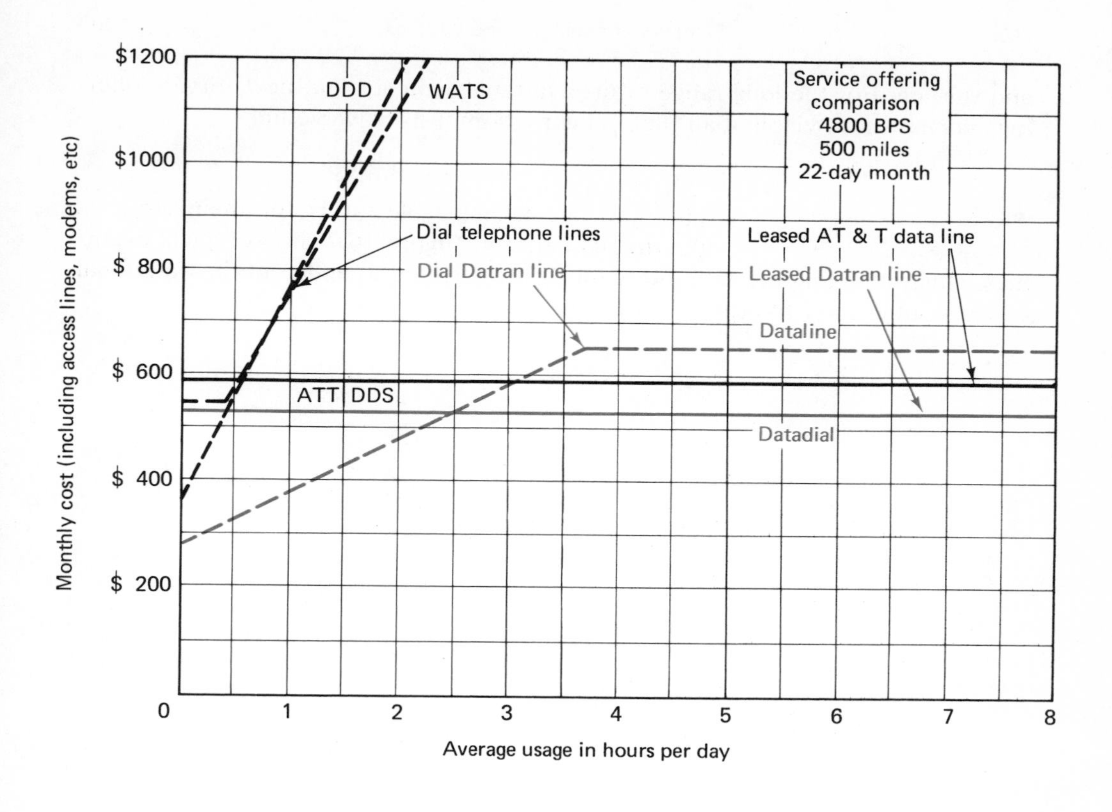

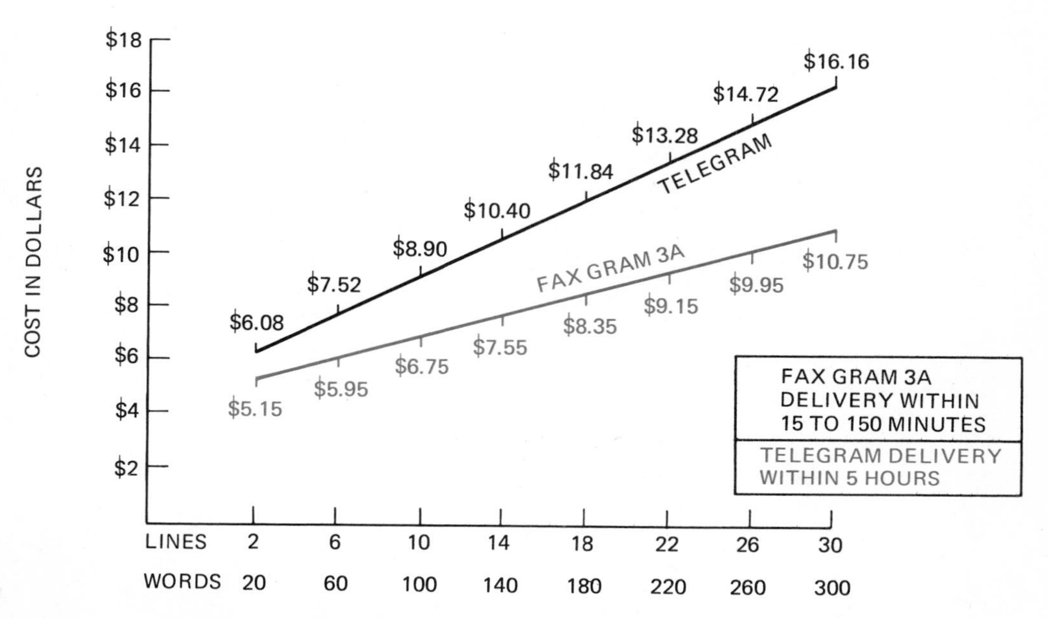

Figure 16.7 Typical cost differences between the new and the traditional common carriers. (Such costs are changing rapidly.)

and service. For the long-range future Datran's switches and new service offerings appear more viable than their construction of microwave links.

TARIFFS

The new carriers must file tariffs for their services, like the old carriers. Figure 16.7 shows typical examples, illustrating that prices in these tariffs are selected to undercut the traditional carriers sufficiently to attract customers.

PART II SYNTHESIS

Scaffolding for wires coming into telephone exchanges, London, 1909. (*Courtesy The Post Office, England.*)

17 CORPORATE COMMUNICATIONS, AN UNMANAGED RESOURCE

Telecommunications have a vital part to play in improving the productivity of industrial nations. They will become more important as industrial processes are increasingly run by machines. A growing proportion of human work will be concerned with handling information rather than operating tools or lugging items around a factory. The relative costs of physical transportation and telecommunications are rapidly changing. Physical transportation is becoming more expensive as fuel costs rise, while long-distance bandwidth is dropping in cost.

Most corporations today spend a substantial proportion of their money and human talent on communication of one form or another. Businessmen spend most of their day communicating. Most white-collar workers spend much of their time communicating with superiors, subordinates, customers, suppliers, secretaries, and computers. The information handling process typically costs from 5% to 30% of an organization's total expenses [1]. Given such a large expenditure, it is desirable to ask: how can corporate communications be made as effective and inexpensive as possible.

If we look at telephone communication alone there is much scope for lowering costs. Table 17.1 shows typical U.S. costs for traditional corporate

Table 17.1 Cost of corporate telecommunications as a percentage of total operating expenses

Industry	Range	Average
Airlines	3%–7%	4%
Banking and Finance	0.6%–4.2%	1.5%
Insurance	1%–3%	2%
Manufacturing	0.3%–2%	0.5%
Securities	8%–12%	10%

telecommunications, most of which is telephone traffic [1]. Large corporations spend many millions of dollars on their telephone facilities.

Surprisingly, about three-fourths of this expenditure relates to *internal* telecommunications. A widespread corporation can therefore profit by having an internal telephone network designed to minimize costs. Such a network is likely to incorporate leased lines, WATS lines (giving a fixed monthly charge to or from specified areas), lines from the specialized common carriers which are lower in cost than the telephone company lines, and switching arrangements to interconnect these lines. Some corporate networks now include leased satellite channels.

MINIMIZING THE NETWORK COST

Designing a minimum-cost corporate telephone network is a complex operation. In the United States it is much more complex now than in the 1960s because of the competitive tariffs and new service configurations. Among the actions today that will lower the telephone bill are:

1. Selection of an optimum mix of leased lines, WATS lines, satellite channels, privately owned microwave and millimeterwave facilities, and public facilities.
2. Selection of appropriate lines and services from the specialized common carriers.
3. Optimum geographic layout of the leased line network.
4. Selection of appropriate tandem dial and common control switching facilities.
5. Selection of equipment which will route *external* long-distance calls over the leased-line network.
6. Use of equipment that will automatically send calls by minimum-cost routes. cost routes.
7. Appropriate PABX selection.
8. Elimination of switchboard operators.
9. Use of the feature on computerized PABXs which prevents specified telephones from making specified long-distance or expensive calls.
10. Use of telephone monitoring facilities.

Surprisingly, in spite of the magnitude of telecommunication costs, most corporations exercise little control over them. There is often no attempt at corporate network optimization; the telephone facilities are acquired by local managers, who are told what to lease by the telephone company. Leasing a 500-line electromechanical PABX often required a million-dollar 5-year contract, but there was no technical study or high management involvement as there would be with other equipment of similar cost, for example, computers. Telephone expenditure has been taken for granted.

The new computerized PABXs and telephone monitoring equipment can do much to lower telephone bills as illustrated in Box 17.1. (Different regulations in different countries prohibit some of the techniques in Box 17.1.)

BOX 17.1 How to cut a corporation's telephone bill

USER MONITORING	• All users' telephone calls are listed so that users can be asked to explain them. Telephone misuse is eliminated. • Refunds are demanded for personal calls. • Excessively long calls are eliminated. • Management is able to promulgate new rules on telephone usage.
EXPENSE ALLOCATION	• Precise allocation of phone costs to departments and budgets. • Department managers are made responsible for their phone budgets. • Users dial a code to identify personal calls. • Users dial codes to indicate subject matter, for billing or budgetary control.
CLIENT BILLING	• Detailed bills for clients are prepared in service organizations such as law firms, architects, accountants, etc. permitting 100% billback of telephone costs. • Call dial codes are used to indicate client and subject matter for client billing.
AUTOMATION	• Elimination of telephone operator positions.
CALL RESTRICTION	• Specified extensions are prevented from dialing certain area codes. Geographical restriction for each employee. • Specified extensions can dial only within specified hours (e.g., 9:00 A.M. to 5:00 P.M.). • Users sent warning tones when calls exceed a specified time.

continued

BOX 17.1 ***Continued***

TRUNK MONITORING	• Quick detection of malfunctioning corporate trunks (which often go undetected for months). • Telecommunication facilities brought to maximum operating efficiency. • Refunds won from carriers for malfunctions. • Console attendant indication of heavily utilized trunk groups.
NETWORK REPORTS	• Computer-produced reports recommending adjustments to network facilities (including WATS lines, specialized common carrier lines, etc.) to handle existing traffic at minimum cost. Such reports are based on trunk monitoring. • User frustration reports summarize uncompleted calls.
LEAST-COST ROUTING	• Each call is routed by the least expensive route at that instant; e.g., first choice: tie-line; second choice: WATS; third choice: DDD. • Takes advantage of time zones to increase bulk facilities' utilization. • Permits a major increase in bulk facilities utilization. System automatically calls back user when a low-cost route is free.
PRIORITY SYSTEM	• Low priority callers are prevented from using expensive routes. • High priority callers are given a high grade of service – almost no busy signals. • Nonreal-time traffic sent when trunks are idle.

continued

BOX 17.1 Continued

OFF-PREMISES ACCESS TO CORPORATE NETWORK	• Use of corporate network gives executives inexpensive calls from home. • Reduces employee credit card calls. • Uses bulk corporate facilities which stand idle after hours. • Off-premises to off-premises calls are possible via corporate network.

DATA NETWORKS

In addition to telephone networks, corporations have networks for transmitting data. The volume of computer data transmitted in many corporations is growing at more than 25% per year, and will increase by a factor of 10 between 1970 and 1980.

Being controlled by the data processing staff, data networks are often well optimized, unlike the telephone network. However there are often multiple data networks forming parts of different computer systems which were designed separately. The combination of all data transmission facilities in a large corporation does not normally form an optimized whole. Some corporations are now in the process of linking together separate data networks to cut cost and increase throughput.

To control data networks, appropriate software and control mechanisms are needed. Unfortunately the different networks often used different line control procedures, and so to combine them the line control software and hardware had to be changed. Often the terminals had to be changed. Users and manufacturers attempted to standardize the network procedures to make future network growth and optimization easier.

Integrating the separate data facilities has several advantages. It can lower the overall cost. It can decrease the network response time if it is done in an appropriate fashion. It can increase the total reliability of connections. Some terminal users can have more remote computers and more data banks available to them.

However optimization of the data transmission facilities *alone* is suboptimization. What is really needed is optimization of *all* of the corporate telecommunications. Data transmission costs vary from 1% to 20% of the total corporate transmission costs. Because of economies of scale, especially with the newer technologies, it can pay to combine the facilities that are leased for voice and data traffic. As we have stressed, it pays to combine real-time and nonreal-time traffic.

In the future as voice digitization techniques spread, the combining of voice and data traffic will become increasingly important.

OTHER COMMUNICATION COSTS

Today, corporate telecommunications is generally thought of as being telephone and computer traffic, possibly with a few small extras such as occasional facsimile messages. The total cost of *communicating* in a corporation, however, is much greater than the cost of telephone and data transmission—typically between 5 and 10 times as much. Two major contributers to this cost are the sending of mail in a corporation and the cost of physical travel for communications. In many large corporations about three-fourths of these expenses are for internal communications (as with telephone expenses).

An average piece of correspondence in the United States has been estimated to cost $10 or more to be conceived, formatted, copied, transported, received, read, and filed [2]. This is much higher than the cost of making a telephone call. A very large quantity of memoranda and letters is sent within most corporations. Modern telecommunications enables us to ask a new question: Should not corporate correspondence be sent in an electronic form rather than in the form of paperwork. Electronic memoranda have two important advantages. First, given appropriate system design they can be cheaper. Second, they can reach their destination faster. The highest priority correspondence can reach its destination in minutes.

ELECTRONIC MEMORANDA

There are a variety of ways in which memoranda can be handled electronically:

1. Telegraphic Message Switching

Memoranda are typed into terminals like telegraph machines and delivered by a store-and-forward system. They are filed in the system, not by filing clerks.

2. Visual Display Message Switching

Memoranda are typed into screen units which permit easy editing of documents. Retrieval on screen units avoids paper handling. Screen units on executives' desks would be used for many other functions.

3. Magnetic Card Systems

Magnetic card typewriters permit easy editing and storage of documents, and may form the input to a message switching system.

4. Facsimile

Facsimile machines permit drawings, signatures, logos, and handwritten notes to be sent. If transmitted or stored digitally, facsimile documents require about ten times as many bits as alphanumeric documents. In some cases use of facsimile can save typing costs.

5. Voice Message Storage

As electronics costs drop and secretarial costs rise, the cheapest way to transmit and store interoffice messages may be in the form of speech. Speech messages could go straight from the sender to the receiver and avoid any intermediate human processing. Widely distributed memoranda can be made available by telephone.

Many of the machines used for message input and output could be machines already existing in offices with communications adaptors added, for example, typewriters, magnetic-card typewriters, and copying machines.

SPEECH MEMORANDA

Executives spend much of their time dictating messages to other people in their corporation, which the secretaries have to type. The most labor-saving, and potentially cost-saving, form in which to deliver such messages is in spoken-voice form. The spoken-voice messages—voicegrams—would be filed in computer storage. Each recipient would be notified when messages were waiting for him, or could check his file periodically from an ordinary telephone. The cost of storing such messages in a large on-line library store would be about 20 cents (U.S.) per year for messages coded with delta modulation, 2 cents per year for messages coded with vocoder techniques, or 0.01 cents per year for messages composed with code-book works (see Fig. 15.1).

The use of speech memoranda instead of typed memoranda would be alien to some executives who cherish their present way of operating. However, it would be more convenient to send many memos by telephone. Whenever an executive cannot reach a person he telephones, he can immediately leave a message with no danger of secretaries misinterpreting it. When an executive is traveling he can telephone his own file, at night if necessary, and listen to the memos that were sent to him. Security controls can be designed to prevent *any* other person listening to it. Such a system could be more convenient and accessible than a memo-typing operation.

Memos are used in corporations to provide a *record* of past instructions

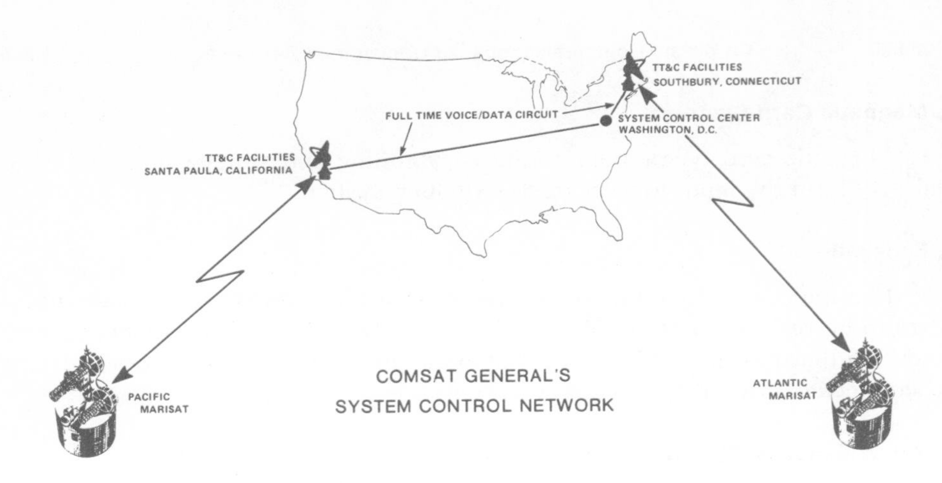

Figure 17.1 The Marisat satellite system enables ships throughout the world to be connected to corporate networks for voice and data transmission. Ships can now have the same telecommunication facilities as a terrestrial office. (*Pictures courtesy Comsat General.*)

and communications, and the memo sender *signs* the document. The voice-print of an individual saying certain words, for example speaking his personnel number, is as good an identification as a signature. The system may be designed to store speech memoranda in long-term files when instructed to do so. When not so instructed, it keeps them for a given period of time after they are delivered (perhaps a month) and then erases them.

Memoranda containing tables, diagrams, or other items not conveniently represented by speech would still be sent on paper. The paper may be delivered by facsimile or alphanumeric coding, and stored in the same system as the speech memoranda.

AN ALTERNATIVE TO TRAVEL

Not only is the total cost of typing and filing memos greater than the telephone bill in most corporations, the total cost of business travel is also greater—and the cost is rising. Some large corporations spend more than $100 million on business travel within the United States. This does not include the cost of the time of the persons traveling or the effects of wear and tear on them, which for some executives are considerable.

The technology has now reached a point when telecommunications can form an effective and cost-saving substitute for some travel. It is necessary to ask the question: What are the *best* ways for people to communicate with people at a distance? This is a complex question and there are many types of answers to it. Many of the answers require transmission at higher rates than that of local telephone loops, often in very brief bursts.

Among the facilities which could be used to improve communications at a distance are the following:

1. Telephone Conference Calls

Conference calls have been infrequently used in business because of the difficulty of setting them up. Some new computerized PABXs give their users the ability to set up multiple-party voice connections without operator intervention. This can be a valuable facility because it is often very useful to consult a third party during a conversation, or include several people in a telephone discussion.

A telephone "meeting" with many parties at separate locations needs a certain discipline imposing upon it to make it effective. It is necessary for each person to know who is speaking and to be able to indicate to the dispersed group that he wants to speak. This can be done if the meeting has a chairman who disciplines the conversation. Another possibility is for each caller to have a small strip with lights. Each participant has one of the lights associated with him, which is on when he is talking, and which he can make flash when he wants to talk. A low bit-rate control channel is derived from the speech channel for operating the lights. The strip may be designed so that a caller can write the names of the parties by the lights assigned to them.

Telephone meetings may also be held between meeting-rooms in different locations equipped with speaker-phones.

2. Picturephone

As we discussed in Chapter 8, Picturephone adds information to a telephone call by permitting the observation of facial expressions. The cost of this extra information is high—many times the cost of a telephone call—and for many corporate calls it is not worth it. Picturephone does not have sufficient resolution for users to read typed documents, contracts, computer printouts, detailed engineering drawings, and so on.

3. Facsimile

Paper documents can be transmitted fairly quickly by facsimile means. Telephone callers may use facsimile transmission to enhance their conversation so that they can exchange sketches or documents and discuss them.

4. Still Video

Callers may employ a screen in conjunction with their telephone call, on which still images can be displayed. If the image is to be displayed and discussed while the conversation proceeds, it is desirable that it should be transmitted fairly quickly—say in 5 seconds. Picturephone transmits one frame every 1/30th of a second. A Picturephone-quality frame transmitted in 5 seconds would need $\frac{1}{30 \times 5}$ times the Picturephone bit rate, or 42,000 bits per second. A black-and-white printed sheet would require about 250,000 bits, or 50,000 bits per second, and a much higher resolution screen than Picturephone. If the telephone call is occupying 56,000 bits per second on a PCM channel, then the image could go over the same channel interrupting the speech in one direction for 5 seconds or less.

To enhance the conversation both parties should be able to look at the same image at the same time and point to it. A moveable arrow may be provided on the screen for this purpose, along with a low speed subchannel for conveying its movements.

It may be desirable to see the caller's face, but as a still image rather than the moving image of Picturephone which takes so much bandwidth. The lens which is used to transmit documents could also be used like a Picturephone lens to record the face of the caller. A system could be designed in which a telephoner could press a button to capture the image of the face he is talking to. He could then have some idea of a person's expression at a selected critical moment in the conversation. This would not convey as much human information as Picturephone, but could be transmitted over a duplex channel of telephone bandwidth.

5. Electronic Flip-Charts

Most major locations in corporations will have visual display units connected to computers. Such screens have been used effectively for enhancing person-to-person communications as well as person-to-machine communications. Two or more individuals talk by telephone and discuss information which resides in a computer storage. All participants to the conversation see the same data displayed, and can modify it.

At its simplest level, the machine is being used merely to display human ideas with clarity. In many corporations today, flip-charts are used for this purpose. An employee has a set of facts or ideas which he must present quickly and efficiently to management or colleagues. He writes the information in a concise form, so that it can be grasped quickly, on large sheets of paper which can be hung on a flip-chart stand. The information in the boxes of this book, for example Box 15.1, is typical of what would be written on flip-charts. Employees travel with a roll of flip-charts to make half-hour presentations to management. The information could be conveyed equally well if the data on the flip-charts were entered into a computer system as a set of single-screen displays, and edited until they were as concise and clear as possible, preferably in color. The persons talk by telephone, using the electronic flip-charts in the same way as paper flip-charts.

Electronic flip-charts have several advantages other than avoiding the need to travel. First, they remain in the computer storage after the conversation. Management rarely admit it, but they probably remember only a portion of the data that are flip-charted at them. It would be useful if they could review the charts again privately at their leisure, and perhaps discuss them with persons other than the original presenter.

The production of paper flip-charts is often made a laborious task. Neat magic-marker lettering takes time to write, and the wording is frequently modified. Computer software could make the entering and editing of screen charts a fast operation.

In a corporation the many flip-chart presentations could be filed and indexed, with appropriate security locks. Many flip-chart presentations are made on the same or related subjects and the indices would permit computer searches to be made for these.

In some corporations flip-charts are one of the main forms of communication, with much money being spent on air travel by persons making flip-chart presentations. It can be a highly efficient form of communication and is susceptible to mechanization. *Computer-assisted flip-charting* has major advantages.

6. Communication via a Data Base

Communication links which handle flip-charts could also handle data which are assimilated and stored by computer data-base systems. Data base technology imposes a measure of precision on the way data is defined and referred to, and data base administrators have often been surprised by how different departments or managers call the same data by different names or different data by the same name. When communication takes place between parties using a common data base there is less chance of imprecision.

In a system at Westinghouse [3], a graphics terminal is used for production scheduling based on sales forecasts of washing machines. This is a complex operation because Westinghouse makes over one hundred models, all available in several colors. Once a month, the production and marketing managers travel to Pittsburgh to work together on the display console. The marketing managers evaluate market forecasts and assist the production managers in working out the production schedule. The use of the terminal permits more options to be explored than were possible before. Before, according to Reference 3, a "seat of the pants" approach was necessary. Now the two groups of managers can communicate with precision. The managers involved, once experienced with the technique "wouldn't want to do their scheduling any other way." [3]

The same type of meeting could take place via telecommunication links, with the parties involved able to discuss data which all can see. In some cases the data will be modified or processed during the conversation.

In reflecting on the ideal forms of man-computer dialogue, it seems that they have much in common with ideal forms of person-to-person dialogue via machines. As person-machine communication improves, and person-to-person communication becomes more precise, the two will increasingly tend to require the same hardware, channels, and features.

7. Teleconference Rooms

Because some of the facilities for improved communications are expensive they may be installed in conference rooms rather than in individuals' offices. A conference room designed for meetings via telecommunications may be equipped with multiple television screens and cameras as in Fig. 8.6, with electronic flip-chart facilities, links to data-base systems, and facsimile equipment. In some cases teleconference rooms are equipped with voice conferencing facilities and the equipment selected can all be used over voice-grade links.

Video equipment and the requisite communication channels are expensive, but in many situations not as expensive as the alternative physical travel. Several major U.S. corporations are now using video conference facilities, in some instances to lessen the cost and strain of air travel, in others to give better human communications. IBM has a video education network in New York. New England has a video network linking hospitals.

8. Radio Paging

Radio paging systems make it possible to contact individuals who are not sitting near a telephone. The individuals wear a small inconspicuous radio receiver which can signal them either with an alarm tone, or with a spoken message. Individuals roaming about a factory floor can be instructed to pick up the nearest telephone. Service personnel miles from anywhere can be instructed to go to a stated customer. Paging is one of the facilities of some computerized PABXs, but is more often done manually. A few corporations make a massive use of radio paging.

9. Two-way Mobile Radio

Two-way radio can be used in the ways discussed in Chapter 12, and may have a major growth ahead. Two-way communication to persons far from conventional telecommunication facilities is being accomplished via satellite. WESTAR links connect to off-shore oil-drilling rigs. The MARISAT satellites connect ships around the world to their head offices. Ships have been million dollar facilities without the transmission capability of a terrestrial office. Now they can be linked into their corporate communications network like any branch office.

Remote facilities on land can also be linked to corporate networks by radio—possibly via satellite. For example, the trucks that service tractors or earth moving equipment in developing countries can have radio links to their offices.

TELECOMMUNICATIONS MANAGEMENT

The design and management of a corporation's telecommunication facilities need to be conducted on a centralized basis, taking all divisions of the corporation and all different users of telecommunications into consideration. Because of the economies of scale and scope for optimization, centralized design and management can save a substantial amount of money, and can provide a corporation with better facilities than it would have otherwise.

Booz-Allen & Hamilton [1] experience shows that organizations carrying out a centralized audit of conventional telephone facilities typically show savings of 8% to 15%. This saving is soon lost if centralized management procedures do not follow up the controls on a permanent basis.

Telecommunications management is rapidly becoming a much more complex job. The complexity is arising first from the new types of tariff structures and services such as those of the specialized common carriers, the interconnect industry, the satellite common carriers, and the value-added common carriers. Second, it comes from an increase in the equipment complexity and diversity, for example computerized exchanges, concentrators, intelligent terminal controllers, data network hardware, satellite demand assignment equipment, radio facilities, and so on. Third, complexity arises from the need to incorporate diverse traffic into the network facilities, such as data transmission, facsimile, video, electronic memoranda, security monitoring, and radio paging. Fourth, it may be difficult to establish the tradeoffs between various reasons for travel and telecommunications, and secretarial services and telecommunications.

Efficient telecommunications management is particularly important today because telecommunications costs in most corporations are rising more rapidly than the inflation rate. Surprisingly few corporations have tight centralized management of their telecommunications resources. Those that do, rarely take the broad perspective of communications tradeoffs including the travel budget, the secretarial costs and data processing budget.

A number of stages can be observed in the development of telecommunications management:

Stage 1. No centralized management. Lines and equipment are leased haphazardly by user departments.

Stage 2. Centralized planning and management of the corporate telephone facilities. Optimization of separate data networks by data processing system designers.

Stage 3. An attempt to combine the various data networks into an integrated data transmission facility, possibly including a message switching system.

Stage 4. An attempt to combine the data and voice facilities into an overall optimized plan.

Stage 5. A study of all forms of corporate communications including internal mail and memo typing, executive travel, and use of information resources.

Stage 6. Cost optimization of all forms of corporate communications.

To achieve Stage 6 is a very complex operation. Few corporations have contemplated it yet, let alone succeeded. The technology becoming available will make it increasingly profitable.

REFERENCES

1. From a draft of an article by Harvey L. Poppel and Anthony G. Ward: *Time to Tame Telecommunications,* Booz-Allen & Hamilton, 1975.

2. The $10 figure is from a Booz-Allen & Hamilton study and is quoted in the above reference.

3. William E. Workman, "Which Color Washer Will They Choose?" *Computer Decisions,* December 1969.

The great telephone tower in Stockholm, 1910. (*Courtesy L. M. Ericsson.*)

18 COMPUTERIZED NETWORKS

There is much scope for computer assistance in using the new transmission networks. We have already discussed the added features in today's computerized telephone exchanges (Chapter 6) and the requirements for locating vehicles or persons with radiotelephones (Chapter 12). When we consider the additional uses of communications, other forms of assistance will be possible—in fact, in many cases, essential.

We are likely to see changes on telephone networks in private organizations before we see them on the public networks. The typical corporation tie-line system leaves much to be desired. It is constantly giving busy signals. Often it is difficult to track down a person on it, and it rarely handles data transmission efficiently. In the years ahead we shall probably see computer-controlled corporate networks, with their own switching and other facilities and, in an increasing number of cases, with international links.

The world's largest private telephone network is that used by the U.S. military—AUTOVON (*AUTO*matic *VO*ice *N*etwork). Its total circuit mileage in 1970 was equivalent to the entire Bell System in the early 1950s. It handles voice and data communications, and calls can be encrypted if necessary. The circuits link U.S. military installations all over the world.

A corporate telephone network does not need to be nuclear-bombproof, but many other features of AUTOVON would be of value. A corporate network needs reliability, not "survivability." Like AUTOVON, it should handle data and voice over the same lines; also, as in the military, there will often be a separate specialized data network. As on AUTOVON, some calls will be highly important and should be connected as quickly as possible, but the majority of calls will be low-priority administrative traffic. It is desirable for certain top executives and their assistants to be able to contact each other immediately. On AUTOVON, all "command and control" calls are set up quickly, and this step could be profitably copied in industry. Today executives commonly receive busy signals from the tie-line network because the lines are flooded with unimportant chitchat.

PRECEDENCE CALLING

AUTOVON uses Touchtone telephones with 4 additional (red) keys to the right of the normal 12 keys (see Fig. 18.1). Pressing the P key designates the call as "priority," in which case it preempts ordinary calls. I (immediate) preempts priority calls. F (flash) and FO (flash override) give the highest levels of precedence. When a general presses FO, it overrides everything. The called party receives a special "precedence" ringing signal. A unique tone informs telephone users when they are being interrupted by a higher-precedence call. If one of these keys is used on a telephone not authorized to use that level of precedence, a prerecorded voice (what rank?) tells the caller that such a call cannot be put through.

Figure 18.1 The AUTOVON telephones have four keys to the right of the conventional keyboard for making priority calls, and if necessary, preempting calls of lower priority. Many of the facilities of this military network would be of value on corporate networks. (*Courtesy AT&T.*)

HOT LINES

In addition to the ability to make precedence calls, the general can have one or more "hot lines." He picks up a red telephone on his desk and is almost immediately connected to a predetermined location. Alternatively, he can press one key on a telephone with the same effect. The top executive in industry might have a "hot line" to his sales manager, distant plant managers, or presidents of subsidiary companies. In the future he is quite likely to have a hot line to an industrial "war room." The hot-line connections on AUTOVON are, in fact, switched. This factor makes them lower in cost and more reliable. A preprogrammed precedence level is automatically assigned to such calls, and service is so fast that a user often does not know that the call is switched. By the time he puts his ear to the handset, the phone of his hot-line associate appears to be ringing.

Another function of AUTOVON is the automatic setting up of conferences. A number representing a preselected list of conferees is keyed; the system looks up their numbers in its memory and switches lines to them. The conferees in this automatic hookup can be located all over the world.

In future private networks, a computerized PABX may be programmed to search for an individual, to try his number repeatedly if it is busy, or to page him by radio. The extension of the hot lines by radio paging could make key individuals permanently accessible.

SECURITY

Most corporate networks today have poor security in that the lines can be easily tapped, especially on the users' premises. Probably the only way to make transmission truly secure is to use cryptography, enciphering the transmission in digital form.

OBTAINING THE REQUIRED COMPUTER SERVICE

Where terminals are used for communicating with computers, obtaining the required service may be simply a matter of dialing the right machine. As the variety of applications available from one terminal multiplies, however, computer assistance will be needed in finding the service. Furthermore, the terminal will often be on a leased line rather than on a dial-up line (at least with today's tariffs). The procedure may begin, then, by using the terminal to select the required service.

IBM has display units in its branch offices throughout North America connected on leased lines to its "Advanced Administrative System." After the user signs on, the computer helps him select the appropriate application programs with a "conversation" such as the following:

TERMINAL: FUNCTIONAL AREAS—ENTER LINE NUMBER:

1. FIELD COMMUNICATIONS SUPPORT
2. SALES AND QUOTA RECORD INQUIRY
3. DP ORDERS AND MOVEMENTS
4. OP ORDERS AND MOVEMENTS
5. SUPPLY ORDERS
6. CARD ORDERS
7. CUSTOMER ORDERS
8. PARTS AND PUBLICATIONS ORDERS
9. ACCOUNTS RECEIVABLE

USER: 3

The user thus indicates that he is interested in the application area of "DP orders and movements." The computer then narrows down his field of interest further as follows:

TERMINAL: TYPE OF EQUIPMENT—ENTER LINE NUMBER:

1. SYSTEM
2. UNIT RECORD

USER: 1

The computer then proceeds to establish what operation he wishes to carry out within the field of system DP orders and movements, as follows:

TERMINAL: TYPES OF TRANSACTION—ENTER LINE NUMBER:

1. PRE-AUDIT
2. ORDER
3. RESCHEDULE
4. ALTERATION
5. SHIPMENT
6. RECEIPT
7. RPQ REQUEST
8. CABLE ORDER
9. CANCELLATION
10. FUTURE RENTAL STOP
11. INSTALLATION
12. RENTAL STOP
13. INQUIRY

USER: 13

The computer has now determined which application programs to use to process the subsequent transactions.

In this system, all the application programs are presently in one of two computer centers. In the future, terminals on leased lines in large corporations will access multiple computer centers. A conversation like the preceding one may be the means by which the switching computer connects the terminal to the processing computer it will use.

INFORMATION CONTROL ROOMS

After another decade or so of development of computer usage, many different machines will be accessible by telecommunications for different functions. It is probable that some large general-purpose machines will offer many functions in the same system, but others will be separate, single-purpose systems, often handling a particular type of data base. Some computers will be highly specialized. There will probably be machines for setting print and assisting in editing, for example; these will be able to handle all the different type fonts found in modern books and will be able to set headings of different sizes. Book or magazine editors will work at terminals adjusting the text and formating the pages.

Gaining access to the different machines and their services may simply be a question of dialing the right number, which might be looked up in data processing "yellow pages." On the other hand, assistance in finding the right machine might itself be obtained from the terminal, and the telephone company might well have a directory-assistance computer for this purpose. A brief conversation using the terminal screen would identify the required computer, and switching to this computer would then take place. The directory computer might serve many of the common types of inquiries itself and then switch the user to another remote computer when unable to help. The remote computer, in turn, might itself decide at some point in the conversation that a different machine was required and thus arrange for switching to take place. Alternatively, it could merely obtain data or a program from the other machine and continue serving the terminal itself.

In some cases, code conversion will be needed in communication between otherwise incompatible machines, and (as discussed in Chapter 10) such conversion may be a function of network interface computers. Clearly, many major developments in our transmission networks are needed in the years ahead.

DIRECTORY COMPUTERS

Obtaining information from the various computers will become a highly complex process in a large corporation. Many systems do not attempt 100% automation but use a judicious mixture of computers and human experience. One way to do so is to gather together the talent and terminals in rooms equipped to answer or redirect management's questions. Such rooms exist in embryo form today in a variety of organizations, and it is clear that they will grow in complexity and diversity of functions. They range from the "war rooms" of military command and control systems to centers for helping eliminate errors on data collection systems. In controlling a complex set of operations, the decisions that must be made are often brought together by telecommunications into one nerve center. A spectacular example is the NASA control center for the space missions at Houston, Texas. Many less-spectacular examples exist in industry.

The showpiece of many data processing installations in the future may be a "war room" in which all types of information can be sought from different remote computers and routed to locations where needed by management, salesmen, shop foremen, and others. The computers and their storage will no longer be glamorous showpieces but hidden in secure locations.

The functions of such information control rooms are likely to include the following:

1. The staff is responsible for the corporate data base from which management obtains much of its decision-making information. Errors and wrong information inevitably find their way into the data base. There are many checks for detecting such errors, and when found, the control room staff is responsible for correcting them.

2. Some management questions are too complex to be answered easily from the terminals in their locations, and these questions are switched to the expert staff in the control room for answering. The control room staff can cause any display to appear on the screen of a manager's office and can monitor it on its own screen.
3. When emergency situations occur in the corporation, these situations may be referred to the control room, from which centralized expediting can be directed and the situation monitored.
4. When local management wishes to override a decision of its local computer, the management may contact the control room for arbitration. This procedure might be followed, for example, in a district sales office when the computer says that a certain order considered important by the sales office cannot be taken or met.
5. In some types of operations, computers should not be permitted to make *all* decisions in controlling them. The machine must be programmed to recognize when the intervention of a man with *experience* is necessary. The expertise for human intervention will be collected and made accessible in the one location.
6. The boardroom may also be equipped with terminals and screens and may perhaps be close to the "war room." This proximity would enable the highly specialized staff in the war room to be called in if needed to help in answering questions that arise during board meetings or to generate appropriate charts on the boardroom screens.
7. A hierarchy of information control rooms may exist. A factory may have a control room relating to the flow of work through the shop floor. A regional sales headquarters may have one relating to its customer and order situations. Subsidiary companies would probably have their own, separate from their parent company. A head office location may have a group of skilled operations research staff capable of dealing with more complex types of questions.

The manager of the future may have a hot line to his local information room. He may pick up a red telephone on his desk and be immediately switched to personnel who know where to find the answers to various types of questions. If possible, they will answer his question verbally, or display relevant information on the screen in his office, or prepare a printout for him. Sometimes they will route the query to another information room, perhaps more specialized, perhaps more expert, or perhaps in a far-distant part of the organization.

PUBLIC INFORMATION SERVICES

Since the earliest days of the computer, a trend from military usage of information systems to industrial usage can be detected. SAGE† brought real-time processing, visual display screens, and light pens; ten

†SAGE means Semi-Automatic Ground Environment, the first of the systems designed to protect the United States from a surprise air attack.

years later these devices were common in industrial systems. Now the technology of military command-and-control systems is being found in industry. Data transmission networks are spanning commercial organizations. The simulations used in "war games" are finding analogs in industrial and civil situations.

There may be a second stage to this trend. The schemes that were first used in the military at enormous cost reach industry when the cost falls substantially, and then in some cases reach the general public when it falls again. The data banks containing information of value to industry today will hold information sought by the public tomorrow. The public will use computers to help find a house, book theater tickets, provide stock information, give advice on optimizing tax returns, and so on. The terminals that spread from the military to industry will also spread from industry to the home. Just as the manager in industry needs help in using his terminal, so, to a greater extent, will the public. The lessons now being learned about the "man-machine interface" will be essential to home use of terminals; they are the key to acceptance.

Just as the information room in industry needs human skills as well as machines, so it would be a mistake in systems serving the public to attempt to automate everything. The public will need not only "directory assistance" but also many forms of more elaborate aid in using their terminals. As in industry, a human operator will often generate an appropriate display and switch it onto the screen of the inquirer.

In the early days of the telephone, the public received much more help from the staff at the telephone exchange than they do today. They could ask their "friendly operator" for the time, the weather forecast, how to get a nurse, or the name of the film at the neighborhood moving-picture house. Some countries still have friendly human help available by telephone. France, whose telephone system is notoriously underdeveloped, has a service called "Q.E.D.," which enables the public to ask questions about almost anything.

In the early days of terminals for public use, it may be desirable to bring back the friendly human operator. When the average person is using the terminal to shop, to book a journey or a vacation, to balance his bank or credit account, to search for literature on a particular subject, or any number of other functions, he will occasionally need assistance. The computer rooms providing such services will need a staff for this purpose. The terminals should have a HELP button. Unmanned bank "branches" should have a telephone to a friendly assistant. Information rooms with skilled personnel similar to those for management assistance will be needed in many other areas of terminal usage. One prime usage will be in obtaining the switched connections to the requisite computers. Let us end this chapter, then (in a book mainly concerned with automation), with a plea for the skilled use of *people* in the systems. The trend of employment toward service industries and away from the manufacturing industries will continue in the United States, and one segment that will need people will be that of the rapidly growing information services.

1922. A picture from the *Illustrated London News* showing a family listening to an early broadcast. One of the pioneers of radio—a man who was later to become very famous—said that he could foresee little use for public broadcasting except for the broadcasting of Sunday morning sermons because that was the only form of mass oratory that the public regularly listened to.

19 TERMINALS IN THE HOME

Now evolving in society are electronic payments networks, police emergency and information networks, networks for travel and theater booking, library networks, and networks for use by shops, supermarkets, hospitals, doctors, real estate agencies, stock brokers, employment agencies, and universities. Particularly important in the future will be networks serving schools, so that children can learn from the terminals and become familiar with the wired society they are going to live in.

A natural extension of the growth of computer networks which serve the public is for the networks to reach into the home. The home subscriber can use his telephone, television set, or any inexpensive terminal, to access a wide variety of different services. Pocket calculators have sold in quantity to the home market, including some with capabilities other than doing arithmetic, such as storing bank balances, giving time reminders, doing bond, mortgage and other financial calculations, and, in some cases, microcomputer programming capability. Some industry spokesmen have been highly enthusiastic about the future market potential for electronic devices in the home.

Various think tanks and research organizations have produced long lists of services which modern telecommunications could provide in the home. Some experimental systems have set about providing them, proving that at least the technology can work. Box 19.1 lists some of the possible home services, omitting ones which seem particularly dubious economically such as menu selection, shopping lists, and so on.

The difficulty with providing such home services is that most of them are unlikely to make a profit for a long time. Packages that incorporate many different services could make a profit if there was a sufficient wide coverage and widespread acceptance. But the wide acceptance will not grow if the initial systems fail due to lack of profit. In other words we have the old telecommunications chicken-and-egg problem: *wide coverage is needed for profitability but initial systems cannot have wide coverage.*

BOX 19.1 Some possible telecommunications services in the home

Passive entertainment
- Radio
- Many television channels
- Pay TV (e.g. Home Box Office—movies, sports, etc.)
- Dial-up music/sound library
- Dial-up movies
- Subscriber-originated programming

People-to-people communications
- Telephone
- Videophones
- Still-picture phones
- Videoconferencing
- Telephone answering service
- Voicegram service
- Message sending service
- Telemedical services
- Psychiatric consultation
- Local ombudsman
- Access to elected officials

Interactive television
- Interactive educational programs
- Interactive television games
- Quiz shows
- Advertizing and sales
- Politics
- Television ratings
- Public opinion polls
- Debates on local issues
- Telemedical applications
- Interactive pornography
- Betting on horse races
- Gambling on other sports

Still-picture interaction
- Computer-assisted instruction
- Shopping
- Catalogue displays
- Advertizing and ordering
- Consumer reports
- Entertainment guide
- City information

- Obtaining travel advice and directions
- Tour information
- Boating/fishing information
- Sports reports
- Weather forecasts
- Hobby information
- Book/literature reviews
- Book library service
- Encyclopedia
- Politics
- Computer dating
- Real estate
- Games for children's entertainment
- Gambling games (bingo)

Monitoring

- Fire alarms on-line to fire service
- Burglar alarms on-line to police
- Remote control of heating and air conditioning (a user sending commands to this equipment from a distant telephone)
- Remote control of cooker
- Water, gas, and electricity meter reading
- Television audience counting

Telephone voice-answerback

- Stock market information
- Weather reports
- Sports information
- Banking
- Medical diagnosis
- Electronic voting

Home printer

- Electronic delivery of newspaper/magazines
- Customized news service
- Stock market ticker
- Electronic mail
- Message delivery
- Text editing; report preparation
- Secretarial assistance
- Customized advertizing
- Consumer guidance
- Information retrieval

Obtaining transportation schedules
Obtaining travel advice/maps

Computer terminals

Income tax preparation
Recording of tax information
Banking
Domestic accounting
Entertainment/sports reservations
Restaurant reservations
Travel planning and reservations
Computer-assisted instruction
Computation
Investment comparison and analysis
Investment monitoring
Work at home
Access to company files
Information retrieval
Library/literature/document searches
Searching for goods to buy
Shopping information; price lists and comparisons
Real estate searching
Job searching
Vocational counseling
Obtaining insurance
Obtaining licenses
Medicare claims
Medical diagnosis
Emergency medical information
Yellow pages
Communications directory assistance
Dictionary/glossary/thesaurus
Address records
Diary, appointments, reminders
Message sending
Christmas card/invitation lists
Housing, health, welfare, and social information
Games (e.g. chess)
Computer dating
Obtaining sports partners

HOME TERMINALS USED BY INDUSTRY

Computer terminals in homes *today* are generally not there at the user's expense. These users, for the most part, are university staffs engaged in research, programmers employed by a large corporation or sometimes a "software house," a handful of executives enthralled by the new technology, and some traveling salesmen who use their terminals when they arrive home to transmit their orders to a computer.

In writing a report, a group of authors could type the text directly into their respective home terminals. The report would reside in the memory of a distant machine. They could then modify it, edit it, restructure it, snip bits out, correct each other's work, add to each other's ideas, and instruct the terminal to type clean copies when they were ready. Possibly magazine editing will be done with such aids in the future.

When the opportunity to use computers from the home becomes common, it seems logical for some employees to spend at least part of their time working at home rather than at an office. The overhead cost of providing staff with offices and desks is very high, especially in large cities; some of this cost will therefore be saved by the use of home terminals. A manager can see what remote staff members are doing by telephoning them, dialing up the computer they are using and examining their work, and sending them instructions on their home terminals. It is possible, indeed, that in the future some companies may have almost no offices. A software company for producing programs can cut its costs significantly if most of its personnel work takes place at home. Some companies are operating highly successfully on this premise. Parents who participate in such a scheme may be relieved of much of the boredom they feel when they are unable to leave their children.

The growth of computerized teaching—from today's experiments to tomorrow's industry—will need a tremendous amount of human thought and development of programs. Much of this endeavor may also take place in the home, just as books are written at home today. Step by step, a teacher can build up his lessons on his home terminal. Occasionally he may dial a colleague to ask him to try out what he had produced on *his* terminal. When the work is near completion, he may try it out in a classroom, study the reactions of students, and return home to make appropriate modifications.

This is not only possible for teaching programs; it is also valuable for many other computer uses now within our grasp. Legal data banks, medical data banks, data banks for all types of professional users, as well as nonprofessional people seeking information, are going to take an enormous amount of intelligent effort to build up. Most of this work is likely to be done with terminals, probably terminals which impose standardized formats on the data being entered. It is as though we have to rewrite all our textbooks and reference documents in a form that permits terminal users to access the data and instruct computers to search for and manipulate the information.

We see here a beginning of a return to "cottage industry"—a trend that will probably increase greatly during the next few decades. Nevertheless, at the moment, a few factors are counteracting this trend. The first is the reluctance of some companies to give their systems analysts or other employees home terminals because such a step seems an unprecedented and potentially unpopular encroachment on leisure time. The second is a feeling that regardless of what they are doing, employees ought to be at their desks from 9 A.M. to 5 P.M. (At home, who knows, they might be watching television!) The third is a feeling that people cannot work at home because the environment is not suitable. Perhaps some homes are unsuitable, but the author of this book had no difficulty in writing it at home with a rate of productivity that would be impossible in his office.

We believe that these three views are not generally held by intelligent and enthusiastic members of the community and that this is an era when we must throw off the mores of tradition and rethink what is applicable to the age of teleprocessing.

Certainly if such "cottage industry" spreads extensively in our society, it will drastically change social patterns.

COMPUTER AMATEURS

Perhaps the next group to help introduce terminals into the home on a large scale will be the computer hobbyists. The equipment that ham-radio enthusiasts and other such groups now have at home is more expensive than basic computer terminals, such as teleprinters or alphabetic screen units. Furthermore, the interest and excitement likely to arise by being able to dial up and work with an ever-growing number of computers will probably be far greater and far more absorbing than the attraction of such hobbies as ham radio and home movies. It is probable that as the number of available computers grows, as education on computers spreads, and as leisure time increases as a result of automation, the computer amateurs will become a growing body. Magazines are now being produced for them. Industry will encourage them and enthusiastically sell to them.

Computer hobbyists may fall into a number of different groups. Some will hope to produce and sell their own programs or make them available to other amateurs. There will be some who are less creative and mainly interested in education; they will dial various instructional and information-retrieval programs. Others will be interested in playing games, doing puzzles, or indulging in mathematical recreations. But perhaps the majority will fall under the narcotic spell of programming. Working on ingenious programs (rather than the routine of commercial programming) is, to a certain type of mind, endlessly captivating.

The computer amateur will have significant contributions to make to the development of this technology. Most technical fields are too complicated or specialized for the amateur to make a name for himself, but programming new computer applications has endless scope. In every direction, new territories await the ingenuity and care of a dedicated amateur.

As noted earlier, *enormous* quantities of ingenuity and programming are essential to this new era—not the work of geniuses but ordinary step-by-step construction and testing. The work requires a high order of craftsmanship. In general, it is creative, enjoyable work, work that persons at home can do, work that disabled and in some cases blind people are doing. It is work to which the hobbyist or the enthusiast making money in his spare time will contribute enormously.

DOMESTIC MASS MARKET

The time will come when the computer terminal is a natural adjunct to daily living. Sooner or later computing will become a mass domestic market and the computer manufacturers' revenue will soar. The airline industry, the automobile industry, telecommunications, and other complex technical industries all spent two decades or more of limited growth but then expanded rapidly when the *general public* accepted and used their product. This will probably happen in the computer industry also with the help of microprocessors and data transmission.

The home user will in time have access to a wide variety of data banks and programs in different machines. He will be able to store financial details for his tax return, learn French, scan the local lending library files, or play games with a computer. If he plays chess with it, he will be able to adjust its level of skill to his own. It has been suggested that news will be presented via terminals in the future; the user will skip quickly through pages or indexes for what he wants to read on his screen. Because the machines' files will be very large, foreign newspapers transmitted by satellite could also reside in his local machine. He may have a machine to *print* newspapers in the home, although use of the screen might be preferable. If he wants a back number, he will be able to call for it, using a computerized index to past information.

As illustrated previously, a man interested in the stock market could dial up a computer holding a file of all stock prices, trading volumes, and relevant ratios for the last 20 years. Possibly stockbrokers would make the information available free and would provide analytical routines to their clients. When a client bought or sold stock, he would be able to give the appropriate orders directly to the stockbroker's computer via his own terminal.

On the other hand, someone lacking the money to buy stock could play at buying it. The computer would calculate the effect of his buy and sell orders, permit him to trade on margin, pretend to give him loans, but no actual cash transfer would take place (apart from the cost of using the machine). When friends visit, he would be able to dial up his records to show them that he started with $100,000 and in six months had made $78,429! Perchance to dream in glorious detail. At least he will have had lots of practice ready for when he does eventually become rich.

A valuable home service may be the preparation of tax returns, which in some countries have become outrageously complicated, with much scope for computer assistance in finding ways to lessen the tax paid. Data for the tax return could be entered into the system throughout the year.

SHOPPING

Although a businessperson or executive might use a home computer terminal for scanning newspapers or for stock market studies, a homemaker is more apt to use it for shopping. In some countries, punched-card supermarkets have come into operation. The shopper at the supermarket picks up a card for each item he wants to buy and these cards are then fed through a tabulator. He pays the bill, and the goods are delivered from the stockroom. The advantages for the supermarket are that less space and less capital outlay are needed and there is less pilferage. Still, why must the shopper come into the store at all? He could scan a list of the available goods and their prices at several different shops on the home terminal and then use the terminal to place the order. An organization selling in this way could cut overhead to a minimum by eliminating stores or lessening their size.

The customer who buys through such an automated catalog avoids the exasperation of fruitless searches for hard-to-find merchandise. Besides replacement bulbs for projectors condemned by planned obsolescence, spare parts for automobiles, or rare phonograph records—all items likely to be found, if at all, in only one store in the city—there are tedious searches for special items like summer houses for rent, theater tickets for a particular show, or boats with a particular specification. Presently these items are found through classified ads in newspapers or through agents.

Much of the work of such agents, and of the advertisement columns, could be done more cheaply and efficiently by a data bank accessible from terminals. Until home terminals are common, it is unlikely that those who now make their living from putting buyers in touch with sellers will cooperate with each other to set up systems that will make themselves redundant.

Many newspapers and magazines, already competing with television for advertising revenue, may go out of business. Printing unions are well organized to prevent the automation of printing itself; thus many newspaper publishers may be unable to avoid closing down when the classified ads emigrate to a computer.

The preceding services may be provided free by the advertisers; many others will have a charge, with the bookkeeping almost certainly being done on a central computer. An automated diary might cost one cent per entry and a secretary or spouse in a different location could update it. Hunting through dictionaries, almanacs, abstracts of literature, and similar sources might cost about one dollar an hour, much less than a trip to a good library, and more fruitful.

As more such services become available, the economic justification for terminals in the home will be strengthened.

SPORTS

Sports, too, will be aided by computer, and the sports fan will be able to obtain all manner of information on his home terminal. In top golf tournaments today, computers are used to keep track of everything that is happening. Observers stationed around the course report information to a computer station by means of walkie-talkies. The machine digests all the information, operates a scoreboard for the clubhouse gallery, press, and television; and displays on a screen hole-by-hole scores and such information as greens reached in par, number of putts on each hole, and lengths of drives on selected holes. Instantaneous comparisons between players can be produced along with all manner of asides, such as remarkable runs of birdies. And golf does not come up to baseball in providing comparisons.

The terminal owner of the future will presumably be able to dial machines giving up-to-the-minute detailed information on any sport instead of being restricted to the one or two items fed by the television channels. One imagines a fan of the future watching pro football television on a Sunday afternoon much as today, but with the teleprinter chattering away at one's side printing commentaries it has been instructed to give on other games.

Using CATV responses as in Fig. 9.7 or Viewdata (Fig. 19.1), sporting events could be watched on television and bets placed, using a computer system, before or while the event is taking place.

HORSE RACING

In the winter of 1967–68 England was swept by a disastrous plague of foot-and-mouth disease that killed many millions of pounds-sterling worth of livestock. Most horse racing was stopped because of the epidemic, and persons who enjoy placing bets with local bookies were highly despondent. Fortunately, *The Evening Standard* of London realized that betting on a horse race does not actually require real horses galloping around a track—indeed, in the age of computers it is inefficient and a waste of manpower (to say nothing of "horsepower"). Consequently, the paper made the following announcement: "*The Evening Standard* today proudly announces that it has devised, and will stage, the World's First Electronic Horse Race." It went on to say that the Massey-Ferguson Gold Cup, canceled because of foot-and-mouth disease, would be run on the London University Atlas computer. A mathematical model of the horse race was programmed with the help of racing experts who provided the details on horses and their form over the previous two years, jockeys, distance between fences, and so on.

The computerized horse race met with the full approval of the National Hunt Committee and the Cheltenham Racecourse Executive. The BBC broad-

Figure 19.1 *The British Post Office Viewdata system.* Viewdata is an interactive information service for the home and office, which links a conventional television set to the telephone local loop and thence to a data network. It is now in pilot operation.

Viewdata employs elegant easy-to-use dialogues. The user can access a wide variety of information for the cost of a *local* telephone call plus a small fee for the data accessed. Two of several Viewdata keyboards:

Page 32342a
WEST END CINEMAS

ABC	THE GODFATHER PART 2 SHAFTESBURY AVE 01 836 8861	2.45 & 8.00
CASINO ...	EARTHQUAKE 01 437 6877	2.30 5.30 8.30
COLUMBIA .	FUNNY LADY SHAFTESBURY AVE 01 734 5414	2.30 5.20 8.00
DOMINION .	WIND AND THE LION TOTTENHAM COURT RD 01 580 9562	2.15 5.00 7.45
ODEON	SHAMPOO HAYMARKET 01 930 2738	1.55 5.30 8.00
PRINCE CHARLES	EMMANUELLE LEICESTER SQ 01 437 8181 LATE SHOW FRI & SAT 11.45	2.45 6.15 9.00

Key *# to return

1.

Any entrepreneur, from a private hobbyist to a big publishing firm, can put information on the Viewdata files. If the public or business persons use it a small royalty will be paid. Programs can also be placed on the files and used by the public or business. A user can send a message to another user.

It has been forecast that by 1985 there will be many millions of Viewdata sets in Britain, but this depends on whether the Post Office invests in the scheme to a large enough extent. Several other countries are preparing plans for Viewdata-like services.

The screens shown here are:
1. Local movie programs
2. Part of a dialogue for obtaining travel information
3. Up-to-the-minute stock market results
4. Start of a dialogue for obtaining education courses via Viewdata

(Photos courtesy British Post Office.)

2.

* INSURANCE * — STOCK EXCHANGE

	OPEN	LATEST	
Britannic Ord	131	[illegible]	
Bowring	66	71	A
Commercial Union Ord	150	151	
Equity & Law Life Ass	152	153	
General Accident Ord	160	161	
Howden Alexander Ord	135	137	
Heath CE Ord	240	240	–
Leslie & Goodwin Ord	101	101	
Legal & General	127	129	
London & Manchester	112	114	XD
Minet Holdings Ord	172	171	
Mathews Wrightson Ord	157	157	
Pearl Assurance Ord	212	[illegible]	
Phoenix Assurance Ord	214	216	XR
Prudential Assurance	128	129	
Royal Insurance	294	297	
Sedgewick Forbes Ord	230	234	

Key *# to return PAGE 42504

3.

Page 35a

EDUCATION

0 COURSES AVAILABLE
-Schools,Playgroups,Universities
-Further Education,Evening Classes

1 PREVIOUS EXAMINATION PAPERS

2 TEACH YOURSELF COURSES

3 EDUCATIONAL INFORMATION
-Information for teachers,Lectures
-Open University,Higher Degrees

4 EXAMINATION REVISION NOTES

5 GRADED QUIZZES

Key number required

4.

cast a full commentary on the race, with commentators sounding no less excited because the horses were not real. The commemorative Gold Cup was awarded to the winner; the jockeys received their normal fee.

Clearly this concept can be extended. With terminals in the home, the racing enthusiast can have a race any time he wishes. It would probably be a fine after-dinner entertainment. He and his guests could use the terminal to ask questions about the various horses' form, and the computer might ask whether these are simulated bets or whether an actual cash transfer will take place. The race should be no less exciting than an actual race. There is no need to have the monotonous voice we sometimes hear from computers. Even today's equipment can reproduce the voices and even the intonations of popular sports commentators. Using cable TV in an interactive fashion, bets could be placed on the simulated race, and in other simulated activities the viewer might be permitted to modify the course of events.

Many other such games will be played with the terminals. Who knows what forms gambling might take in the computerized society, with the home-gambler's bank balance being automatically added to or depleted. Perhaps in America one will be able to telephone the local Mafia computer.

HOUSEHOLD APPLIANCES

The telephone line, in addition to its normal use, could be used for activating household appliances. A family, driving home after a few days vacation, may be able to telephone their home and then key some digits on the Touchtone telephone that switch on the heat or air-conditioning unit. Before leaving for work a person will preprogram the kitchen equipment to cook a meal. He will then phone at the appropriate time and have the meal prepared. Or possibly a computer might telephone the equipment and instruct it step by step to perform a sequence given to the computer the night before—for example, to switch on the oven at 3 P.M. to cook the roast that had been placed in it and to move some vegetables out of the freezer compartment, leaving them to thaw. At the appropriate time, the vegetables would be heated in their aluminum foils, and the dish-warmer would be switched on.

Cable television gives the added possibility, as we discussed in Chapter 9, that *pictures* can be received, in color, as responses in conversations with computers. This process will have many appealing applications. It will be particularly valuable with teaching programs for children. Children usually find computer terminals a fascinating toy and want to play with them endlessly. They will have no difficulty in learning to operate the keyboard that enables them to converse with this new medium.

In the advertiser's paradise of America, all kinds of highly colored catalogs will become available the moment the new medium arrives, and there will be varied enticements for exploring them. Very elaborate presentations of products will become possible. The Sears-Roebuck catalog, for example, might include film sequences, although the user would still be free to "turn the pages," to use the index, to select and reject. As with American television, advertising would help to pay for the new medium. Perhaps critical consumer guides will also become automated to aid product exploration. Having scanned the relevant catalogs and inspected pictures of the goods in detail, the shopper could then use the same terminals to order items, with the money being automatically deducted from his bank account.

HOW SOON?

Probably before the year 2000 many of the services in Box 19.1, and others not thought of, will be available in homes. The technology is available today but setting up the services will take time, talent, and money. The reader might glance again at the curves in Fig. 2.1 and reflect what the curve for home information services, interactive television or dial-up movies will be like. We suspect that the growth will be faster than that of the telephone or automobile, but the take-off may not be as sudden as that of television. On the other hand the take-off of CB radio into a large consumer market in the U.S. occurred suddenly and rapidly, and a service like Viewdata *could* take off rapidly.

The home market is very fickle and it may be some fad that first brings *Subscriber Response Services* into the home. The ability to select any of the current top 20 pop records for home playing might have great appeal. Off-track betting in the home could suddenly become fashionable. To watch the horses before a race on color television, and be able to place a bet using the same set, could have enormous attractions. The off-track betting computers already exist. Such a service might surge rapidly to betting on all types of sport, the equipment installation cost subsidized in part by bookies and advertisers. Television game shows would probably join the bandwagon, and a bingo-happy nation would acquire the facilities that would later give it university courses in the home and encyclopedic information sources.

City wiring at the turn of the century.

20 THE WIRED CITY

The term "wired city" is used to describe a city in which homes, shops, and offices have telecommunications facilities. Today's cities are wired for telephone usage, so the term normally implies more advanced facilities such as CATV, interactive television, and data networks. Often it is used as a futuristic term meaning a city in which all services possible via telecommunications can be obtained (at a price). Exponents of the term often cite the traffic congestion and pollution of dense areas and envision a city without pollution where telecommunications substitutes for much of the use of vehicles.

CITY SCENARIO

Imagine a city with parks and flowers and lakes, where the air is crystal clear and cars are kept in large parking lots on the outskirts. The houses and apartments have wall-size television screens and computer terminals. Banking can be done from the home and so can as much shopping as is desirable. The high-rise buildings are not too close, so they all have good views, and all persons living in the city can walk through the gardens or rain-free pedestrian malls to shops, restaurants, and pubs. There is a good delivery service, and the shopping carts can be taken home if desired. There is fast public transportation to the theaters, university, places of work, and car parks. Working at home is encouraged and made easy for some by the videophones and computer facilities. The switching of the telephone and video channels is organized so that conferences can take place.

There is excellent computer-assisted instruction, and a large library of educational films is accessible from the home, leaving the teachers free to concentrate on the more human and creative aspects of education. Information retrieval facilities give the inhabitants access via their screens to news, sports information, stock market figures, weather reports, encyclopedias, and vast storages of reports and documents. However, far from acting as a substitute for books, the screens allow the citizens to browse in excellent local libraries from which books or magazines can be delivered or collected in person.

There is almost no robbery in the streets because most persons carry little cash. Restaurants and stores all have bank-card terminals, and the fund transfer system is designed so that the bank cards cannot be used except by their owner. Citizens can wear radio devices for calling police or ambulances if they wish. The homes have burglar and fire alarms which are wired to the police and fire stations. Some buildings use television cameras for security purposes, wired to centralized guard stations. All utility billing is done automatically with wired meters.

The screens in the home provide excellent telemedical facilities, some computerized and some via video connections to nurses and doctors. There are emergency communications in case of the need for ambulances. Some homes have printers or facsimile machines with which they can obtain business documents, custom-selected news items, financial or stock market reports, electronic mail, bank statements, airline schedules, and so on. They can use a computer to edit documents they write and to transmit these documents to the people they work with.

The city we have described is very different to big cities of today. It is possible that today's smaller cities like Geneva, say, or Ottawa could evolve into pleasant pollution-free wired cities. Two questions are relevant. What mix of telecommunication facilities should the ideal wired city have? And how do we get there from here?

TODAY'S WIRING

Many, but not all, of the wired-city uses of telecommunications can be achieved with today's wiring—with a combination of telephone loops and CATV cables. Few of the services listed in Box 19.1 need two-way video facilities, although two-way video gives some of the most appealing applications. Many of the services need switched channels; however many can operate with no switching, or only rudimentary switching on the video channel. Many of the applications can operate with only telephone wiring.

Today we have complete interconnectability in telephone switching, and no switching on the video channels although rudimentary switches can be placed at the cable head. This mix of capabilities means that for some applications both telephone and CATV cable are used. To obtain dial-up movies in schools, for example, the telephone is used to dial a movie library. The requested movie is loaded onto a machine which is manually switched to the requisite video channel. This operation is satisfactory for schools but would not be an adequate way to provide a dial-up movie service for the home because too many homes are connected to one cable. Dial-up movie service to the home is expensive in bandwidth and the bandwidth likely to be available for the next ten years is inadequate for this application.

Most CATV cables carry 12 to 40 video channels to more than a thousand subscribers, so any services in which more than a handful of subscribers want a video channel *to themselves* cannot be handled with adaptations of today's cable systems. On the other hand, many subscribers could have a speech or music channel or a data channel to themselves. Music library service in the home appears economically viable; film library service in the home does not.

DEFICIENCIES IN TODAY'S WIRING

If future cities are conceived as having two-way switched video facilities, the present types of wiring into the home are inadequate. Considering this, some authorities have commented that the current forms of wiring constitute a dead-end line of development, and new wiring layouts should be planned and encouraged. Box 20.1 summarizes the four types of channel into the home today. The following could be regarded as deficiencies in these forms of wiring when viewed from the wired-city point of view:

1. Telephone cabling into the home and office is too low in bandwidth for most video applications.
2. CATV cables have too many homes on one cable for switched video applications. The tree-structured layout of the cable is optimized for one-way broadcasting of signals to many homes, possibly with brief digital responses, and is generally inappropriate for switched video applications.
3. The capacity of the CATV cable is inappropriate for video library services into the home.
4. Picturephone cabling with two extra wire-pairs into one subscriber's premises is expensive. In most cases existing wire pairs cannot be used, first because shielding is often necessary, and second because amplifiers and equalization are required and these are difficult to add at a later time, especially because most telephone companies cannot always locate and identify the individual wires. Wire-pair videophone cabling is straining a dead-end technology to the limit.
5. Today's CATV cables can carry a diverse mix of signals but in some cases severe interference problems have been encountered (intermodulation between different types of signals on the cable). These can be corrected but the cost of the cable is substantially increased.
6. Some form of trunking and nationwide switching network is needed for CATV.
7. Continuous-channel switching and multiplexing (like that on the telephone networks) is very inefficient for data networks carrying brief bursts of data. Some form of switched data network is needed with burst switching (Fig. 10.2), in order to allow data to be handled at a suitably low cost.
8. Most telephone local loops and the intended Picturephone loops are not shared. The average home telephone is in use for less than half an hour per day. For the rest of the time the telephone cable into the home stands idle. Furthermore that cable *could* carry many simultaneous calls; it only carries one. The efficiency of usage of this form of wiring is thus substantially less than 1%.

BOX 20.1 The telecommunication channels into the home

Telephone	*CATV*	*Radio Broadcasting*	*Picturephone*
Moderate cost (Typical figure: $500 investment cost per subscriber)	Lower cost than telephone (Typical figure: $150 investment cost per subscriber)	Minimal cost	High cost (Too high for most subscribers for a long time hence)
1 subscriber per channel	Many subscribers per channel (Typical figure: 2000)	Unlimited subscribers	1 subscriber per channel
Switching (complete interconnectability)	No switching (or rudimentary channel switching)	No switching	Switching (eventually like telephone switching)
2-way	Full video: 1-way Some cables have limited reverse transmission capability.	1-way (future possibility of low capacity reverse channel using packet radio [q.v.])	2-way
Fully interactive	Some cables have limited interactive capability	Not interactive (pseudointeractive capability using techniques like CEEFAX [q.v.])	Fully interactive
Low capacity	High capacity	High capacity	Medium capacity
(2-wire, 3 KHz channel. More bandwidth – up to 1 MHz – if the subscriber loop is specially adapted)	(12 to 40 2-wire channels of 4.6 MHz; lower bandwidth reverse channel)	(Many MHz available)	(4-wire, 1 MHz)
Nationwide network	No nationwide network yet	Mostly localized with nationwide networking	Multi-city network planned

NEW WIRING LAYOUTS

What types of wiring layout might be used instead of today's layouts for future wired cities?

Several principles seem relevant to answering that question. First, telecommunications is characterized by *major* economies of scale. Figure 20.1 illustrates the economies of scale of telephone trunks. The cost per voice channel is *much* lower on a large trunk group than on a small one.

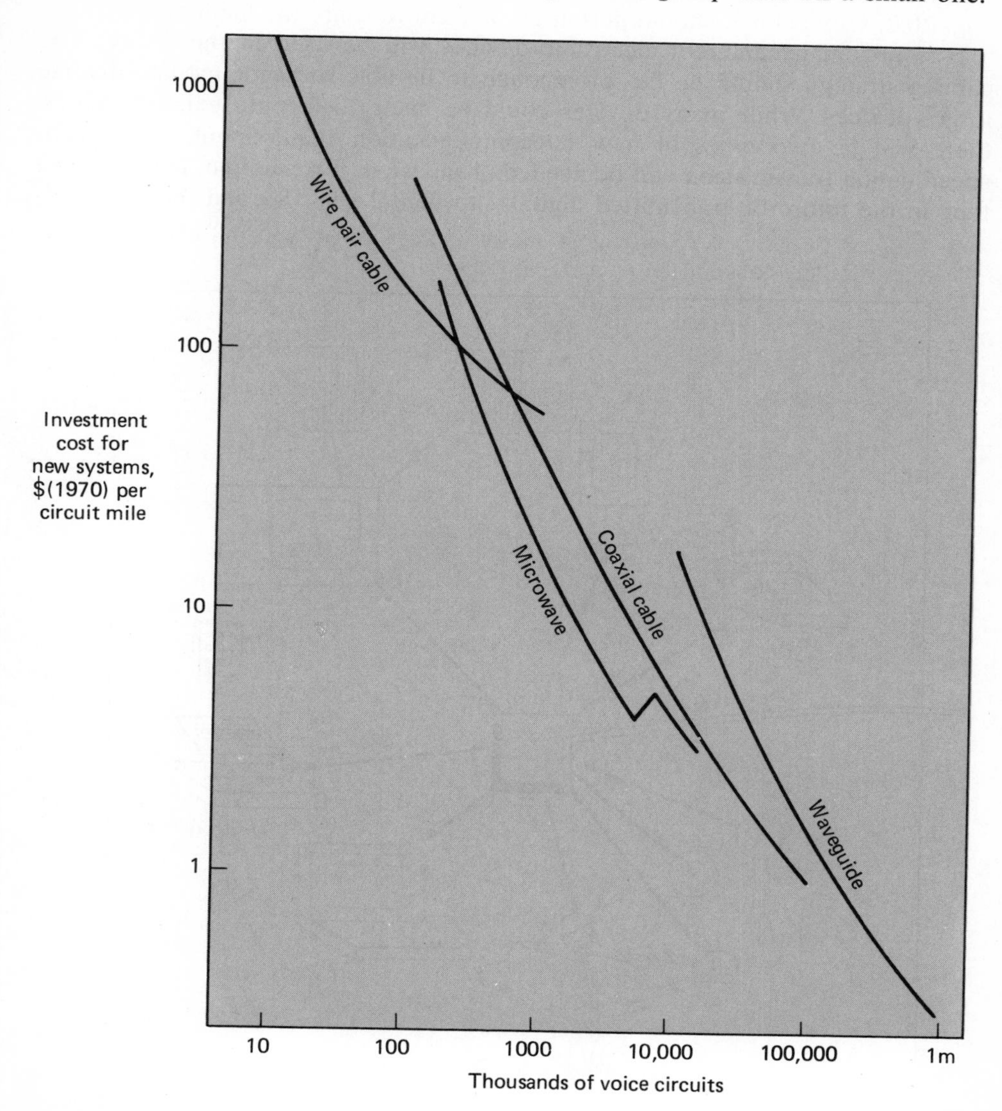

Figure 20.1 Telecommunications is characterized by *major* economies of scale. This chart shows such economies in AT&T trunks. Similar economies could relate to wired-city trunks.

Economies of scale will apply in future city wiring. One television channel is equivalent to about a thousand voice channels. Many subscribers should share city trunks of high capacity.

Second, it is unnecessarily expensive to have separate television and telephone wiring into the home. They could both be carried on the same cable. Future homes will have a variety of telecommunicating devices such as those in Fig. 20.2. City cabling should permit all of them to share the facility.

Third, many different electronic devices will be used in the future. The cabling strategy should be flexible enough to be able to accommodate diverse future devices. While many devices could be made to fit onto existing cables, there will be two types of new telecommunication requirement. First high-speed digital transmission will be needed. Television, for example, may at some time in the future be transmitted digitally to digital sets. Second, higher band-

Figure 20.2 One coaxial cable into the home could serve all of its future transmission needs.

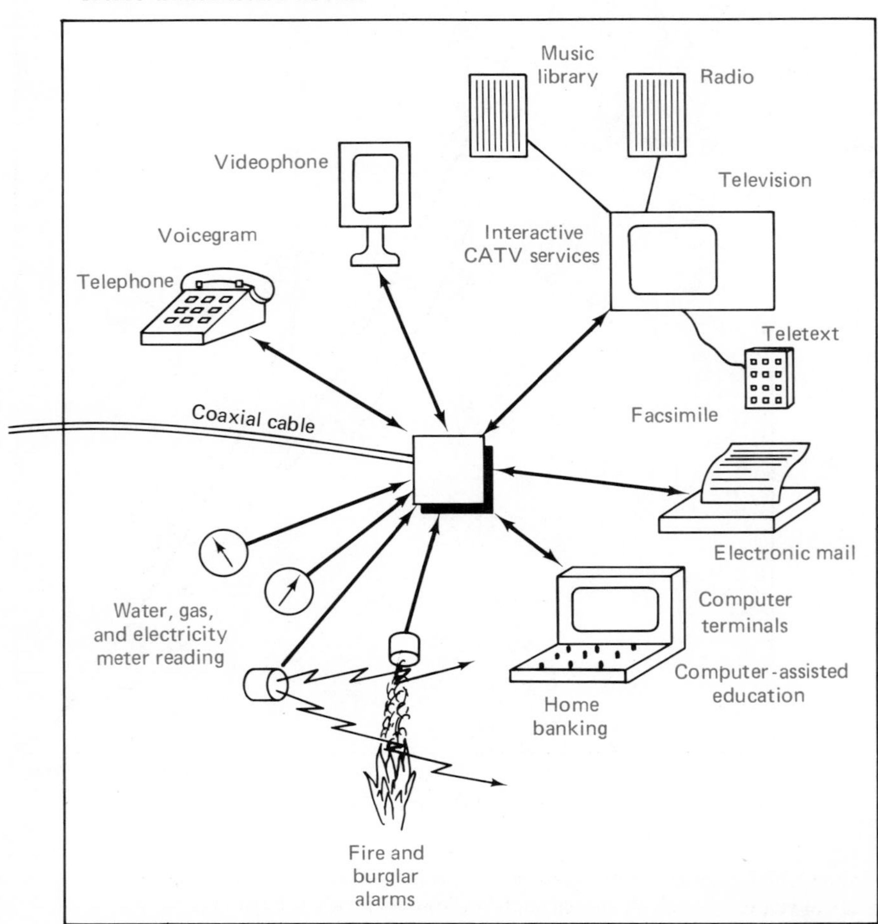

widths than today will be used. High-fidelity television with 2000 lines to the screen would need substantially higher bandwidth than today's television.

Fourth, coaxial cables are not the ultimate in cable development. At some time in the not too distant future, optical fiber cables will become competitive with coaxial cables. Optical fiber cables offer major advantages. They offer a very high capacity in a thin flexible cable, and many such cables can be enclosed in one sheath. There is no radiation from the cable and so the separate cables in the sheath do not interfere with one another. The world copper supplies are being expended rapidly; there will be a major and lasting copper shortage before the end of the century. By then optical fiber ought to have replaced copper for new local telecommunication cabling.

Fifth, the cabling design should be such that growth, both in bandwidth and numbers of subscribers, can be achieved economically over a wide range. Such growth in interactive cable uses cannot be achieved with today's tree-structured CATV cable layouts.

There is a variety of possible city-cabling layouts which could grow to meet the needs of future wired cities. Some of the systems that have been proposed use switching units that are much smaller than telephone exchanges and much more numerous. The use of small switch units close to the subscribers is referred to as *distributed switching*.

Distributed switching is used in two existing types of cable TV system, the *Rediffusion* system, and the *Discade* system. CATV systems with switching facilities located near subscriber locations are called *remote selection CATV systems*. In the Rediffusion system short wire pairs link a subscriber's home to the CATV city trunk. With the Discade system each home has a small coaxial cable to the nearby trunk switches. This coaxial drop would have enough capacity to carry out any of the signals that are envisaged for a wired city, but a major expansion of the city trunking facilities would be needed.

An extension of CATV remote selection systems has great appeal as the city cabling layout of the future. The unshared coaxial drop into the home is economically justifiable today on the Discade system. The city trunks and their remote switches or concentrators could be upgraded a stage at a time as new services are added and the subscribing population grows.

Figure 20.3 shows a possible wired city layout. A high-capacity wideband trunking system spans the city, connecting the remote switches and concentrators to a central wideband switching office. Such trunks would be bundles of coaxial tubes today, bundles of optical fibers in the future when that technology has matured. They may be analog today; digital in the future when digital technology has further dropped in cost. The city trunks would serve the needs of business as well as homes, providing the means for video conferencing, high-speed facsimile transmission, electronic payments, computer interconnection, and so on.

The end office of the wideband trunking system may be separate from the local telephone switching offices, as shown in Fig. 20.3. Telephone signals in a new city might reach the home on its coaxial drop. In an old city many of the

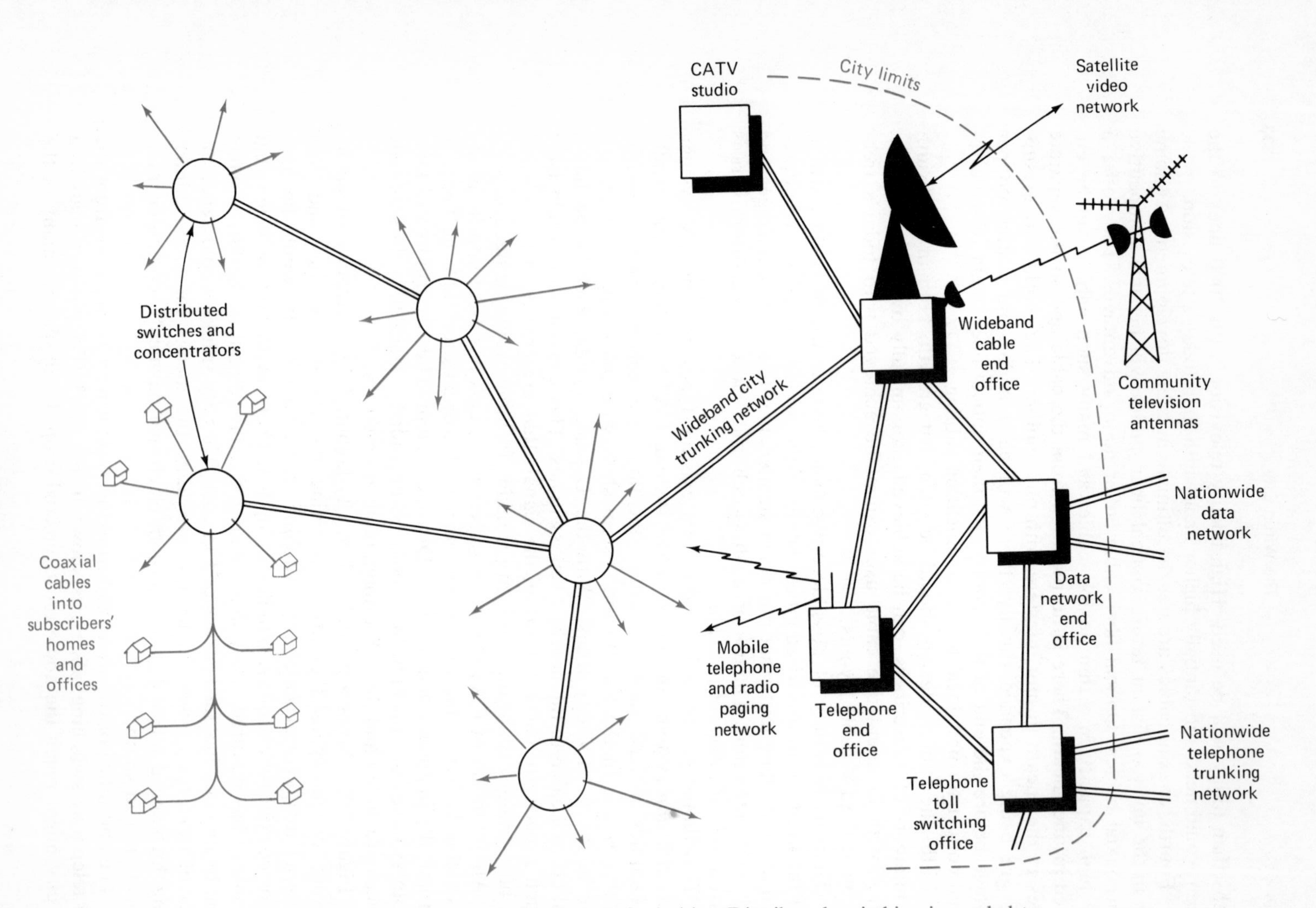

Figure 20.3 Possible wiring for future wired cities. Distributed switching is needed to handle the wideband distribution of signals. Any wired-city proposal presents severe problems of ownership and regulation.

existing telephone wire-pair loops connecting homes to the telephone office will remain. These are not shown on Fig. 20.3. Ideally, one cable into the home is enough. A major new function of the telephone office will be to provide control and switching for radio telephones, dispatching systems, and radio paging units.

The coaxial drops to subscribers from the remote trunk switches could be shared by several subscribers or could each serve only one subscriber. In the cable connecting several houses in Fig. 20.3 each house could be allocated a different frequency band or different time slots, or each house could use a different coaxial tube in a cable consisting of a bundle of tubes (frequency-division, time-division, or space-division multiplexing).

Fig. 20.3 shows three types of intercity linkages. The telephone trunking network is a natural evolution of today's telephone system. The data network is an evolution of today's packet-switching or fast-connect networks. The wide-band or CATV facilities are shown being interconnected via communication satellites.

NONTECHNICAL PROBLEMS

A variety of city cabling strategies could be devised which could grow to provide the services of a future wired city. The technology could be made to work; however there are severe implementation problems which are nontechnical. First, most such schemes are unlikely to be profitable in the near future. The market for the new services will build slowly and the economies of scale will not be quickly realizable. Second there are problems of ownership and regulation. Today's regulation in the United States places constraints on both the telephone and CATV companies that would prevent *either* constructing wired city facilities that would handle both telephone and television. Either industry would fight against integrated cabling owned by the other. Third, there is a vast and growing investment in existing telephone and CATV cabling. This investment cannot be written off quickly by commercial corporations.

Because of these problems it is, perhaps, unlikely that fully wired cities will come into existence without government involvement. Government involvement can take a variety of different forms. A likely form is the implementation of pilot systems in selected areas, such as new towns or university areas. It may be difficult, however, to transfer the technology from a pilot system to nationwide reality. In spite of the technological leadership of the United States, it might be more difficult to achieve wired cities in the United States than in some other countries because of legal and regulatory entanglements.

WIRED BUILDINGS

While there are massive economic and regulatory problems concerned with the concept of the wired city, multifunction wiring of a *building* or associated group of buildings is more immediately practicable. In a large office building, shopping center, university campus, or factory cluster, many telecommunication facilities are needed. All can benefit if they are planned in an integrated fashion.

A modern shopping center contains buildings which might include the following: shops, a supermarket, a bank, a restaurant, a post office, a broker's office, a travel agent, a real estate and insurance office, and possibly a local government office. When the owners of the shopping center plan its services, telecommunications services such as the following may be planned:

1. Telephone—a common PABX.
2. Credit authorization terminals.
3. Bank-card terminals.
4. Cash-dispensing machines.
5. On-line cash registers.

   } Banking services for the group may be handled by the on-site bank. (items 2–5)

6. On-line supermarket checkout stations.
7. Piped music.
8. On-line fire detectors.
9. On-line burglar alarms.
10. Police hot-line.
11. Closed-circuit television for security.
12. Security patrol stations.
13. Centralized guard console.

    } All security services may be handled centrally. (items 8–13)

14. Centralized reading of gas, water, and electricity meters.
15. Computer control of heating, air-conditioning, and lighting to minimize fuel bills.
16. Computer terminals in the insurance office.
17. Computer terminals in the broker's office.
18. Stock broker information services.
19. Computer terminals in the post office.
20. Airline reservation terminals in the travel agent.
21. Other travel-booking terminals.
22. Terminals in the real-estate office.
23. Terminals in the government office.

    } All connected to distant computers. (items 16–23)

A similarly long list could be drawn up for other buildings or sites such as office blocks, hospitals, hotels or apartment blocks with shopping arcades, or industrial sites. There is much to be said for having common wiring, common switching or control mechanisms, common line disciplines when possible, and in general the sharing of telecommunication facilities.

SEPARATE TRANSMISSION TECHNOLOGIES

The transmission facilities of the world have long been dominated by telephone requirements. There were telephone circuits within buildings, across cities, and spanning nations, and all formed part of an integrated telephone architecture. Today the situation is beginning to appear

BOX 20.2 Separate technologies are emerging for in-plant links, city links, nationwide links, and worldwide links

Transoceanic Links	*Nationwide Links*	*Citywide Links*	*In-Plant Links*
• 4/6 GHz satellites	• 4/6 GHz satellites and 12/14 GHz satellites with rooftop antennas	• Nonmultiplexed telephone wire pairs	• Telephone wiring
• Transoceanic coaxial cable	• Microwave systems • Coaxial carrier systems	• Multichannel CATV cables with thousands of drops	• Video cabling
	• Waveguide systems	• Two-way switching video cables	• High-speed digital loops
	• Optical fiber trunks		
	• MF broadcasting	• HF, VHF, and UHF broadcasting	
		• HF, VHF, and UHF mobile radio • Packet radio	• VHF radio paging

more fragmented, and separate technologies are emerging for the wired building, the wired city, and the wired nation.

For nationwide transmission the specialized common carriers are building microwave routes and using satellite circuits. AT&T estimates that its lowest cost per channel will come from waveguide systems and possibly from optical fiber systems in the 1980s. In the long run the most economical technology may be satellites.

Citywide transmission is still dominated by telephone wire pairs. Tree-structured CATV cables serving thousands of locations are spreading fast, and there are various plans for 2-way interaction using the cables. Cities have HF, VHF, and UHF broadcasting and these frequency bands are also used for mobile radio.

Most building communications consists of ducts filled with telephone wire pairs. Now, high-speed digital loops are coming into use for data transmission, and could carry voice also. Some buildings are being wired with video channels. VHF radio paging is used within buildings.

Box 20.2 lists the dominant transmission technologies for building, citywide, nationwide, and worldwide, links.

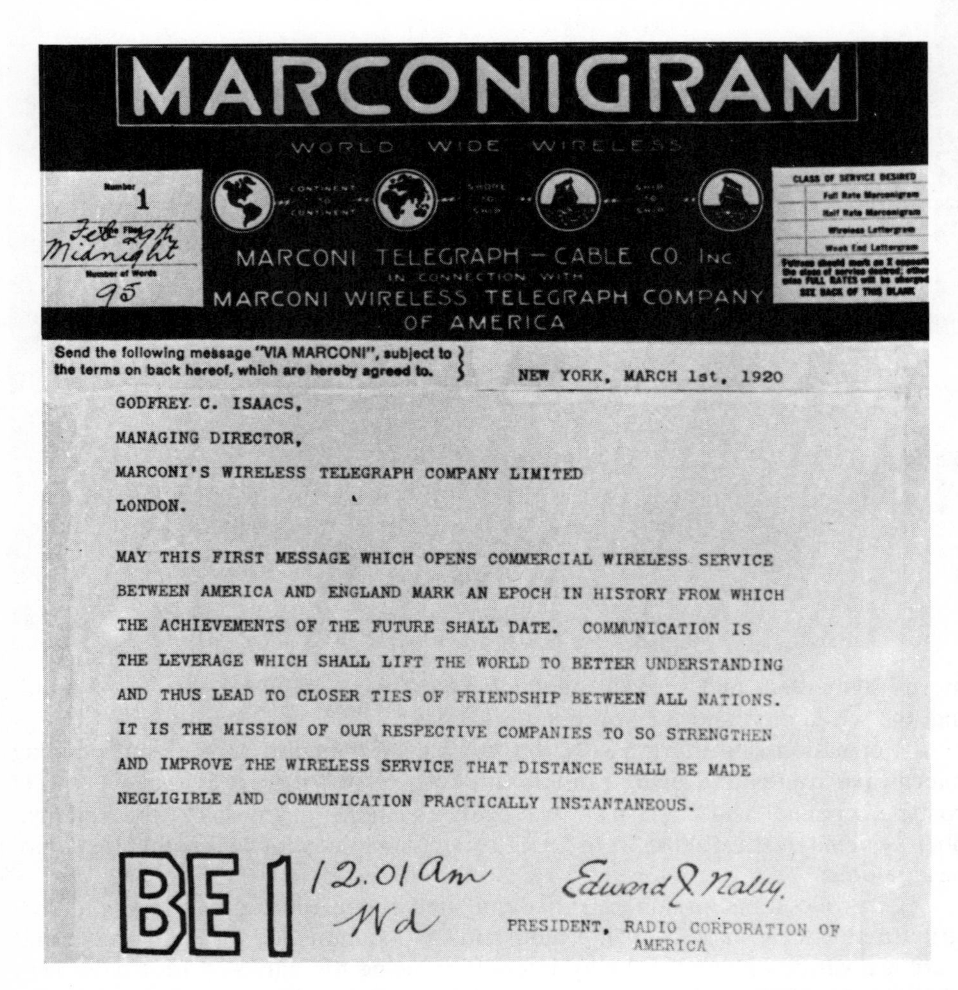

MARCONIGRAM

WORLD WIDE WIRELESS

CONTINENT TO CONTINENT — SHORE TO SHIP — SHIP TO SHIP

Number 1

Time Filed Feb 29th Midnight

Number of Words 95

MARCONI TELEGRAPH — CABLE CO. INC

IN CONNECTION WITH

MARCONI WIRELESS TELEGRAPH COMPANY OF AMERICA

CLASS OF SERVICE DESIRED

	Full Rate Marconigram
	Half Rate Marconigram
	Wireless Lettergram
	Week End Lettergram

SEE BACK OF THIS BLANK

Send the following message "VIA MARCONI", subject to the terms on back hereof, which are hereby agreed to.

NEW YORK, MARCH 1st, 1920

GODFREY C. ISAACS,

MANAGING DIRECTOR,

MARCONI'S WIRELESS TELEGRAPH COMPANY LIMITED

LONDON.

MAY THIS FIRST MESSAGE WHICH OPENS COMMERCIAL WIRELESS SERVICE BETWEEN AMERICA AND ENGLAND MARK AN EPOCH IN HISTORY FROM WHICH THE ACHIEVEMENTS OF THE FUTURE SHALL DATE. COMMUNICATION IS THE LEVERAGE WHICH SHALL LIFT THE WORLD TO BETTER UNDERSTANDING AND THUS LEAD TO CLOSER TIES OF FRIENDSHIP BETWEEN ALL NATIONS. IT IS THE MISSION OF OUR RESPECTIVE COMPANIES TO SO STRENGTHEN AND IMPROVE THE WIRELESS SERVICE THAT DISTANCE SHALL BE MADE NEGLIGIBLE AND COMMUNICATION PRACTICALLY INSTANTANEOUS.

BE 1 12.01 am Wd

Edward J. Nally.

PRESIDENT, RADIO CORPORATION OF AMERICA

A copy of the first commercial transatlantic wireless message, 1920. (*Courtesy Radio Corporation of America.*)

21 GLOBAL REACH

The news of Lincoln's assassination took twelve days to reach London. Now, telecommunications signals pass around the world in a fraction of a second, and Marshall McLuhan's phrase "global village" is used to describe the effect of television in linking distant peoples.

Because of its global nature and the dropping cost of long distance links, telecommunications will increasingly have worldwide, rather than nationwide or citywide, effects. We spoke in the last chapter of a wired city. Not far behind is a wired world.

The most important technology in worldwide communications is the satellite, and it is difficult to escape the conclusion that today's communication satellites are but an elemental beginning of a technology with tremendous growth ahead of it. One day readers of this book will look back at WESTAR, the first U.S. domestic satellite, with the amused but admiring sense of history that we have when we look back at Explorer I, America's first grapefruit-sized response to Sputnik. WESTAR weighs 300 pounds. At some time during the 1980s it will become economically reasonable to take hardware weighing many tons into geosynchronous orbit.

A new era in space will begin when the space shuttle is operating. We will look back at today's throw-away rockets as an incredible waste that was necessary for man to make his first steps into space. Imagine having to throw away a 747 every time it was used! Combined with the space shuttle some means of ferrying hardware into geosynchronous orbit will be needed.

Today we look upon the geosynchronous orbit as a limited resource. Using today's microwave frequencies (4/6 GHz) the satellites must be kept almost 2000 miles apart to prevent signals transmitted to adjacent satellites from interfering with one another. The limitations can be removed however. Satellites of different frequencies can be employed without interfering, and higher frequency satellites can be spaced more closely. Work on high-power lasers is bearing fruit, partly because of massive military expenditure. Some up-links may one day be laser beams, and satellites using these can be very close together without interference. Lasers operate at more than 10,000 times the frequency of today's up-links, and hence have a much higher potential bandwidth. They may not be used on the down-links because of possible dangers to persons on earth. Highly directional millimeterwave spot beams (10–40 GHz and possibly higher) could carry the signals down. With such a technology it is difficult to see any limit on satellites other than man's own capability to use the information transmitted.

As we have stressed the main key to advanced satellite design is weight. It is breathtaking to reflect what could be done with a satellite several times the weight of NASA's ATS-6.

Two parameters are critical in the design of satellites, the diameter of the satellite antenna, and the on-board power of the satellite. The strength of the signal received from the satellite is approximately proportional to the fourth power of the diameter of its antenna. Most of today's communications satellites have an antenna about five feet in diameter. In the utter stillness of space, with no gravitational force or breath of wind, a large antenna can be a fine gossamer-like structure, impossible on earth. The 100-feet diameter aluminized mylar sphere of the Pageos satellite weighed only 120 lbs. NASA design studies have described antennas for astronomical telescopes 10 *miles* in diameter, deployed in orbit as a fine spinning mesh pulled into position by centrifugal force.

The WESTAR satellite uses about 250 watts, generated by its solar cells. Satellites which spread out sails of solar cells like the Skylab satellite could produce much more electric power. The Skylab satellite sails generated 12,000 watts each.

There is no restriction other than cost to the amount of energy that could be generated in the satellite orbit. The total solar energy intercepted by a 10 mile wide strip at the height of the satellites is hundreds of times greater than the total amount of electricity consumed on earth. Indeed there have been serious proposals from NASA [1] and Arthur D. Little [2] for building a solar station in geosynchronous orbit generating 10,000 megawatts or so—enough power to supply the whole of New York City. The biggest problem would be getting the power down to earth on a microwave beam sufficiently diffuse not to cause harm. The total cost of power generation in geosynchronous orbit in the 1980s (including the launch) could be as low as $2 per watt if it were done on a large enough scale. Nuclear generators in orbit have also been proposed.

Before the year 2000, mankind could have immensely powerful satellites with large antennas beaming as much information as we are capable of using to our rooftops.

At some time this century a vast industry will grow up placing massive hardware in orbit in a ring around the earth 22,300 miles above the Equator. Eventually, it will become economical to have service vehicles in geosynchronous orbit, repairing, refueling, or assisting in the deployment of the satellite equipment. It has even been suggested the geosynchronous orbit will become the right place for certain new manufacturing processes which need the intense vacuum of outer space and perhaps the absence of gravity. The vacuum of geosynchronous orbit is more perfect than in low orbit and, being stationary above the earth, the production process could be constantly linked with high bandwidth telecommunications to terrestrial computers, control rooms and video monitors. One can imagine future solid-state logic circuitry or computer memory being fabricated in the utter purity of space with microscopic components thousands of layers deep being deposited on silicon.

It is fascinating to read the original article in *Wireless World* in which Arthur C. Clarke first proposed communication satellites in 1945 [3]. He argued that it would be prohibitively expensive to build a *terrestrial* network for trunking television or wideband signals:

> The service area of a television station, even on a very good site, is only about a hundred miles across. To cover a small country would require a network of transmitters connected by coaxial lines, wave guides, or VHF relay links. A recent theoretical study has shown that such a system would require repeaters at intervals of fifty miles or less. A system of this kind could provide television coverage, at a very considerable cost, over a whole of a small country. It would be out of the question to provide a large continent with such a service, and only the main centers of population could be included in the network.

Clarke suggested that satellites would eventually be far less expensive than terrestrial links, and most of the details he gave of satellites were surprisingly accurate.

Clarke underestimated the money that would be spent. High bandwidth terrestrial trunks *were* built across some continents before satellites could be used. Nevertheless, the simple logic of his argument is valid. Satellites, now that they are practical, are vastly cheaper than continental networks of coaxial cable, microwave, and the new waveguide systems. Hence the wiring of much of the world, which does not yet have the vast investments in terrestrial links, may come from satellite systems.

Certain developing nations which do not have the advanced terrestrial networks of Europe or America are planning satellites. Some countries which had major difficulties in deploying telephone cables because the villagers steal the wire to tie up their goats, are now planning satellite antennas on the roofs of telephone exchanges. It seems probable that certain of these nations will have more advanced satellite systems than Europe in ten years time, and their capital development costs for telecommunications will be less.

Not only should the long-distance links be of a different technology in the developing nations, but also the local distribution may be different. Such nations do not need a vast acreage of UHF spectrum to be allocated to television; VHF television is enough. That leaves plenty of UHF spectrum free for allocation to radio telephones, packet radio, paging, and broadcast data. These services are relatively more valuable in areas where telephone cabling is not as ubiquitous and high quality as in the major industrial countries. A businessman may find a radio telephone invaluable in places where conventional telephones are poor or nonexistent. Packet radio units may be attached to computer terminals instead of modems—and cost about the same. They may be linked to communication satellite relays, as in the Hawaii ALOHA system, thus permitting the spread of teleprocessing and data-base usage in areas where telephone links are bad.

Much of the developing world has less than one telephone per 100 people, and the telephones that do exist are mainly in business and government. (North America has about 65 telephones per 100 people) Slowly, telephone service to houses will be provided. The best approach to providing such service will not be to copy the local loops of the West. They represent the technology of an earlier era when copper was cheap and electronic equipment was expensive. Future telephone wiring design should consider:

1. Many homes using the same cable by multiplexing.
2. Multiple services, including CATV, sharing the same cable.
3. Coaxial and optical fiber cables.
4. Distributed switching and concentration.
5. Incorporation of radio links.

In the developing world, communication facilities are built in a different sequence from that in the West. The West had railroads, then roads, and later the airplane. In Brazil many towns can only be reached by air; later roads are built leading to them, and later, if at all, railroads. Similarly satellites will bring the links of culture and commerce to towns without telephone trunks. Data radio will be available to locations which cannot send data by telephone. Cable television will reach some homes before the telephone.

Rather than duplicating the facilities of the West, developing nations should use the new technology in new ways related to their own special problems and opportunities.

One of the great potentials of world telecommunications is education. Almost all countries of the world have television, but in much of the developing world only a small proportion of the people see television as yet, and the programming is very limited. A satellite such as NASA's ATS-6 can beam television down to relatively low-cost antennas. In an experiment in India television was received from ATS-6 in several thousand villages. Now India plans to have its own satellite. In a decade or so hundreds of millions of people, currently isolated from world communications, will join the world's television-watching hordes. This immensely powerful medium can enlighten, educate, entertain, spread literacy, spread better farming methods, or foster greed, provoke antagonisms, and spread themes of chaos. There is little evidence as yet that it will be used to spread the best of man's culture. But whatever its use will be, the world can never be the same again.

Medical assistance can be brought by telecommunications to areas of the world where doctors or specialists are in short supply. Experiments in telemedicine in the United States have been highly successful. In some the patient and doctor have been miles apart, connected by a television link. In some, instrumentation has been connected to the patient to transmit heart or other signals to a computer or specialist at the other end of a satellite link. ATS-6, the same satellite that brought television to India, was used to bring medical help to remote areas in the United States.

Satellites can transport other forms of expertize also, making available crop disease experts, consulting engineers, or world authorities, at remote locations where their special skills are needed.

In terms of the technology, if not the tariff, a communication link across the United States costs the same as a communication link across the world. As satellites become more powerful and rooftop antenna technology matures, corporations (in those countries which permit it) will use satellites for relatively short-distance links. The investment cost of a circuit from New York to Boston will be the same as a circuit from New York to Buenos Aires. Corporations will be able to tie together their worldwide operations in the same way that they tie together their national operations with telecommunications. However, although the channels are similar, the cost of international satellite circuits may be kept artificially higher than national satellite circuits.

Worldwide travel is more onerous than national travel, so there will possibly be a greater incentive to devise effective man-to-man telecommunications globally than nationally if the cost becomes similar. The leased-line networks, electronic mail, videoconferencing, voicegram systems, and computer links, discussed in Chapter 17, will become an essential facility of global corporations, vital to their fast efficient functioning.

Ships are becoming linked into the corporate telecommunication networks with the *Marisat* satellite system. Optimal fleet scheduling is a computer application which saves much money. Inventories can be maintained and resources moved around on a worldwide basis. Funds can be electronically transferred from one country to another, and switched to different currencies.

Multinational corporate networks, like national ones before them, will raise severe arguments about centralization and decentralization. The local staff should handle local situations, but there are some functions that arguably might benefit from centralization—bulk buying, inventory control, product design, capital management, some computer services, guidance in software development, etc. Video conference rooms and computerized information systems increase the degree to which head-office executives might attempt to guide or interfere with management in other countries.

The American military has superb worldwide communications. The effect of such fast and well-integrated information flow is that it involves high-level commanders in decisions that formerly would have been made at some lower level: the buck is passed upward at great speed. Often the President is involved in what were previously field decisions. In the Vietnam war the press reported how President Johnson became involved in such command decisions; for example, he personally reviewed each day's bombing targets. Field commanders, it is sometimes reported, resent such interference from on high, but the technology now permits bringing the best brains together on critical situations. To an increasing degree, command will be exercised from head-office locations and involve persons and computers in different locations connected via terminal screens.

The expenditure for air travel is very high in some worldwide corporations. Some of this travel is unavoidable, but for a large part of it telecommunications could form a substitute if superbly effective video, facsimile, and computer output facilities were available, and all capable of being hooked into conference calls. To do so it is necessary to bypass today's telephone loops.

Newsweek [4] estimated that the West needs to spend $500 billion over a period of ten years on the development of new energy sources if it is to become independent of the OPEC cartel. When new energy sources are listed, telecommunications does not normally appear on the list, yet satellite facilities appropriately used could bring a massive reduction in the consumption of petroleum products for human transportation. Roughly $100 billion will be spent on capital improvement of terrestrial telecommunications facilities in the West over the next ten years. The cost of superb satellite facilities is small compared with these figures.

As the Club of Rome and other models show (however inaccurately) many of the vital resources necessary for running today's industrial society are going to run perilously short. Petroleum is merely one of these resources. At the same time satellites will be spreading television to a vast world population which does not yet see television. In spite of all efforts to control it, the world population will have grown to 6 billion by the year 2000 and by then will be growing by a larger number per year than today. If the impact of satellite broadcasting is to make this vastly growing population demand the way of life of the industrial countries with their cars, plastics, steaks, and heated homes, then the world is in for a traumatic period.

An overly rapid introduction of advanced communications into societies with primitive communications is a formula for chaos. There seems little doubt that mankind is heading into a period of great upheaval. Satellites, as much as any technology, will contribute to this.

Satellites will give multinational corporations instantaneous global communications. Electronic fund transfer networks will be able to move vast sums of money instantaneously from one country to another. A coup d'état in Saudi Arabia could be seen immediately on the screens in the Pentagon where military leaders are in satellite communication with the American fleets and have the ability to jam the Arab satellites. Satellite receive-only antennas will be used at the heads of the CATV cables; high fidelity television will become a new electronics market with the bandwidths available to have wall-sized screens. World shortages of raw materials will become common as developing nations assume that their only way to wealth is to combine in cartels to increase the prices of the raw materials the West does not have. The spread of affluence will be highly selective with a few nations and groups, like the Arabs, becoming rich enough to cause economic chaos, while much of the burgeoning world population is starving. The bewildered public will watch the effects of economic and political turmoil and world starvation on the high fidelity wall screens.

All powerful technologies can be used for good or for ill. It is imperative that those persons who influence society should understand the new potentials that are emerging. Western society will change so that the products we consume are different from today—geared to new shortages and to new technological riches. There are limits to growth in many of our existing consumer patterns. But there are no limits near in the consumption of information, the growth of culture, or the development of the mind of man. The new information channels, which can access the world's data banks, film libraries, computer-assisted instruction programs, digitized encyclopedias with built-in film clips, will become available to mankind everywhere.

REFERENCES

1. P. E. Glaser, O. E. Maynard, J. Mackovcisk Jr., and E. L. Ralph, "Feasibility Study of a Satellite Solar Power Station," NASA Contractor Report CR-2357, 1974.
2. "Satellite Solar Power Station: An Option for Power Generation," briefing before the Task Force on Energy of the Committee of Science and Astronautics, U.S. House of Representatives, 92nd Congress. Second Session, Vol. II, U.S. Government Printing Office, 72-902-0.
3. Arthur C. Clarke, "Extraterrestrial Relays," *Wireless World,* London, October 1945.
4. *Newsweek,* February 17, 1975, page 25.

The International Telegraph Union and related bodies have done much to achieve international standardization. This photograph shows the heads of national delegations attending the first conference in Paris, 1865.

22 THE LAW AND POLITICS

Technological change has reached a fast and furious pace and is rapidly becoming faster. Most of society's institutions, on the other hand, can only change slowly, and in many cases avert their eyes from the frightening onrush of science. Herein lies a dilemma that will make itself felt in chaotic ways in the decades ahead.

Some uses of technology in society will involve increasingly large expenditures of money and exceedingly elaborate systems design. This prospect is true for the new schemes for urban transportation, for example, air traffic control, medicine, educational facilities, civil engineering, urban renewal, city center design, pollution control, environmental design, ecological problems, and many computer uses. It is especially true for telecommunications.

Where vast expenditures and public issues are involved, the arguments can rapidly become ideological rather than pragmatic. Some countries will insist on government ownership and civil service control. In others, ownership design and management by the private sector are an inviolate creed.

The trouble with telecommunications, as with some other public services, is that certain segments of the industry can be more cost effective if they are monopolistic. Having more than one telephone company operate in the same city with competing lines and switching offices is too wasteful and is likely to result in inconvenience for the users. In telegraphy some American subscribers used to pay a high price for having Telex machines on Western Union lines side by side with TWX machines which used to be on AT&T lines, solely so that they could reach different subscribers. Furthermore, where limited physical resources are involved, as with uses of the radio spectrum and satellites, it is necessary to employ these resources reasonably efficiently. It is argued that such factors make telecommunications a "natural" monopoly, at least in certain aspects, and hence there is state ownership in some countries and tight government regulation in others. Unfortunately, both state ownership and regulation have led to inefficiencies and to the sapping of incentive, sometimes to a degree that has been highly destructive. They have led to pricing practices and "reward systems" which encourage inefficient uses of technology and inhibit technical change.

Left to itself, the regulated monopoly is going to find many reasons for avoiding change. The telephone companies around the world are prepared to keep their switching offices for 40 years, although in an era of intercommunicating computers and real-time terminals the design may already be hopelessly inadequate. And so we have a problem.

Today, with the diverse new technologies discussed in this book, the ideal approach would be that of a massive piece of systems engineering. The telephone lines, electronic mail service, satellites, data links, mobile radio devices, broadcasting, and cable TV could all be interlinked in one gigantic system that would use the different facilities to their best advantage. Tight radio-spectrum engineering would ensue, and there would be an optimum balance between use of radio and use of cable for different functions. The old systems and plant would be integrated into the new. Computers would control the multipurpose switching and the tracking of mobile system elements. The result would be massive engineering design on a scale comparable with that of the moon landing.

It is a version of this dream that lies behind the state ownership and control of telecommunications in many countries. Another aspect of the subject is that the telecommunication facilities are vital for national defense and security.

The 1969 bill in England creating the British Post Office Corporation gives it a monopoly over all potential uses of telecommunications, including the provision of channels for broadcasting, data transmission, mobile radio, remote reading of electricity meters, burglar alarms, and so on. The corporation has the "exclusive privilege" of

> running systems for the conveyance, through the agency of electric, magnetic, electromagnetic, electro-chemical and electro-mechanical energy . . . speech, sounds, pictures, signals to control apparatus and signals serving for the importation of any matter otherwise than in the form of sound or visual images.

This privilege includes the regulation of communication between "persons and persons, things and things, and persons and things." As the London *Economist* said when the corporation was formed, that seems to include everything up to love at first sight.

The corporation even notified the companies renting television sets for cable connection that it could, conceivably, take over their rental in the future.

The British government's approach to the future of telecommunications was summed up in a statement to the House of Commons by John Stonehouse, then Postmaster General. In view of our comments in Chapter 2 about technological surprises and forecasting failures, his statement seems a classic of doctrinaire socialism:

> We must ensure that all possible technical developments are caught within the monopoly . . . *before they are invented* and before they become competitive with the monopoly, for that would be a very embarrassing position.

One wonders how many inventors will save the British government embarrasment by bringing their invention to the United States.

The United States also has its dogma. An essential part of the American creed is that public communications networks should be run by private industry. The opening section of the 1968 President's Task Force on Communications Policy (Final Report) said,

> We have taken pains to protect society against the risks of concentrated power, in the hands of either government or of the communication companies.

And later,

> In approaching these problems, we have been guided by the basic premise underlying the law and policy affecting American industry and commerce: that, unless clearly inimical to the public interest, free market competition affords the most reliable incentives for innovation, cost reduction, and efficient resource allocation. Hence *competition should be the rule and monopoly the exception.*

An extensive section on "Future Opportunities for Television" lamented, as indeed it should, the lack of diversity in American television. The average family has its television set switched on for more than 6 hours per day. Most sets can obtain many times more channels than sets in most other countries. Yet a foreigner visiting the United States is frequently amazed that such an affluent country has such poor programming on radio and television, strangely lacking in both diversity and quality. The opportunities offered by the medium have been gravely missed. The Task Force report discussed several possible solutions to this problem, including Pay-TV, cable TV, and various sources of educational TV. However, it made *no mention whatsoever* of the only solution that has given diverse and intelligent programs in some other countries—a nationwide government-operated network like the BBC in Britain or the NHK in Japan. This omission seems to suggest that the topic is taboo in the United States.

If private enterprise has failed in culture, it has certainly succeeded in technology. One might judge the relative claims of government versus private ownership of telecommunications, or of competition versus absolute monopoly, by grading the performance of the two systems. Most government-controlled monopolies would not score very well. Indeed, most would fail abysmally on many counts. Countries with such systems generally have long waiting lists for telephones, woefully inadequate data transmission facilities, out-of-date switching offices, crude PABXs, expensive long-distance calls, hardly any portable radio transceivers in public use, and regard the Picturephone as science fiction.

Broadband data transmission is virtually unknown. One hears horror stories of people waiting two years for a telephone to be installed.

At the end of the 1940s America began to replace step-by-step (Strowger) switch telephone exchanges with the faster, more flexible crossbar switch. In the late 1960s it began to replace the crossbar-switch exchanges with vastly superior computerized switching. Europe is still installing Strowger-switch exchanges in many areas.

The laboratories in Europe produce excellent theoretical concepts relating to telecommunications and a few working prototypes, but the public networks remain old-fashioned and under-capitalized.

It seems that competitive forces are essential to keep the telecommunication companies on their toes. Monopoly becomes moribund. The problem, then, is how to preserve the integrity of the vast network, take advantage of economies of scale, avoid excess duplication, and conserve scarce natural resources such as spectrum space, without stifling invention and entrepreneurship.

Since 1968 in the United States there have been substantial changes in the policy and rules which regulate telecommunications. The changes have almost all been of a nature that makes the telecommunications industry more competitive. All such changes were fiercely fought by the established common carriers, but nevertheless resulted in new product designs and tariffs that were highly competitive, for example AT&T's Dimension PBX, Hi-Lo tariff, Design Line phones, and Dataphone Digital Service.

Box 22.1 summarizes the main events in this period relating to telecommunications regulation. They gave rise to the growth of major new industries each of which has the potential of becoming a multi-billion-dollar industry:

1. *The interconnect industry* which makes devices which can be connected to the telephone network such as decorator telephones, private switching offices, telephone answering machines, key telephones with new facilities, modems, computer terminals designed for direct interconnection, radio telephones, and so on.
2. *The specialized common carrier industry* which constructs telecommunications links for specialized markets, mainly microwave channels for corporate voice and data circuits.
3. *The value-added network (VAN) industry* which leases channels from the carriers, connects computers to the channels and provides new service at new prices, such as packet-switched terminal-to-computer interconnections, message delivery, and facsimile transmission.
4. *The domestic satellite industry* which manufactures and operates satellites and earth station equipment and has the potential of becoming a major growth industry when many corporate locations have their own satellite antennas and demand-assignment equipment.

In addition the new legislation gave a major boost to the *cable television industry,* protecting it on the one hand from the big three television networks

BOX 22.1 A summary of the major regulatory developments in the United States since 1968, almost all of which have the thrust of making telecommunications more competitive

1968 (December): *PRESIDENT'S TASK FORCE ON TELECOMMUNICATIONS POLICY*

In recommending telecommunications policy:

> "we have been guided by the basic premise underlying the law and policy affecting American industry and commerce: that, unless clearly inimical to the public interest, free market competition affords the most reliable incentives for innovation, cost reduction, and efficient resource allocation. Hence competition should be the rule and monopoly the exception."

1970: *OFFICE OF TELECOMMUNICATIONS POLICY* was established as an executive branch of government to study in depth long-range policy alternatives and make recommendations to the FCC and Congress.

DEVELOPMENTS LEADING TO THE GROWTH OF THE INTERCONNECT INDUSTRY

Prior to 1968 ATT Tariff No. 263 was in force:

> "No equipment, apparatus, circuit, or device, not furnished by the Telephone Company shall be attached to or connected with the facilities furnished by the Telephone Company, whether physically, by induction or otherwise (with specified exceptions for police, hospitals, etc.)."

1968: *CARTERFONE DECISION*

A landmark decision followed a long antitrust action by the Carter Electronics Corporation who wanted to couple their mobile radio system to the telephone network. The FCC ruled that the above and related paragraphs were unreasonable and should be stricken from the tariff. The ruling which gave birth to a new industry said,

> "A customer desiring to use an interconnect device . . . should be able to do so, so long as the interconnection does not adversely affect the telephone company's operations or the telephone system's utility for others."

continued

BOX 22.1 *Continued*

1969: An AT&T revised tariff came into effect permitting the direct interconnection of customer equipment via a "direct access arrangement" provided by AT&T, which protected the network. The device limited the input signal strengths and performed all network control signaling functions.

1969 onwards: Rapid growth of the "interconnect industry" making PABXs, modems, decorator telephones, radio phones, and other devices for interconnection to the telephone network.

DEVELOPMENTS LEADING TO THE GROWTH OF THE SPECIALIZED CARRIER INDUSTRY

1963: Microwave Communications Inc. (MCI) files for permission to construct a common-carrier microwave system from St. Louis to Chicago.

1969: A landmark decision by the FCC permits MCI to commence construction (6 years after their original application).

1969, 1970: Many firms including multiple MCI affiliates apply for permission to become specialized common carriers. (There were over 1900 new microwave station applications.) The established carriers petition the FCC to reverse its MCI ruling.

1971: After lengthy hearings the FCC gives an overall policy approval to the specialized common carrier concept (Docket #18920).

1971 onwards: Rapid growth of the specialized common carrier in dustry installing systems for specific business-oriented uses.

DEVELOPMENTS LEADING TO THE GROWTH OF THE CABLE TELEVISION (CATV) INDUSTRY

1968: Justice Department Antitrust Decision urges the FCC to allow CATV to develop as a competitive medium and to permit CATV program origination and advertising.

1969: FCC permits all CATV systems to originate their own programming (program origination was originally made mandatory, then optional).

1970: FCC permits CATV operators to import distant signals, and substitute commercials on them.

continued

BOX 22.1 *Continued*

1970: FCC prohibits phone companies from operating CATV systems in markets where they have phone facilities.

1972: FCC ruled that:

1. Cables must have at least 20 channels.
2. Cables must have built-in capacity for 2-way communication.
3. For each broadcast channel carried, there must be an equivalent bandwidth for nonbroadcast users.
4. There must be one free, dedicated, noncommercial, uncensored, *public-access channel* available on a non-discriminatory basis.
5. In addition, one channel for educational and one channel for local government use, free of charge, must be set aside for 5 years.
6. Minimal production facilities (a television studio) for public use must be maintained.

DEVELOPMENTS CONCERNING THE DATA PROCESSING INDUSTRY

1966: The FCC initiated a *Computer Inquiry* to resolve the "Regulatory and Policy Problems Presented by the Interdependence of Computer and Communication Services and Facilities."

1973: The Computer Inquiry terminated, defining the following services involving computers and communications (FCC Docket #16979):

1. Local data processing. 2. Remote access data processing. 3. Hybrid data processing.	NOT TO BE REGULATED.
4. Hybrid communications. 5. Message switching (and packet switching). 6. Pure telecommunications.	REGULATED BY FCC.

Hybrid data processing was defined as "a hybrid service offering wherein the message switching capability is incidental to the data processing function or purpose"—Not regulated.

continued

BOX 22.1 *Continued*

Hybrid communications was defined as "a hybrid service offering wherein the data processing capability is incidental to the message switching function or purpose"—Regulated by the FCC. There is a grey area between hybrid data processing and hybrid communications.

The decision not to regulate data processing services leaves the computer service bureau industry free and competitive.

Common carriers may not offer data processing services (hybrid or otherwise) except through a separate corporation with separate facilities, officers, and accounting, and circuits obtained like any other corporation.

DEVELOPMENTS LEADING TO THE GROWTH OF THE DOMESTIC SATELLITE INDUSTRY

1962: The Communications Satellite Act, which established Comsat, effectively prevented the use of satellites for transmission within the United States.

1970: The new Office of Telecommunications Policy recommended that any financially and technically qualified entity should be permitted to establish and operate domestic satellite facilities.

1971: The FCC formulated its "open skies" policy:

> "We will consider applications by all legally, technically, and financially qualified entities proposing the establishment and operation of domestic communications satellite systems designed to provide the capability for multiple or specialized communications services. Applicants may propose the rendition of such services directly to the public on a common carrier basis or by the lease of facilities to other common carriers, or any combination of such arrangements. Applicants may also propose private ownership and use or the joint cooperative use of the system by the several owners thereof. Applicants may further propose the shared use of some facilities by different systems, or a division in the ownership of various system components (e.g., user ownership of earth stations to afford direct access to the space segment of a common carrier or cooperative system)."

1972: Eight major applications were received. The FCC responded favorably to the applications with the exception of those from

continued

BOX 22.1 *Continued*

AT&T and COMSAT. Restrictions placed on the latter were "to minimize the effects that AT&T's economic strength and common carrier dominance and COMSAT's role in Intelsat and relation to AT&T might have on multiple entry by new carriers during the difficult start-up periods." AT&T was permitted to lease transponders from COMSAT for an initial period of three years but use them only for MTS, WATS, AUTOVON, emergency restoration in the event of terrestrial outage, and possibly services to Alaska, Hawaii, Puerto Rico, and the Virgin Islands.

1974: Western Union launched WESTAR I and II, the first U.S. domestic satellites.

1975: A joint IBM-COMSAT application was disallowed, but the participants informed that modifications of the application more likely to encourage competition would be allowed. A new IBM-COMSAT-Aetna application in which IBM had a 40% interest was encouraged. SBS (Satellite Business Systems) was formed and given permission to operate a demand-assigned multiple-access satellite system.

The FCC filing said: "With the proposed SBS system, the distinction between central and remote computing is virtually eliminated. Central computers will be able to communicate with remote computers at virtually the *same high data rates at which they process data internally*. A company's data base can, in effect, be moved out to remote processors at all traffic concentration points, and thus be much closer to the company's most remote operations."

DEVELOPMENTS LEADING TO THE GROWTH OF THE VALUE-ADDED NETWORKS

1971: The Office of Telecommunications Policy recommended that "second-tier" common carriers should be permitted to develop and offer new services derived from "first-tier" carrier channels.

1973: PCI, Graphnet, and Telenet applied for permission to operate value-added networks using channels leased from other carriers. PCI and Telenet intended to provide packet-switched connections for computing; Graphnet intended to provide message and facsimile transmission. Permission was granted and other such applications followed.

continued

It is clear that the potential for "second-tier" carriers extends far beyond packet switched networks for data.

1975: First customer transmissions on Graphnet, and later Telenet, networks.

1976 ITT files for its COMPAK value-added carrier.

1974: The Justice Department Antitrust Division recommended that AT&T should be broken up, with equipment procurement being done on a competitive basis rather than from a wholly-owned subsidiary (Western Electric), and interstate transmission being operated by a separate corporation (now AT&T Long Lines).

1976: AT&T drafted the Consumer Communications Reform Act which would declare that actions to encourage competition for long distance and interstate private line services are contrary to the public interest if such competition does not provide innovative service and establish new markets. The act would affirm the authority of states rather than the FCC to regulate interconnection to the telephone network. The FCC prior to authorizing a specialized common carrier must examine all evidence to assure that the authorization will not be a cause for established telephone companies to increase charges for local telephone service. If the act became law it would eliminate most telecommunications competition in the U.S.A. and hence be the end of an era represented by the other developments in this Box.

and on the other hand from the big telephone companies. It has also opened up a vast new potential for mobile radio applications that will also give rise to a multi-billion-dollar industry.

There is no question that the new competition has injected a vitality into the telecommunications industry that was absent prior to 1968. The type of entrepreneur who made America great has something to be excited about. A few Americans have already become multimillionaires through telecommunications and more may join them (including antitrust lawyers). However AT&T drafted an Act in 1976, the Consumer Communications Reform Act, (See Box 22.1) designed to limit further competition.

ARGUMENTS AGAINST COMPETITION

Some strenuous arguments are voiced *against* competition. First, economies of scale exist in telecommunications highways. Figure 22.1 illustrates the approximate investment cost of links of different capacity. The cost *per voice channel* of a 100,000-channel link (like the Bell L5 carrier) is about one hundredth that of a 100-channel link (a small bundle of

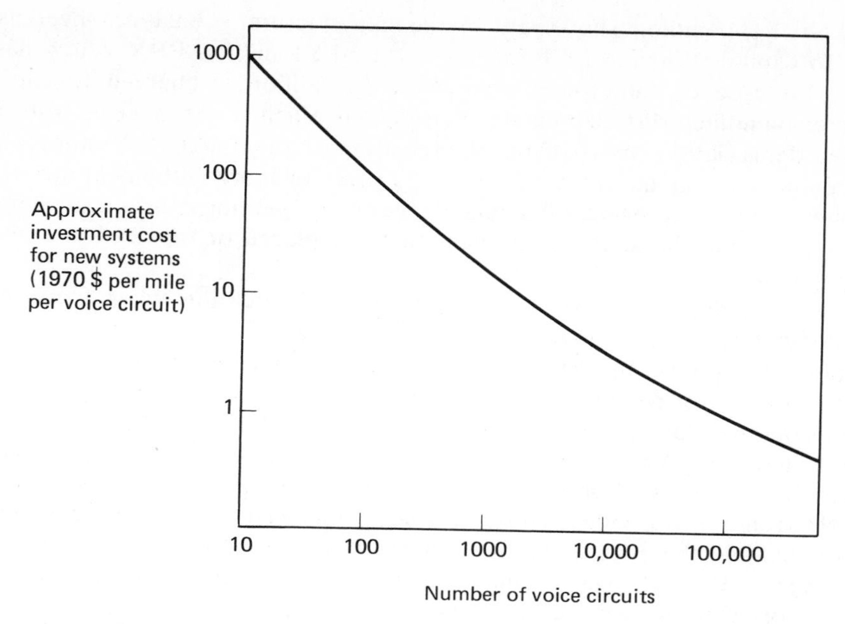

Figure 22.1 Economies of scale in terrestrial transmission.

wire pairs). Economies of scale are also pronounced in satellite systems. Only large common carriers can take full advantage of the economies of scale.

Second, some telecommunication facilities are a scarce natural resource, like the radio spectrum and the geosynchronous satellite orbit. Because these are uniquely valuable to mankind they should be used in an efficient manner. Competition tends to lock us into wasteful uses with unnecessary interference or duplication. A carefully planned nationwide system could make better use of resources.

Third, competing systems are wasteful in some areas, and cause interconnect problems. If several telephone companies laid wires down the same street to give the subscriber a competitive choice, the subscriber would have to pay more for his telephone service. Indeed, it is wasteful for him to have separate telephone and television cables. One coaxial cable (in the future an optical fiber cable) could carry all the signals to his home including signals not transmitted today such as electronic mail and hi-fi music.

An argument against competition sometimes heard in the United States is that the accounting practices of the telephone companies were not designed for the pressures of competitive industry. The telephone companies capitalize much that equivalent competitive corporations would charge to income. AT&T for example capitalizes the full purchase price of equipment including the manufacturing profit and much of the labor cost of installation. Large competitive high-technology corporations charge most of such expense against income. When a computer manufacturer installs a leased $1 million computer for a customer only about $1/4 million shows up on the manufacturer's balance sheet as a charge to capital. When AT&T installs a leased $1 million PBX which has a similar obsolescence rate today, the whole $1 million is charged to capital. Computer manufacturers depreciate their rented machines in 5 years with accelerated depreciation (most of the depreciation at the start). Telephone companies depreciate similar equipment over 20 years or more with linear depreciation. These figures represent a financial danger if competition causes a substantial fraction of the telephone company plant to be displaced or replaced with modern equipment.

The U.S. railroads were in a similar financial situation when faced with rising competition from highways and airlines. Theodore F. Brophy, president of General Telephone and Electronics, commented at a regulatory conference, "I hope I will never be able to write a book on the new competitive philosophy in the telephone industry and entitle it 'From That Wonderful Regulatory Philosophy That Gave You Penn Central.' " The cost of what Brophy fears could be high. But the cost of denying American industry satellite antennas on the roof, electronic mail, video conferencing, high-speed burst computer interconnections, and the other potential benefits of competition would be immeasurably higher. It would be like having prevented the growth of highways and airlines in order to protect the railroads.

NO NATURAL MONOPOLY

The term "natural monopoly" has been applied both to long haul and local telecommunications. It is a very questionable term.

For long-haul transmission there are clearly competing alternatives, thus:

Public versus leased links.

Competing cable systems.

Cable versus microwave systems.

Terrestrial versus satellite systems.

General-purpose versus specialized systems.

Since the spread of long-haul competition the telephone companies have claimed that at least for *local* distribution there are no competing alternatives and that their local wire pair loops are definitely a "natural" monopoly. In reality there are many possible alternatives to wire pair loops for different types of transmission, thus:

CATV cables.

Packet radio.

Private microwave or millimeterwave radio. } for corporate or government use.

Infrared transmission.

Optical fibers.

Satellite antennas.

Pseudo-interactive data broadcasting (like Ceefax but with a higher data rate).

The telephone loops are not suitable for some types of transmission such as video uses and high-speed data bursts. Their limited bandwidth has held back the development of some types of computing. They sometimes have poor reliability and the user usually has no standby facility because only one cable serves a building. If the telephone loops were regarded not as a "natural monopoly," but one of several computing technologies, system designers would select the standby facilities or channel characteristics they need from the alternative.

Today's telephone loops use a massive amount of copper, and the world is running short of copper. A multidropped coaxial loop going to hundreds of subscribers (or in the future an optical fiber loop) can carry signals to subscribers at a fraction of the cost of a wire pair going to only one subscriber.

U.S. REGULATORY ISSUES

Let us comment further on some of the regulatory issues

1. Rate of Return Regulation

Some U.S. common carriers have their profits regulated by the state and federal governments in a system known as the "reward system." A formula is used, which to a large extent determines the attitudes of the carriers. The choice of formula can either impede or encourage technical progress. Several aspects of today's reward system seem to be impeding progress.

The regulation of the major common carriers operates on the concept of "rate of return." In essence the formula says that for every dollar's worth of plant and equipment in the telephone system, the telephone company is allowed to charge rates which give it a return after all expenses of approximately 20 cents before tax. This has led to a practice which some observers of the industry refer to as "gold plating" of telephone company plant. A Bell System microwave tower, for example, is *much* more expensive than a microwave tower of the same capacity from the new specialized common carriers to which rate of return regulation does not apply.

Furthermore the formula encourages the telephone companies to depreciate their plant as slowly as possible in order to make the book value of the plant as high as possible. This is the reason that there is such a large difference between the depreciation schedules of the telephone companies and, say, the computer manufacturers. But this form of accounting is highly dangerous when the rate of change of technology is as great as today and the threat of competition is present.

Telephone companies not only write off their equipment over a long period; they build it to be long lasting. It is normal to design a telecommunications plant for an operating life of 40 years. Long-lived plant with little maintenance cost earns the maximum "reward."

A 40-year replacement cycle might have appeared reasonable in the 1930s, but in today's electronics world the rate of change is so fast that equipment is obsolete almost before it is installed. Few persons in data processing expect to keep a computer longer than 6 years. An IBM 7070 computer which cost nearly a million dollars in 1964 was sold by the Parke-Bernet Antique Galleries for $2250 in 1970. The purchase price of computing equipment is commonly equivalent to about 3 or 4 years' rental; yet most commercial machines are rented. The Bell System ESS computers for switching, however, were planned to have a 40-year operating life.

The promise of today's technology and the demands of the data processing world would imply a rapid installation of PCM equipment, digital channels, satellite links, computerized exchanges, Touchtone dialing, and so on. The reward system, however, encourages carriers to replace existing equipment only slowly.

Fortunately, *expansion* of the networks is rapid, and therefore new types of plant *are* being installed fast. If the carriers double their circuit mileage every 5 years then in 20 years equipment in use today will account for only about one-sixteenth of the total network. In this case, the problem is one of interfacing the old equipment with the new. The compatibility problem is a constant drag on new development. The telecommunications circuits of the 1980s will have to be compatible with those installed in World War II if the rules of the game do not change.

The rate-of-return formula also encourages low maintenance expenditures. Thus we not only have obsolete plant but also low expenditure on maintenance. Some of the major U.S. cities have had serious problems with their telephone service because of this and because of shortage of attention to the new user class who transmit data. Under the streets of New York there are old-fashioned cables in enormous quantities and many of them are in very poor condition, rotted by old age and by the steam and gases under the streets. The regulations give little encouragement to replace them, but they are storing up future problems for the subscribers.

The worst aspect of the rate-of-return regulation is that it tends to discourage projects which could bring a massive saving in capital equipment costs, as could the use of large telephone company satellites today.

2. Domestic Satellites

The domestic satellite issue is politically explosive because satellites could bypass the established telephone trunks, carrying a large nation's toll traffic at much lower cost. More than that, they can be the key to new types of corporate services and computer usage because they can handle much higher bit rates than telephone lines. Burst modems transmitting 60 million bits per second per transponder are in use.

The Communications Satellite Act of 1962 prevented domestic satellites being launched in the United States until the FCC formulated its "open skies" policy in 1971. In that policy the FCC stated that it would consider any reasonable application from organizations wishing to operate satellites. However it then rejected applications coming from *big* organizations such as AT&T and Comsat.

Competition will almost certainly give rise to a greater diversity of schemes than if one monopoly were responsible for the satellite and earth stations. Already we are seeing such diverse uses of satellites as the WESTAR leased channels, the Musak antennas, the ALOHA system, the MARISAT ship-borne antennas, the ATS-6 chicken-wire earth stations for TV reception, and the General Electric corporate earth stations. However, the economies of scale are great in satellites and we would be served better by a few very large satellites than many little ones like WESTAR. There are problems with radio interference, and industry might become locked into a less-than-optimal system with political pressures preventing it backing off into a more integrated approach when the technology evolves.

NASA would possibly be the best organization to develop pilot satellite schemes. Its communication satellites to date, and especially the spectacular ATS-6, have been the world's most advanced. At the time of writing its budget has been cut, some say tragically, so that ATS-6 is the end of that activity. With the cutback in space travel, communication satellite development seems a highly desirable way to use NASA's talent and facilities.

The debate could continue endlessly: which is best, free competition or integrated systems *à la* NASA?

A possible way to obtain the best of both worlds would be to have one organization, perhaps COMSAT or NASA, be responsible for the space segment, and completely free competition in the ground segment. Economies of scale would be achieved by launching satellites weighing many tons, filled with commercially useful transponders (unlike ATS-6 which is a research vehicle) and giving voice channels for one tenth of the cost of terrestrial transmission.

3. International Circuits

There are two main ways to provide a circuit today across the ocean, satellite and submarine cable. The cost per circuit is far lower with satellite than with cable; yet suboceanic cables are still being designed and laid. Why?

The satellite delay time of a quarter of a second has been cited as a reason by the cable proponents. The added military security that comes from having more than one transmission medium is another reason. However, the most important fact is that today no single firm in the United States is permitted to operate *both* cable and satellite or whichever it chooses. We therefore have vast vested interests lobbying for cable on the one side and satellite on the other. The present investments are likely to be regulated on the basis of compromises that seem fair to industry claimants but not designed to reduce costs by maximizing efficiency.

One INTELSAT IVA satellite could handle all the present Atlantic basin traffic, and a second one could handle all the Pacific traffic. Yet the FCC authorizes new trans-Atlantic cables of higher cost than satellites and a small fraction of the traffic capacity. It then insists that the public should be charged the same whether their call goes on the expensive cable circuit or the cheap satellite circuit. A cable to the Virgin Islands was also authorized to operate in parallel with the satellite. The cables have only a small fraction of the satellite capacity, yet the FCC stipulated that new traffic on the Virgin Islands cable was to be allocated on a ratio of 50:50 between satellite and cable. The cost to users of such decisions will be high.

Satellites could drop the cost of international calls to a small fraction of their present cost. With the large satellites ahead, this prospect is likely. However, if satellites are forced to compete with cable on a 50-50 basis, the user benefit will not materialize.

The President's Task Force, reviewing this situation, argued strongly for the formation of a single entity for all U.S. international transmission. This entity would operate cable and satellite circuits, or any other that seemed appropriate.

The Task Force also considered the possibility of unregulated competition in this area but concluded that such competition was unreasonable because it might eventually result in the elimination of cables; it also argued that it would be disadvantageous to have more than one organization launching the satellites. Furthermore, establishing conditions for effective competition would be very difficult. The report concluded that "a natural monopoly is rarely encountered in the real world" but that the transmission segment of international communications appeared to be one.

The creation of the "single entity" should rationalize this industry. The Task Force went on to recommend, however, that the entity be subject to conditions limiting the monopolistic disadvantages, as follows [1]:

1. It should be limited to the function of providing the transmission links and should sell only transmission capacity. It is possible also that the earth stations could be user-owned. The transmission capacity would be sold to competing common carriers. The entity would thus be a "carrier's carrier."
2. It should not engage in manufacturing or have manufacturing affiliations. These factors should be provided by the competitive market place. Satellite and earth station manufacturing would remain a highly competitive industry, thus ensuring vigorous and rapid technological development.
3. It would not provide domestic U.S. transmission or have affiliations with domestic carriers. This prohibition would ensure diverse development of domestic satellite potentials.
4. It would be subject to strengthened government regulation.

In spite of the Task Force's well-developed arguments, it is far from likely that the "single entity" will be formed by Congress. Existing telephone companies will fight to avoid losing their cables.

4. Foreign Attachments

Prior to 1969, U.S. telephone company tariffs prohibited the connection of any device or circuit to the telephone lines, unless it was provided by the telephone company. Devices could not even be linked to the telephone by induction or other nonphysical means.

The Carter Electronics Corporation produced 4500 Carterfones between 1959 and 1966. A Carterfone was an accoustical coupling device for interconnecting the base station of a mobile radio system with the public telephone network. A person using a mobile radio telephone clearly benefits greatly from its interconnection to the telephone system. However AT&T claimed that Carterfones violated the tariffs. In 1966 Carter brought an antitrust action against AT&T. The Texas court held that the FCC should judge the "justness, reasonableness, validity, application and effect of the tariff."

In 1968, after bitter controversy, the FCC pronounced a historic decision that permitted a new industry to be born. They declared the tariff "unreasonable and unreasonably discriminatory" and ruled that the paragraphs prohibiting interconnection be stricken from the tariff. The ruling said that "a customer desiring to use an interconnect device . . . should be able to do so, so long as the interconnection does not adversely affect the telephone company's operations or the telephone system's utility for others."

After the ruling AT&T designed an interconnection device called a *Direct Access Arrangement* (DAA). This is, in essence, an isolation transformer which protects the telephone network from high voltages or signals of too great an amplitude. It attenuates the customer signal if it is too strong. It is illustrated in Fig. 22.2, and is installed on the user's premises in a small plastic box on the wall. It must be used with a telephone company dial and switchhook—usually a conventional telephone. Slightly different DAA's are used for machines which automatically originate and answer calls.

The DAA was leased from the telephone company at a price that seemed out of proportion to its manufacturing cost. Five years later the FCC ruled that machines following certain standards could be attached directly to the telephone lines without a DAA.

After the DAA and its associated tariff were introduced the electronics magazines soon contained advertisements for new devices that could be attached to the telephone system. These included computer terminals with built-in modems, decorator telephones of all shapes and sizes, PABXs, new devices for accessing PABXs, telephone-connected burglar and fire alarms, telephone answering machines, remote control equipment, radio telephones which you could take out to the swimming pool, and so on. Yankee ingenuity had been set free.

One important use is the direct interconnection of corporate tie-line and radio networks to the public network. In some systems a corporate user at home can dial a local call to an interface with his corporate network, and on this obtain a nationwide call which may terminate with another public network link. A connection has then been made between two public telephones at a lower cost than a toll telephone call.

For home use, the telephone link could be used to operate equipment. To save heating fuel, for example, a home could be left at a low temperature when it is empty and the occupant may telephone before he returns to instruct a control unit to switch on the heat. Automatic cookers could be switched on similarly.

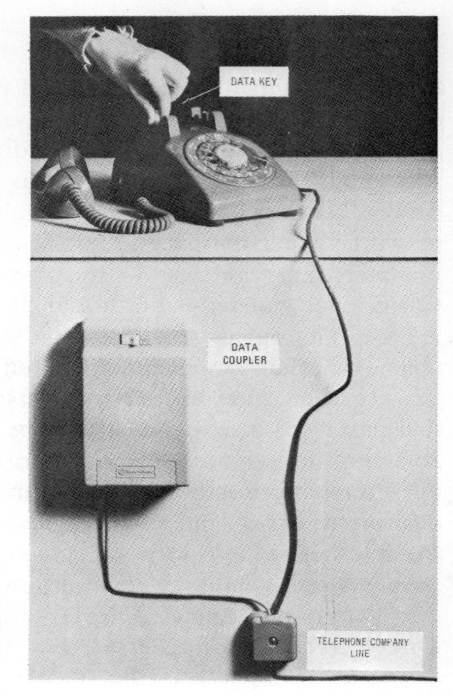

Figure 22.2 The Data Access Arrangement, DAA, gives users in North America freedom to design *foreign attachment* data handling machines and connect them to the telephone network.

NOTE: Devices of approved design can now be attached directly to the telephone wires without a DAA.

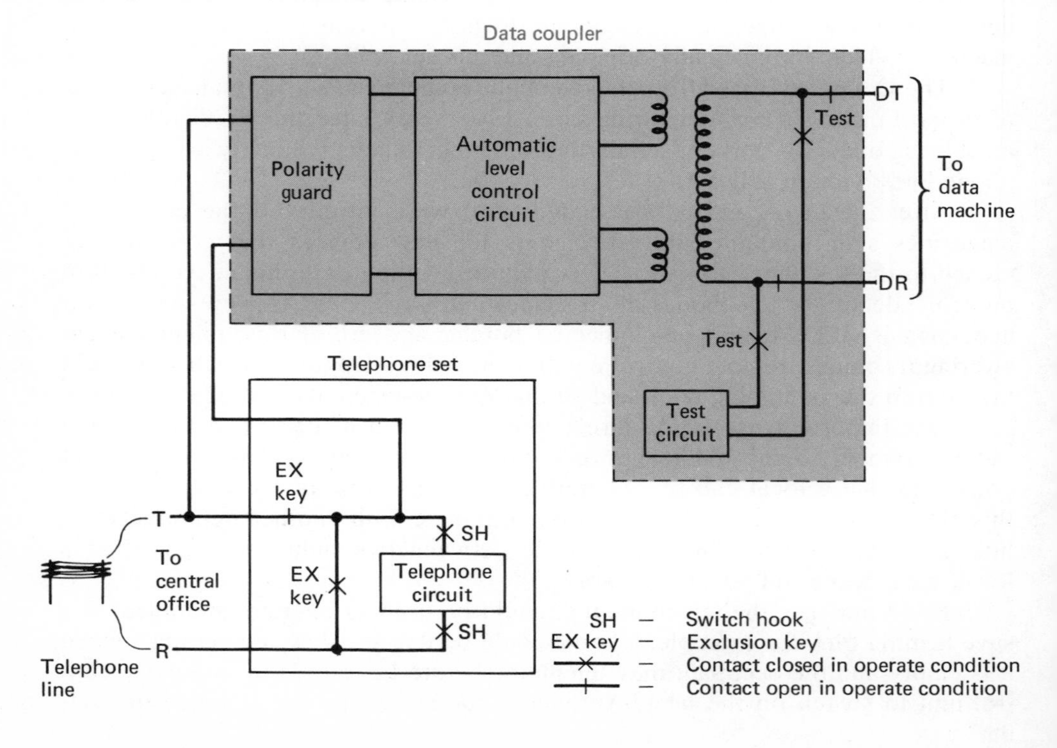

Figure 22.3 The *foreign attachments* ruling in the USA led to new consumer products as well as products for corporate and computer networks. Above is a *Rovafone* portable telephone, which can be carried indoors or outdoors, and linked to the telephone loop by a short-distance radio connection. Particularly useful in the home are telephone answering machines, costing $100 to $150. (*Photo from Fonetron, Inc.*)

5. Specialized Common Carriers

After the Carterfone decision, a second precedent-shattering development took place in the United States in August 1969 when the FCC authorized a small company, Microwave Communications, Inc. to set up a public microwave system between St. Louis and Chicago, paralleling AT&T's routes. John D. Goeken, the 37-year-old president and founder of MCI, had spent 6 years and $400,000 fighting a legal battle with the telephone company to obtain this right. He said he could offer a wider choice of bandwidths than AT&T and prices as much as 94% cheaper.

After the FCC decision, Goeken suddenly found massive sources of finance available to him and applied to extend the system nationwide. Within a few months the FCC was deluged with other applications for new networks, in-

BOX 22.2 Government regulations in the United States which have had a harmful effect on the development of telecommunications

- Rate-of-return regulation discourages the replacement of obsolete equipment and encourages the construction of unnecessarily expensive plant.
- Regulation has permitted a financial structuring of the traditional common carriers which makes them highly vulnerable to the growth of badly-needed competition.
- The Communication Satellite Act and other regulation discouraged the launching of domestic satellites in the United States until 1974. America should have had domestic satellites in 1969 when the military had TACSAT I.
- Regulation has encouraged the use of expensive transoceanic cables rather than permitting free competition from the much cheaper satellite technology.
- Because of the economies of scale in satellite transmission the United States could and should by now have domestic satellites weighing several tons and carrying more than 100,000 voice channels each. Regulation has prevented this and instead AT&T's terrestrial network is being expanded at a cost of several billion dollars per year.
- NASA has created man's most advanced communications satellites and has been responsible for the best research and development in this area. But now, following the success of ATS-6, NASA is prevented from developing further such satellites.
- Regulation has discouraged the development of lower-cost alternatives to today's local loops.
- The T1 and T2 carriers are not being taken into business users' premises where they would have such a major effect on the development of data transmission.
- Throughout the petroleum crisis there has been a failure to perceive that advanced telecommunications form a substitute for the use of petroleum in physical travel. Government should have been encouraging the use of videoconferencing, Picturephone, and electronic mail.
- In spite of the Post Office's massive ongoing losses there has been little or no government encouragement of electronic mail. The economies of scale of the Post Office traffic volume would make transmission via satellite much lower in cost than the physical movement and sorting of encodable mail.

BOX 22.2 *Continued*

- The interconnect industry and specialized common carriers were prevented from coming into existence until 1968. There was a *six year* delay in giving MCI its approval.
- The use of cable television has enormous potential, yet the FCC crippled the growth of this industry until 1970.
- Television is the most powerful mass communication medium in history yet the government failed to encourage its use for socially responsible purposes until the token and inadequate introduction of educational television in 1967.
- The UHF radio band is uniquely valuable for mobile radio applications for which spectrum space is so short, yet the FCC inappropriately allocated 420 MHz of UHF frequencies to television where it has been mostly unused. CATV technologies are the long-term answer to expanding television facilities.
- The frequencies allocated the satellite transmission in the early 1960s were exactly those frequencies which would cause the maximum problem with terrestrial microwave interference, thereby preventing the installation of satellite antennas where they could be of the most commercial value.
- Radio spectrum allocation has been done in an out-of-date fashion, preventing many potentially valuable users of radio.
- A better quality television image is desirable, and necessary for new uses of television such as electronic library facilities in the home, yet standards have not been set for high-fidelity television.
- In many areas the United States has not conformed with international telecommunications standards.
- AT&T was permitted to grow into a massive vertical monopoly preventing competition in manufacturing telephone equipment for the Bell System. For 30 years subscribers were given little choice and almost no new equipment such as telephone answering machines.
- Western Electric, manufacturing much of the world's best telecommunications equipment, is allowed to follow a ruling of not selling it abroad, thereby negatively affecting the U.S. balance of payments.
- In general, the native ingenuity of Americans and the drive of U.S. entrepreneurs would have created a richer assortment of telecommunication uses if not prevented by regulations.

NOTE: There have been some more encouraging regulatory developments since 1968, summarized in Box 22.1.

cluding an application from the Data Transmission Company (Datran) to build a nationwide switched *digital* microwave system. The established common carriers protested the new applications and fought unsuccessfully to obtain a reversal of the MCI decision. Within a short time the FCC received applications to build 1900 new microwave stations and 40,000 route miles of transmission.

The FCC conducted a lengthy inquiry into the new common carriers applications, and in 1971 gave an overall policy approval of the *specialized common carriers*. In December 1971 the first MCI customer went on-line and within five years specialized common carriers had more than $400 million worth of plant installed.

6. Cream Skimming

A major concern of AT&T about the specialized common carriers was that the new companies wanted to "skim the cream" of telecommunications, operating in only the most profitable market segments and locations. The large common carriers are constrained to provide their services at similar prices on a nondiscriminatory basis. The pricing of telephone service in the U.S. has, until very recently, been based on the concept of *value* to the subscriber as determined by the telephone companies. This is the reason why businesses pay more than homes for equivalent service. Such discrepancies in pricing may vanish in coming years if the industry moves to pricing based on cost of service.

It is expensive to run telephone lines to sparsely settled regions. The high profits from densely settled regions help to pay for the telephones in the remote regions. If a new carrier operates in only the high-profit regions then it can naturally charge lower prices than the established carriers. The eventual outcome of unregulated competition would be that the telephone charges in remote areas would become high. Some would contend that this situation should be permitted to occur and that an artificial price structure is wrong: a remote farmstead should pay heavily for its telephone—or Picturephone. Are we going to do the same with cable TV when it provides services that over-the-air TV does not? Should the TV cable to the remote farmstead be subsidized? Should Picturephone and data transmission be cheaper in the 50 or so largest cities?

For the time being it is thought desirable to inject new competition into telecommunications, and new companies *have to* build up their operations starting with a small high-demand segment of the market. Hence new carriers are permitted to cream-skim.

AT&T's answer was a two-price-level tariff for leased lines, the Hi-Lo tariff, and its DDS tariff for data transmission operating to only the most profitable cities.

7. Public Use of Remote Computing

Telecommunication systems use computers in different ways. Some use them for switching, some for storing messages which are transmitted, some for processing the data transmitted. At one extreme the computer merely switches

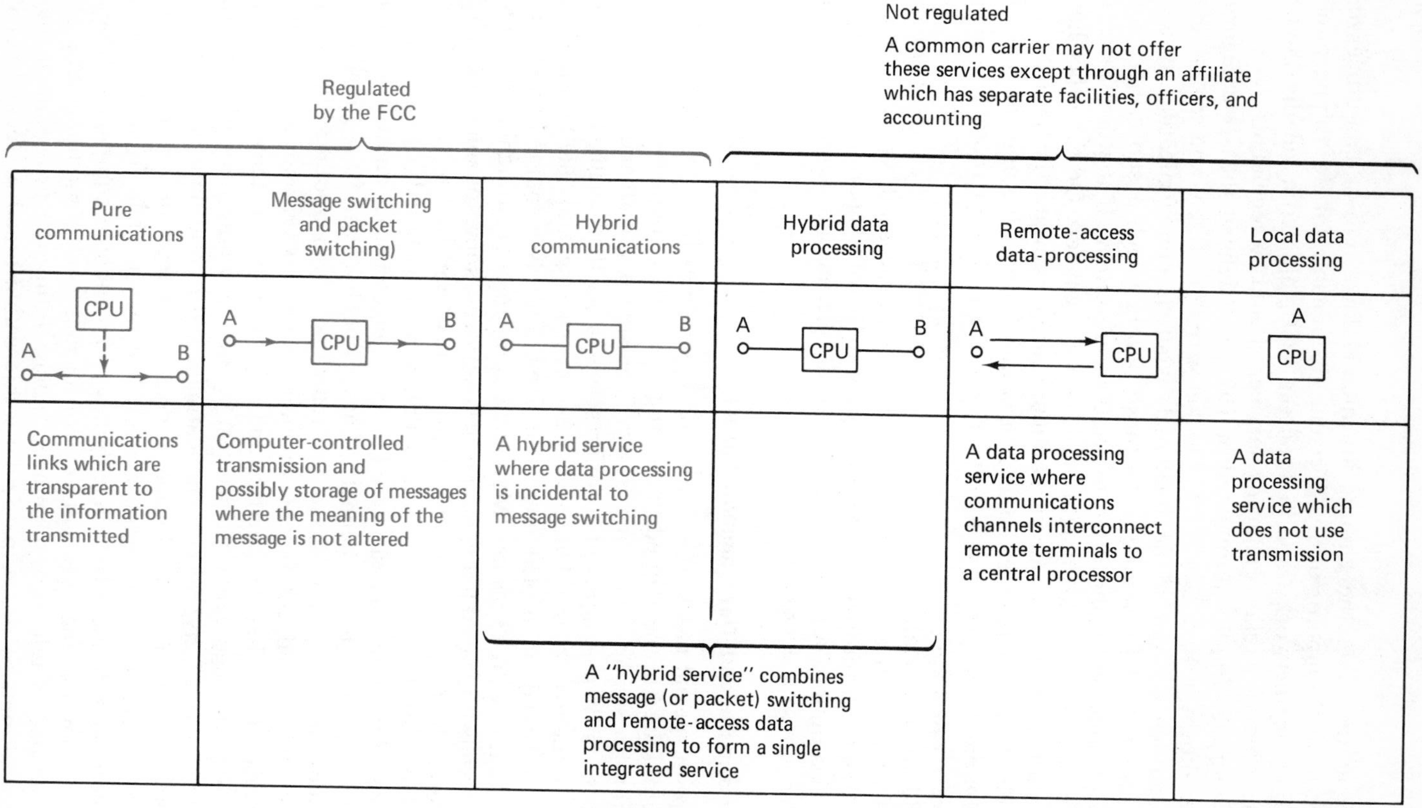

Regulated by the FCC			Not regulated. A common carrier may not offer these services except through an affiliate which has separate facilities, officers, and accounting		
Pure communications	Message switching and packet switching)	Hybrid communications	Hybrid data processing	Remote-access data-processing	Local data processing
A — B (CPU)	A → CPU → B	A — CPU — B	A — CPU — B	A ⇄ CPU	A CPU
Communications links which are transparent to the information transmitted	Computer-controlled transmission and possibly storage of messages where the meaning of the message is not altered	A hybrid service where data processing is incidental to message switching		A data processing service where communications channels interconnect remote terminals to a central processor	A data processing service which does not use transmission
		A "hybrid service" combines message (or packet) switching and remote-access data processing to form a single integrated service			

Figure 22.4 Range of services defined by the FCC computer inquiry final decision.

the circuits; at the other the circuits are merely links into a data-processing system. The term "computer utility" became fashionable for describing public access to computer networks, and in 1966 the FCC initiated a lengthy inquiry to determine whether public computing services should be regulated.

The *Computer Inquiry* terminated in 1973 and defined the six categories of operation shown in Fig. 22.4 (FCC Docket #16979). Local and remote data processing service are not to be regulated, whereas communication systems are. There is a *hybrid service* between these two in which a subscriber sends data, it is processed and transmitted to another subscriber. If the data processing is the primary part of this operation, it is not regulated. On the other hand if the operation is primarily one of communication between the parties, it *is* regulated. The former is referred to as *hybrid data processing* and the latter as *hybrid communications*. There is a grey area in the middle of these two about which lawyers will argue.

Hybrid communication services must be completely tariffed and regulated by the FCC. Common carriers may not offer data processing services (hybrid or otherwise) except through a separate corporation with separate facilities, officers, and accounting. AT&T is excluded from offering any such services under an earlier consent decree.

8. First-Tier and Second-Tier Common Carriers

Except in special cases, the sharing of leased lines by different customers has been prohibited in the United States and most other countries. It has rarely been possible to set up as a "line broker," leasing a broadband channel and employing one's own equipment to offer services on that channel to different customers. Recently in the United States the sharing and line-broking rules have been removed on leased interstate lines. This is a major breakthrough towards more innovative services. Many new companies are attaching computers and other devices to leased lines to offer new services, some for small specialized markets which the bigger telephone companies may not want to serve.

The Office of Telecommunications Policy recommended a policy of *first-tier* and *second-tier* common carriers. The first-tier construct and own telecommunications links and lease channels. They typically own 50%–100% of the channel miles in service and lease the remainder from another carrier. The second-tier carriers add equipment, including multiplexors and computers, to channels leased from first-tier carriers and sell services that they create in this way, including message-delivery services, packet-switching networks, information retrieval services, hybrid communications, and hybrid data processing services. Taking this view to its logical conclusion, telephone service could be provided by a second-tier common carrier, leasing channels from the specialized common carriers and adding telephones and switching equipment. It seems likely that second-tier markets will develop in many telecommunications areas including the use of television channels, mobile radio, the provision of music, electronic mail, and so on. A particularly interesting breed of second-tier carrier may arise leasing satellite channels and installing small customer earth stations. The second-tier carrier often minimizes his investment in terminals by letting the customer provide these.

Legislation in favor of second-tier carriers increases the diversity and competitiveness of the telecommunications industry.

9. Value-Added Networks

In 1973 Packet Communications Inc. (PCI) filed for permission to operate a packet-switched network similar to the ARPA network but with a slightly different proposed architecture. Shortly afterwards Graphnet filed for permission to operate a facsimile network delivering messages to facsimile machines from computers, terminals, or other facsimile machines. Telenet filed to operate a packet-switching network like the ARPA network, delivering data packets between data-processing machines or terminals in a fraction of a second. All three received permission. Graphnet and then Telenet became operational in a limited initial form in 1975.

This genre of second-tier became known as a *value-added* carrier, and their networks were known as *value-added* networks. Value-added common carriers are often thought of as operating packet-switched networks. However, their networks do not have to be packet-switched; other structures are possible. The Graphnet operation does not use packet-switching.

In 1976 ITT announced that it would start a value-added carrier, called COMPAK, which would transmit data and facsimile. The speed, code, and format conversion possible with a value-added, computer-controlled network are valuable for facsimile documents because many facsimile machines are incompatible. Incompatible machines can only transmit to one another when a value-added network is used. MCI also commenced a value-added service. It is likely that we shall see value-added networks being operated by the specialized and traditional common carriers as well as by the new second-tier carriers.

It is possible that value-added networks will become a major industry shipping bursts of data among computer users at a data rate far higher than that possible on public telephone lines and a cost far lower.

10. Cable TV Development

As indicated in Chapter 9, cable TV offers a wide variety of exciting prospects. The existing TV broadcasters, however, have opposed it so vehemently that in the 1960s its development was seriously retarded. Considerable money can be commanded for such fights. In 1966 the FCC was persuaded to impose rules so stringent that they virtually stopped extension of cable service. In the top 100 markets (about 89% of all TV sets), they required operators who wanted to import distant signals to prove that doing so would not harm any existing stations or *any that might later be established.*

While the cable casters were under crippling attack from the established networks on the one side, the telephone companies attacked them on the other.

Cable TV would have fitted naturally into the telephone companies' service pattern, and it has the great appeal for them of being a business without rate regulation. The cable TV companies needed to have space on the telephone company poles, and often the telephone company manufactured the cable equipment. In leasing the equipment to the cable TV companies, telephone company contracts frequently prevented them from originating programming and from engaging in two-way communications. The prevention of two-way communications rules out many of the interesting applications of cable discussed in Chapter 9. Telephone traffic could be sent more economically on the CATV cable then on telephone loops so the telephone companies would naturally be wary of cable usage developing too freely.

From 1969 to 1972 the regulatory climate changed and breathed vigorous new life into cable TV. CATV operators were allowed to import distant signals and substitute commercials on them. They were instructed to build two-way capability into their cables, and the phone companies were forbidden to operate CATV systems in the same area they operated telephone facilities. Short haul microwave usage was permitted for distributing CATV signals, bypassing phone company facilities. The FCC ruled that the cables must have at least 20 channels and that for each channel received from broadcasts and carried on the cable, there must be another channel for non-broadcast users. The cable must carry one free, dedicated, noncommercial, *public-access channel* available to users on a nondiscriminatory basis, and a rudimentary television studio must be available for public use. The CATV operators were encouraged to originate programs. Thus instead of protecting the major television networks, the FCC was now strongly encouraging competition with them.

In hope of encouraging more socially beneficial uses of television the FCC ruled that one channel for educational and one channel for local government use, free of charge, must be set aside for five years.

11. The UHF Television Channels

As we will discuss in Chapter 24, there is a major problem in radio spectrum usage. Many valuable new uses of radio are being prevented because of spectrum shortage, particularly in land mobile radio applications.

The frequencies usable for such applications are from 5 to 890 MHz. In the United States, the federal government reserves about 300 MHz of this amount for itself in total. Of the remaining 585 MHz, no less than 420 MHz are allocated UHF television broadcasting—the active band from 470 to 890 MHz. This figure is in addition to the 72 MHz of VHF television frequencies (channels 2 to 13) in normal use.

This allocation was made in 1949. Unfortunately, UHF television in the United States was very slow to take off. Few commercial stations broadcast UHF at all. One reason is that the transmitting equipment is more expensive than VHF. Until 1964 very few sets were able to pick up UHF, and then a law was passed saying that all new sets must be able to pick it up. Still, many sets today are rarely used for this purpose because there are no clearly marked UHF channels that the channel-selection knob clicks to.

The rapid growth of cable TV at present makes it seem even more unlikely that all the UHF spectrum space allocated to television will be used. In view of "the silent crisis" of the radio spectrum, which has so many valuable potential uses, it seems appropriate to distribute new television services by cable rather than radio, at least in the denser areas of the country where cable TV is economic. Some, and possibly all, of the unused UHF frequencies should be freed for other purposes.

The powerful television industry, however, is fiercely opposed to this view and is capable of great political pressure. The UHF allocation costs broadcasters nothing, and even if they do not use it they do not want to give any of it up to other types of users.

When regulatory authorities yield to the powerful vested interests, new uses of technology can be impeded.

12. Radio Spectrum Allocation

One of the most pressing problems is the allocation of the radio spectrum. It seems clear that flexibility is needed in spectrum management. The present method of allocating frequency bands nationwide is out of date. Spectrum engineering and allocation by TAS (time-area-spectrum) packages (Chapter 24) are needed. Transferable "property rights" in spectrum may be a solution. The President's Task Force recommended an "eclectic" approach introducing major—if incremental—modifications of existing administrative, economic, and engineering practices.

It summarized its conclusions as follows [2]:

A. As a basic guideline, we should seek that combination of spectrum uses which offer maximum social and economic contribution to the national welfare and security.

Accordingly, the following principles emerge:

1. We should seek the continuing substitution of higher-valued spectrum uses for lower-valued uses and the *addition* of uses whose net effect is to increase overall benefits, with due consideration of all imbedded capital investments.
2. Unused spectrum resources should be employed to meet any legitimate need provided that this does not cause excessive interference to existing uses, conforms with established standards and international agreements, and does not interfere with established plans for higher-valued uses.
3. Comprehensive coordination of all spectrum use is required, under a continuing framework of public administration.

B. Greater consideration of economic factors is necessary.

1. An improved schedule of fees for spectrum licenses should be developed, which reflects the extent of spectrum use (e.g., bandwidth, power, service area, time availability and the level of demand for spectrum rights). And intensive studies should be conducted of other means to account for economic value, including adjustable license fees, spectrum leasing, and taxation.
2. License privileges should clearly be stated for each class of spectrum use (e.g., land mobile, radio relay, etc.) in terms of interference probability, channel loading, service quality, and other appropriate factors.
3. Administrative procedures should be modified to permit greater transferability of licenses among legitimate spectrum users within broad service classifications, subject to all relevant conditions of the initial license, including the requirement that all exchanges or transfers be registered and approved by the spectrum management authority.
4. Procedures should be developed whereby a prospective spectrum user may obtain a license even though this would represent a potential source of harmful interference to an established clear channel user, provided that prior arrangements are concluded between all affected parties, including adequate compensation or indemnification by the new user.

C. Greater attention to individual spectrum uses should be achieved through "spectrum engineering" and related technical considerations.
1. A more flexible approach to spectrum management should be adopted, under which the National Table of Frequency Allocations is transformed over time from a fixed allocation by user category to a basic planning guide by service classification.
2. A comprehensive spectrum engineering capability for individualized planning and engineering of spectrum uses should be developed, charged with continuing improvement in technical design and operating standards for all transmitting and receiving equipment and other devices that materially affect spectrum use.

D. Enhanced management capabilities and a restructuring of responsibility and authority are required.
1. Legislation should be considered which would vest in an Executive Branch agency overall responsibility for ensuring efficient spectrum use for all government and non-government uses; this legislation should contain appropriate guidance as to coordination required between the spectrum manager and the FCC in areas of mutual interest and concern.
2. The agency should be given the resources needed to develop a strong interdisciplinary capability embracing technical, economic, social, and legal skills, to support its spectrum planning, management, and coordination responsibilities as described in this Report.
3. In particular, the agency should: (a) determine and continually update the division of spectrum among various classes of users, and adminis-

ter its use on the basis of detailed planning and engineering, at local and national levels; (b) establish and enforce technical standards applicable to all transmitting and receiving equipment and other devices that materially affect spectrum use; (c) coordinate federal R&D activities oriented toward spectrum management and use, except those directed to fulfill a specific mission of another agency; and (d) administer any user fee system now existing or later established.

4. In the interim, to meet existing spectrum management problems and to prepare for the future, resources should be provided to begin effectively to implement the general and specific recommendations of this report.

E. Specific recommendations in selected problem areas.
   1. Land mobile radio services
      (a) Land mobile radio services should be authorized to use spectrum resources now within the national allocations for UHF television broadcasting which are unusable by television stations under the present TV station allotment plans; subject to operating criteria which will avoid harmful interference to television broadcasting on adjacent channels or in adjacent geographic areas.
      (b) Equipment and operating standards should be established for engineering future land mobile services to permit closer spacing of base stations sharing the same frequency assignment: the use of multi-channel radio equipment should be encouraged wherever this would economically provide more efficient spectrum use.
      (c) Development and use of common-user and common-carrier mobile radio systems—including those employing wire-line trunking between individual base stations—should be encouraged, particularly for users with intermittent service requirements.
      (d) A range of channel loading criteria should be established to encourage effective frequency sharing among complementary uses and to provide a satisfactory and well defined quality of service to each user.
      (e) The sub-allocation of land mobile spectrum bands by user class should be substantially discontinued. Any remaining sub-allocations should be flexibly administered within each geographic area.
      (f) Procedures should be established whereby members of the general public now restricted to the Citizens Radio classification may be licensed to use certain land mobile spectrum resources subject to compliance with reasonable technical and operating standards and appropriate channel-loading criteria.
   2. Public safety radio services
      (a) The public safety radio services, in particular, should be incorporated into the government spectrum allocation and management framework.
      (b) Operating standards requiring greater time and geographic frequency sharing among public safety agencies should be established.

   (c) Development of common-user mobile radio systems for public safety services should be encouraged.

3. Television broadcasting
   (a) Spectrum resources presently allocated for broadcasting which are unusable for that purpose under existing station allotment plans should be made available for land mobile and other uses.
   (b) Studies of improved techniques for television broadcasting should be carried out on a continuing basis, with respect to alternative distribution methods, channel bandwidth reductions, and reduction in total spectrum allocations.
4. Microwave service (1000 – 10,000 MHz)
   (a) Improved operating standards (e.g., modulation, antenna directivity, space diversity, etc.) should be established to achieve greater spectrum re-use and interference protection between terrestrial facilities sharing the same frequency ranges.
   (b) The criteria for satellite/terrestrial sharing of all spectrum allocations below 10,000 MHz should be re-evaluated, giving due consideration to the significant technical differences between domestic and international satellite systems and to improvement in technical data since the existing criteria were established.
   (c) Experimental programs should be conducted to ascertain the probability of interference between satellite earth stations and terrestrial radio relay stations, in shared frequency bands below 10,000 MHz.
   (d) Improved criteria and coordination procedures should be developed for efficient sharing of spectrum allocations and orbital locations among domestic and international satellite systems, both government and non-government.
5. Millimeter-wave bands (above 10,000 MHz)
   (a) Continuing research and development activities needed to bring about effective and efficient use of these spectrum bands should be encouraged, through federal R&D programs and flexible policies with regard to the potential uses of these bands.
   (b) Existing domestic allocations of all millimeter-wave bands should be reviewed to determine the feasibility of inter-service sharing of these bands as an alternative to exclusive domestic allocations to satellite and terrestrial services.

GOVERNMENT FUNDING

Looking at the eventual future of a telecommunications plant, it seems certain that it will become almost fully digital, with digital switching. Analog signals will be carried in digital form, and intricate solid-state logic will be used throughout the networks. The cost of transmission with such networks will become a fraction of today's cost, and the capability will be much greater.

A problem exists in how to progress from today's analog plant to tomorrow's digital plant. To make the transition it may be necessary to make investments that will make the operation of today's telephone plant temporarily more

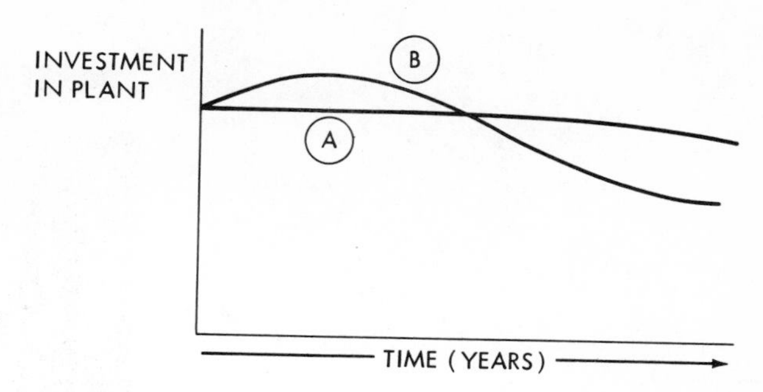

Figure 22.5 Which nations will take path B?

expensive, as shown in Fig. 22.5. Taking course B rather than course A will result in lower costs in years ahead, at the expense of higher costs in the immediate future. However, if course B represents the building of a digital network with telephone, data, and other facilities integrated, it will provide much greater capability than course A in the years ahead. The value of this to the economy can be estimated only very approximately but is likely to be immense, even if one discounts intangible benefits such as computer-assisted instruction in the home (which might become technology's most powerful societal force).

Future value to the economy, especially when filled with so many intangibles, rarely motivates businessmen to open their purses. If investment bankers select the course, it may not be course B. Failure of imagination is absolute if there is not a high return on venture capital in a few years.

To develop highways, massive government funding was necessary, and a car without roads is like a videophone set without communication lines. U.S. federal funding alone on highways exceeded $70 billion in one 10-year period. A similar expenditure on communication satellite facilities would work miracles (a spectacular satellite costs less than $70 million). The payoff for using government funds to push digital telecommunications over the hump in course B will be far greater than the payoff from many uses of government funds.

Some nations probably will spend the money to take course B. Many governments, however, will lack the imagination and the courage to invest sufficiently in this vital future resource. Europe has no interstate telecommunications authority today. It is fascinating to reflect what a Common Market NASA-like agency could do for Europe, building the satellite, video, electronic mail, and value-added networks that will be such an important part of our future.

THE BATTLES AHEAD

Great battles are yet to be fought in telecommunications. The stakes are high and the changes needed are revolutionary. The fighting will be fierce.

When the battles are over, the vast interlocking networks will look very different from today. The condition of man, living in a maze of electronic signals, will become profoundly changed.

There is a danger that politics, lobbying, monopolistic sloth, regulatory ignorance, or destructive competition will rob us of part of the riches that the

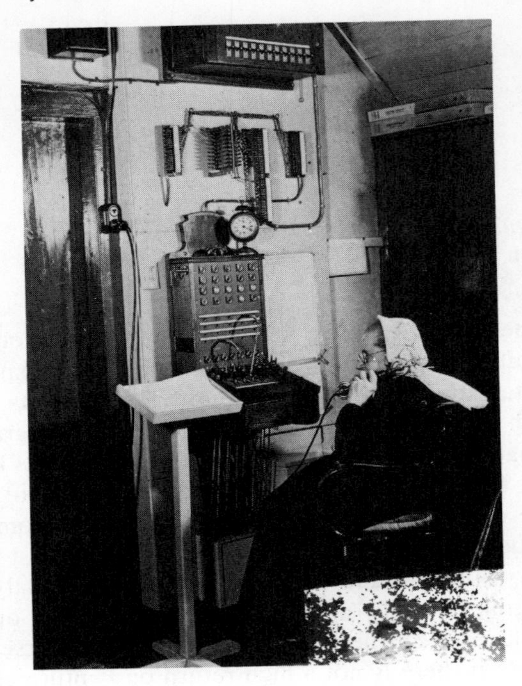

A typical rural telephone exchange in the Netherlands (about 1930). As new telecommunication facilities come into operation, they have to coexist with the old. (*Courtesy Dutch Posts, Telegraph and Telephone Administration, The Hague.*)

technology could bring. It behooves our politicians and regulators to understand fully the many facets of possible future developments in telecommunications. However, it seems sadly possible (as with some other public uses of technology) that through lack of such understanding, we shall fail to make full use of the new opportunities for enriching man's world.

REFERENCES

1. Final Report of the President's Task Force on Communications Policy, Chapter 2, page 9, Washington, D.C., December 1968, Chapter 2, pages 38–47.

2. "The Use and Management of the Electromagnetic Spectrum," Chapter 8 of the Final Report of the President's Task Force on Communications Policy, Washington, D.C., December 1968.

23 A FUTURE SCENARIO

This chapter attempts to summarize the possible development of the technologies we have described by outlining a telecommunications scenario for the rest of the century.

The straight line in Fig. 2.5 has been rising by a factor of 10 every 17 years. It appears that its rise will be at least as great in the future, but that many of the channels will become digital, thereby giving new opportunities for using telecommunications. Interestingly, the number of radio transmitters in use is also increasing exponentially and has been going up by a factor of 10 every 17 years. In 1950 there were 3 transmitters per 1000 persons in the United States. In 1967 there were 30 transmitters per 1000 persons, somewhat less than half of which were mobile two-way radios. In the mid-1970's there was a major upward surge in citizens' band radio transmitters. The President's Task Force on Communications Policy forecast a continuation of the exponential trend, in which case there would be 300 transmitters per 1000 persons by 1984. Most of these transmitters would be mobile radios.

In this chapter we assume that the line in Fig. 2.5 will continue upwards for the rest of the century, which is indicated by current research. We will also assume that computer logic circuitry continues to become cheaper, smaller, more reliable, less power consuming, and less heat dissipating. We will assume that the size of on-line file systems continues to increase and that their cost continues to decrease.

Given these assumptions, we will outline a table of events that should be regarded not as a *forecast* but as a statement of what is likely to be *possible.* Whether or not discrete events in the table come into being will depend on the entrepreneurs of the day, on laws, and on the societal structure. The reader might choose to slide some of the events listed backwards or forwards in time. Sooner or later the telecommunications developments listed will occur. They may occur later than on this table because of shortage of capital, or because of harmful government regulations. Those nations which have the advanced facilities first will be in a position to achieve higher industrial productivity, faster economic growth, and greater societal benefits.

NOT SO MUCH A FORECAST AS A STATEMENT OF POTENTIAL

Late 1970s

Semiconductor logic and memory circuitry continues its rapid drop in cost. Several manufacturers set up highly automated mass-production lines to manufacture LSI wafers. Instead of dicing the wafers into chips, techniques are devised for using entire wafers and bypassing faulty segments. LSI *design* is further automated so that with the automated production facilities LSI wafers or chips can be designed somewhat more quickly. Charge-couple devices, magnetic-bubble devices, and semiconductor lasers, join the family of cheap mass-produced miniature electronic circuits.

The cost of microprocessors drops below $20.00 each. Nonvolatile semiconductor memory becomes available at low cost. Large charge-couple and magnetic-bubble memories become available. Pocket machines which had little or no memory in the early 1970s become available with substantial memories. The program libraries for pocket programmable machines grow impressively. Designers everywhere are faced with an imposing challenge in how to use this new hardware.

Mass on-line storage devices are marketed which hold a trillion (10^{12}) bits. These are referred to as *terabit* stores. These machines are initially perceived as being equivalent to on-line tape libraries, but later become general-purpose on-line computer storage units. The generation of computers which appears in the late 1970s is oriented towards the use of telecommunications, interactive terminals and mass storage units.

Data processing continues its prodigious growth, with some of the fastest growing segments being related to data transmission. Minicomputers spread faster than larger computers and microcomputers spread faster than minis. Small, fast, and inexpensive files units become available for use by the minis, and solid state storage units become available for the microcomputers.

Much more logic and memory (loosely termed "intelligence") is built into computer terminals. Many terminals have their own files. Data processing systems evolve towards distributed intelligence, distributed processing, multi-computer networks, distributed files and finally distributed data base systems. In some systems there are strong reasons for centralization rather than distribution of data and processing. So the industry advances along multiple paths: large centralized systems, systems with distributed storage and processing, stand-alone minicomputers and minicomputer networks, and portable microcomputers.

Data networks are installed in many countries, initially serving only the dense cities. There is a high level of incompatibility between the different networks. Some are leased line networks. Some use packet switching. Some use fast-connect circuit-switching and some use slow conventional telephone switching. Some use satellites. New CCITT recommendations for *virtual call* and *permanent virtual circuit* protocols promise to lessen the incompatibility problem.

The U.S. specialized common-carrier industry grows turbulently, with takeovers and financial suspense stories. The recession tends to help in that corporations seeking to cut their telephone bills lease the cheaper channels. The weakest link is the local distribution. Major changes occur in the telephone company tariffs as a result of the new competition. AT&T attempts by legislation to recover its monopoly position.

The value-added carriers in the United States grow, providing tariffs that are of value for interactive computing, short-batch data transmission, and message delivery. Telenet in the United States, and the Datapac service in Canada serve increased numbers of computer users and provide tariffs appropriate for small users such as locations connected to electronic fund transfer networks.

Some companies become hybrids between a specialized common carrier and a value-added carrier. They lease an increasing number of digital channels from AT&T to supplement their own and connect them to their switching computers. At least one major telephone company announces a packet-switching offering. In other countries a variety of equivalent facilities are built, providing the data networks mentioned above.

Facsimile mail and document transmission increases in popularity and drops in cost. Conventional mail delivery increases in cost and unreliability. It becomes clear that facsimile gives the only way to deliver mail rapidly and hence a large market grows. Electronic mail and message services become an important field for the value-added carriers.

The effect of distance on transmission drops, and some tariffs are independent of distance. This has a major effect on the organization of data processing in some nationwide corporations, leading to increased consolidation of large systems. At the same time, the spread of minicomputers leads to increased fragmentation in other corporations, and to the spread of distributed systems.

Privacy and security of data transmitted and stored become a major concern. Several large-scale crimes and scandals occur. A variety of legislation is passed related to the use and control of data in computers. The technology evolves so that it becomes possible to build very secure systems.

Users with direct access to the data networks do not need modems, and can often transmit at higher speeds than those on telephone circuits. However, many users have to be connected to the data networks via a telephone link. The local telephone loops act as a bottleneck preventing high-speed connection. Attempts to bypass the local loops are made with optical transmission, cable TV and millimeterwave radio. Optical transmission is affected by bad weather; cable TV does not go into many business premises; millimeterwave transmission is used but is often too expensive. There are some cases of AT&T installing T1 carrier links into users' premises, carrying 1.344 mbps.

Several domestic satellites join the existing WESTAR and RCA satellites in North America. There is continuing argument about the regulation of satellites, most of it carried out in ignorance of the technology potential. The first satellites with commercial transponders at 12–14 GHz are launched. These permit earth stations on city rooftops and users' premises anywhere, and also permit smaller and cheaper satellite antennas. Modems operating at 60 million bits per second are used with existing satellite transponders (such as WESTAR and RCA).

Domestic satellites are launched for Japan, Australia, Indonesia, Iran, and Arab countries. Europe uses its small *Symphonie* satellite but not to enhance Europe's telephone or data transmission networks.

Frequency-division multiple-access (FDMA) is employed to enable users in geographically scattered locations to share the satellite transponders on a demand basis, i.e., in a manner which varies with their traffic requirements. Later as digital techniques come to the fore with satellites, time-division multiple-access (TDMA) is used for the same purpose. TDMA enables users to request voice or video channels, and data channels of widely varying data rates. These demand-assigned multiple-access techniques make it economical for areas without a high traffic volume to have a satellite earth station.

TDMA is also demonstrated on cable television channels and high-speed PCM channels.

Satellite usage is fragmenting into two types; first augmentation of existing common carrier networks by means of a small number of large earth stations, and second, bypassing common carrier networks with large numbers of small earth stations. Systems in the latter category include those which provide multiple-access corporate networks and those which provide the broadcasting of music (Musak) and television. Experimental satellites such as Canada's CTS demonstrate that television can be received from satellites by low-cost antennas if the satellite generates enough power. NASA's ATS-6 satellite was moved along the geosynchronous orbit to India where it broadcast television to thousands of villages using cheap antennas made with chicken wire. It is clear that satellites like ATS-6 can have a major impact on education, medical treatment and farming methods, in the developing nations. Many satellite earth stations are used by the CATV industry for interlinking their systems and distributing programs.

INTELSAT IVA is launched and INTELSAT V is planned. International telecommunications traffic increases rapidly; most developing nations now have telecommunication links via satellite. Regulation artificially supports the subocean cables with the result the satellite channels remain expensive and traffic does not grow as fast as it otherwise would.

The stress on PCM transmission in telephone networks grows. The T1 carrier spreads until most wire trunks are digital. Data under voice (DUV) on microwave links enables nationwide data transmission at T1 rates (1.544 mbps). The T2 carrier also spreads rapidly (6.312 mbps).

A variety of links operating at T4 speed (274 mbps) are built into the Bell System. Some suburban coaxial cables have 274 mbps sent down each coaxial tube in the cable. A digital radio system comes into use carrying multiple channels of 274 mbps at a frequency of 18 GHz. These radio links are of particular value in areas with conventional microwave frequency congestion. The Bell WT4 waveguide system is installed on a pilot basis and transmits 60 full duplex channels of 274 mbps. 274 mbps is transmitted via a prototype satellite transponder, and also over optical fibers developed at Bell Laboratories. Much development work is done on optical fibers which appear to be one of the most important of future telecommunications media. Public use of optical fiber cables is demonstrated at T3 speed.

It seems clear that computer-controlled time-division switches are the right way to interconnect the time-division digital telephone trunks. Bell's ESS 4 trunk switch operates in this way and is installed successfully. Large numbers of ESS 1 and ESS 2 computerized central offices are installed. The Bell System is on its way to becoming a vast computer network.

CCITT recommends standards for digital trunks of a range of speeds. They are different from the U.S. T-carrier standards.

The telephone companies transmit speech in PCM form using a full duplex 64,000 bps channel for one telephone conversation (including signaling). A major thrust in telecommunications development is the encoding of speech into a smaller number of bits. *Codec* design improves greatly, so that 96 or more conversations of excellent quality can be sent over a T1 speed channel (instead of 24). Codecs are marketed which can send more than one conversation over a circuit operating at 9600 bps, with recognizable but somewhat distorted speech.

Interconnections are permitted to U.S. telephone channels without the direct access arrangement (DAA) that was required earlier.

Computerized PBXs spread very rapidly, especially in the U.S. with AT&T's Dimension system. The new PBXs substantially improve the service to corporate telephone users and are often linked into corporate tie-line systems.

Touchtone telephones are installed extensively throughout North America. Only sets with 12 keys are installed.

Telephone voice answerback by computers spreads rapidly. A wide variety of terminals employing voice answerback are marketed. There is general recognition of its usefulness in creating an effective man-machine interface, especially in applications where data are being entered by the terminal user as in factory data collection and sales-order entry systems.

Use of the Touchtone telephone as a computer terminal spreads for diverse applications. A variety of devices are marketed to connect to the telephone for this system, including additional keyboards, cheap small printers, a terminal identification device for security purposes, and an attachment for displaying responses on the screen of a domestic TV set. "Voice print" recognition is used on an experimental basis on some systems for identifying a terminal user. A few systems with speech input of a very small vocabulary are used.

A revolutionary change begins in the handling of financial transactions—electronic fund transfer. Transactions are transmitted between banks in electronic form. Payrolls are distributed to employees' banks electronically. Bank cards with machine-readable stripes are disseminated to the public and machine-readable cards threaten to replace nonmachine-readable credit cards. The cards at first have two *readable* stripes and then a third *writable* stripe. The public can obtain cash and automated teller services from terminals in the streets without going into banks. Terminals are placed in supermarkets, stores and restaurants which enable bank card users to obtain goods without cash. The spread of such techniques seems likely to be rapid, at least in North America, and threatens to change the entire structure of the banking and consumer-finance industry. It needs new forms of data transmission networks.

Some countries introduce a limited form of interactive television in experimental systems, in which text, news, stock market figures, weather maps, etc., can be displayed on the screen. Some such systems employ data continuously broadcast over the television channels, for example, Britain's Ceefax system. Others use cable television.

Britain's Open University grows to become the largest university in Europe in terms of numbers of students and degrees issued. It has a large campus with professors, studios, and television production facilities, but no students. The students study at home with a television set, taking tests each week. The Open University changes the patterns of adult education. It has a higher proportion of women students than other European universities.

Cable television continues to spread rapidly. To attract customers it needs to offer more programs than through-the-air television, and searches for programming with a high sales appeal, such as sports and first-run movies. In many cities additional charges are made for certain sets of programs, and especially for advertisement-free movies and pornographic programs. Most new cables have a capacity of about 40 channels. Some channels carry locally-originated programs. Sometimes these are of high community interest, but often they are of very poor quality.

The initial marketing of Picturephone brings few subscribers. It is clear that Picturephone needs to be redesigned to give additional facilities and to be of lower cost. Bell Laboratories works on this, and demonstrates Picturephone sets that can operate digitally using a fraction of the bit rate of the earlier Picturephone Model II. Slow-scan videophone sets usable on voice-grade lines are marketed by RCA and others.

A number of organizations install video-conferencing facilities using television sets. Some countries offer a service with leasable video-conference meeting rooms in major cities.

Video transmission is generally difficult over the telephone network because of the cost of long distance circuits and the lack of local distribution circuits. Video links via satellites and rooftop antennas are demonstrated. As gasoline continues to rise in cost it is clear that telecommunications substitutes for travel are needed.

There is a spectacular boom in Citizens Band radio in North America. By the end of the 1970's more than half of all new cars in the U.S. are equipped with it. The enjoyment of communication with strangers via CB becomes a major sociological factor. Lonely people go out in their car with their CB radio on. Many restaurants and motels take bookings by CB. Adaptors are used so that calls may be placed on the telephone system via a CB link. Devices are built which allow CB users to remotely switch on home heating and air conditioning, thereby conserving fuel. The frequencies available for CB are expanded but many enthusiasts claim that they are still not enough.

AT&T implements a *cellular* mobile radio telephone system in Chicago. The system works well and demonstrates that all American cities could have a high coverage of mobile radio telephones. Other manufacturers demonstrate portable radio telephones which give high quality reception.

ARPA demonstrates a portable packet radio system in a California city. It is clear that a city suitably equipped could have pocket radio terminals little larger than an HP65 pocket computer.

A manufacturer demonstrates an LSI wristwatch which picks up the Dow Jones from a broadcast radio signal. Several corporations apply for FCC permission to use certain narrow bandwidth UHF frequencies for data broadcasting.

Early 1980s

The technology of long-distance calls has dropped still further, in cost, though little of this drop is passed on to the telephone subscriber.

There is a widespread use of telephone answering machines and other facilities for recording spoken messages. Some systems for relaying spoken messages (voicegrams) are used.

The rapid spread of computerized ("intelligent") PABXs continues, and more facilities are added to such machines. Corporate telephone users become accustomed to using the new facilities such as automatically transferring calls, setting up conference calls, leaving spoken messages, paging, and so on.

The cost of international calls has fallen dramatically. The satellite links are used to handle data, facsimile, and television with equal facility. This fact, together with the world-wide spread of identical-looking hotel chains, bars, multinational advertising, and multinational corporations, makes the planet seem much smaller. The drop in cost of long-distance transmission in the United States had a major effect on the organization of national corporations and their information processing. Now the same effect is being felt on international corporations. Data banks of international corporate data are used by large firms in the United States, Europe, and Japan. As in military establishments a decade before, decisions in the field could be instantly flashed back to centralized command posts. Factories are sited where labor costs are low. Laboratories and programming centers are set up where talent is plentiful and cheap. Administrative offices are located in countries with favorable tax laws. All are linked with leased lines, and now data dial-up costs are becoming favorable because of international data networks.

Data transmission costs a small fraction of what it cost ten years earlier. There are three powerful reasons for this. First is the widespread use of PCM and other digital telephone channels. A speech channel is often equivalent to 64,000 bits per second, instead of the 4,800 bits per second of the early 1970's. Second is the widespread use of burst multiplexing and burst switching techniques, including packet-switching networks, electronic fast-connect circuit switching, highly flexible time-division multiplex networks, powerful concentrators, communication satellite data-network architectures, and in some areas, pocket radio. These techniques permit a large-scale sharing of the transmission and switching facilities. Third is the widespread use of microprocessors in terminals or cluster controllers, providing many "distributed intelligence" functions which permit powerful use of the machines with far fewer messages transmitted and fewer bits per message.

This drop in cost has led to a massive growth in data transmission applications, with many new types of applications, vast terminal networks, and in some cases many thousands of terminals connected to one computer center. The mass market for terminals resulted in devices of much lower cost, which in turn encouraged their ubiquitous use.

The use of computers as a hobby has by now become widespread, and a major section of industry has grown up to cater to the computer amateurs. A wide variety of microcomputers, microfiles, and solid-state displays are available at low prices for the computer amateurs to use. It is one of mankind's most captivating hobbies and there are a rapidly growing number of *computer bums,* who for a period do little else. Many computer amateurs use cheap satellite antennas designed for digital reception. Several satellite transponders are used for data broadcasting, relaying in total several hundred million bits per second. Some amateurs have more expensive earth stations which transmit as well as receive. There is a widespread use of packet radio by the computer amateurs. Many systems and data banks have been set up primarily for amateur usage, but many amateurs also gain access to commercial data banks. Some amateurs exhibit remarkable skill in bypassing security procedures and gaining unauthorized access to computer systems.

Magazines for the amateur market have greatly increased their circulation.

Amateurs can obtain newsletters at their terminals and can register a "profile" to determine what categories of information they receive. Computing spreads like a drug to a large number of people, and once hooked they cannot let the machines alone. Computer amateurs, much more so than the radio amateurs of an earlier era, are able to make significant and original contributions to the industry, especially in programs and in data bank contents.

Data banks with 10^{13} bits of directly accessible storage are fairly common. Such storage is used for photographs, drawings, and documents in image forms, as well as for digital data. Much telecommunication usage is for access to the numerous data banks rather than merely access to processing power which could be obtained from local minicomputers. The cost of storing alphanumeric data in large electronic storage units is now much cheaper than storing the data on paper in filing cabinets, or even in the form of printed books. And the cost per bit continues to fall.

Information retrieval systems permitting a fast and efficient search of library data bases—books, reports, corporate data, patents, legal documents, etc., are now in common use.

Major advances occur in the digitization of facsimile images. (Character recognition techniques are employed for print, and other techniques for corporate logos and signatures.) Typical facsimile pages which used to be compressed into 200,000 bits can now be compressed into 20,000. LSI chips become available for this compression. Hence many documents are stored and transmitted in "non-coded" image form. A terabit (10^{12} bit) storage can store 50 million pages of documents in image form. Hence massive information retrieval and library systems come into existence, many in government, which permit their users to carry out computerized searches for information.

A "war room" in business is now conventional. It takes many forms and is given many different names. Sometimes it is the showpiece of a firm's data processing. Many offices of top management have video links to the firm's information center. There is now (after some bitter failures) a general recognition that the human element in the information center is as important as the machine element. Experienced and highly professional staffs operate with an array of terminals and wall screens that often rivals a NASA Mission Control Center in appearance. Although some managers like to demonstrate their prowess at operating their own terminal, many have an assistant for this task, or else they use their video link to a local information room, which in turn may route some questions on a remote or central information room.

Computer voice input systems permit a user to speak to a computer over the telephone, using a very limited vocabulary of clearly separated words. The computer responds with spoken voice words.

The computer manufacturers begin to perceive, a little late, a future saturation in the market for simple data processing. This is highly dangerous for them when the cost of digital electronics is dropping so rapidly. Major funding is therefore applied to "artificial intelligence" techniques, which have been worked on in universities, with low funding for two decades. These lead to "intelligent" industrial robots, human speech recognition, recognition of significant patterns in intelligence data, medical diagnosis, military techniques for directing unmanned planes and missiles to targets, intelligent compilers, attractive uses of graphics, intelligent data base systems which a user can employ without programming and man-computer dialogues in which the machine can comprehend and react to miscellaneous human input.

Telecommunications is extensively used in medicine. Information from all manner of patient instrumentation is transmitted to specialists or computers. "Prediagnosis" interviews are carried out between patient and distant computer, often to determine whether the patient should see a doctor or not, or visit a hospital.

Automatic monitoring of chronically sick patients is done by computer, sometimes with the automatic administering of drugs. Sometimes patients are monitored during normal daily activities by means of miniature instrumentation connected to radio transmitters (as with the astronauts). In some cases their readings are recorded by a tiny machine that they can later link to the telephone and transmit the readings to the hospital computer. Remote diagnostic studios are used with powerful television lenses. With the help of a nurse in the studio, a distant doctor or specialist can examine a patient as though the patient were in his office. The patient can see him and talk to him. The doctor can fill the whole of his color screen with the pupil of the patient's eye, or tongue or skin rash. He can listen to a distant stethoscope and can see both instrument readings and computer analyses of them. Patients in remote areas have "telemedicine" access to highly qualified and specialized doctors and facilities when they need them. Many doctors are resistant to telemedicine but detailed studies of working systems demonstrate its success. Many hospitals have remote access to highly expensive and specialized computer facilities.

Picturephone gains some acceptance mainly because of its much lower cost. Vital to its usage in industry and government is its new ability to display still images of documents in a clearly readable manner and in such a way that viewers at both ends of the line can point to items on the screen which they are discussing.

Video channels of various types become more widely used as a substitute for physical travel. As transportation costs continue to increase video channels are increasingly regarded as a cost-saving mechanism rather than as a luxury or status symbol. Equipment using satellites to give video communication drops in cost partly because of signal compression and partly because it is designed so that the image (but not the speech or data channel) degrades badly during very heavy storms. This degradation permits much cheaper engineering.

AT&T put a new optical fiber cable system into service on a trial basis. The separate fibers in the cable each transmit at the T4 rate of 274 mbps. There is every indication that such systems will be a great success and that by including many fibers in a cable an extremely high information-carrying capacity can be built up. A small flexible optical-fiber cable is also installed on a trial basis for local loop transmission.

The CATV organizations install an experimental optical fiber cable television system, delivering television in digital form. A multifiber cable could carry a large number of television channels or a prodigious data rate into the home.

A "high-fidelity" television service is started with more lines to the screen. Seven-foot wall screens are marketed and the sets operate in a digital fashion. The digital bit stream using a form of differential modulation reaches the home over a coaxial cable with frequent digital repeaters. It is not planned to transmit high-fidelity television over VHF or UHF radio. It now seems clear that optical fiber rather than coaxial cable is the technology of the future for CATV. To carry CATV signals to remote locations, satellites are employed.

The hi-fi music industry is in upheaval with the spread of digital recording techniques. It is clear that digital hi-fi can give more distortion-free and noise-free sound than analog systems and is also cheaper. Digital hi-fi quadraphonic recordings are marketed which have great brilliance and clarity. A receive-only antenna at a remote farmhouse is much cheaper than a CATV cable. Digital storage units of a trillion (10^{12}) bits can hold a thousand hours of hi-fi recordings or advertisements so that they are randomly accessible, and not excessively expensive. Public music library services are started, allowing the public to request hi-fi recordings over their CATV cables for a fee.

There is extensive use of millimeterwave radio links in the suburbs and cities for distributing all types of signals. Most such systems operate digitally and some use digital repeaters spaced at intervals down the streets or on the city rooftops. A very high channel capacity with a low level of noise and distortion is made possible. The highly directed beams eliminate interference between different transmitters, and it is clear that a large number of such systems using the same frequency bands will be employed in a city.

There is growing use of *time-division multiple-access* (TDMA) to enable digital channels to be shared by different geographical locations in a manner which varies with the demand for channels.

Satellite technology spreads rapidly especially for private corporate and government networks. These networks share transponders by using multiple-access techniques (FDMA or TDMA). More powerful satellites are launched using the 12–14 GHz frequencies, and the cost of private earth stations drops greatly. For data-only channels the earth station equipment is almost as cheap as the channel control equipment used on terrestrial telephone channels.

Data appear cheap compared with speech on the satellite channels because speech is transmitted digitally, usually at rates higher than 20,000 bps.

The space shuttle is operating successfully—ferrying loads into earth orbit at a small fraction of the cost of doing this in the 1970s. Additional hardware has enabled NASA to use the shuttle in positioning a communications satellite of much greater weight than any previous geosynchronous satellite. This research vehicle demonstrates that earth station costs can be brought down to a very low level, and that satellite broadcasting directly to the home is economically viable.

The number of satellites broadcasting to the lesser-developed nations increases. The cost of appropriate television sets and antennas has dropped substantially. The sets are used for education, entertainment, and propaganda. Although television in the United States continues to focus on entertainment programs, because advertising revenue supported it, television in the poorer nations often does not, for it is largely under governmental control. There is little or no advertising revenue, and thus the entertainment programs are broadcast mainly to lure viewers into watching the education or propaganda broadcasts. Many thousands of villages in India, the Middle East, Africa, Asia, and South America are seeing television for the first time.

The value-added common carriers make extensive use of satellite channels, expanding their services in many directions. In addition to handling computer traffic they widely handle electronic mail, messages, voicegrams, electronic funds transfer, library searches, medical instrumentation data, credit checks, and so on. They provide services such as adding corporate letterheads to facsimile documents, filing audit trails, storing vital records in secure locations, and interfacing with international networks and message carriers.

Electronic payment mechanisms are in widespread use in some countries. The financial entities such as banks have undergone major structural changes. 3-stripe bank cards have largely replaced credit cards, and many payments are made directly from the bank-card machines. Terminals in stores, supermarkets and restaurants are on-line to the financial networks. Customer-operated bank terminals are available everywhere in some countries, and customers far from their own bank can still use some of its services. Some banks have set up unmanned "branches," which provide customers with desks, telephone and terminals behind a locked door. The employment of bank tellers has dramatically declined.

Cash transfers take place within the electronic systems. Electronic fund transfers take place within one computer, or between two different computers, holding the accounts of the persons concerned. They are sometimes initiated by "preauthorization"—that is, instructions in advance of the payment being given to the bank's computer for the routine payment of salary, rents, dues, etc. They are sometimes initiated by the use of a bank card. The cost per transaction is substantially lower than with checks or credit cards. There is no longer any talk of a "checkless" society. The number of checks in the U.S. has risen to 50 billion per year and bankers desperately hope that electronic fund transfer will lessen the deluge of check processing. The EFT terminals have done much to prevent the high level of crime that became associated with credit cards.

The use of mobile communications equipment has risen greatly. *Cellular* mobile telephone systems come into existence in many cities. The cost of vehicle telephones has dropped and in suitably equipped cities anyone who applies for a mobile telephone can have one. The pushbuttons and sound quality are similar to regular Touchtone telephones.

There are now millions of paging devices in use. A nationwide paging service is in operation.

Data broadcasting is used in many locations distributing high speed streams of data on UHF channels. Broadcast data can be picked up on a variety of machines. A popular machine is a pocket calculator designed to carry out a variety of financial functions and equipped to receive broadcast stock market prices, and other financial data. Probably the most common device used to receive broadcast data is the home television set. A very large number of data pages can be received on the screens. Some of these are in the form of a broadcast "magazine" of which the user can request the page he wants to see, or can initiate page turning. Some are organized into "dialogues" in which the user's keyboard response causes the set to select a particular broadcast frame.

Britain's Open University is now being emulated in many countries. There is a major international exchange of program material. Publishers and television networks compete in the market for programs and related texts. This market is becoming highly lucrative. Interactive facilities in the home are now used in addition to television for automating student tests and for computer-assisted instruction.

There is increasing concern about public literacy. The proportion of people who have difficulty reading has grown with the growth of the video media.

The use of packet radio is growing, mainly for specialized applications such as automatic meter reading, police work, delivery vans, military uses, burglar alarms in vehicles, etc. There is substantial use of packet radio in some developing nations which have inadequate local telephone loops for connecting conventional computer terminals to computers. A packet radio connection for a terminal costs about the same as a conventional telephone modem. In North America pocket calculators with packet radio capabilities have been demonstrated which could have an extremely versatile range of applications.

Television newscasters become fond of taking "real-time public opinion polls," in which the public are asked to respond to an issue or comment on an interview by pressing keys on their telephone. A computer accumulates and summarizes details of the responses, which are then broadcast. Scandals occur due to rigged results and greater security is built into the system. Some authorities advocate that real-time public opinion polls should evolve into real-time referenda for government—real-time democracy. Others claim that any such form of government would spell the end of the Western economic system.

The year 1984 comes and goes without anything similar to George Orwell's predictions coming to pass, at least in North America. Nevertheless it is clear that a technology exists which is incomparably more powerful than anything Orwell thought of. (Orwell did not imagine computers or data banks, or packet radio.) In spite of all of the new legislation, personal privacy exists only because the government is essentially benevolent. No legislation would prevent a totalitarian government from using the technology for controlling individuals if it wanted to. Many of the world's governments are now authoritarian and some would question their benevolence.

Late 1980s

People, at least those who live in North America, are becoming accustomed to a society in which many functions of life are carried out by telecommunications. Numerous persons now work at home at least part of the time. Tax deductions for home facilities needed for work are standard and include computer terminals, videophones, and video-conference screens. Many firms install these devices at their own expense in the homes of employed persons who need them. Many homes are now built with a childproof, spouseproof office. America's traditional antipathy to soundproofing diminishes. Curiously, the companies most reluctant to allow their employees to work at home are the giant, conservative, computer and telephone companies which make it possible.

There are now many millions of fairly inexpensive electronic fund transfer terminals in use. Some persons have them in their homes.

A new generation (of people) is now dominant which can communicate with the computers with ease over the various transmission links. Programming is taught at an early age in schools, and most well-educated persons under 30 can use one programming language fluently. The computer and software industries have spent much time and money developing the "man-machine interface" so that the ubiquitous terminals are usable by the greatest number of people. Nevertheless, some minds seem naturally at home with the new technology, whereas for others it is a struggle. Some persons seem to have a built-in hostility to this form of communication, which is becoming so vital in society.

A person who is well-adapted to the technology can carry out an amazing number of different functions from his home terminals. An ever-increasing world of computers, data banks, sound, film, and picture libraries is there to explore. Many authorities, however, still believe that the technology is only in its infancy. Certainly a vast amount of work lies ahead in building up the data banks, writing teaching programs, improving computer-assisted medical diagnosis, and so on. Many data-bank uses that met with initial skepticism from the professional men they were designed for, are now gaining wide acceptance, but the work required to make them comprehensive is enormous.

Much television interviewing takes place remotely with the subject being in his office or home, and interviewer being in the studio many miles away, often in a different country. Small, unmanned news studios exist in many cities, using a variety of remote-controlled cameras, zoom lenses, and back-projection facilities.

"High-fidelity" television proves to be popular, and cables for it are laid down in large numbers. Some affluent homes have wall-sized screens. The vividness of the large color picture provides a more "hot" medium in McLuhan's sense of the word than the earlier small TV screen.

Three dimensional television is demonstrated using large wall screens.

Commercial satellites launched with the space shuttle are in operation. They weigh several tons and generate many kilowatts of power. New satellite designs could generate megawatts of power. There is an excess of satellite bandwidth for current applications, and this forces the cost per channel down. However it is clear that the projected growth in videoconferencing and two-way video communications will need prodigious quantities of bandwidth. NASA launches an experimental satellite using laser communications to the satellite. It works well and holds out the promise of extremely high bandwidths. It is thought that laser communications is necessary for the exploration of the planets.

AT&T extend their use of optical fiber cables both for long-haul and urban trunks. Cables are installed with more glass fibers than currently needed so that rapidly rising demand for video channels can be accommodated. AT&T and other carriers make extensive use of digital radio operating at 18 and 39 GHz in the suburbs and cities. Many different types of channels now operate at the T4 speed of 274 million bits per second. Computerized concentrators, exchanges, and packet switching systems interlink these digital channels.

Optical fibers are increasingly used for local loops, with new legislation permitting telephone and television companies to share the same local channels. Many corporate locations and a small percentage of American homes now have optical cables serving them. The wired city is coming together.

The satellites now permit home pickup of worldwide television with rooftop antennas. Some television programs are dubbed in many languages. Because a television sound channel requires only about one thousandth of the bandwidth of the picture, this does not substantially increase the overall bandwidth requirements. A rooftop antenna can now pick up almost as many television programs as there were sound stations on the shortwave radio of the 1960s.

Telephones in vehicles are widespread. Cars are manufactured with the option of a Touchtone keyboard on the dial, alongside all the other electronic equipment. Discussion exists on whether a car telephone, plus a loudspeaker that is never switched off while the vehicle is on the public roads, should be a legal requirement. The loudspeaker would receive tones and voices that interrupt the car radio or music player and that are concerned with safety, accidents, parking, toll paying, route-finding, and so on. Improvements in the nationwide dialing system for vehicles are needed. In many parts of the country, a car still cannot be reached by a long-distance telephone call.

Personal portable telephones have been in use for some time by the military forces, who have achieved almost worldwide dialing of key persons. In the cities, fire, police, and other personnel have portable radiotelephones. Many corporations use portable telephones within factories or office buildings, and some are now setting up a nationwide corporate dialing network. A public service that is an extension of the car telephone service and that permits small transceivers to be carried anywhere is now clearly practicable.

A wide variety of microcomputers are marketed with artificial intelligence functions.

Major strides are made in industrial automation, with the widespread use of "intelligent" robot machines. Some factories contain thousands of inexpensive computers controlling machines and feeding information to larger systems. The automated production lines operate round the clock and over weekends with little human attendance. Nowhere is automation more fully used than in the manufacture of the computers themselves. Immensely intricate and high-speed production lines produce the wafers of LSI circuits, charge-couple, and magnetic-bubble devices, semiconductor lasers, etc., all in vacuum and all untouched by human hand.

Digital library storages are now available which store 100 terabits (10^{14} bits) on-line. These stores are used in data processing applications, for storing vast libraries of documents (hundreds of millions of facsimile pages), for large on-line music libraries, or for holding up to a thousand hours of randomly accessible television programming. Intelligence and police agencies use such stores for holding up to a million hours of recorded telephone conversations or conversations from bugging devices.

Every issue of newspapers such as *The New York Times, The Times of London, Le Monde,* and *The Wall Street Journal,* are stored in information retrieval systems with automatic indexing originally pioneered by *The New York Times* so that a user can make a computerized search for past news stories on any subject. Similar information retrieval systems exist for magazines, technical reports, legal documents, United Nations documents, patents, etc. These systems are accessible from terminals anywhere in the world via the satellite data networks.

"Personalized" newspapers come into operation. Instead of being presented with an impersonal and often superficial selection of all the news, as today, a subscriber may register his news-requirement "profile." He then receives detailed news on topics that interest him. This information may be printed out on his home terminal or stored for him so that he can display it on his screen when he wishes. All types of categories can be registered—for example, local news about a district other than where he lives, news about a particular industry, company, or stock, scientific news, movie reviews by particular reviewers, news about crime, sex, war, or business, and foreign news.

Some of the dire predictions of the Club of Rome are beginning to be regarded as self-evident. World population now exceeds five billion and a larger number of people per year is being added to it than ever before. As petroleum becomes scarcer it is becoming very expensive, and some of the main minerals needed by industry are also rising in cost precipitously as their supply shrinks. Because of the relative costs consumer patterns are changing. Communications and computer-oriented goods are increasing in capability by leaps and bounds without increasing in cost, whereas goods requiring a major use of scarce raw materials are rising in cost.

Early 1990s

On-line library storages for computers now exceed 10^{15} bits where this can be used. Such a store can hold 10,000 hours of video programming which can be accessed at random.

Mass production and mass marketing have given most homes wall screens. Laser-driven optical-fiber channels are being laid into homes to give a very large number of high-fidelity TV channels. Dial-up channels (as opposed to cable TV channels) are being installed in limited numbers to carry the large-screen pictures. The latter are being sold mainly to industry and, to a minor extent as yet, to affluent homes. The office of a top ranking executive is now likely to have a wall-sized screen facing his desk. It can either be connected in its entirety to another location or fragmented into several smaller screens for conference or multimedia use. All such facilities are digital. Office and apartment blocks have small computers that act as concentrators, and control screen fragmenting and switching.

The wall screens are now frequently used by lecturers. A lecturer may conduct a class of 16 or 32 people, all in different locations and all using videophone sets. They may be in their homes, they may be in worldwide business locations. The lecturer can see all of their faces, and they can see either his face or diagrams, objects, slides, or film clippings, which he switches on to their screens. He can speak to any of the "class" individually and they can speak to him. They cannot see their fellow students and so tend to ask questions with little embarrassment from possible class reaction. The teacher may occasionally let his students see the rest of the class. If he wants, he may switch the face of a questioner onto the class screens. On the other hand, he can address any one student without the others hearing. Similar facilities are used for sales meetings, management briefings, and by a manager addressing his employees who work at home. Some such links in industry are international.

Commercial satellites weighing many tons are now in geosynchronous orbit, operating at frequencies of 12–14 and 20–30 GHz. Higher frequencies are in experimental use and are clearly needed because of the large demand for video channels. Transmissions to satellites with laser beams are in experimental use. Because of the high satellite power, antennas for transmitting and receiving video channels are much less expensive than a decade before, and are installed at many corporate and government locations. Receive-only video antennas are on sale at Sears Roebuck. Cryptograph is employed in most corporate and government satellite transmission to ensure security.

Telephone speech transmission by satellite costs a fraction of long distance terrestrial transmission. Nevertheless *public* telephone calls cost the same whether routed over satellites or terrestrial trunks. The government considers it essential to keep a high proportion of terrestrial telephone channels in operation for security reasons—an enemy might attack the satellites. A highly expensive military system is in operation to protect the communications satellites.

The video and speech channels are all digital. By comparison text transmission or alphanumeric data transmission is very inexpensive. An hour of video transmission is equivalent to 30 billion bits of data transmission (10 million Western Union telegrams!). Electronic mail is a small fraction of the cost of any type of mail operation involving manual pickup, delivery, or sorting. Nevertheless the U.S. Post Office still has several hundred thousand employees. A nationwide voicegram service in in operation.

Copper wire pair local loops are now generally regarded as an anachronism. However the cost of replacing them is high and wherever possible telephone companies allow them to remain. In some cities and suburban areas the coverage of the new wideband digital local loops is high. Many of these use optical fibers with time-division multiple-access. The remaining copper wire cables in some cities have deteriorated seriously, and give a proportion of catastrophic failures which some users claim to be intolerable.

The information retrieval systems which have been highly successful for searching text libraries are now being extended to video libraries. It is possible for researchers or program directors to search abstracts of news broadcasts or other video programs and then examine segments of the retrieved program at a computer controlled screen.

Computer assisted editing is generally used in video program production, and often video segments stored in distant libraries are edited into programs.

In order to maximize the utilization of the exceedingly expensive machinery, many factories and government departments operate with 3 1/2-day work weeks. Employees alternate 3- and 4-day weeks with 3- and 4-day weekends, and the machinery works 7 days a week, round the clock. To fill the long weekends, many employees have second homes, boats, hi-fis, television, and elaborate hobbies. Many take university courses, usually in telecommunications. Many have second jobs.

The movement to the cities of earlier decades has been partially reversed, with much house building in country areas. This is caused by the spread of second homes, the availability of telecommunication services such as television via satellite, and data networks, the spread of rural information-related industries, working at home, and the worsening crime and conditions in the cities.

A directory the size of the "yellow pages" is published, listing all the telecommunication services available and giving the codes necessary for communicating with the innumerable computers. Most shopping is done from the home screens. With increasing automation, employment in the service industries has become much greater than in the manufacturing industries, and now many persons are employed in giving specialized assistance in the information networks. The HELP key on the Touchtone telephone overlays and other devices are used frequently. As in the information centers of industry a decade earlier, it is found that much more useful facilities can be provided if specialized human assistance and human expertise are built into the system. A most useful attribute on many computer systems is an on-line human being. Almost 10% of the United States Gross National Product is spent on the communication services discussed in this chapter (including broadcasting).

Leisure time has increased greatly because of the increased automation in manufacturing and clerical processes. Major growth industries are those relating to the use of leisure time, especially uses which do not involve substantial expenditures of fossil fuels. Few factory employees work a 5-day week. Workers in the information and entertainment industries, entrepreneurs, system designers, and other creative people frequently more dedicated to their jobs, often work 60-hour weeks.

Printed newspapers cease publication production in the United States except for a minor intellectual press, a few picture newspapers for low-IQ readers, and some local newspapers. Some weekly news magazines survive.

Television news has become exceedingly vivid with worldwide satellite coverage. The news from some parts of the world seems to be increasingly disastrous as populations grow, aspirations are inflamed to unfulfillable levels, and raw materials, fuel, and fertilizers are in increasingly short supply. At least one cable TV channel carries big-screen news all day. Very high quality photography on the wall screen shows riots and world catastrophes in fine detail, including local wars and terrorism and a new wave of famines in certain underdeveloped and overpopulated countries.

The unrest in the underdeveloped world is undoubtedly inflamed by their television window to the affluent nations.

Late 1990s

Industry is to a major extent run by machines. Automated production lines and industrial robots carry out much of the physical work and data processing systems carry out much of the administrative work. There is almost no machine tool that does not contain a computer. Paperwork is largely avoided by having computers send orders and invoices directly to other computers, and by using electronic fund transfer for most payments. The jobs for human beings tend to be more difficult than they were in an earlier era. Maintenance is more complex. Production planning and cost optimization is more complex. Management and entrepreneurship are now more complex. The easy jobs have been automated.

The wideband, rapidly-switched communication networks, and the communications satellites weighing many tons, form an infrastructure vital to the running of industry. Nations which do not have this expensive infrastructure find that their industrial productivity and hence economic strength is falling far behind the nations with advanced telecommunications. Some developing nations such as Brazil and Iran have advanced satellite channels which are vital both to their industry and to social services such as education and medicine. The telecommunications facilities of Brazil are in some ways more advanced than those of Europe because Brazil had to invest heavily in satellites and TDMA, and built few wire-pair trunks.

Multinational corporations make extensive use of satellite networks and other telecommunications facilities to tie together their far-flung operations. A salesman in Africa uses terminals, sometimes radio terminals, on-line to New York. Worldwide data-base systems are used for corporate control and planning. The satellites permit video conferences to be held between staff in many countries.

Substantial industry now takes place in space. The manufacturing experiments launched 15 years earlier by the space shuttle were highly successful and led to remote controlled factory units in orbit. Many industrial processes require the absence of gravity or the high vacuum and utter purity of space. Not least of these is the production of three-dimensional solid-state logic and memory modules in which hundreds of millions of transistors are grown expitaxially, layer upon layer, to form modules the size of matchboxes. Fashionable young ladies wear wrist computers made in orbit. Solid-state wall screen components are also made in space. Some factories operate in low earth orbit, but many are in geosynchronous orbit so that their vast solar power supplies operate 24 hours a day. Most factory modules do not have men permanently in space but are manned from earth via multiple video and telemetry links. Telecommunications control is easier when the factory is in geosynchronous orbit.

The first retrieval of minerals for industrial use has begun on the moon. Several of the planets and their moons are now being explored and photographed by roving laboratory modules controlled by laser communications from earth. The search for life on the planets continues to create new scientific mysteries.

The first permanent space colony has been set up at the stable position between the earth and moon, called L5. The future of space colonies is not clear but the view is growing that they should become self-sufficient, highly populated industrial units. As with new frontiers of earlier ages, they attract tough and restless individuals. The persons in space have ready access to the movie libraries, quadraphonic music, and data banks of earth. Earth-like environments are simulated for them by telecommunications. Television watchers on earth view the breathtaking new sports that are played in space.

The population of earth passes six billion and is still growing almost exponentially. It is often commented that space cities will absorb some of the excess population but population is growing far too fast for space to relieve the problem in the plannable future.

Most people now carry a portable radio transceiver with a telephone keyboard. They have a wallet full of credit-card size overlays. When an individual is dialed, he can be reached in most parts of the country. The zones of radio inaccessibility are diminishing. It has been suggested that the public should be issued transceivers that transmit their national identification number, even when switched off. These devices would help in controlling crime. They would also be used in most financial transactions. CB and cellular mobile telephone technologies have merged, giving the users the option of cheap party lines or private lines via radio, or both. Some telephone systems have a facility which enables a person to be dialed using his national or corporate identification number rather than a telephone number. The telephone system routes this to a computer which can convert it to a telephone number which will operate the requisite switches.

The magnetic-stripe cards used originally for banking are now carried by almost everyone and have long been used for many applications other than banking, including paying for goods, operating ticket-vending machines, operating parking lot gates and auto toll booths, maintaining security, operating office and house doors, and so on. Now devices are marketed which could replace these plastic cards by identifying a person directly by means of some physical characteristic of that person such as hand geometry. Such devices in combination with a spoken or keyed number provide higher security and lesser possibility of crime. Just as the plastic cards once gave rise to talk of a cashless society, so the new devices cause talk of a "plasticless society."

Some affluent homes are replacing their Picturephone with a dial-up service that uses the digital wall screens. The wall screen can be linked either to the TV cable or to the new dial-up videophone network. With the latter, one apartment or one room in a house can be connected to another via the wall screens. Large-screen videotape recorders are inexpensive, and families often dial up relatives to play them videotapes of the children.

The dial-up channel makes it possible to request movies for individual playing in the home.

"Personalized" video news services are set up. In the same way that printed news became individually selected for each subscriber "profile," now video big-screen newscasts are similarly assembled.

In addition to being able to select what he wants instead of passively watching what is fed to the masses, the home viewer now has a channel on which he can "converse" with the medium. This channel is often used for teaching. "Computer-assisted instruction" thus progressed from terminals giving an alphanumeric response from the computer, to terminals giving a color slide response and presentation, and then to terminals giving movie sequences. All these facilities in turn were initially used in institutions and later with a dial-up or cable TV service to the home. The interactive media are used for many functions other than teaching, but much "passive" programming is still used also.

Computerized, video dating gains popularity with customers examining computer-selected prospective dates on their home screens. Many other video services using computers become popular, the video image adding a dimension and a popularity absent with earlier computer services.

The ability to program and communicate with computers is now widespread. Among persons under 30, the inability to program is regarded as a form of illiteracy. There is a serious gap between the capabilities of young and old people in communicating with the all-pervading machines. Computers and terminals are now almost as widespread and interconnected as the telephone network was twenty years ago. A single query on the data networks might invoke the use of dozens of computers. A major emphasis has been placed on building "knowledge" systems rather than merely data systems, so that users concerned with a given subject automatically gain access to multiple knowledge banks and relevant programs.

In large cities, the movie libraries are located close to the telephone (videophone) central offices; and as the channel is permanently connected from the central office to the home, the telecommunication cost of dialing for movies or news is not high. In rural districts, however, it is very high because of the long-distance transmission to the movie library. Here the distribution of movies in cartridges is likely to continue. In spite of the vast growth of telecommunications, there are thus still advantages in metropolitan living. The high bandwidth, two-way, telecommunication services are much cheaper in the cities. We have not yet reached the point when the drop in long-distance transmission costs can overtake man's ability to consume increasing bandwidth.

Public movie theaters have declined under the competition from the home entertainment media and now cater to two main markets. First, they show movies with a degree of sex and obscenity not permitted on the home media and catering to a market too poor to afford the wall screens. Second, other theaters show spectacular movies on screens occupying 180° or 360° of the field of vision; these screens can create an impact greater than the home wall screens.

Circular-domed rooms come into use in which the entire ceiling and walls are a three-dimensional color TV screen. These can be linked by a communication line to any appropriate camera system, videotape or transmitter, or they can be operated from their own cartridge, which can generate an environment of sylvan tranquility, spring in the Andes, earth orbit, or the Kilauea volcano erupting.

There is experimental use of drugs administered under electronic control in conjunction with entertainment media, largely to heighten and "edit" emotional reaction.

It seems clear to many authorities that the staggering advances in molecular biology are going to merge with the electronic technology. This prospect is dismaying to many older people (those who read this book in the 1970s), but strangely enough, a new generation of students is emerging that appears to welcome it.

PART III TECHNOLOGY

Valve panel at the first high-power radio station to use thermionic vacuum tubes for transmitting, 1923. (*Courtesy The Marconi Company Ltd., England.*)

24 RADIO CHANNELS

Information can be transmitted by two different types of medium; cable and radio. In the former, different transmissions do not interfere with one another in an uncontrollable manner; they can travel over different cables and any number of cables can be used. In the latter, they do interfere and the radio frequencies available must be carefully allocated and rationed. The radio spectrum forms a natural resource that is in limited supply. As we move to high frequencies, the resource becomes more plentiful but has new problems, which we shall discuss.

In both cable and 'wireless' systems there is a trend toward increasingly high frequencies. It is this advance of the frontiers of technology into higher-frequency zones that gives us the increasing bandwidth and channel capacity discussed earlier.

THE SPECTRUM

Figure 24.1 shows the portion of the electromagnetic spectrum that is used for telecommunications today. Everyone is familiar with the rainbow-colored spectrum of light created by separating the visible frequencies into individual colors; this is but a small portion of a vast spectrum laying out all of the radiation frequencies of physics. Those shown in the figure are a fraction of the frequencies known to physicists, who discuss rays of 10^{25} Hz and higher. The scale of frequencies in Fig. 24.1 is logarithmic. If we drew this diagram with a linear scale so that that part of the spectrum used today for commercial communication links occupied the width of this page, then the part of the known electromagnetic spectrum as yet unused would require a sheet of paper stretching from earth to a point a hundred times the distance of the sun! Theoretically, the quantity of information the medium can carry is proportional to the distance in this scale.

Information-carrying capacity is approximately proportional to bandwidth (difference between highest and lowest frequency in a band of frequencies).

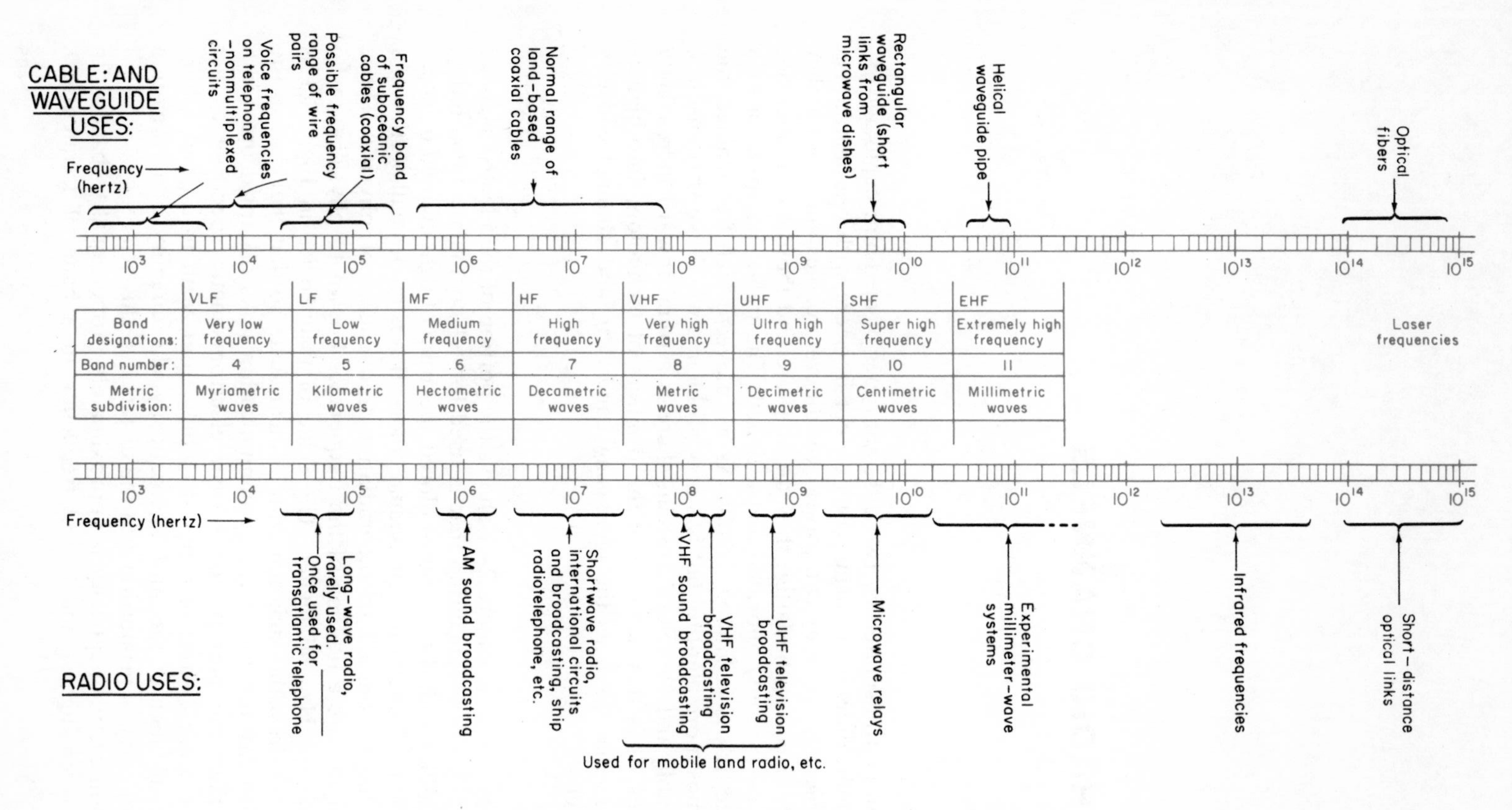

Figure 24.1 Telecommunication uses of the electromagnetic spectrum.

The higher we go up the electromagnetic spectrum, the more plentiful is bandwidth. If we can extend the upper limits of radio usage from centimetric to millimetric waves (Fig. 24.1), we will increase the total bandwidth available by a factor of 10. These new frequencies have their own special problems, however, as will be discussed shortly.

RADIO WAVES

This chapter discusses the use of radio; the next will discuss cables and waveguides.

Box 24.1 summarizes the designations given to radio frequency bands. Radio waves of different frequencies differ in their properties, which may be briefly summarized as follows:

1. Long-wave Radio, 30 to 300 kHz

The original radio telegraphy frequencies. Once used for trans-Atlantic telephone. Little used today.

2. Medium-wave Radio, 300 kHz to 3 MHz

Mainly used for the AM radio broadcasting frequencies. The familiar dial of the home AM radio tunes over a band from about 500 to 1500 kHz. The upper end of the band is used for mobile land transmitters and for amateur use.

3. Shortwave Radio, 3 to 30 MHz

Used for shortwave radio broadcasting, international overseas ship radiotelephone, for aircraft, for much amateur use, and for mobile land transmitters.

Below 30 MHz, the spectrum is principally suited to very long distance transmission. Hence most of this region is used for international communication and broadcasting, as well as for maritime and aviation use. Government and nongovernment users share it almost equally. These frequencies are reflected by the ionosphere and therefore can bounce around the world. Being strongly scattered by the atmosphere, they are able to pass around large obstacles. Thus they are more easily used in car broadcast receivers than the higher frequencies. Tuning a domestic AM or shortwave radio set, the user can pick up stations from far away if the set is sensitive. This fact is especially true at night, for the ionosphere has better reflecting properties when not in sunlight. On the other hand, there is much atmospheric noise—noise largely absent at higher frequencies.

4. The VHF Band, 30 to 300 MHz

A large part of this band is also used for radio broadcasting; it is familiar to the domestic user as the FM band on his radio set. The user now must generally be within a hundred miles or so of the transmitter. The VHF band is also used for television broadcasting. Television needs a bandwidth many hundreds of times greater than sound broadcasting, and therefore the frequencies below this band could not be used for television.

The VHF band and the frequencies up to 1000 MHz are the most suitable for mobile communications on land. They are used for police, fire, and ambulance services, taxis, delivery fleets, car radiotelephones, personal paging systems, and so on. The demand for mobile radios is increasing rapidly, which means that this part of the spectrum is becoming overcrowded and there is an ever-growing shortage of these frequencies. Furthermore, technology has now brought us to a point where we could like to have a major increase in the numbers of portable transmitters and receivers, as discussed in Chapter 12, if control of use of the spectrum permits. Fortunately, there *are* ways to achieve much more efficient allocation and use of the spectrum.

Atmospheric noise decreases with increasing operating frequency and eventually falls below the level of the thermal noise that is always present in electronic circuits. Radio circuits above 100 MHz are as noise-free as cable circuits. VHF radio can have an almost noise-free background.

Propagation is not reflected by the ionosphere. It takes place in the lower atmosphere and becomes closer to a straight-line path as the frequency increases. The radiation is increasingly reflected by large objects. High buildings can cause "shadows" on television, and the helicopters over cities interfere with VHF radio and television in some locations.

5. The UHF Band, 300 to 3000 MHz

The UHF band again has a capacity ten times greater than the VHF band. A major block of it has been allocated for television. This band provides more channels than VHF, but these were allocated more recently and only a few are used as yet. Below about 1000 MHz, this band is used for mobile land radio, as is VHF, and shares its congestion problems.

Above 1000 MHz, the UHF band joins the SHF frequencies in being mainly used for radar and navigational devices and the point-to-point communications.

6. The SHF Band, 3 to 30 GHz

At these high frequencies, point-to-point transmission is generally necessary. The frequencies are thus too high for general mobile land radios. Atmospheric scattering is slight. Some circuits are used in which waves from powerful transmitters are scattered by the troposphere, but normally line-of-sight transmission is employed.

Frequencies from 1 to 10 GHz are referred to as *microwave*. Microwave radio is one of the chief means of transmitting long-distance telephone calls and television programs. As with coaxial cable, the high bandwidth of microwave channels can transmit many thousands of telephone calls, and many television programs, simultaneously. Microwave antennas, within line-of-sight of each other, are erected on tall towers and rooftops. These antennas and towers form chains across the country, each such location amplifying and retransmitting signals. They are usually spaced about 30 miles apart. If they were farther apart, the curvature of the earth and varying retractive index of the atmosphere would necessitate very high towers, absorption due to rain and snow would become severe, and large, expensive antennas would be needed.

Many cities throughout the world now have skylines that are dominated by a tower carrying microwave antennas. Tokyo has a tower like the Eiffel Tower but 40 feet higher. East Berlin has one 1185 ft. high. One of London's most expensive dinners can be eaten in a revolving restaurant just above the Post Office tower microwave antennas, and Moscow possesses a tower that is 250 ft. higher than the Empire State Building.

Such relays today use frequencies from about 1 to 12 GHz. The higher the frequency, the easier it is to direct a narrow beam that does not interfere with other nearby transmitters using the same frequencies. A 12 GHz beam spreads over an angle of about 1°.

Different moisture and temperature layers can cause the beam to bend and vary in amplitude, just as we sometimes see light shimmering over a hot surface or causing minor mirages along a road surface in the sun. Occasionally bending effects can cause fading. Rain can increase the attenuation slightly, especially at the higher microwave frequencies, and occasionally trouble is caused by reflection from unanticipated objects, such as helicopters or new skyscrapers in a city. To a limited extent, automatic compensation for changes in the radio attenuation is built into the repeaters.

7. Millimeter Waves Above 10 GHz

The new frontier of radio is in millimeter waves. The bandwidth is enormous and so is the payoff in discovering how to use it.

Waves of these frequencies suffer severe attenuation in rain and snow. Below 6 GHz, the effects of rain play no part in the spacing of microwave relay towers. Above 10 GHz, the effects become severe, limiting the distance between repeaters. Nevertheless, radio systems operating at frequencies above 12 GHz are coming into use in cities and suburban areas.

The higher the frequency, the smaller the antenna needed to make a narrow-angle beam. Small antennas on tall posts in city streets can relay millimeter-wave signals of extremely high bandwidth. Because of the short range and narrow angle beams, different city pathways can be prevented from interfering with one another.

8. Infrared and Optical Transmission

At the frequencies of light, beams travel in straight lines and can be sharply focused with lenses. Consequently, there is no problem with interference. On the contrary, there may be a problem with vibration of the light source because the beam is so accurately focused. No license is needed from the FCC to operate at these frequencies. The big snag with these frequencies is that they are heav-

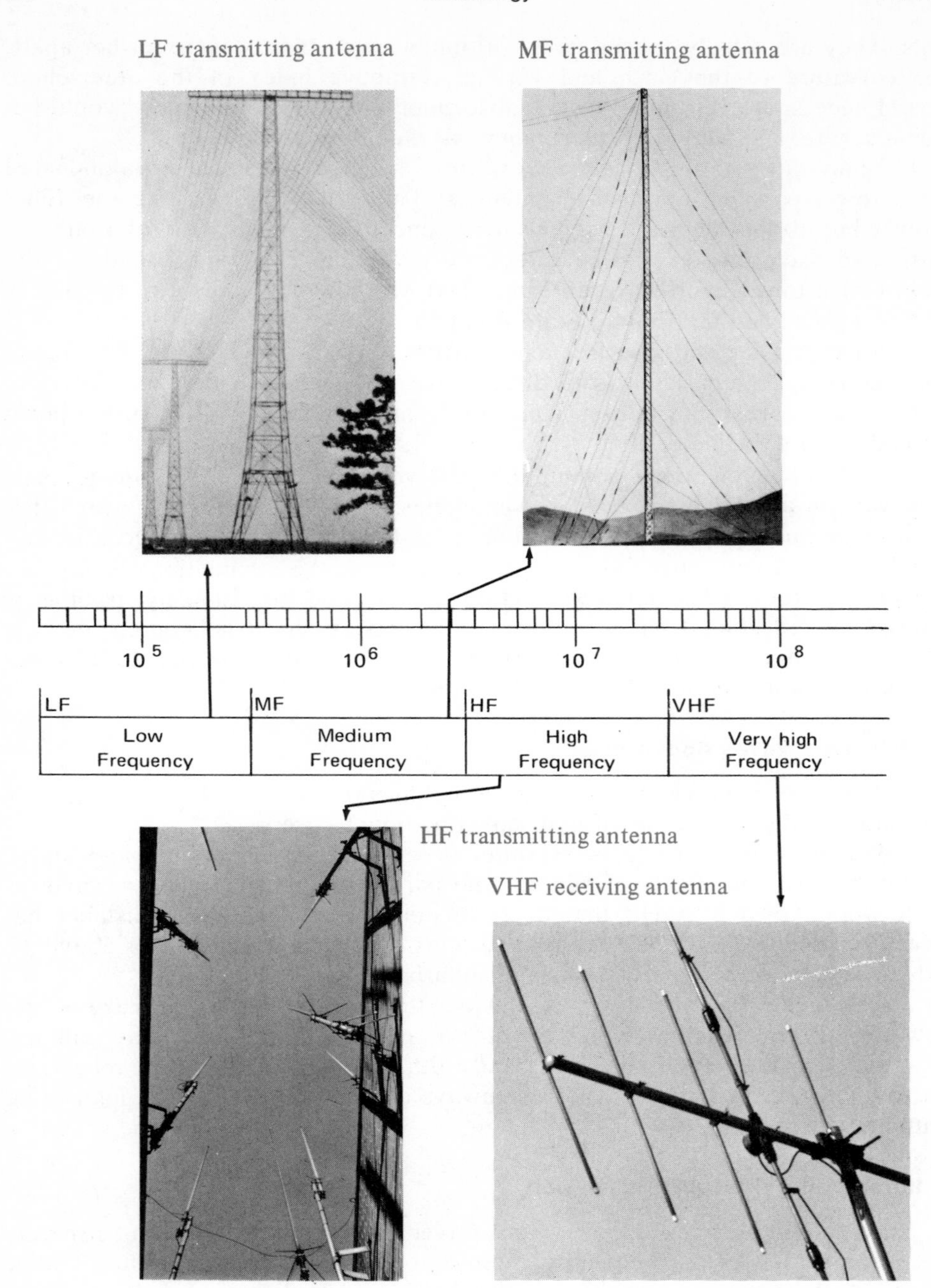

Figure 24.2 Antennas for different radio frequencies.

Microwave transmitting and receiving antenna

UHF receiving antenna

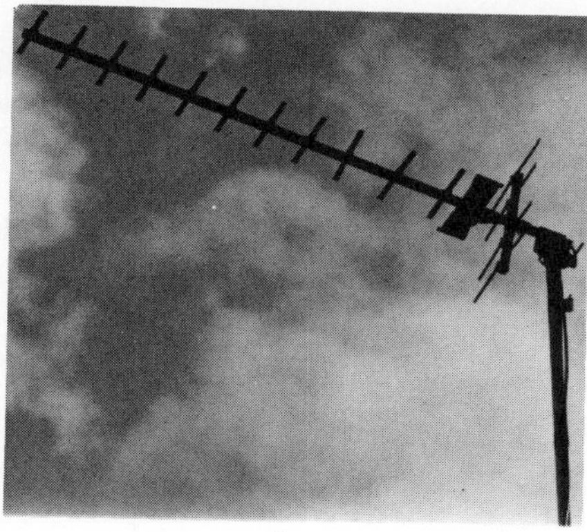

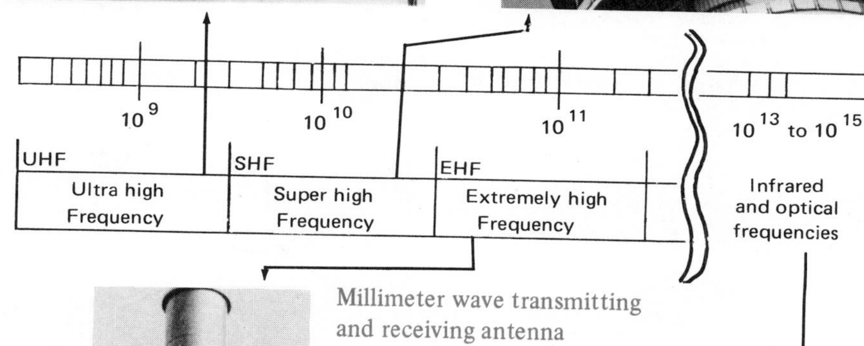

Millimeter wave transmitting and receiving antenna

Infrared transmitting and receiving antenna

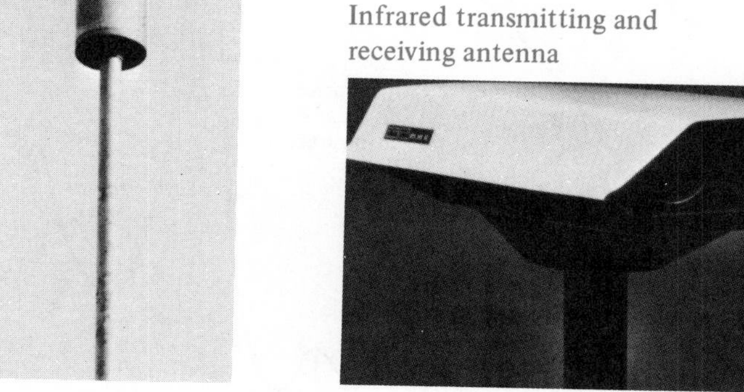

ily absorbed in thick fog, as we can see with the unaided eye. The signal can also be blotted out with very intense rain. Because of these effects of bad weather, optical and the lower frequency infrared transmission is normally limited to short distance links.

Figure 24.2 shows the antennas used for different radio frequencies.

RADIO ABOVE 10 GHz

Much of the new technology in radio systems relates to the use of frequencies above 10 GHz. Frequencies from 10.9 to 36 GHz are referred to as the K band. The advantages of such frequencies are that the available bandwidths are high and the radio congestion of lower frequencies is avoided. The equipment can be small and can give a relatively low cost per channel. However,

Figure 24.3 Microwave and millimeterwave attenuation in the atmosphere.

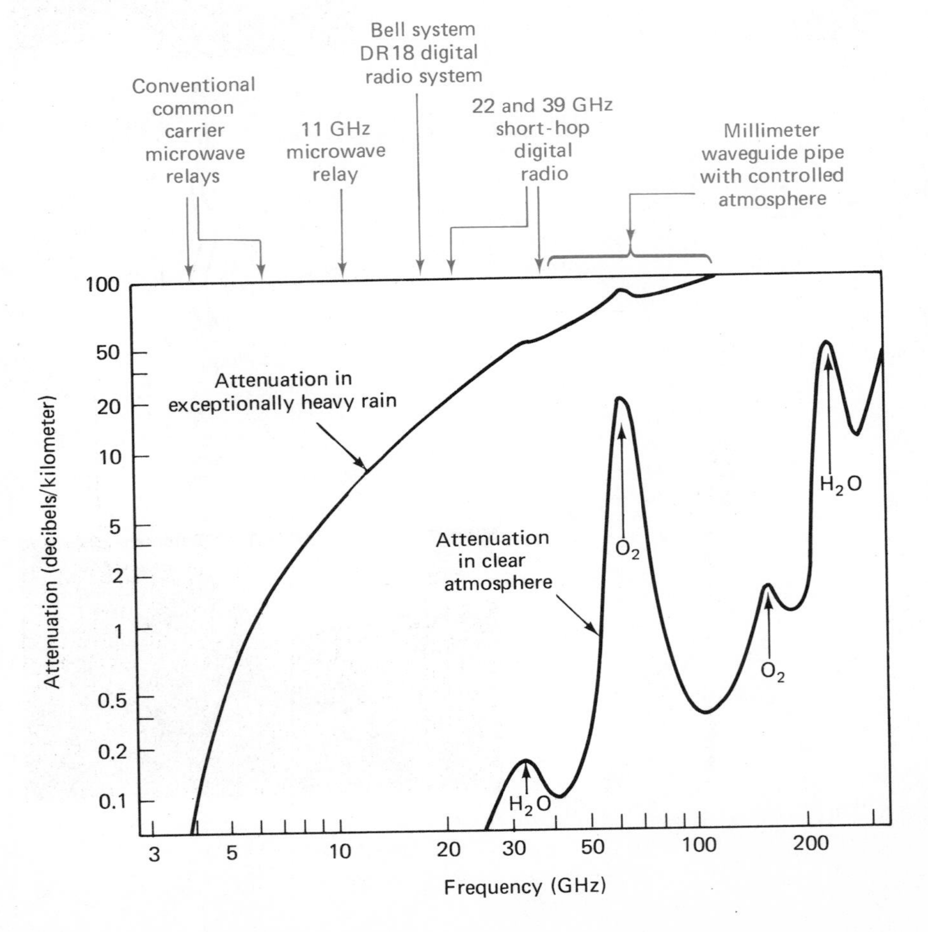

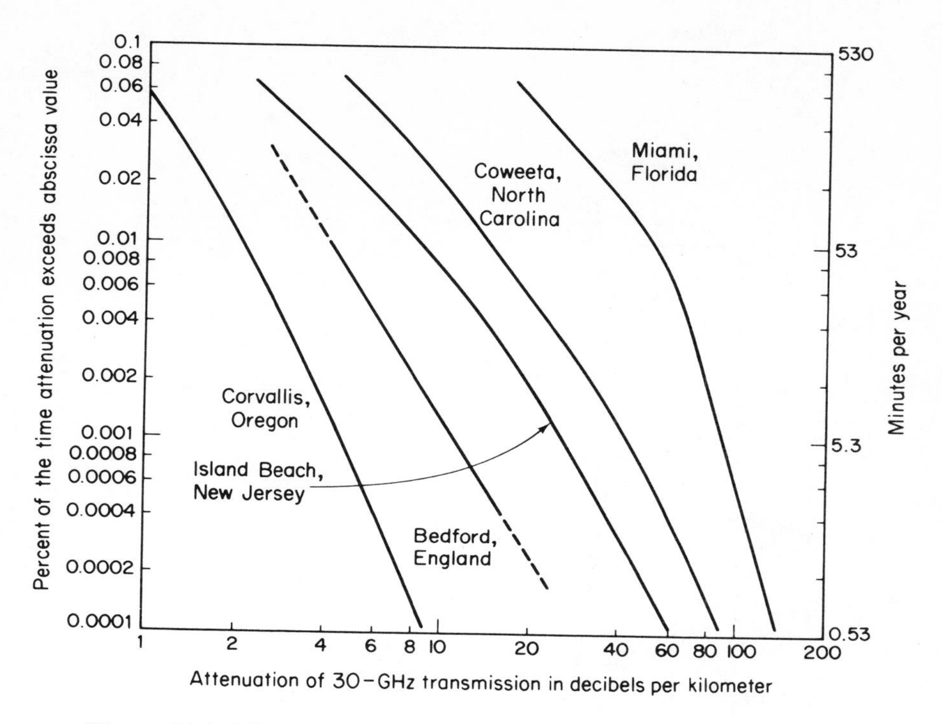

Figure 24.4 The number of minutes per year that a given level of attenuation at 30 GHz is exceeded for parts of the U.S. and England. (*Reproduced from D. C. Hogg, "Millimeter-Wave Communication Through the Atmosphere," Science No. 3810, January 5, 1968, with permission.*)

there is one major disadvantage. The signals are heavily attenuated by rain, and some frequencies are attenuated by the atmosphere.

The lower curve of Fig. 24.3 shows the attenuation of microwave and millimeter frequencies in the atmosphere when the weather is good. The peaks in the curve are caused by the absorbtion of specific frequencies by water and oxygen molecules. Oxygen causes severe absorption at 60 GHz. When millimeter waves of this frequency are sent through a waveguide pipe, the pipe is filled with a gas such as nitrogen to avoid the oxygen absorption.

The upper curve shows the attenuation caused by exceedingly heavy rain—rain at a rate of 100 mm (4 inches) per hour. Rain this heavy occurs only infrequently in most locations. One option therefore is to build a radio system which works well *most of the time,* but will occasionally be blotted out. A common carrier might use this option if the calls affected can be rerouted automatically via other types of circuit. Millimeterwave radio engineers can be found eyeing the weather like yachtsmen.

The amount of heavy rain varies widely from location to location. Figure 24.4 shows the number of minutes per year that the attenuation on a 30-GHz radio path exceeds certain limits. The result is not directly related to the

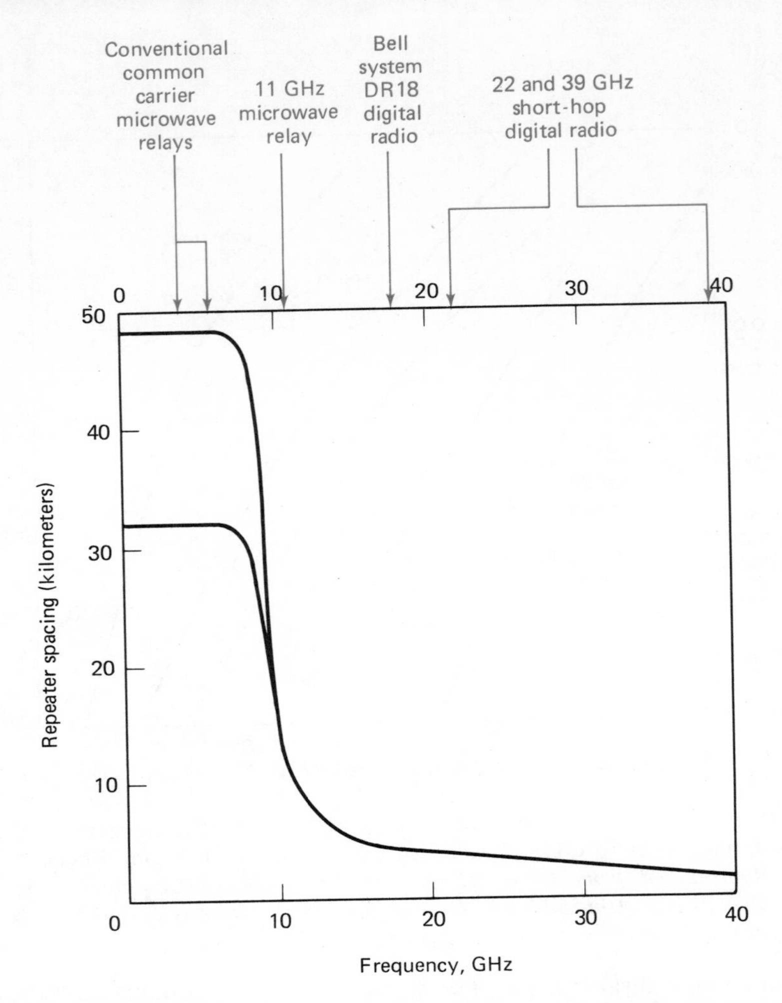

Figure 24.5 Repeater spacing as determined by a rain rate of 100 mm (4 inches) per hour (the upper curve of Fig. 24.3), and a 40 dB fade margin. This assumes uniform rain over the entire path. (*Data from Bell Laboratories.*)

amount of time it rains because it depends upon the intensity of the rain. London, where one rarely goes out without a raincoat, does better than Miami, because when it does rain in Miami it rains cats and dogs. Very heavy rain is often isolated geographically into localized storms. The most intense storms are the most limited in area, which raises the possibility of automatically switching transmission to alternate paths when storms occur.

The high attenuation, however, means that short hops with many repeaters will have to be used. The repeater, therefore, must be of low cost and high reliability. It must require little or no manual attention or maintenance. Consequently, solid-state circuitry must be used. The repeaters may stand on poles in the city streets; in this case, their appearance must not be unattractive.

Figure 24.5 shows a Bell Laboratories diagram showing the distance between repeaters necessary for the typical weather of the United States. The weather has no effect on repeater spacing below 7 GHz—the frequencies of common carrier microwave links. The typical spacing of 40 kilometers is determined by the curvature of the earth, the spreading of the microwave signal, and the effect that these factors have on tower and antenna costs. At 11 GHz—frequencies at which some specialized common carriers have planned local distribution—the repeaters should be within 10 kilometers of one another. At 18 GHz the repeaters need to be within about 5 kilometers, and at 39 GHz, within 2 or 3 kilometers. These figures are based on the gloomiest of weather assumptions: rain at 100 mm (4 inches) per hour over the entire hop.

FAIR-WEATHER RADIO

Because very heavy rain occurs only infrequently, one option with millimeterwave radio is to build channels which cease to function in very bad storms, or which severely degrade their mode of operation. There are two types of situation in which such an option is attractive. First when important traffic can be routed by an alternate means, and second when the traffic is not time-critical.

Electronic mail or other forms of message delivery can be delayed until after storms are over. Batch data transmission can be delayed. These applications may make use of the high bit rates of millimeterwave radio. In certain interactive uses of computers a link unavailability of .02, or even higher, can be tolerated. Users often face this today because of hardware and software problems. Video telephones or teleconferences might be regarded as non-time-critical in that loss of video can be accepted for a while provided that the speech remains clear. Speech can be routed by conventional telephone channels when the video link fails.

Another approach to surviving bad weather is to design channels which decrease their transmission capacity when the signal-to-noise ratio becomes poor. With digital transmission error-correcting codes can be used which may halve the transmission rate but compensate for much of the effects of increased noise. On FM systems a wide-ranging trade-off can be exercised between signal bandwidth and tolerance of a bad signal-to-noise ratio. This is illustrated in Fig. 24.6 which shows the modulation of an IF carrier of 70 MHz with a sine wave of frequency 1 MHz, by five different circuits. The top one uses amplitude modulation, and the two sidebands can be seen on either side of the non-information carrying component at 70 MHz. The other four use frequency modulation with different modulation indices. When the modulation index is low the energy tends to be clustered around the carrier frequency—though not so much as with amplitude modulation. When the modulation index is high the energy can be spread over a wide range of frequencies.

In practice the signal transmitted is not composed of one sine wave, but of many or of a continuous band of frequencies. Looking at the bottom diagram of Fig. 24.6 the reader may imagine a band of frequencies up to 1 MHz being

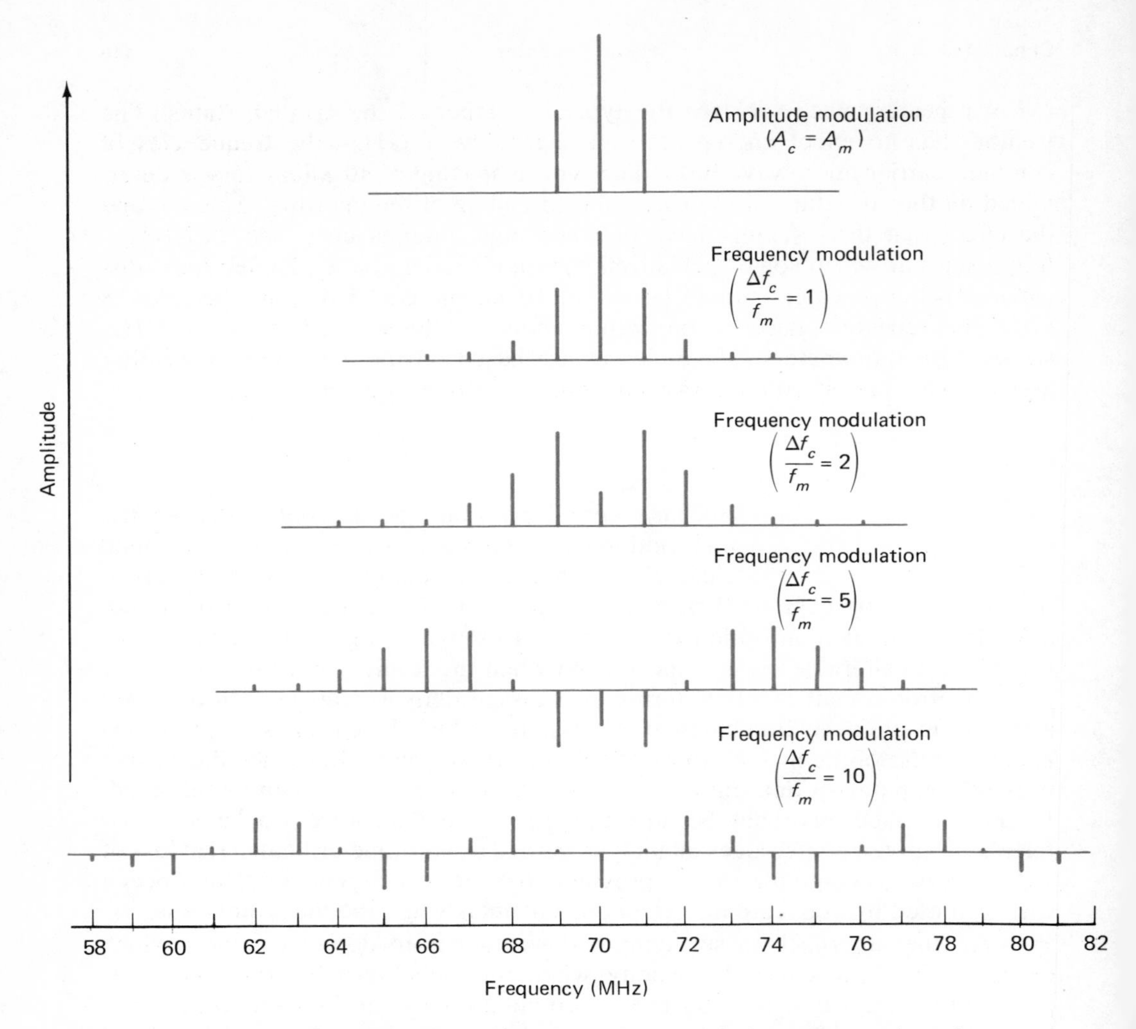

Figure 24.6 Spectra resulting from the modulation of a carrier of frequency 70 MHz with a sine wave of frequency 1 MHz. Frequency modulation can spread the energy across an entire transponder bandwidth.

smeared across the bandwidth of a radio carrier—from 52 to 88 MHz. In some millimeterwave systems a signal is smeared across a bandwidth *far* bigger than the signal's bandwidth. Such a signal can be recovered from the transmission even when the signal-to-noise ratio is very poor.

AT&T'S DR-18 SYSTEM

To avoid the congestion in frequency bands below 12 GHz, the FCC proposed the allocation of frequency bands at 18, 22, and 39 GHz for common carrier and private communications.

In the mid-1970s, AT&T introduced a radio system operating at 18 GHz into its telephone network. It is called DR 18, DR meaning *digital radio*. The repeaters are $2\frac{1}{2}$ to $6\frac{1}{2}$ kilometers apart. The antennas are housed in canisters, shown in Fig. 24.7, on slim steel masts or on building roofs. The canisters also contain the system's solid-state receivers, transmitters, and associated interconnecting equipment.

Figure 24.7 AT&T's DR18 digital radio system operating at 18 GHz, designed for major communications arteries in or near cities. A relay station such as that shown could be along highways or on city roofs. It relays eight two-way channels of 274 mb/s each. Seven of these are working channels and one spare. The system can transmit 28,224 simultaneous PCM telephone calls. The relays are spaced 1½ to 4 miles apart.

This radio system is designed to act as a major communications artery in metropolitan areas. It will carry up to 28,224 simultaneous telephone conversations, and bypasses the radio congestion that was illustrated in Fig. 13.7. Because of the short spacing necessary between antennas, it will not be used to haul signals over long distances.

The transmission is entirely digital and is designed to be resistant to the effects of bad weather. Each relay station has 16 receiver and transmitter parts, giving 8 two-way channels. There are 7 working channels and one held in reserve for automatic switchover in case one of the working channels should fail. Each of the channels transmits 274 million bits per second in each direction (totaling 1918 mb/s). The AT&T T4 standard for 274 mb/s is used. The 274

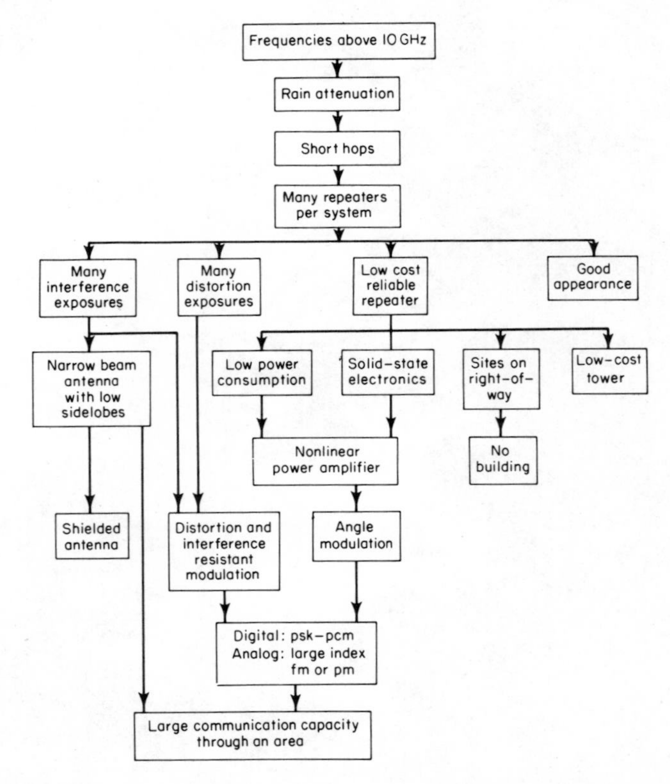

Figure 24.8 Characteristics of radio systems operating at frequencies above 10 GHz. (*Redrawn with permission from the Bell System Technical Journal.*)

mb/s bit stream carries up to 4032 telephone calls in PCM form (64,000 b/s per call). The radio system can thus handle $7 \times 4032 = 28{,}224$ calls. The 274 mb/s channel is intended to interconnect directly to other AT&T channels, discussed later, which also transmit 274 million bits per second, and which will eventually form nationwide links.

Because there are many hops in such a trunk system there are many exposures to distortion. A form of modulation that is resistant to the distortion is needed. Digital operation with frequency or phase shift keying gives a rugged form of operation, and it is likely that most K-band radio systems will be digital.

Figure 24.8 summarizes design characteristics of common carrier radio systems operating above 10 GHz.

SYSTEMS FOR PRIVATE USE

While common carrier systems such as the Bell DR-18 are engineered for high traffic volumes, another breed of system is intended to provide relatively low-cost links to private organizations.

A corporation, government body, or university can use a small millimeter-wave antenna on the roof to obtain a high-speed data link to a computer center, a link from a PABX to a satellite earth station, or a number of private lines between buildings. A specialized common carrier can use similar antennas to distribute its signals in cities.

Figure 24.10 shows a small millimeterwave radio system made by OKI Electronics. It operates at 39 GHz and is small—diameter 18 inches. It is designed to be mounted with a bracket to a 4-inch pole.

The system is designed to transmit 1.544 or 6.312 million bits/second, compatible with the T1 and T2 carriers respectively, and can therefore carry 24 or 96 voice channels—very different to the 28,224 voice channels of the Bell DR18 system. A range of interfacing and multiplexing equipment operates with such systems which provides the types of voice and data channels that are used with the T1 and T2 carriers.

Such systems give a maximum error rate of one bit error in 10^8. They are highly resistant to noise because high index frequency modulation is used to smear a signal with a signal of 1.544 Mb/s over a bandwidth of 50 MHz.

The FCC has assigned two groups of 14 channels with a spacing of 50 MHz, between 38.6 and 40 GHz as shown in Box 24.1. The spacing between antennas could be 10 miles if some unavailability due to rain can be tolerated, say 0.03%. With closer spacing the unavailability is less.

Figure 24.9 An experimental repeater for relaying radio signals of frequencies above 11 GHz down city streets. Built at the Bell Laboratories. (*Photographs courtesy of Bell Telephone Laboratories.*)

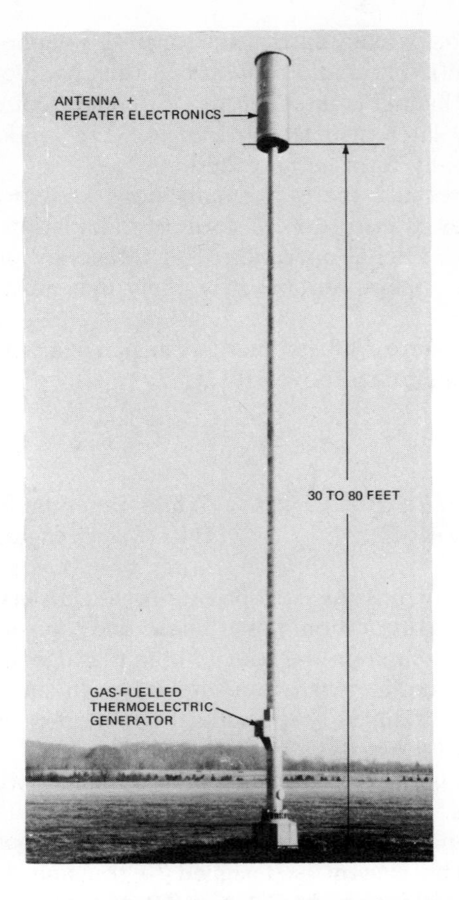

Receptors such as this may be sited every one to two miles.

A front view without the weather cover or inner absorbing liner.

Figure 24.10 A low cost millimeterwave antenna capable of transmitting 1.544 or 6.312 Mb/s (T1 or T2 carrier) for distances of a few miles, with an error rate of 10^{-8}. The radome cover is 18 in.; it is mounted on a 4 in. pole. It uses the frequencies listed in Box 24.1. This is the MACT-1 System manufactured by OKI Electronics.

The system is designed for the mass market of voice and data channels. It is likely that when a market develops for higher bandwidth links for applications such as videoconferencing, higher capacity millimeterwave radio systems will become available.

Going still higher in frequency, the absorption caused by the atmosphere itself becomes significant (see Fig. 24.3). At 60 GHz, the first 0_2 absorption band causes a level of absorption comparable to that of heavy rain. It has, however, been suggested that wideband signals such as television might be distributed at such frequencies by using small relay units every 300 ft. along the

BOX 24.1 FCC channel allocation for 39 GHz millimeterwave radio. Two groups of 14 channels each used by devices such as that in Fig. 24.10, for urban signal distribution

Channel Group A			Channel Group B		
Channel Number	*Frequency Band (MHz)*		*Channel Number*	*Frequency Band (MHz)*	
1–A	38,600	38,650	1–B	39,300	39,350
2–A	38,650	38,700	2–B	39,350	39,400
3–A	38,700	38,750	3–B	39,400	39,450
4–A	38,750	38,800	4–B	39,450	39,500
5–A	38,800	38,850	5–B	39,500	39,550
6–A	38,850	38,900	6–B	39,550	39,600
7–A	38,900	38,950	7–B	39,600	39,650
8–A	38,950	39,000	8–B	39,650	39,700
9–A	39,000	39,050	9–B	39,700	39,750
10–A	39,050	39,100	10–B	39,750	39,800
11–A	39,100	39,150	11–B	39,800	39,850
12–A	39,150	39,200	12–B	39,850	39,900
13–A	39,200	39,250	13–B	39,900	39,950
14–A	39,250	39,300	14–B	39,950	40,000

street and at intersections. The householder would have a small receiving antenna on a windowsill or rooftop, pointed at the nearest relay. Whether this method would be a cheaper way than cable to distribute television remains to be seen.

INFRARED AND OPTICAL TRANSMISSION

Much higher in the spectrum, infrared and optical systems can be used. A variety of somewhat experimental infrared and optical systems have been marketed and installed. They transmit only a minute fraction of the available bandwidth at those frequencies, but are relatively inexpensive. The weather is a worse problem than with millimeterwave radio. Fog as well as rain seriously attenuates the signal.

IBM used infrared transmission at Expo '67 in Montreal. The links are shown in Fig. 24.11. One carried two television channels a distance of almost two miles. The other was a data link carrying 1.3 million bits per second over a distance of about half a mile.

An optical transmission system can work in two possible ways. First, a laser can be used and the waves transmitted can be coherent—that is, the same phase relationship exists throughout the beam. The receiver could then use heterodyne detection, as in radio reception, in which a fixed frequency is added to the received signal to produce fluctuations (or beats) or a frequency equal to the difference between the two signals. The second approach is cruder. It uses fluctuating noncoherent light (or infrared signals) and detection is accomplished by photon-counting schemes—that is, direct detection of the incoming light with some form of photodiode. Coherent transmission and heterodyne detection was expected to be the better scheme. Background noise is not amplified with the signal and a larger information bandwidth is possible. However, noncoherent transmission has, so far, proved to be less inexpensive and it is less perturbed by atmospheric path fluctuations.

The success of noncoherent optical transmission has largely been due to the *light-emitting diode* (LED). This is a small semiconductor diode that emits radiation when a current is passed through it. The frequency of the emission depends upon the material of which the diode is made. It is desirable to select a material that emits at a frequency to which a radiation detector is most sensitive. Galium arsenide is commonly used as this emits at the sensitivity peak of a silicon detector. This tiny diode emits light approximately proportional in intensity to the current used.

The information rate that can be transmitted with a light-emitting diode is limited by the fact that the device takes a certain time to turn on. Typical diodes in use today take from 2 nanoseconds (2×10^{-9} second) to 100 nanoseconds to rise to full emitting power. Such devices have been used to transmit two TV channels or data rates up to about 200 mbps.

Figure 24.12 shows an inexpensive optical data transmission unit in operation. This device, the CTS 1815 OPTRAN transceiver can transmit and receive up to 250,000 bits per second. It is typically sited on the roof of a building, where it is inconspicuous because of its small size (19″ × 12″ × 5″). It can transmit more than a mile in clear weather, however, it is more likely to be used for distances of 500 to 300 ft, depending on local weather conditions.

Figure 24.13 shows laser systems, which can transmit a high bandwidth signal over a short distance, and a relatively low bandwidth signal many miles. Such systems are installed today in areas where cabling is a problem, i.e. across urban areas, freeways, airports, etc.

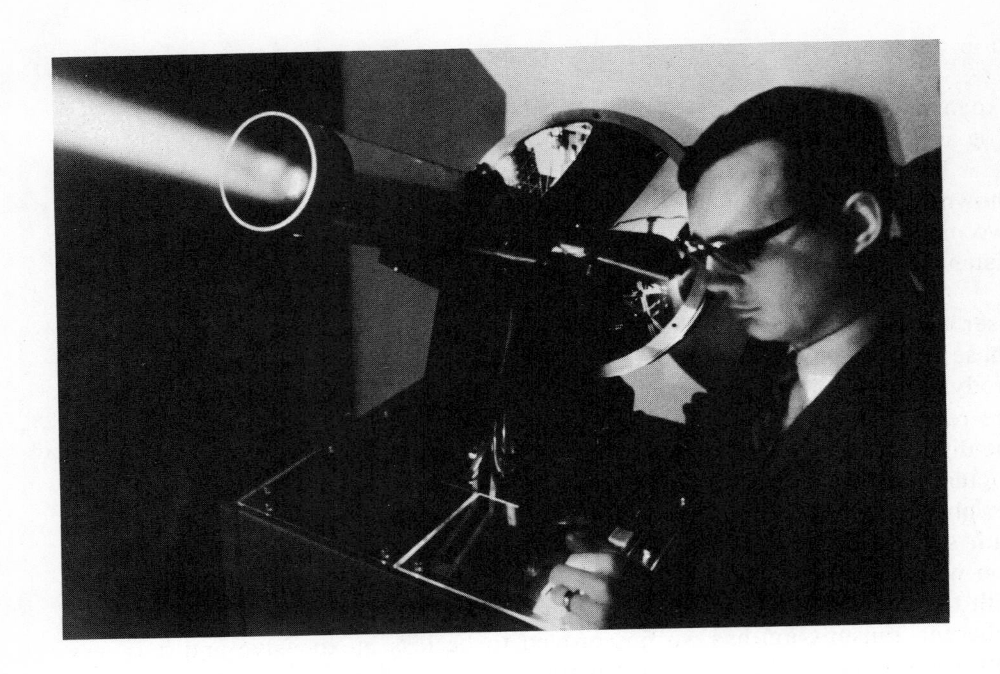

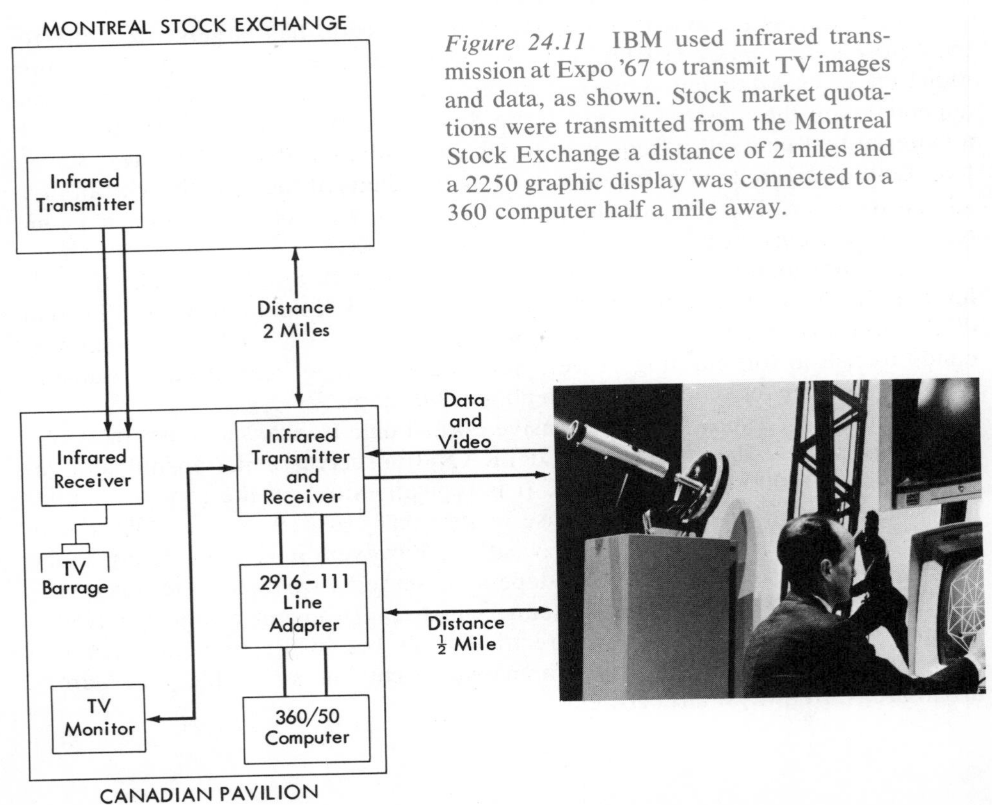

Figure 24.11 IBM used infrared transmission at Expo '67 to transmit TV images and data, as shown. Stock market quotations were transmitted from the Montreal Stock Exchange a distance of 2 miles and a 2250 graphic display was connected to a 360 computer half a mile away.

Figure 24.12 Optical and infrared transmission may form the basis of inexpensive signal distribution schemes in the cities. Part (a) is the Optran transceiver that transmits and receives, using noncoherent optical transmission, at a rate of 250,000 bits per second.

(a)

(b)

Part (b) is an installation of the Optran device. The unit on the left is transmitting to a similar device on a roof hidden in the mist behind the church steeple, half a mile away.

Figure 24.13 Three inexpensive laser communication devices manufactured by American Laser Systems, Inc.

ALS Model 736
GaAs Laser Voice & Data Communicator
Data rate: 10 Kb/s
Highly secure voice communications
Range: 15 miles
Peak output power: 10 watts
Optical pulsewidth: 100 msec

ALS Model 741
GaAs Laser Voice & Data Communicator
Data rate: 1.544 Mb/s
24 channel PCM telephone circuits
Range: 15 miles
Peak output power: 1 watt
Optical pulsewidth: 10 msec

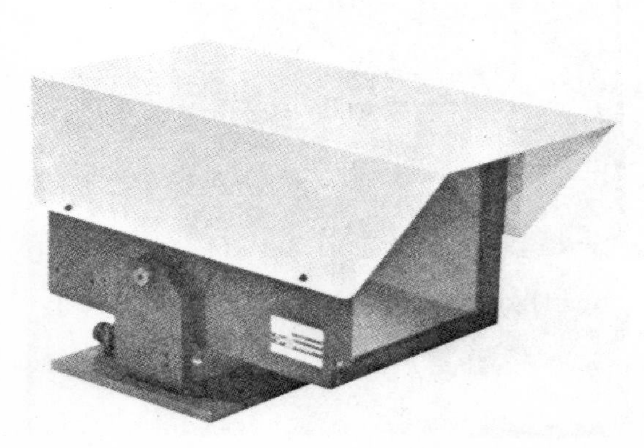

ALS Model 747
Television Transmission Set
High resolution video: 800 line
Video signal bandwidth: 15 MHz
Range: 2000 ft
Signal-to-noise ratio at 2000 ft 48 db
Peak output power: 4 milliwatts

Optical transmission has a great future in optical cables which we discuss in the next chapter. For through-the-air usage its application is limited.

ALLOCATION OF THE RADIO SPECTRUM

There are both international agreements and more detailed national regulations on the usage of the radio spectrum. In some countries the national frequency allocations are "classified." In the United States they are not, and the U.S. National Table of Frequency Allocations extends from 10 kHz to 90 GHz. It has hundreds of bands of frequencies allocated to different uses—government, military, and civilian. Bands allocated to one service are often separated by bands allocated to another. Some bands are allocated to two or more services on a shared basis.

Figure 24.14 summarizes the way the bands are allocated. The frequencies from 30 to 1000 MHz are the most congested, these being the frequencies needed for television and radio broadcasting and for mobile radio services. The FCC [1] has described the situation in the land mobile services as being one of "congestion" (1958), "extreme congestion" (1962), "acute frequency shortage" (1964), and a "disaster" (1970). The frequency shortage has seriously impeded progress in developing mobile radio services.

The congested area is shown in more detail in Fig. 24.15, this time the international frequency allocation chart is given. Usage of frequencies is more internationally uniform below 30 MHz because it is in these lower frequencies that worldwide transmission is possible.

WAYS TO ALLEVIATE THE FREQUENCY SHORTAGE

Fortunately, there are many ways in which the spectrum congestion can be relieved. These methods involve not merely a better sharing of the spectrum but better uses of today's technology. They cannot be achieved without national and, in some cases, international cooperation and control. It is highly desirable for government to continually review and reallocate a nation's electromagnetic resources. New problems will arise with the efficient use of satellites, which must be prevented from interfering with one another and with other land resources. It is necessary to consider these problems along with other spectrum problems.

The following are steps that can be taken to alleviate the spectrum problems:

1. Better Allocation of the Frequencies

Some services for which blocks of frequencies are allocated are underutilized while others are bursting at the seams. Bands allocated for forest conservation, forest products, and highway maintenance are underutilized, for example, whereas those for business radio and radio telephone are so crowded

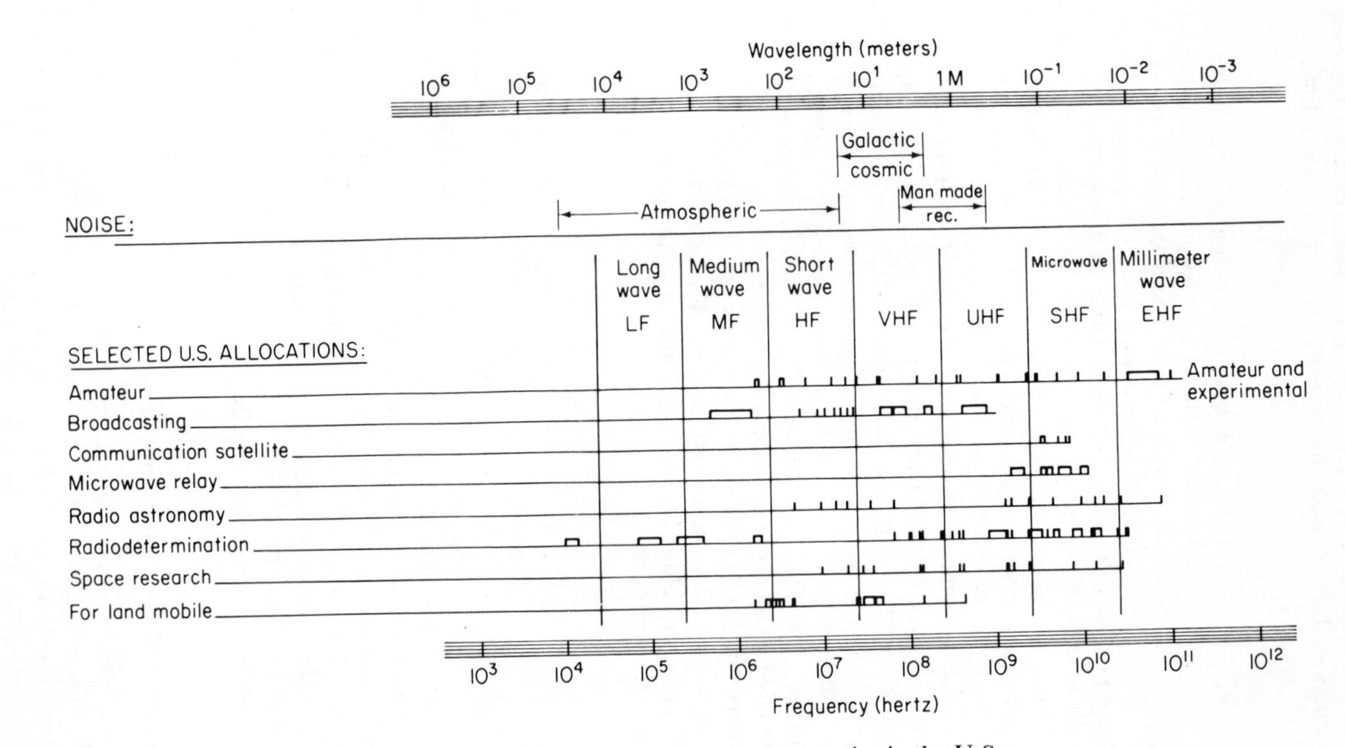

Figure 24.14 The allocation of radio frequencies in the U.S.

that they are holding up progress which could have a valuable effect on the economy. The wide band allocated for UHF television channels is little utilized today, and the unused portion could carry many thousands of radio telephone channels. In the United States, as discussed in Chapter 12, the FCC reallocated a large area of UHF spectrum to mobile radio uses. Some military allocations are also underutilized. Some parts of the spectrum are not allocated at all because of possible future needs.

Spectrum allocations must be changed as the needs and technology change. The spectrum should be more finely apportioned, in smaller blocks, and the allocation should be much more dynamic.

2. Frequencies Should Be Shared

It no longer seems appropriate to allocate frequencies in nationwide blocks. In the early 1970s the band allocated to the petroleum industry is the same in New York City as in the oil fields of Texas. Taxis have the same number of channels in the Oregon forests as they do in Los Angeles, and frequencies allocated to the forestry service in Oregon cannot be used, except for forestry in Los Angeles.

A few decades ago this situation would have made more sense because longer wavelengths were used, which caused long-range interference. The farther we move into higher-frequency bands, the shorter the range of interference and the greater the need for *localized* frequency allocation.

Again, different services do not have the same utilization at different times of the day or different times of the year. Most business use of frequencies in mobile land transmitters, paging schemes, vehicle control and scheduling, radiotelephones, and so on will take place between the hours of 8 A.M. and 6 P.M. Peak television watching is from 6 P.M. to midnight. Automatic machines, transmitting computer data, or sending mail in facsimile form could operate at night. Would it not make sense to allocate some of the television bandwidth to other functions outside the hours of 6 P.M. to midnight? Again certain users needing mobile radio may be able to confine their activities to just a few hours a day.

Various authorities [2] have advocated that market forces should apply to spectrum allocation on a basis of "packages" of frequencies covering a given area and a given time. "TAS packages" (time, area, and spectrum) should be used rather than the National Table of Frequency Allocations (Fig. 24.13).

3. Narrower Transmitter Bands

Some equipment using the spectrum is designed in a way that wastes its frequencies—after all, spectrum use is free. Many mobile voice transmitters used to use 100 kHz per telephone channel—the same telephone channel which receives only 4 kHz of equivalent bandwidth on the Bell System. Radio systems need much wider spacing than cable systems to separate different transmissions; however, with today's state-of-the-art, they do not need 100

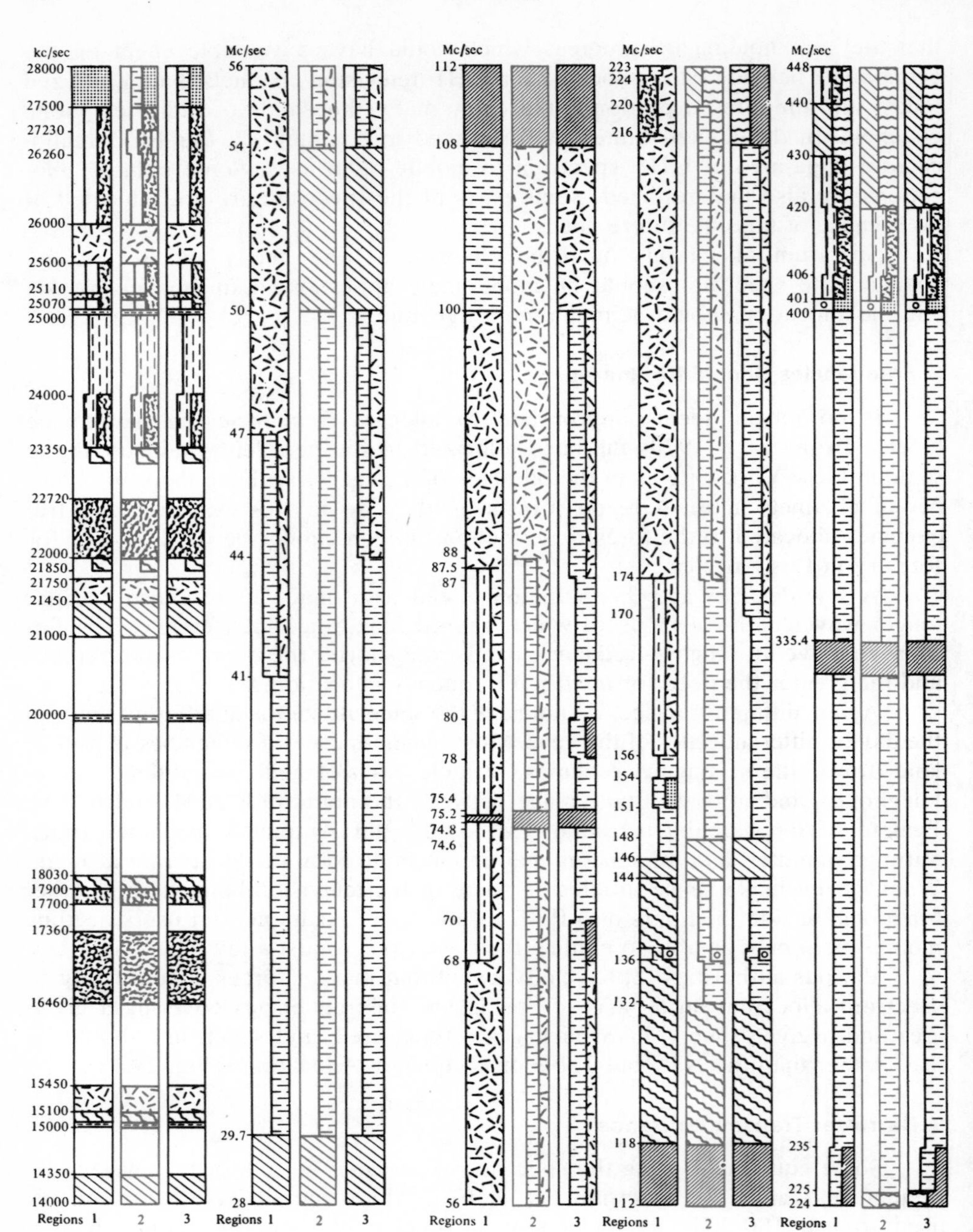

Figure 24.15 International frequency allocations, showing the most crowded part of the spectrum. (*By permission of the Radio Corporation of America.*)

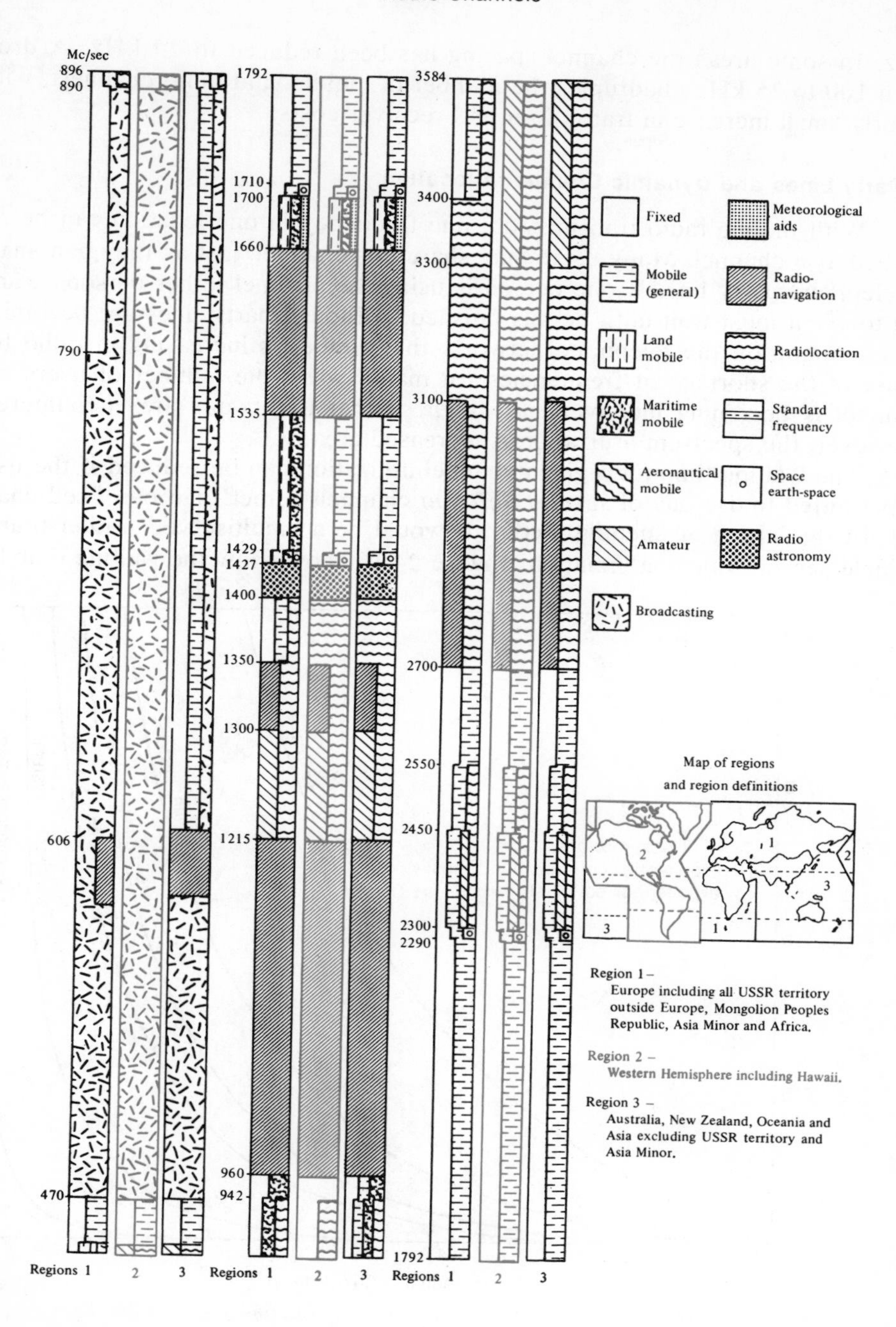

Figure 24.15 (cont.)

kHz. In some areas the channel spacing has been reduced to 25 kHz. A drop from 100 to 25 kHz quadruples the number of mobile land radio channels, with a fairly small increase in transmitter and receiver costs.

4. Party Lines and Dynamic Channel Allocation

With mobile radio transmission, one frequency in one location can be regarded as a channel. Many users may share this channel, just as they can share a telephone party line. If one person is using the channel, other persons wanting to use it must wait until he has finished. Although party lines are becoming less common on the telephone network, the number is increasing in radio because of the shortage of frequencies. In many cases, the number of users per channel is becoming inconveniently high so that long waits are encountered. However, the spectrum utilization is increased greatly.

The waiting time for a given channel utilization can be reduced if the user is permitted to use one of *several different* channels rather than one fixed channel. In the jargon of queuing theory, it would be a "multiserver" rather than a "single server" queuing situation. Figure 24.16 shows how the waiting time for

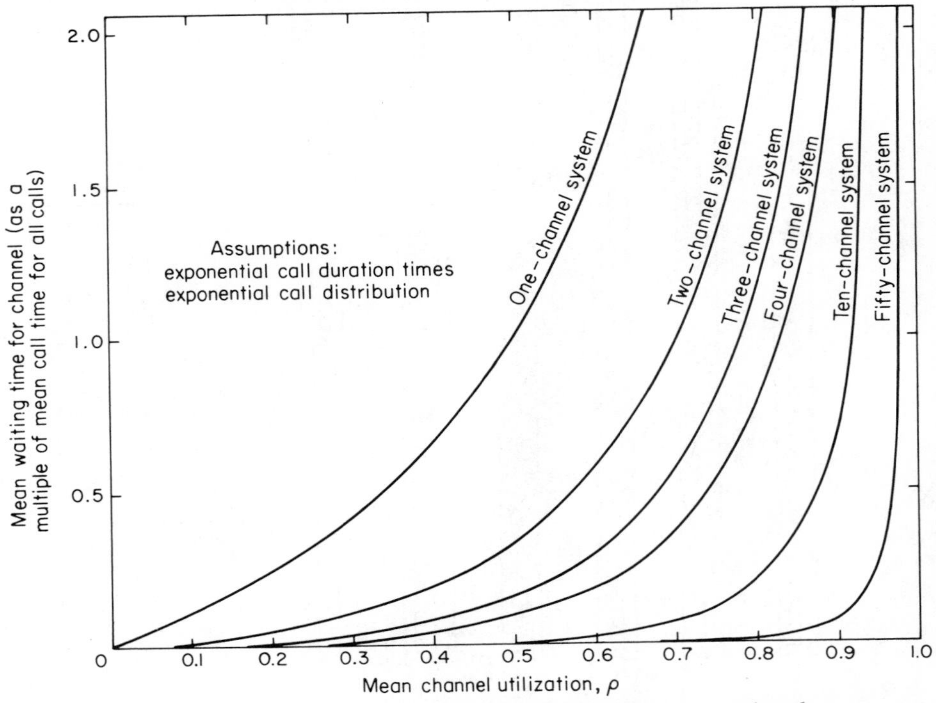

Figure 24.16 A system which automatically queues users so that they can be allocated any one of many channels, will increase the permissible channel utilization and decrease the user waiting time.

queues varies with the numbers of channels (or servers). These curves assume an exponential distribution of the channel-holding times and calls arriving at random—that is, a Poisson distribution of number of calls per unit time. These conditions approximate the situation on typical mobile radio channels. They also assume that callers wait until a channel becomes free.

It will be seen that the queuing time is less for the multichannel queues. With a channel utilization of, say 0.5, there is a substantial mean waiting time if only one channel is used. It is less if two channels are used. If ten are used, the mean waiting time is very low indeed.

To take advantage of this a mobile radio transmitter and receiver operate on several different channels. When it comes into operation, it searches for a free channel and then uses it. This process adds to the cost of the transmitter and receiver, but, once again, in an age of large-scale integration, circuit logic for the searching operation will be inexpensive.

5. Low-Power Transmitters

A television transmitter has high power and beams its signal over distances of 100 miles or so. Nothing else over a wide area can use its frequencies. A taxi fleet operating within a city does not need such a powerful transmitter. Separate taxi fleets in nearby cities could use the same frequencies if the transmitters employed were suitably low in power.

In any area over which continuous coverage is needed, it is often desirable to reuse the frequencies for different transmissions in different locations. A vehicle fleet in one area should not be prevented from reusing the same frequencies as a vehicle fleet in another area. This is the principle of the *cellular* radio telephone systems discussed in Chapter 12.

In order to serve as large a number of mobile radios as possible, the zones should clearly be small. In other words, the transmitters should be low in power. It will be necessary in cellular systems for vehicles to travel from zones of one frequency band to zones of another and for calls to and from them to be appropriately and automatically switched.

If mobile transmitters cover a distance of 4 miles and same-frequency zones are separated by one zone, then each frequency used must be unique within an area of about 113 square miles. For television transmitters covering a distance of up to 100 miles, the equivalent area will be 70,000 square miles. The frequencies used by one television channel could be used for over a thousand mobile radio channels. But because of its wide area coverage, one television broadcast uses frequencies that could be reused by several hundred thousand mobile radio channels with appropriately low power transmitters.

6. Directional Antennas and Polarization

Transmitting antennas can be built to transmit in one particular direction, and receivers can be built to receive from that direction only. The higher the transmission frequency, the narrower the transmitted cone. Microwave relay antennas can use a very narrow beam. The beam has a somewhat wider spread with UHF and VHF bands. However, a receiver at one location could separate signals of the same frequency beamed to it from four different directions at right angles to each other. Vertical and horizontal polarization similarly enables the separation of same-frequency signals and may be used in conjunction with directional antennas to increase the number of signals that can be distributed in a city.

Comparing today's radio links operating above 10 GHz with today's television antennas is like comparing a searchlight beam with a floodlight. In the diverse needs of our cities for the distribution of many types of signals "floodlight" transmission is far too wasteful. Much radio distribution in the future will be in the form of criss-crossing "searchlight" beams spanning the rooftops; the narrower the beams the better. To achieve these narrow beams higher frequencies are needed and hence radio techniques pushing up beyond the microwave frequencies into millimeter waves.

7. Maximum Use of Cables Rather Than Radio

Television distribution of the future will commonly use the "searchlight beam" microwave or millimeterwave radio, or satellite transmission, to carry signals to the head of cable TV systems. The flooding of large areas with very high power transmission could be replaced by a judicious combination of cable distribution and small localized transmitters, leaving the maximum "time-area-spectrum packages" for other use. If these steps are taken, large slices of bandwidth reserved for television today could be freed for other services, at least in certain cities. The broadcasting companies, however will remain reluctant to give up its frequency allocations.

In general, cable and satellite distribution of very high bandwidth signals is preferable to VHF or UHF broadcasting.

8. Use of Higher Frequencies

If systems can be designed to utilize frequencies above 10 GHz, the bandwidth available for radio transmission will become several times the total of that used today. Millimeterwaves permit highly directional antennas and will be used mainly for "searchlight beam" distribution of signals within cities. Still higher in the spectrum are infrared and optical frequencies.

These higher frequencies offer the prospect of relief from the microwave congestion in major cities. Within the next ten years radio systems for trunking telephone and television channels into metropolitan areas will make a major use of frequencies above 10 GHz.

9. Use of Data Systems

Data can be carried far more economically than voice. Large numbers of data machines can share a relatively small spectrum space, as has been demonstrated by the packet radio systems. Many radio users employ voice channels today, or even television channels, when data channels would suffice or be preferable. The social or business uses of television could be expanded enormously by data broadcasting.

SPECTRUM ENGINEERING

To allocate the radio spectrum mainly on the basis of a *National Table of Frequency Allocations* is inadequate to the complex interleaving of data technologies that is now possible. The term *spectrum engineering* refers to more flexible spectrum management techniques in which the allocation of blocks of frequency, time, and area will be tailored to particular situations.

Much of today's equipment and many of today's operating practices result in severe spectrum wastage. Spectrum engineering will tighten the efficiency of spectrum usage with all methods that prove economical, including those just discussed. Today's users commonly employ broader bandwidths than are necessary, as well as higher power, less-directional antennas, less-effective modulation techniques, less than optimum-antenna locations, and no polarization. Large acreages of spectrum are today allocated to the distribution of signals that would be better sent by cable. Spectrum engineering attempts to apply standards and operating criteria that would reduce the wastage.

A detailed study of spectrum engineering requirements was carried out by the IEE/EIA Joint Technical Advisory Committee (JTAC) [8]. According to the committee, two characteristics should dominate future spectrum engineering.

First, the system must be evolutionary. The present system's functioning is understood, in varying degrees, by many thousands of people—millions, if licenses' operators and Amateur and Citizens Band Services licenses are included. Their loyal, if often exasperated and occasionally imperfect, support of the system is in a way the bedrock of the nation's spectrum usage structure. Changes in the system must be well thought out and amply foreshadowed, or the continued cooperation of those concerned will simply not be possible. Also, recovery of capital investment in equipment must be considered in planning for changes. But none of this means that evolution must continue at the snail's pace of the past.

Second, spectrum engineering and management thinking must continue to move away from the concept of controlling spectrum usage through simple but rather restrictive and rigid administrative rules. The movement must be in the direction of increasingly individualized technical assessment of applications under explicitly formulated priority criteria, and under a reduced and more flexible employment of block allocation concepts. This will require much heavier use of analytical and data processing capabilities in the nation's spectrum engineering than at present; but it would result in stronger spectrum engineering capable of supporting the more flexible and effective management needed if fuller spectrum utilization is to be achieved.

Under the scheme advocated by the JTAC, user's spectrum allocations would expire periodically and have to be renewed. The needs of would-be users and users whose authorization to operate is expiring would be considered together. An attempt would be made to fit all users in by employing the latest developments in technology.

Under the JTAC scheme, computers would maintain a detailed data base of current spectrum users and their equipment characteristics. They would also keep a data base of available spectrum acreage, and the computers would be used to assist in the selection of frequency assignments. The usage of assigned frequencies would be carefully monitored.

CHARGING FOR SPECTRUM USAGE

The merits of better spectrum engineering are indisputable. However, it can be argued that rather than having an elaborate government agency for the allocation of user priorities, market mechanisms should be allowed to play a part. Spectrum acreage, like land acreage, this argument says, should be bought and sold, the price reflecting the relative strengths of supply and demand in different parts of the spectrum.

Where spectrum space is short, the price will rise. It is unlikely that there would be large, unused tracts of prime acreage like the present UHF television channels. If the cost of the large bandwidth of a television channel becomes too high, the result will be a big incentive for cable TV.

A fairly difficult technical problem exists in how to define marketable "plots" in such a way that their owner cannot interfere with other users. The time-area-spectrum (TAS) packages mentioned earlier seem suitable at VHF and lower frequencies. At higher frequencies, however, directivity of antennas, polarization, reflection off buildings, and other factors make the demarcation of electromagnetic "plots" more difficult.

If such problems can be solved, perhaps the main objection to a market mechanism is the conflict between what is profitable for businessmen in radio usage and what is socially desirable. The maximizing of advertisement revenue, for example, makes for poor television. Radio astronomers would probably suffer if there were a free market in spectrum space. Educational uses of the media would be held back if the spectrum price was high. Spectrum allocation is likely to remain a highly controversial topic.

SPACE TELECOMMUNICATIONS

An area of spectrum allocation that has assumed major importance in the 1970s is the use of radio by satellites. In the UHF and SHF bands, satellite systems and terrestrial systems can interfere with one another. It is desirable to achieve an appropriate mix of frequency allocations that will permit the full flowering of satellite usage.

As space transmissions transcend national borders it is very important that there should be international agreement on the use of space frequencies. Such agreements originate at Administrative Radio Conferences held under the auspices of the International Telecommunication Union (ITU) and comprise the Radio Regulations. These Regulations have the force of a treaty to which each signatory is bound under international law. Two conferences have allocated frequencies for space usage: the Extraordinary Administrative Radio Conference of 1963 (EARC) and the World Administrative Radio Conference for Space Telecommunications of 1971 (WARC).

The frequencies of interest for satellites are the higher UHF frequencies, the SHF band, and the lower EHF frequencies. These are the frequencies of the radio "window" the satellites can employ.

The lowest frequency allocated for geosynchronous communication satellites is 2.5 GHz. Figure 24.17 shows the allocation of radio frequencies of 2.5 GHz and above, made at the EARC and WARC conferences.

The radio regulations divide the world into three regions as shown in the map at the top of Fig. 24.16. Region 1 consists of Europe, Asia Minor, Africa, all USSR territory outside Europe, and the Mongolian Peoples Republic. Region 2 consists of the Western Hemisphere: the Americas including Hawaii and Greenland. Region 3 consists of Australia, New Zealand, Oceania, and the parts of Asia not in Region 1. It will be observed that there are some differences between the allocations for the three regions.

There are many frequency allocations shown in Fig. 24.16 for geosynchronous (fixed) *communications* satellites, and a few for *broadcasting* satellites. Various other uses of space are allocated.

Various allocations for radio in space are made for frequencies below 2.5 GHz, and these are listed in Fig. 24.18. Most of them are for mobile satellites or satellites having specific functions. Most are for bandwidths which are low compared with the bandwidths of today's main communication satellites. Some communication satellites low in the UHF band have been used, with low bandwidths, and can be operated with relatively inexpensive earth stations. Other UHF frequencies are used for sending commands to geosynchronous satellites. Some military satellites use UHF frequencies not allocated internationally for space purposes.

Frequencies in the lower part of the UHF band have the advantage that the electronic equipment is relatively efficient and inexpensive. The military

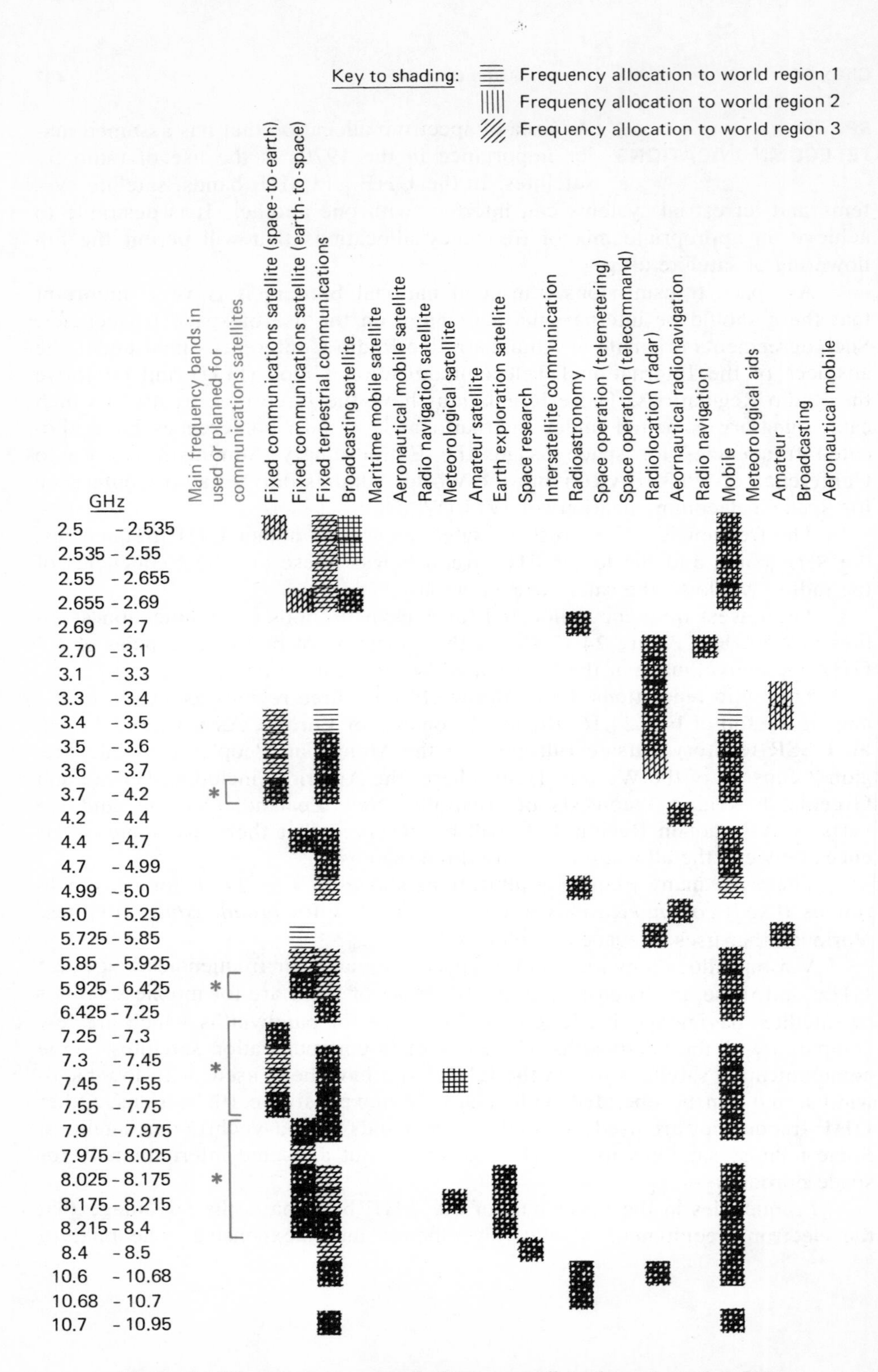

Figure 24.17 International radio frequency allocations above 2.5 GHz.

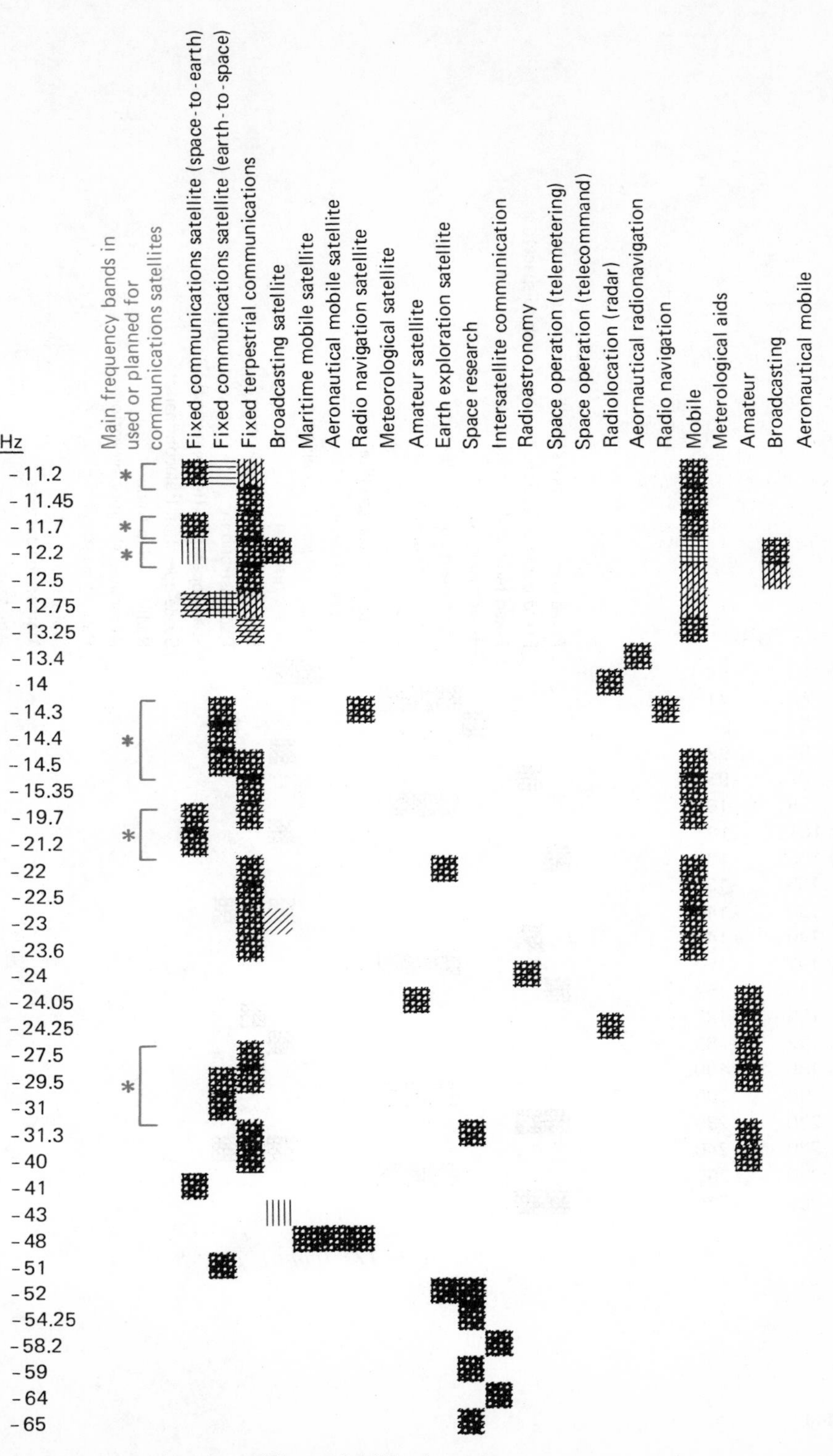

Figure 24.17 (cont.)

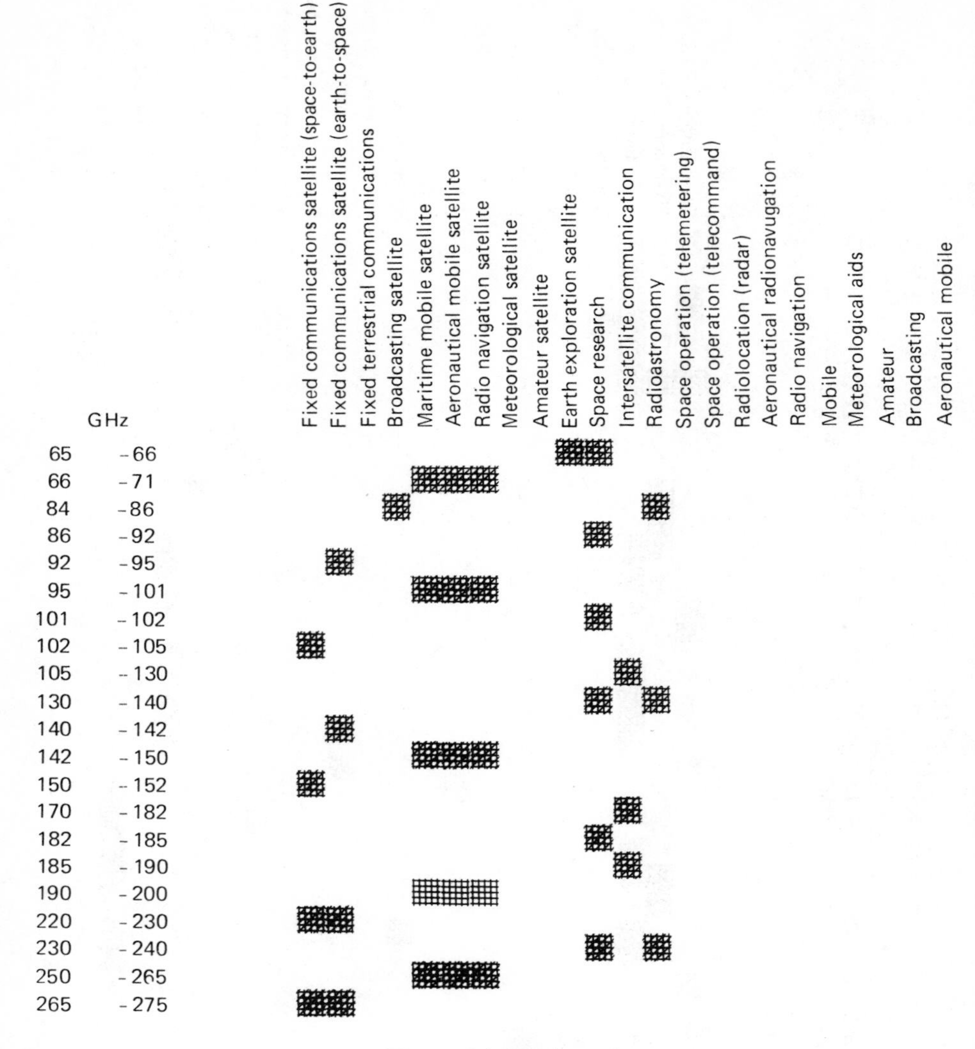

Figure 24.17 (cont.)

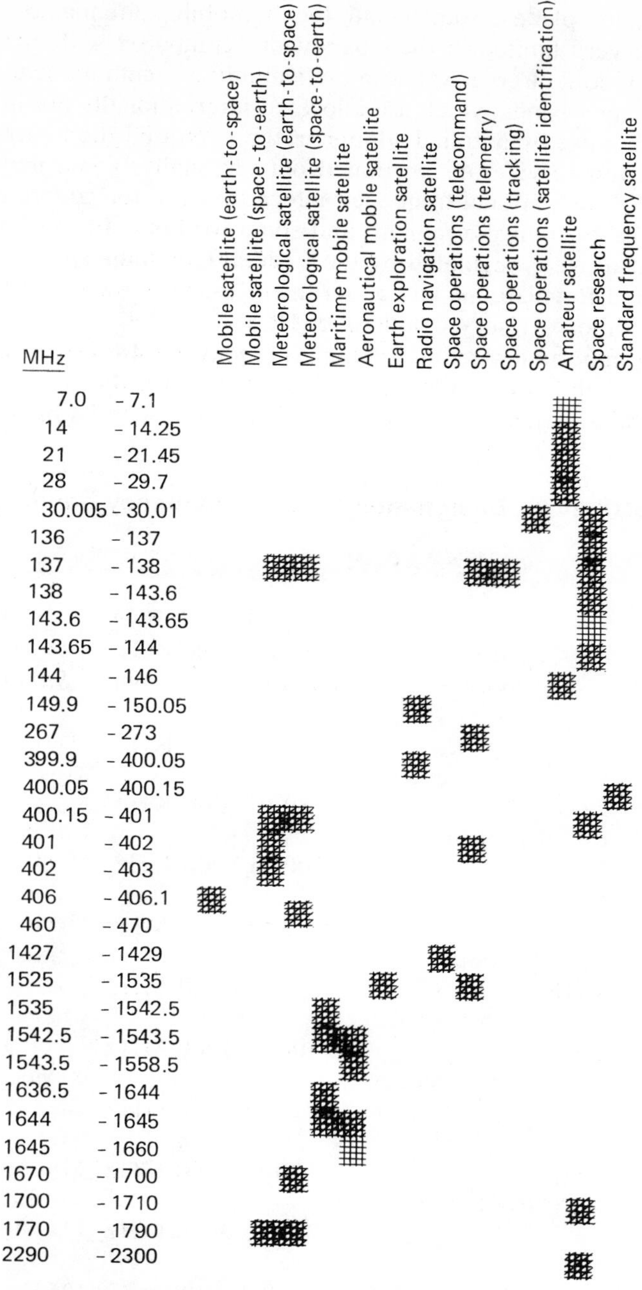

Figure 24.18 Radio frequencies below 2.5 GHz which are allocated to space uses.

uses UHF satellites primarily for tactical communications from equipment on vehicles, ships, and planes. Some small, highly mobile, antennas on vehicles are used. Although very limited in their bandwidth compared with SHF satellites, they are nevertheless of great value in tactical military communications.

The frequency bands which are allocated internationally are in turn reallocated for national use by national administrations. Not all the international allocations for a given service are portioned out nationally; some may be kept in reserve. Some of the national frequency allocations are for government or military usage, some for nongovernment, and some for both. In the United States, nongovernment usage is regulated by the Federal Communications Commission (FCC) Rules and Regulations, and government usage is regulated by the Interdepartmental Radio Advisory Committee (IRAC).

The most important satellite frequencies today are 4/6 GHz. In a few years the 12/14 GHz band will assume major importance. Further in the future there will probably be a major use of the 20/30 GHz band with its higher bandwidth allocation.

BOX 24.2 Designation of radio frequency bands.

*Band Number**	*Band Name*	*Frequency Range (including lower figure, excluding higher figure)*	*Metric Subdivision*
4	VLF, Very Low Frequency	3 to 30 KHz	Myriametric waves
5	LF, Low Frequency	30 to 300 KHz	Kilometric waves
6	MF, Medium Frequency	300 to 3000 KHz	Hectometric waves
7	HF, High Frequency	3 to 30 MHz	Decametric waves
8	VHF, Very High Frequency	30 to 300 MHz	Metric waves
9	UHF, Ultra High Frequency	300 to 3000 MHz	Decimetric waves
10	SHF, Super High Frequency	3 to 30 GHz	Centimetric waves
11	EHF, Extra High Frequency	30 to 300 GHz	Millimetric waves
12		300 to 3000 GHz	Decimillimetric waves

*Band Number N extends from $0{\cdot}3 \times 10^N$ to 3×10^N Hz

BOX 24.2 *Continued*

High frequency bands are also given letter designations as follows:

Band	*Frequency Range, GHz*
P	0.225– 0.39
J	0.35– 0.53
L	0.39 – 1.55
S	1.55 – 5.2
C	3.9 – 6.2
X	5.2 –10.9
K	10.9 –36.0
Ku	15.35 –17.25
Q	36–46
V	46–56
W	56–100

1909. Before the use of multiplexing or underground cables, masses of wires filled the streets. On Broadway (New York) some telephone poles were 90 feet high and carried as many as 50 cross arms. (*Courtesy AT&T.*)

25 CABLES & WAVEGUIDES

In view of the limited availability of radio spectrum space, any signal that can economically travel by cable rather than by radio (below 60 GHz) should be made to do so. This is especially true in the VHF and UHF regions, which are needed for mobile transceivers. Television sets can receive their signals via cable, but a taxi, delivery truck, or automobile cannot drag an umbilical cord around the street after it.

The information-carrying capacity of cables has been steadily increasing ever since Gauss and Weber first strung wire over the roofs of Göttingen in 1834. The highest-capacity cables in public use today are the coaxial cables used by the telephone companies for their main telecommunication highways. The next step up the scale is the helical waveguide, which AT&T is beginning to use. Then comes the prospect of optical channels. This chapter discusses these very high capacity facilities.

As cables of larger capacity are employed, so increasing numbers of voice, television, or other signals can be sent over one cable, a factor that will lower the cost per channel-mile. The investment cost of adding a voice channel-mile to Bell System trunks has dropped dramatically and will continue to drop, although the drop is not always reflected in long-distance telephone charges.

COAXIAL CABLES

The common-carrier cables at the end of World War I consisted of many twisted pairs of wires grouped together in protective sheaths. These cables presented problems because of cross talk between separate pairs of wires and because of high signal losses. They could not carry a high frequency and so were unable to transmit broadband signals. There was no need to transmit broadband signals in the late 1940s, but the dawn of television was on the horizon. By 1950 several thousands of miles of coaxial cable were in use for telephone transmission in the United States.

As the signal frequency is increased on a conducting wire the current tends to flow increasingly on the outside surface of the wire. The current uses an increasingly small cross section of the wire, and thus the effective resistance of the wire increases. This is called the *skin effect*. Furthermore, at higher frequencies, an increasing amount of energy is lost by radiation from the wire. Nevertheless, it is desirable to transmit at as high a frequency as possible so that as many separate signals as possible can be sent over the same cable. The skin effect limits the upper frequencies. Cross talk also imposes limitations.

A coaxial cable can transmit much higher frequencies than a wire pair. It consists of a hollow copper cylinder, or other cylindrical conductor, surrounding a single-wire conductor having a common axis (hence coaxial). The space between the cylindrical shell and the inner conductor is filled with an insulator, which may be plastic or mostly air, with supports separating the shell and the inner conductor every inch or so. A coaxial cable is shown in Fig. 25.1.

Several coaxial tubes are normally bound together in one large cable (see Fig. 25.2). At higher frequencies, there is virtually no cross talk between the separate coaxial cables in such a link because the current now tends to flow on the inside of the outer shell. Because of this shielding from noise and cross talk, the signal strength can be allowed to fall to a lower level before amplification.

The coaxial cables of the late 1950s used 0.375-in. tubes with a top frequency of about 3 MHz. A cable typically carried four such tubes, occasionally six. The engineers of the day used to question how anybody could use more bandwidth than that!

Before long, however, the top frequency was raised from 3 to 9 MHz, and the number of one-way voice signals that a single tube could carry rose from 600 to 1800. Now the top frequency of this cable is about 70 MHz and

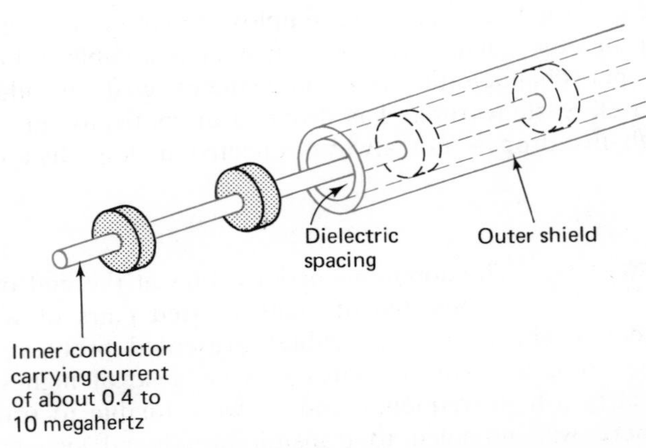

Figure 25.1 Coaxial cable construction.

Figure 25.2 Thick cables containing multiple coaxial tubes, each transmitting thousands of telephone calls, form the highest capacity trunk routes of North America. (*Courtesy of AT&T.*)

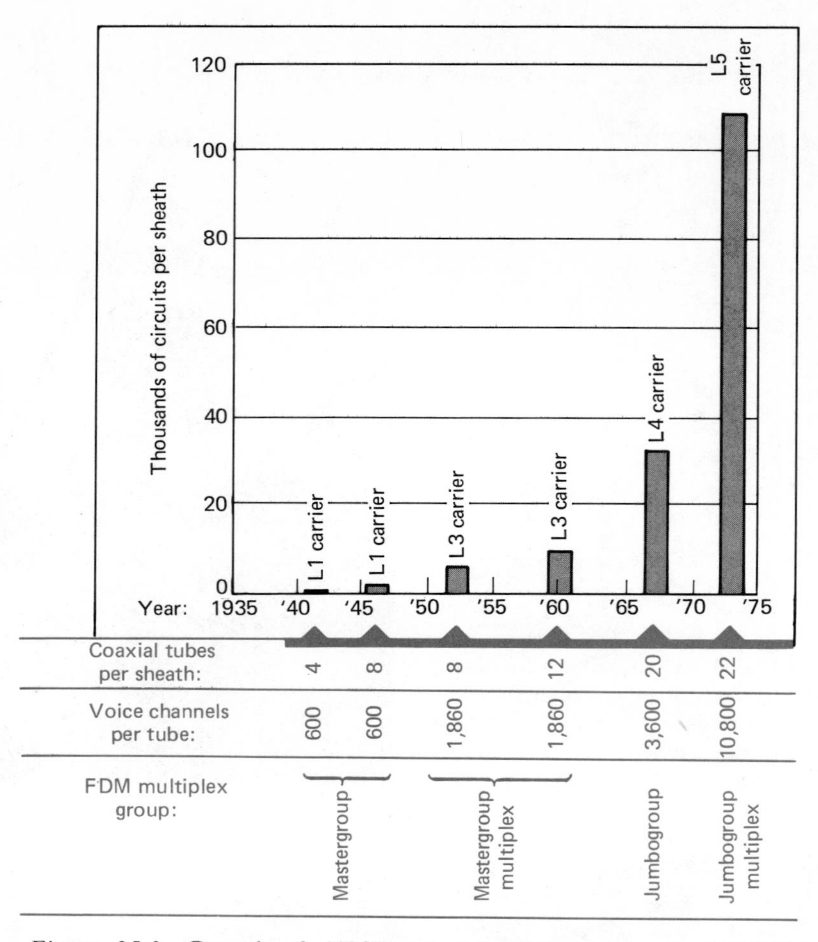

Figure 25.3 Growth of AT&T coaxial cable systems—a 225-fold increase in the capacity of a cable since 1942 and a 10-fold drop in cost per voice channel. There are now more coaxial tubes per cable and higher multiplex groups. (*Source AT&T.*)

one tube can carry 10,800 voice signals. Two tubes are needed for 10,800 two-way telephone conversations. The cable shown in Fig. 25.2 has 22 such tubes; two of these are reserve links in case of failure; the remaining 20 can carry $10 \times 10{,}800 = 108{,}000$ two-way conversations.

Figure 25.3 shows the growth in capacity of coaxial cable systems used by AT&T. There has been a growth in both the number of tubes per cable and the number of channels frequency-division multiplexed in one tube. AT&T's coaxial cables are referred to as "L" systems. L1 was the first. L5 is the latest and greatest, carrying 108,000 two-way telephone calls.

WAVEGUIDES

After the L5 coaxial cable system, AT&T's next major step in expanding its telecommunications arteries is the use of waveguide pipes.

A waveguide is, in essence, a metal tube in which radio waves of very high frequency travel. There are two main types of waveguide, rectangular and circular. *Rectangular waveguides* have been in use for some time as the feed between microwave antennas and their associated electronic equipment. They are not used for long-distance communication and are rarely employed for distances over a few thousand feet. They consist of a rectangular copper or brass tube, (Fig. 25.4). Radiation at microwave frequencies passes down this tube.

Circular waveguides are pipes about 2 in. in diameter. They are constructed with precision and are capable of transmitting frequencies much higher than rectangular waveguides. They are commonly referred to as *millimeter waveguides* because they transmit radiation in the millimeter waveband, typically frequencies from 40 to 110 GHz which would be heavily attenuated if transmitted through the earth's atmosphere (see Fig. 24.3). The atmosphere inside the pipe is carefully controlled to avoid the oxygen absorption peak shown in Fig. 24.3.

Figure 25.5 shows one form of waveguide construction referred to as a helical waveguide because a fine enameled copper wire is wound tightly around the

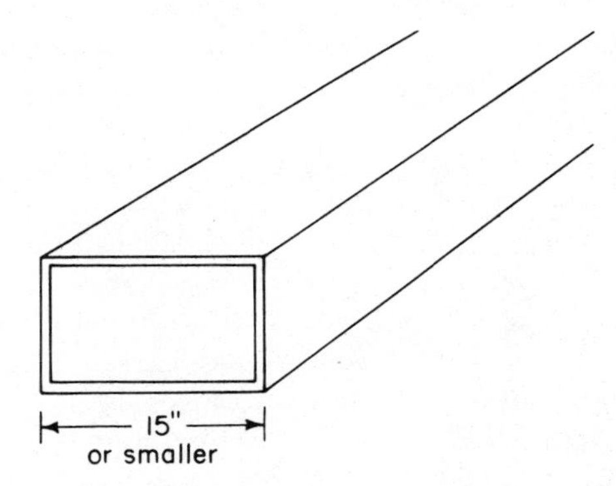

Figure 25.4 Rectangular waveguide.

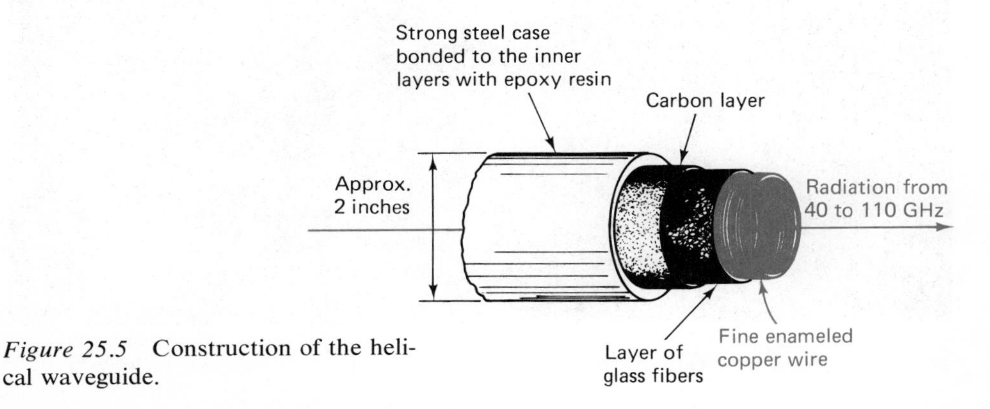

Figure 25.5 Construction of the helical waveguide.

Figure 25.6 AT&T advertisement showing their helical waveguide.

A two-inch empty pipe
can carry 230,000 telephone conversations.

The pipe is no bigger than your wrist.

Yet what really makes it news is that there's absolutely nothing inside.

Except room for 230,000 simultaneous telephone conversations.

In the years to come, millimeter waveguide pipe will be buried four feet underground. In a larger cradling pipe to give it protection and support.

It'll also have its own amplifying system about every 20 miles. So your voice will stay loud and clear.

Even after 3,000 miles.

Yet this little pipe is capable of carrying a lot more than just conversations.

It can also carry TV shows. Picturephone® pictures. Electrocardiograms. And data between thousands of computers.

All at once.

The American Telephone and Telegraph Company and your local Bell Company are always looking for new ways to improve your telephone service.

Sometimes that means developing a better way to use two inches of empty space.

Figure 25.6 (cont.)

inside in a helix. The wire is surrounded by a layer of thin glass fibers and then by a carbon layer. The whole is encased in a strong steel case and bonded to it with epoxy resin. The purpose of this construction is to attenuate undesired modes of wave propagation which would cause excessive dispersion of the signal and hence smear its information content. In another form of construction the pipe is internally copper plated and lined with a 0.25 millimeter layer of polythylene dialectric which also serves to attenuate unwanted modes of propagation. Fig. 25.6 illustrates the Bell System waveguide.

Waveguides cannot have sharp bends in them; however, they can have large radius bends like a railroad track [3]. These gentle bends cause little loss of signal strength. Sharp corners need a junction between two waveguides with a repeater. Because of the need to avoid bends, construction of a nationwide waveguide is an expensive civil engineering operation. It has been estimated that laying down the Bell System waveguide (including right-of-way acquisition) costs from two to four times the cost of the waveguide itself and its electronics. Figure 25.7 illustrates the operation. A protective sheath is buried about four feet underground. This is then opened up at one mile intervals and the waveguide tube slid inside it, cushioned with roller spring supports.

Figure 25.7 The Bell System millimeter waveguide is inserted into a protective pipe which is normally buried in the earth. The pipe is opened up at one mile intervals, and the waveguide tube is slid into it, cushioned with the roller spring supports which can be seen attached to it in this photograph.

The bandwidth available for transmission through the waveguide, 40 to 110 GHz, is a greater bandwidth than all today's through-the-air radio bandwidths added together. The bottom diagram of Fig. 25.8 shows the signal attenuation at different frequencies. This may be compared with coaxial cable and wire pairs, shown also in Fig. 25.8. Theoretically, the attenuation in waveguides should continue to lessen indefinitely as frequency increases. The upper limit shown in Fig. 25.8 is set by today's engineering.

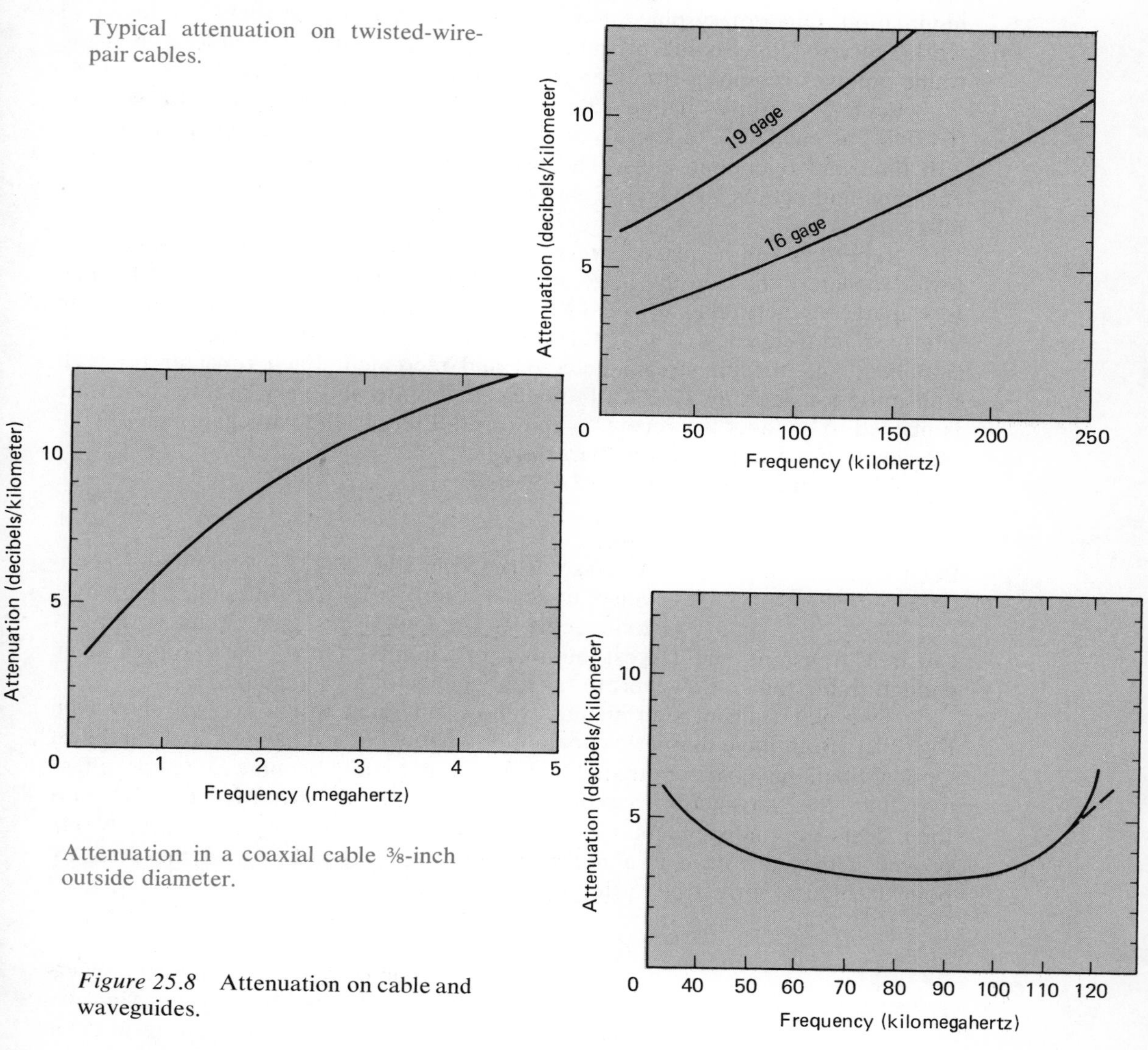

Figure 25.8 Attenuation on cable and waveguides.

Like other media a waveguide could transmit either analog or digital signals. In its initial operation the Bell System waveguide will transmit 60 digital bit streams of 274 million bits per second each (1.644×10^{10} bps in total), in each direction. The lower half of the bandwidth (40 to 75 GHz) is used for transmission in one direction and the upper half (75 to 110 GHz) for the opposite direction. 60 sine wave carriers spaced 550 MHz apart are used in each half of the band. Each is modulated at a rate of 274 mbps, using binary phase modulation. Quaternary phase modulation has been used in the laboratory and works successfully, so the bit rate of the system could be doubled when the traffic volumes require it (to 3.288×10^{10} bps in each direction).

Each 274 mbps channel can carry 4032 telephone calls in PCM form (64,000 bps each). 57 of the channels will be used for traffic giving a total of 230 thousand telephone calls. The other three channels are used as control channels and spares to which a switchover occurs automatically if a channel fails.

Repeaters will be placed every 30 to 40 kilometers on the waveguide—far fewer repeaters than on the large coaxial systems on which they are 2 kilometers apart. At each repeater the 120 received signals are down-converted to an intermediate frequency of 1.35 GHz. At this frequency the signal is equalized, amplified, and the bit stream detected and recreated. Each repeater has 120 transmitters using tiny IMPATT diodes. Elaborate and very precise hardware is needed to connect the receivers and transmitters to the waveguide pipe without affecting the transmission characteristics.

This facility is called the *WT4 System.*

ECONOMIES OF SCALE

As was illustrated in Fig. 20.1, there are major economies of scale in terrestrial transmission. The waveguide achieves the lowest cost per circuit so far because it transmits the largest number of circuits. Only a few routes have enough traffic today, however, to take advantage of a waveguide.

Two new transmission media are now emerging which are not shown on Fig. 20.1. Both have dramatic capabilities—optical fiber channels and satellites. Optical fibers have such potential that they may rapidly become more cost effective than the waveguide. Satellites cannot be fitted onto Fig. 20.1 because their costs are independent of distance and their economies are greatly improved if they are used in a multiple-access fashion rather than the point-to-point fashion of terrestrial systems.

OPTICAL TRANSMISSION

The increases in capacity of telecommunication cables have come from increasing the transmission frequencies that can be used. The next step up in cable technology uses glass fibers with signals at or close to the frequencies of light.

The frequency of light is 10 million times greater than the frequencies used on coaxial cables. Furthermore, a coaxial tube is 10 thousand times the cross-sectional area of the glass fibers being used for transmission at Bell Laboratories, so many more fibers than tubes can be packed into one cable. A long period of development is ahead before optical techniques reach maturity, but the above figures are sufficiently great to indicate an enormous future growth in man's telecommunications capability. The first optical fiber cables, transmitting at only a fraction of the available capacity, will be in use soon.

There are three ways to transmit signals of the frequency of light. *First,* they can be shone through the air as a pencil-thin beam of light, as with the optical systems we discussed in the previous chapter. Such a scheme has many disadvantages. It can be interfered with by anything that will disturb ordinary light—fog, snow, boys with kites. A bird flying through the beam could obliterate 10 million bits. A grave problem in making such a system function correctly would be the variations in the atmospheric path used. Through-the-air optical systems are likely to be used only for short distance links of capacity higher than local telephone loops, but low in comparison with the available optical bandwidths.

Second, light pipes could be used, rigidly constructed from straight line segments. Many experiments were done with light pipes using lenses and carrying multiple laser beams. It became clear that a light pipe system could have an information-carrying capacity much higher than the helical waveguide, but would incur high civil engineering costs. It might be vulnerable to the vibration and twisting of the earth's surface.

Third, light could be transmitted through glass fibers. In the 1960s a glass fiber system seemed impractical except for very short distances because of the absorption that occurred in the glass. In the first half of the 1970s, however, phenomenal progress occurred in glass fiber manufacture. It is now clear optical fibers are one of the most important future technologies of telecommunications.

GLASS FIBERS

By removing almost all the impurities and making glass of a high silica content, fibers with a low transmission loss have been manufactured. Amazingly, some of the Bell Laboratories fibers have a loss 1 to 4 decibels per kilometer—less than the losses in conventional cables which can be as high as 28 decibels per kilometer. Fibers with a loss of less than 1 decibel per kilometer have been produced experimentally. If such fibers could be mass-produced, the repeaters on optical cables could be farther apart and fewer than on conventional cables. There is every indication that these fibers will be mass-produced—eventually in enormous quantity. The raw materials for making glass fibers, unlike copper, are among the world's most plentiful substances.

When light travels 1 meter through ordinary window glass, pyrex, or water, it loses about two thirds of its power (a 5 decibel loss). Through good qual-

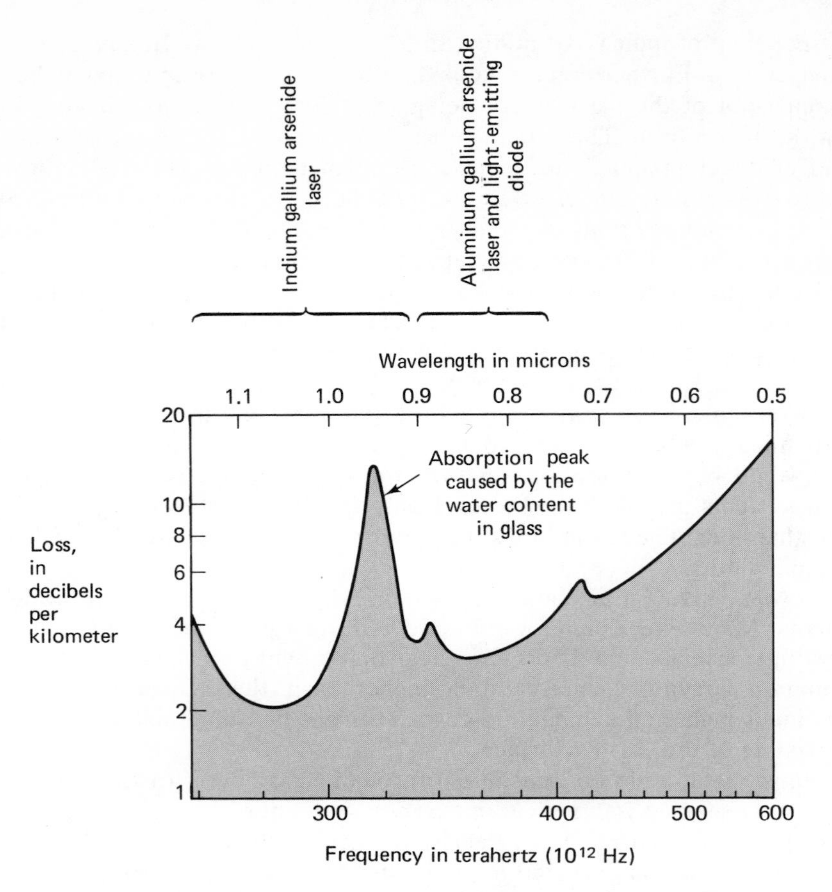

Figure 25.9 Transmission windows of a typical low-loss silica glass fiber. The potentially usable bandwidths are very high. (Compare with Fig. 25.8.)

ity optical glass it can travel 5 meters and suffer the same loss. Through the glass fibers now being manufactured it can travel between one and two kilometers with that loss. Digital repeaters on such cables could be more than 10 kilometers apart.

Why does the light stay inside the fiber? It travels down a cylinder of glass which is surrounded by a substance, usually also glass, of low refractive index. When the beam strikes the edge of its cylinder it is totally reflected and remains inside the cylinder as shown in Fig. 25.9. This total internal reflection occurs in a similar manner at the surface of a pond. If you put your head under the water and look at the surface some distance away, it will appear to be a totally reflecting mirror. The ray of light is not refracted out of the pond at all because of its low striking angle but is totally reflected back into the water. The light beam travels down a fiber and is confined within the fiber by total internal reflection. It will be absorbed somewhat by the fiber, and it will have to be periodically amplified. Light beams of different frequencies could travel together down the same fiber, and a bundle of such fibers thin enough to be highly flexible would be bound together to form one cable.

DISPERSION

As can be seen from Fig. 25.9 the potentially usable bandwidths of glass fibers are very high. If the windows permitting a signal loss of, say, less than 5 decibels in Fig. 25.9 could be fully utilized, one fiber could transmit on the order of 10^{14} bits per second. The transmission rate of today's fibers is far below this theoretical capacity. One factor which limits the transmission capacity is *dispersion* of the signal as it travels down the fiber.

Rays of light travelling down a fiber can be transmitted by different paths as shown in Fig. 25.10. A ray travelling straight down the axis of the fiber will reach its destination before a ray which bounces down the fiber with many reflections. A very short pulse transmitted down a fiber will therefore be spread

Light rays are reflected down a glass fiber with total internal reflection at the surface:

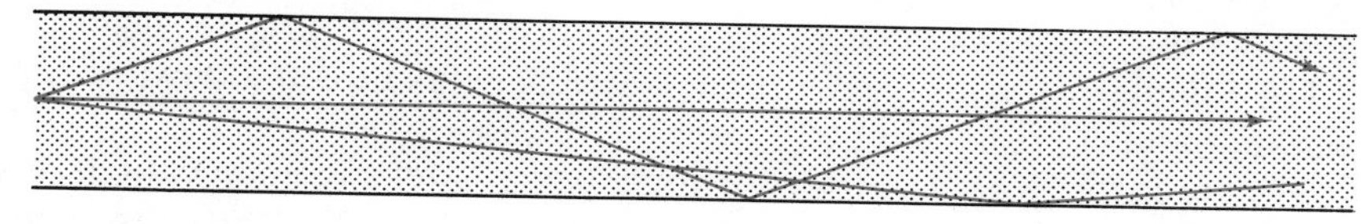

However some light rays travel by shorter paths than others, This causes the signal to become increasingly dispersed:

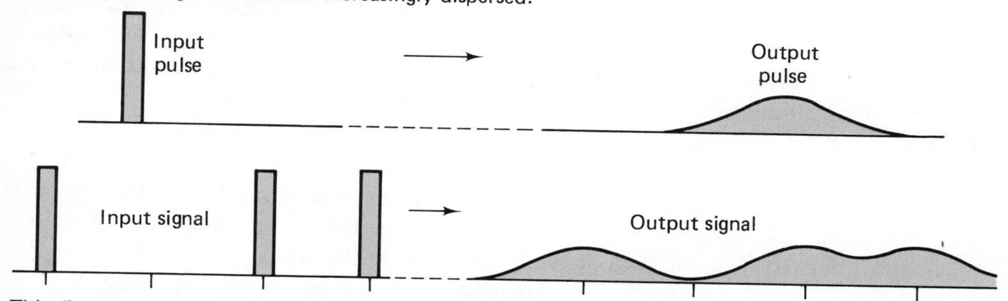

This dispersion puts an upper limit on the signaling rate of a glass fiber. Fibers can be constructed with very low dispersion as shown in Fig. 25.11 and such fibers can carry a very high information rate.

Figure 25.10

out in time, as shown. The further it is transmitted the more it will be spread out. In a typical fiber of 100 microns (0.1 millimeter) diameter, the axial ray is transmitted 1 kilometer in several nanoseconds less time than the ray which takes the longest path. Pulses less than a few nanoseconds apart will interface with one another and become indistinguishable. If the repeaters on the fiber are to be 10 kilometers apart, this pulse spreading limits the transmission rate to well below 100 million bits per second per fiber.

Another property of fibers which causes dispersion is the nonlinear variation of its refractive index with frequency. Different frequency components have different velocities in the glass. If a pulse being transmitted is truly monochromatic (i.e., of a single frequency or color) the variation of refractive index has no effect on it. With fibers employing light from LED's (light-emitting diodes) the pulse contains a mix of frequencies and so is spread out after it has travelled several kilometers. Lasers, on the other hand, transmit single-frequency (monochromatic) pulses.

THREE TYPES OF FIBER

Dispersion can be lessened in several ways. If the fiber was extremely thin there would be little dispersion. Such a fiber, however, would break very easily. Optical cables have therefore been constructed in which many extremely thin fibers are bundled together in a plastic sheath like an electric lamp cord. The bundle is flexible and strong, and a signal is directed down the group of fibers in parallel.

An approach which is better for long-distance telecommunications uses a glass fiber with a fine central core surrounded by glass of slightly lower refractive index, as shown in the center of Fig. 25.11. This is referred to as a single-mode fiber whereas the fiber with the multiple rays at the top of Fig. 25.11 is called a multimode fiber. The electromagnetic waves travel down the fiber with a fine core in a single electromagnetic mode, in effect a single ray travelling down the axis of the fiber. The information-carrying capacity of a single-mode fiber can, in theory, be very high. The difficulty with a single-mode fiber is how to inject a powerful enough signal into the glass core which may have a diameter of only 5 microns (0.005 millimeters). For this a tiny laser is needed.

Another approach shown at the bottom of Fig. 25.11 uses a fiber with a variable refractive index. The refractive index varies from being high at the center to lower at the outside. Rays straying from the center of the fiber are deflected back, the fiber acting as a sort of lens. The refractive index is carefully graded so that the beam is continuously refocused as it travels down the fiber. Rays speed up as they pass into the lower refractive index glass and the result is that all rays arrive at the end of the fiber at approximately the same time. Dispersion is thus kept low.

The Nippon Corporation first put a fiber on the market with a graded index. They called it Selfoc (for self-focusing). It could transmit more than 1 bil-

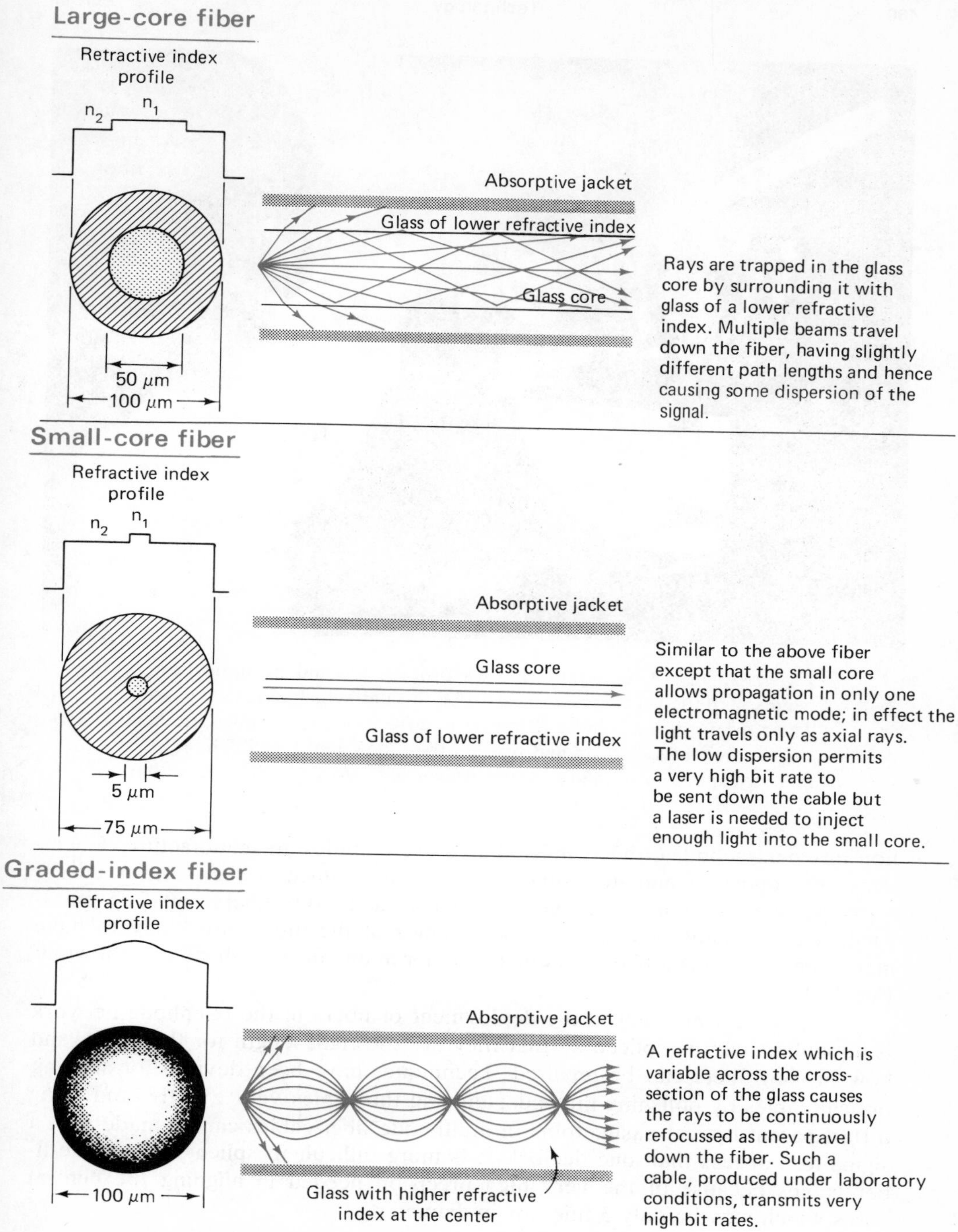

Large-core fiber

Retractive index profile

n_2 n_1

Absorptive jacket

Glass of lower refractive index

Glass core

50 μm

100 μm

Rays are trapped in the glass core by surrounding it with glass of a lower refractive index. Multiple beams travel down the fiber, having slightly different path lengths and hence causing some dispersion of the signal.

Small-core fiber

Refractive index profile

n_2 n_1

Absorptive jacket

Glass core

Glass of lower refractive index

5 μm

75 μm

Similar to the above fiber except that the small core allows propagation in only one electromagnetic mode; in effect the light travels only as axial rays. The low dispersion permits a very high bit rate to be sent down the cable but a laser is needed to inject enough light into the small core.

Graded-index fiber

Refractive index profile

Absorptive jacket

100 μm

Glass with higher refractive index at the center

A refractive index which is variable across the cross-section of the glass causes the rays to be continuously refocussed as they travel down the fiber. Such a cable, produced under laboratory conditions, transmits very high bit rates.

Figure 25.11 Three types of glass fiber propagation of light.

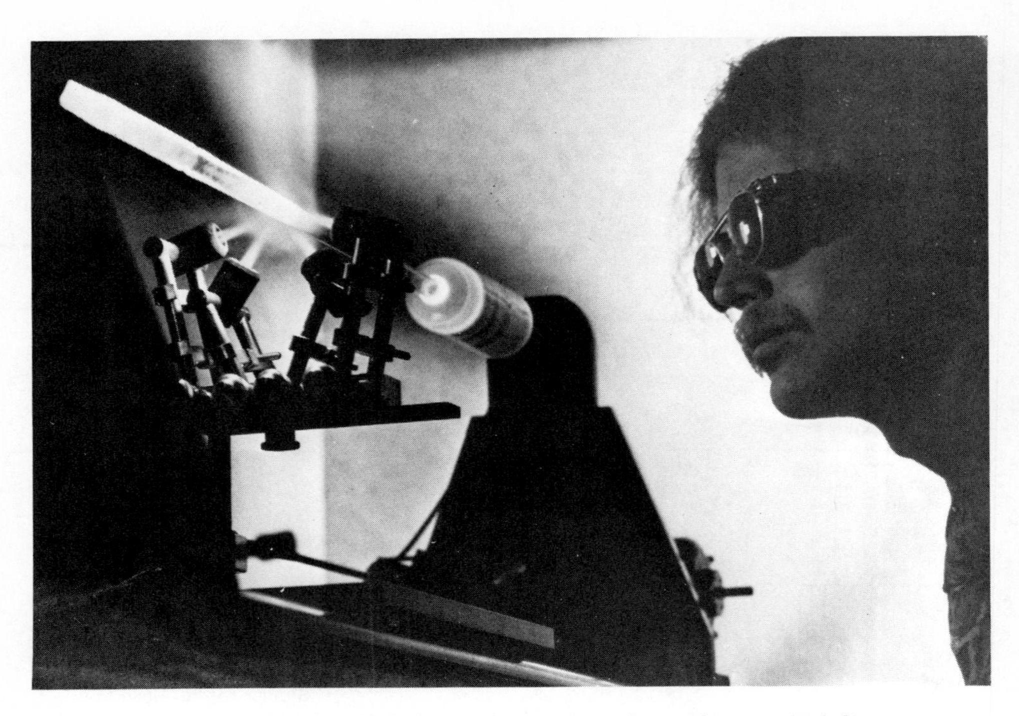

Figure 25.12 A Bell Laboratories process for making optical fibers begins with the heating of a quartz tube through which gases are passed. The tube turns on a lathe as burners move along its length. The heat causes the gases to react and fuse on the tube's inner wall. The tube is then collapsed to a solid rod and drawn out into a lengthy, very thin, fiber.

lion bits per second. Graded index fibers are complex to manufacture. Chemicals are vaporized and deposited on a cylinder which is then drawn out to form the fiber. It is not clear yet whether graded-index fibers could be mass-produced at suitably low cost. Optical cables of the future are likely to have many separate optical fibers grouped together in one flexible sheath, as shown in Fig. 25.13.

A practical problem in the deployment of fibers in the telephone network is how fibers can be spliced so that they are the right length for the ducts, and how broken fibers can be repaired. Techniques have been devised for splicing optical cables by polishing the ends, aligning the cables very exactly, and using a transparent optical paste. Joins in multimode fiber cables can be made with a signal loss of less than one decibel. It is more difficult to splice single-mode fiber cables because of the very high precision needed in aligning the central cores which may be only 5 microns in diameter.

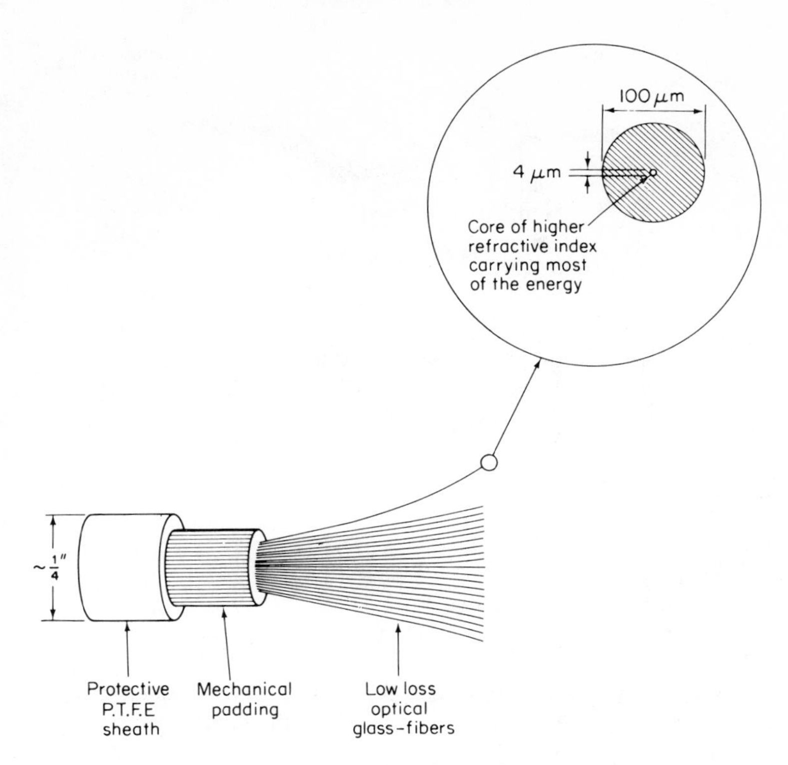

Figure 25.13 Future optical cables will contain hundreds or thousands of optical fibers of one of the three types shown in Fig. 25.11. This illustration shows a cable containing single-mode fibers. Such a cable could carry more telephone calls than all of today's trunk calls, world wide.

LEDS

It is necessary to have a source of light with a long enough life to act as the transmitter, and which can be modulated to carry a high information rate. Glass fibers can transmit either light or laser beams. Many of the early experimental fiber systems used *light-emitting diodes* (LEDs) as a signal source.

Figure 25.14 shows a light-emitting diode, manufactured in the form of a semiconductor chip less than one millimeter across, and bonded with epoxy resin to a glass fiber. As the current reaching the LED is modulated, the light emitted varies accordingly, so no external optical modulator is needed.

The signal transmitted can be received by a photodiode, also a tiny semiconductor chip bonded with the fiber as shown in Fig. 25.15. Using devices like those in Figs. 25.13 and 25.14, experimental transmission systems have been built in Bell Laboratories with relatively inexpensive multimode fibers. Digital repeaters have been built operating at 6.3 and 50 mb/s. Such transmis-

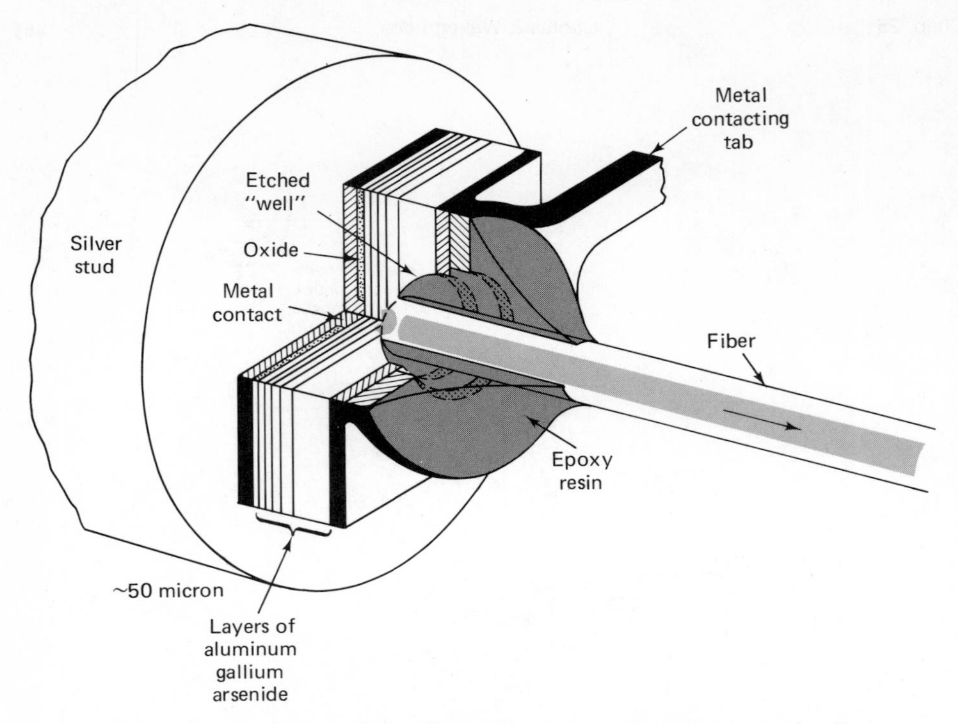

Figure 25.14 A high-brightness light-emitting diode, bonded to a glass transmission fiber. The light emitted varies as the input current is modulated. (*Courtesy AT&T.*)

Figure 25.15 A silicon PIN photodiode receives the light transmitted by the LED in Fig. 25.13 and converts it into current. Photodiodes have operating bandwidths of several hundred megahertz. (*Courtesy AT&T.*)

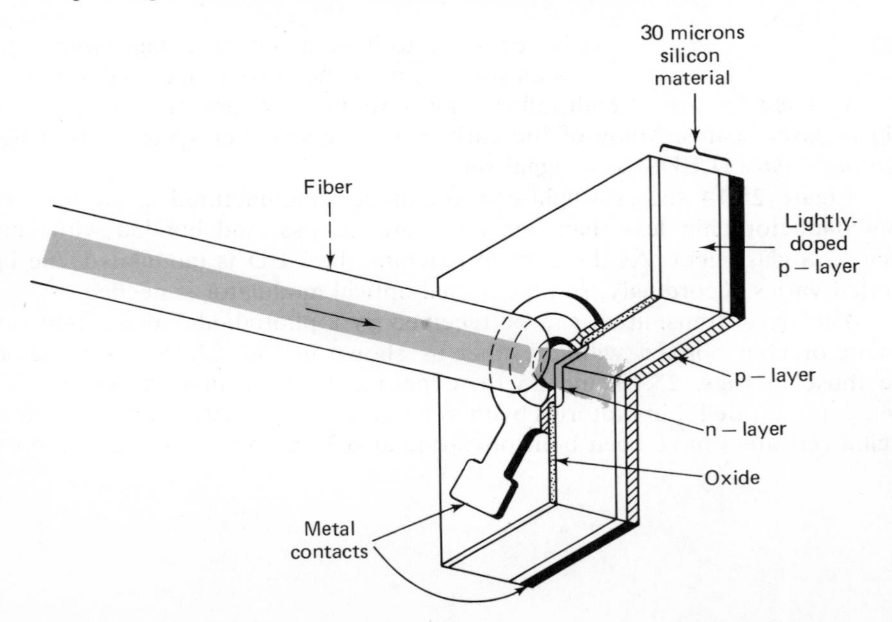

sion systems could either be used for short-haul or intracity trunks. They could also be used as loops into subscriber premises carrying wideband data, Picturephone signals, or groups of telephone channels.

LASERS

For long-distance high-capacity applications the laser shows more promise than the LED.

Laser stands for "Light Amplification by Stimulated Emission of Radiation" and was preceded by *maser,* meaning "Microwave Amplification by Stimulated Emission of Radiation." A laser produces a narrow beam of light that is coherent (all of the waves travel in unison like the waves travelling away from a stone dropped in a pond), and is sharply monochromatic (occupies a single color or frequency) or consists of multiple monochromatic emissions.

An analogy with sound waves is somewhat inexact, but it will help the reader to visualize the difference between a laser beam and an ordinary light beam. The sound from a tuning fork consists of waves that are of one frequency and that are reasonably coherent. On the other hand, if I put a hammer through my office window, the sound waves would be neither monochromatic nor coherent. The former may be compared with a laser beam, the latter with ordinary light. The laser or maser beam is formed by a molecular process somewhat analogous to the tuning fork. It is possible to make certain molecules oscillate with a fixed frequency like a tuning fork.

The electrons in an atom can only move in certain fixed orbits. Associated with each orbit is a particular energy level. The electrons can sometimes be induced to change orbits, and when this step happens, the total energy associated with the atom changes. The atom can therefore take on a number of discrete energy levels—a fact that is well known today from quantum mechanics. Certain processes can induce the electron to jump from one orbit to another or, to state it another way, to induce the atom to switch from one energy level to another. When this happens, the atom either absorbs or emits a quantum of energy. In this way, radio waves, or other electromagnetic radiation is emitted in discrete quantums.

When ordinary light is emitted, the mass of molecules switch their energy levels at random. A random jumble of noncoherent waves is produced. Under the lasing action, however, the molecules are induced to emit in unison; the substance oscillates at a given frequency, and a stream of coherent waves at this single frequency results. This could range from a microwave frequency (maser) to a light frequency (laser), or higher—recently an X-ray laser was produced.

When a laser beam produced by certain lasing molecules falls on other molecules of that type, it can induce oscillation in them. A form of resonance is set up. The reader might imagine a huge pendulum much too heavy to move far by a single hard push. If he gives a series of relatively gentle pushes, however, he can set it swinging. He may go on pushing at just the right point in the swing, and the length of the swing increases until the pendulum builds up great power. This is resonance. His gentle pushes have built up massive oscillations. In a similar manner (and again the analogy is helpful but not exact), a weak laser beam can fall on a lasing substance and cause resonance in it. It sets the molecules oscillating so that a powerful laser beam is emitted. The laser beam has thus been amplified. In this way, a very intense beam of a single frequency or several single frequencies can be emitted.

A beam of ordinary light, even a beam that we describe as monochromatic, actually consists of a small spread of frequencies, each of which would be bent slightly differently by a prism or lens. A laser beam, however, is not dispersed by a prism and optical arrangements can be built for it so precisely that a beam of laser light can be shone onto the moon and illuminate only a small portion of its surface. A beam can be concentrated with a lens into a minute area, and the intense concentration of energy into such a small area causes very localized heating to occur. A cutting or welding tool is provided with miniature precision far beyond the dreams of Swiss watchmakers. The surgeon has a microscopic scalpel; the general a potential death ray.

To summarize, laser light differs from ordinary light in several characteristics. The wide variety of applications of the laser are based on one or more of these factors.

1. Laser light is sharply monochromatic. It may, however, have more than one monochromatic frequency (color). By using filters, a single frequency can be separated from the others.
2. The light is coherent, the waves being regularly arranged and in phase with each other.
3. It can have very great intensity, so great that it could blind a person instantly.
4. Lasers emit a light in a parallel beam rather than in all directions like a bulb filament.
5. Because a laser beam is parallel and monochromatic it can have low dispersion; it can be accurately directed by lenses and prisms. Pulses suffer little spreading when they pass down a suitable fiber.

For telecommunications we can have a highly controllable beam of great intensity, which can be amplified, and which has a frequency 100,000 times higher than today's microwave signals. Its potential information-carrying capacity is thousands of times greater than microwave. It has been said that lasers portend a revolution in telecommunications as fundamental as the invention of radio.

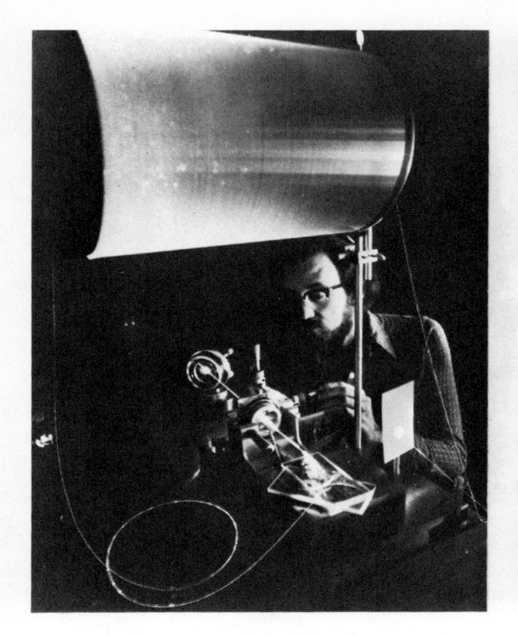

Figure 25.16 The drum at the top left contains half a mile of hair-thin glass fiber. Laser light enters the fiber at the lens in the center of the left photograph, travels half a mile, and then illuminates the card on the right. (*Courtesy Bell Laboratories.*)

LASERS WITH OPTICAL FIBERS

A problem with optical fibers, especially the single-mode fiber shown in Fig. 25.11 which may be the best bet for long-haul transmission, is how to get a powerful signal inside such a tiny fiber. The beam for a single-mode fiber must be shone up a tube about 5 microns (0.005 millimeters) in diameter. Lenses and mirrors could be used but much of the light from an incoherent source would be lost. The laser appears to be the answer. It can provide a very tiny but intensely bright source which emits light in a narrow parallel beam and can be attached directly to the fiber.

A particularly interesting type of laser for this purpose is the semiconductor laser. It is tiny and potentially inexpensive to manufacture in quantity. It can be manufactured by epitaxial methods similar to those which are used for producing LSI circuits like those in pocket calculators and digital wristwatches. Crystals are grown and chemicals are deposited in layers, the outer layers being metallic current-carrying contacts.

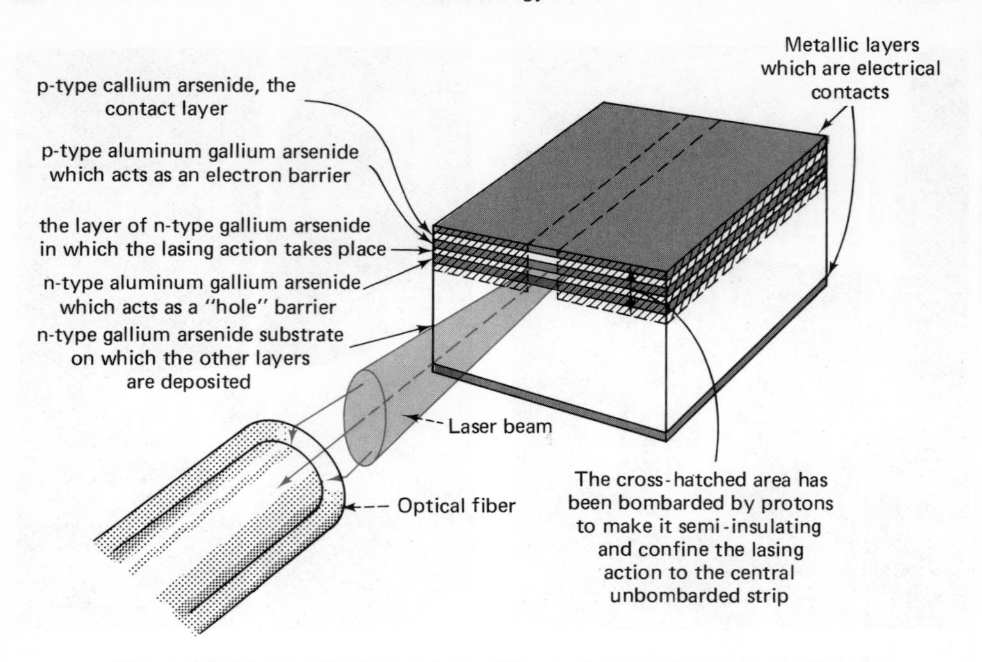

Figure 25.17 A tiny laser chip which emits a narrow intense beam, small enough to enter an optical fiber. Such chips could be mass-produced on a gallium arsenide wafer rather like the mass-production of LSI chips for pocket calculators or wrist watches.

Figure 25.17 shows a semiconductor laser. It is a tiny rectangular slab of gallium arsenide coated with alternating layers of aluminum gallium arsenide and gallium arsenide. At the top and bottom are layers of metal which act as electrical contacts. The layers consist of n-type and p-type regions as shown in Fig. 25.17. The n-type regions contain mobile electrons—negative current carriers. The p-type regions contain positive current carriers which are in effect the absence of electrons from the molecular structure, referred to as "holes." When an electrical current, for example from a battery, is applied to the contacts, electrons from the n-type layers are injected into the p-type layers. In the central p-type gallium arsenide layer, electrons combine with the "holes" and the excess energy is emitted as light. The lasing action takes place in this central layer. The light generated is mostly confined to that layer because the adjoining aluminum gallium arsenide layers are of lower refractive index. The edges of the block are mirror-like so that they reflect the light back into the layer. The light then stimulates the generation of more light. This stimulates yet more light until an intense beam of coherent radiation is emitted.

The lasing action is confined to the central thin layer of gallium arsenide. To produce a pencil-thin beam, all but the central strip of that layer is made semiinsulating by bombarding it with protons during the manufacturing process. The laser chip then emits an intense narrow beam, small enough to enter an optical fiber.

As in the production of LSI circuitry, quantities of such chips are produced simultaneously by depositing the various chemical layers on a wafer of the substrate material, and then dicing the wafer into many chips.

A major problem with the early semiconductor lasers was that their lifetime was short. Research indicated that the processes which eventually led to failure began at defects in the crystalline structure. The defects might be there after manufacture and might be introduced by strain on the laser. The lifetime of lasers has been increased from minutes to years by careful avoidance of crystalline defects. It now appears the semiconductor lasers will have a long enough life for wide application in telecommunications.

MULTIPLEXING LASER SIGNALS

Laser signals can be multiplexed within a glass fiber just as electrical signals can be multiplexed within a wire or coaxial cable. Both frequency and time division multiplexing has been used experimentally.

By varying the semiconductor composition lasers can be made to emit at different frequencies. Different frequency laser beams can be modulated separately and transmitted through the same glass fiber.

To achieve time-division multiplexing a laser can be made to produce a series of very narrow pulses of light. When a laser emits different frequencies, each monochromatic and coherent in phase, these frequencies interfere with one another, in some places cancelling each other out and in certain places reinforcing each other to form a narrow high-intensity pulse. The larger the number of frequencies used, the narrower the pulses can be. Exceedingly narrow pulses have been produced, spaced relatively far apart in time. These pulse trains can be made to carry digital signals and hence can be used for pulse code modulation. Because the pulses are relatively far apart, there is time, using present-day techniques, to modulate them; and because the pulses are very narrow, several pulse trains can be interleaved (time-division multiplexed). Such a process starts with a single pulse stream and slits it into several identical pulse streams. The separate pulse streams are then delayed by different amounts by passing them through optical components, for example fiber coils of different lengths. The pulse streams are then separately modulated and recombined. The timing is such that the combined stream of pulses will interleave the short-duration pulses without any two pulses coinciding.

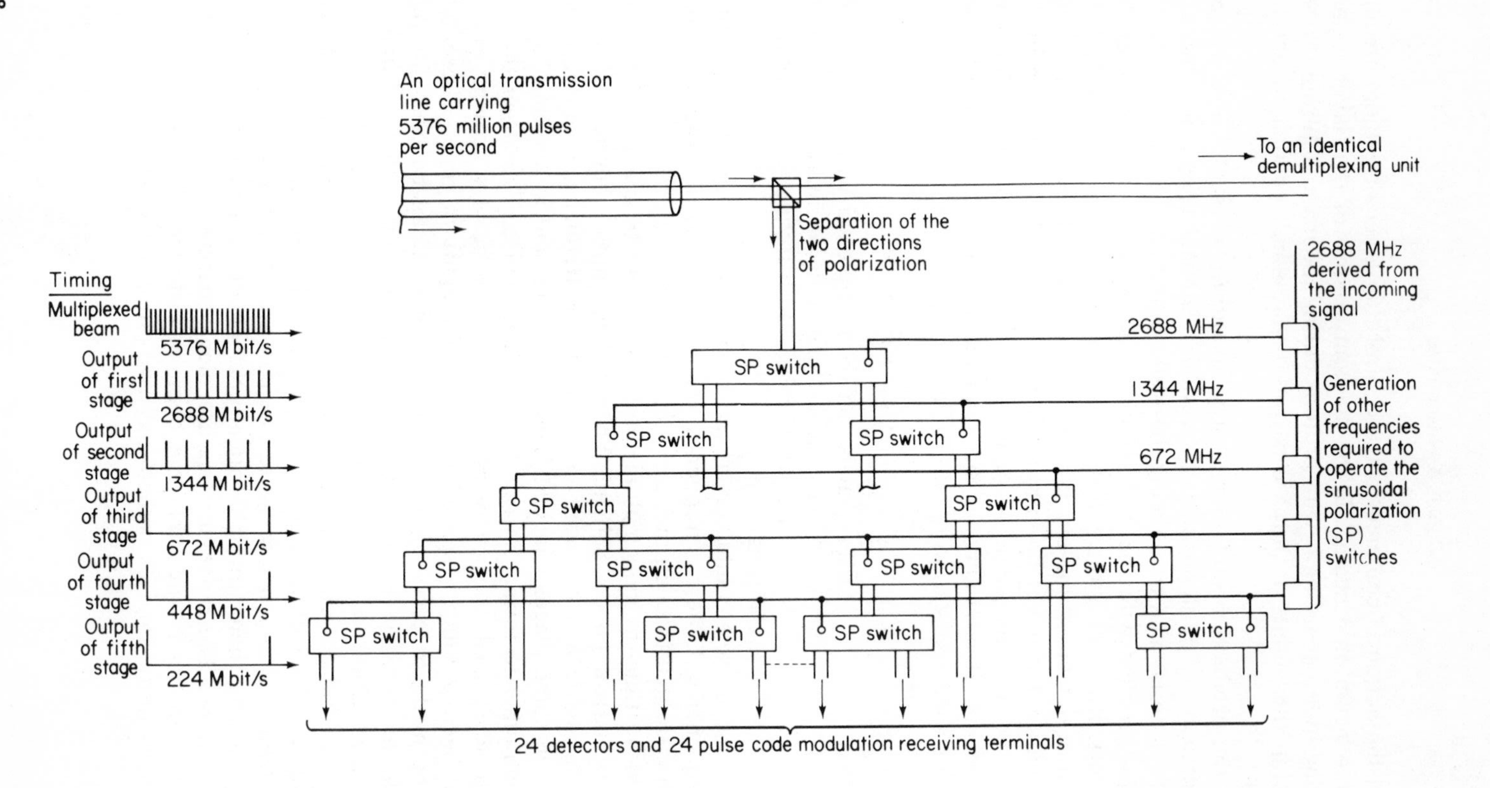

Figure 25.18 Demultiplexing a laser pulse stream in which forty-eight 224 million bit/second signals are sent as pulses to the same frequency.

The first experimental system at the Bell Laboratories used a gaseous helium-neon laser that produced a pulse stream of 224 million pulses per second. A laboratory terminal was built capable of interleaving and transmitting four such pulse streams, giving a total of 896 million bits per second. The pulses are narrow enough, however, to interleave 24 such pulse streams. Furthermore, two light beams polarized at right angles to one another can be transmitted at once, giving 48×224 million = 5.376 billion bits per second.

Figure 25.18 shows a terminal designed to demultiplex a 5.376-billion-bits-per-second pulse stream. The multiplexing process is the converse of what happens in the figure. A polarizing filter at the top of the figure splits the stream into two 2688-million-bits-per-second streams, which then enter two identical demultiplexing units. The demultiplexing units consist of a cascade of polarization switches. The switches consist of crystals that rotate the plane of polarization of light, depending on the voltage applied to them. They are used for selectively removing the pulses of unwanted channels and are operated by a sinusoidal voltage. There must be very exact synchronization and so timing information is derived from the input pulse stream.

Other lasers today can give much higher pulse rates than the above gas laser. Solid state lasers, in general, can give much narrower pulses and there are many types of solid state lasers. A promising one uses yttrium aluminum garnet (YAG) doped with neodymium and generates pulses of 35 picoseconds (35×10^{12} seconds) duration. This laser may be used to give multiplexed PCM channels at rates as high as 1.5×10^{10} bits per second.

Experimental lasers have been produced which give pulse streams in which the pulses are as short as 0.4 picoseconds (0.4×10^{12} seconds—light travels about one tenth of a millimeter in this time!). Theoretically such lasers could give data transmission at more than 10^{12} bits per second through today's optical fibers, and many thousands of such fibers could be packed into one flexible cable.

Today there is no application for such cables. There is no market for transmission media which could handle 10^{12} bits per second. If at some future time today's trunk calls become calls between subscribers with wall screens, or CATV subscribers want the freedom to dial vast movie libraries and video education systems offering a wide choice, such cables will be needed.

LOCAL LOOPS

As we have seen, impressive technical development is under way in the building of telecommunications trunks. However, trunks link only the telephone company offices. There has been little attention to local loops which go from the telephone offices to subscribers' premises. This does not matter, perhaps, if the subscribers want signals of only telephone bandwidths. However, much of the future promise of telecommunications comes from users employing high bandwidth signals.

We should therefore be asking: what alternatives are there to the local telephone loop. Box 25.1 summarizes the possible alternatives. For corporate telecommunication systems one of the most promising alternatives is the bottom one—satellite antennas at corporate locations. This is discussed further in the next chapter.

BOX 25.1 Possible techniques for local distribution of signals (other than the use of analog telephone loops)

- Digital telephone loops carrying baseband digital signals from signal conditioning units (AT&T's DDS technology).
- Digital telephone loops with repeaters (like the T1 carrier).
- Cable TV channels organized to carry signals other than TV.
- Cable TV channels organized with demand-assignment multiple-access techniques.
- Future optical fiber cables (which promise very high bandwidths at relatively low cost).
- Pseudo-interactive VHF or UHF broadcasting (Like Ceefax but carrying a high data rate).
- Packet radio (for interactive data transmission or monitoring).
- Private microwave radio links.
- Private millimeterwave radio links.
- Infrared links.
- Through-the-air optical links.
- Satellite earth stations on user premises.

REFERENCES

1. *Lasers for Communications,* Bell Laboratories Record, April 1975.

2. Tingye Li, *Optical Transmission Research Moves Ahead,* Bell Laboratories Record, Sept. 1975.

3. Merle D. Rigternink, *Better Glass Fibers for Optical Transmission,* Bell Laboratories Record, Sept. 1975.

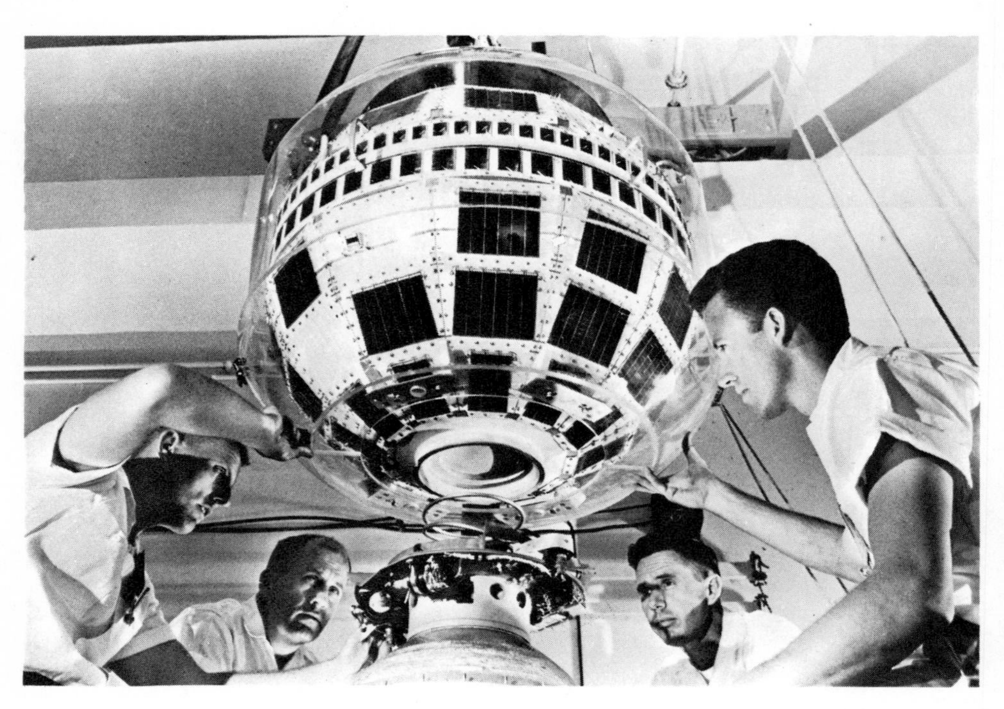

The first communications satellite to relay telephone traffic was not in geosynchronous orbit. It was AT&T's Telstar, 1962. It required massive earth station equipment to track it for its brief time within sight. Telstar earth stations cost about $50 million. (*Courtesy AT&T.*)

26 COMMUNICATIONS SATELLITES

Of all of the technologies discussed in this book several seem to stand head and shoulders above the others in their growth potential. One is microcomputers and associated digital technology. Another is optical fibers. A third is satellites. The first is in the hands of aggressive competitive private enterprise, free of government bondage. Satellite technology, however, is entangled in various ways with government regulation, and it is only too easy to believe that governments in some countries will prevent the full flowering of the technology and the full benefit to the public.

The enthralling potential of satellites comes from the following technical developments:

1. *Use of higher frequencies.* High frequencies are needed, as we discussed in Chapter 13, to avoid mutual interference between satellites and existing terrestrial microwave. Higher frequencies allow more satellites in the geosynchronous orbit, and for some applications cheaper and smaller earth stations.
2. *Use of small, cheap earth stations, at user premises.* New satellite design can permit earth stations cheap enough to receive signals in the home, or permit powerful two-way links from most corporate locations. The microwave devices needed for such earth stations used to be expensive to manufacture. New technology such as the NHK (Japan) planar circuits, mass-produced by etching techniques, make possible very low cost earth stations for installation in the home. These will be used with the Japanese broadcast satellite.
3. *Larger satellites.* We have not yet used large rockets like Saturn I or Saturn V for launching communications satellites. The space shuttle will be operating by 1980. In the mid-1980s satellites a hundred times the weight of the present domestic communications satellites could be launched. In general the larger the satellite, the cheaper the earth stations.

4. *Greater on-board power generation.* Experimental solar cells now exist which are much more efficient than those on today's satellites, and a much larger area of solar cells can be deployed. The solar sails of the Skylab satellite generated 12,000 watts *each,* whereas the WESTAR satellite generates only 250 watts. The more powerful the satellite, the cheaper the earth station.
5. *Multiple spot beams.* The total gain, up and down, of a satellite is proportional to the fourth power of the diameter of its antenna. ATS-6 demonstrated that large umbrella-like antennas can be opened and perform well in space. A large antenna makes possible multiple spot beams, like searchlight beams, from the satellite to dense subscriber zones on earth. Signals can be switched from one spot beam to another.
6. *Demand-assigned multiple-access (DAMA) techniques.* DAMA techniques allow transponder capacity to be shared by many geographically scattered locations such that the capacity they use varies with their instantaneous traffic demands. The sharing permits complete interconnectability between the earth stations so that switching like that on terrestrial trunks is not necessary between the earth stations.
7. *Digital techniques.* It has been demonstrated that more voice channels can be sent by satellite in a digital than in a analog form. All types of signals can be digitized. The multiplexing, concentration, and DAMA, techniques can then be digital, with each earth station using computer-like circuitry.
8. *Burst circuit control.* Satellite circuits are highly effective for interactive computer users, but only if line control disciplines are employed which are different to those conventional on land lines. Such disciplines have been demonstrated. They offer enormous channel capacity for the computer industry.
9. *Lasers.* The attenuation of electromagnetic waves in the atmosphere tends to increase with frequency as shown in Fig. 24.3, but there are certain "windows" of reduced attenuation which could be useful for satellite transmission. The very powerful carbon dioxide laser happens to transmit at the frequency of one of these windows in the infrared part of the spectrum. The useful role of satellites will be greatly extended when laser transmission can be employed, but much development work is needed to resolve the practical problems.
10. *Earth stations which can access multiple satellites.* The earth station in Fig. 26.1 can access several satellites simultaneously.

SATELLITES FOR TELEPHONE TRAFFIC

One of the most ambitious satellite studies ever published was, not surprisingly, from AT&T [1], who launched the world's first real-time communications satellite in 1962. In the regulatory climate of the 1970s this study is not being advertised; nevertheless it illustrates the potential.

The part of the Bell System which handles long distance transmission represents about 17% of its total cost. By 1980 the cost will probably have grown to $20 billion. This figure does not include the switching costs. By 1980 the volume of long distance calls being made *simultaneously* on the Bell System in a peak period will have grown to almost a million. There will be about 20 bil-

Figure 26.1 Comsat has developed an economical earth station which can transmit to and receive from several satellites at the same time. It uses a fixed reflector with multiple moveable antenna feeds. The feeds are inside the hut on the left of the photograph.

This multiple-beam torus antenna (MBTA) is affected less by environmental interference (from terrestrial microwave or other satellites) than a conventional earth station because there is no blockage of the dish aperture, the dominant factor affecting the side-lobe level. The antenna can obtain a high degree of isolation between any two orthogonally polarized beams. The station is unattended and has been used with three Intelsat IV satellites simultaneously.

lion long distance telephone calls per year. A few satellites with multiple spot-beams could handle *almost all of this traffic at a small fraction of the cost of equivalent terrestrial facilities.*

The Bell Laboratories study [1] described how 50 satellites could provide about 100 million voice trunk circuits or equivalent—a hundred times more than Americans will actually use by 1980. The report was published when the enthusiasm for Picturephone was still strong, and if 220 million Americans could not be expected to make 100 million simultaneous telephone calls, Picturephone's gluttonous appetite for bandwidth still made the proposals appear interesting. *Two* such satellites, rather than 50, could make a major impact on today's long-distance telecommunication costs.

The satellites in the study used bands 4 GHz wide at frequencies of 20/30 GHz. The bandwidth was divided into eight bands, each of which carried a pulse rate of 315 million pulses per second. Each pulse carried 2 bits using 4-level pulse code modulation. The satellite therefore had eight transponders for each 4 GHz beam, carrying eight channels of 630 million bits per second each.

Each satellite would be stabilized to $\pm 0.01°$ and would carry a 10 meter multibeam antenna. Such an antenna gives a high gain at 30 GHz. Multiple antenna feeds would give multiple spot beams, each aimed at a specific earth station. Each beam would be able to carry the same frequency band without the beams interfering. In the maximum configuration there would be 50 beams aimed at 50 earth stations, and the satellite would carry 50 sets of 8 transponders. The total throughput of the satellite would thus be $50 \times 8 \times 630$ million $= 2.52 \times 10^{11}$ bits per second. It was estimated that such a satellite would require 7.2 kilowatts of power, and weigh 11,250 pounds. A launch vehicle configuration using Titan rockets or the new space shuttle could be developed that had the launch capability. It is well below Saturn V capability.

If the satellite had fewer than 50 spot beams its weight and power requirements would be less. It could be anywhere from INTELSAT IV size upwards as shown in Table 26.1.

The ground station antennas in the proposal would be designed to point at each satellite. If there were N channels per transponder, S satellites, and G ground stations, the total capacity of the system would be $8N \times S \times G$. In the proposal the 630 megabit throughput of the transponder would be used to carry 10,000 voice channels. A single satellite spot beam would carry $8 \times 10{,}000$ voice channels. A satellite with 50 spot beams to 50 earth stations would carry $8 \times 50 \times 10{,}000 = 4$ million voice channels. The total system capacity would be $8 \times 10{,}000 \times 50 \times 50 = 200$ million channels of which not all would be utilized at any one time because of uneven and fluctuating load distribution.

An elaborate switching system would be a necessary component of such a network, as on the rest of the Bell System. The system must be designed so that the load would be reasonably well balanced between the ground stations. Many of the stations would be close to major metropolitan areas. Switching

Table 26.1 Alternative satellites in the Bell Laboratories proposal [1]

	Number of ground stations served			
	8	16	32	50
Weight of satellite (pounds)	1640	3280	6560	11,250
DC power (kw)	1.15	2.30	4.60	7.20
Transponders	64	128	196	400
Gigabits through satellite	40	80	160	252
One-way equivalent voice circuits through satellite (thousands) or	640	1280	1960	4000
One-way equivalent TV channels through satellite	640	1280	1960	4000

must be designed so that faulty circuits can be bypassed, as with today's terrestrial network. With a suitable switching matrix, a substantial number of failed repeaters in the satellites can be tolerated without inacceptable system degradation. The switching would make the system adaptable to changing traffic patterns and would permit ground stations affected by severe storms to be bypassed. Because of the transmission delay, the switching would be organized so that not more than one satellite link is switched into any circuit.

How does the Bell Laboratories study compare with the state of today's technology? Remarkably well. The ATS-6 satellite launched by NASA *did* have a 9-meter multibeam antenna stabilized to $\pm$ 0.01° (Fig. 26.2). Digital channels now *exist* in commercial operation carrying hundreds of millions of bits per second. If AT&T designed such a satellite today, it might operate at 274 mbps to be compatible with T4 channels now in use. Frequencies at 20/30 GHz *have* been allocated internationally for satellite use with a 3.5 GHz bandwidth. Space shuttle plans for launching satellites into geosynchronous orbits by firing a perigee kick motor in low earth orbit *could* handle satellites of the requisite weight at a moderate cost.

The ATS-6 satellite launched as a research vehicle in 1974 performed remarkably well in orbit and Fairchild, its manufacturers, claimed that they could build a version of the same satellite with 120 transponders designed for commercial use. Each transponder would be capable of relaying about 1800 telephone calls digitally using today's modems and codecs. Improved codecs could double or quadruple this throughput. Six satellites like ATS-6 (1974 technology) could carry as much traffic as the peak traffic on the Bell System toll network (i.e., not including *local* telephone calls). Such satellites should be buildable and launchable for less than $150 million each. (Both this and the other costs should be assumed to be in 1975 dollars.) Let us suppose that 50 cities have satellite antennas at, or near, their toll offices, and that the total cost for each of these earth stations is $2 million. Certain earth stations which

Figure 26.2 NASA's ATS-6 satellite had a 9-meter antenna reflector which opened like an umbrella in space. At the top of the electronics compartment underneath the reflector are multiple antenna feeds which can give multiple spot beams to earth (Fig. 26.3). Each spot beam illuminates an area about 200 miles wide on earth. Signals can be switched from one beam to another electronically. To hold the beams in position the satellite is stabilized to $\pm 0.01°$.

monitor and control the system would be more costly. The total costs, using these figures, are of the order of $1 billion.

AT&T alone is spending $9 to $10 billion *per year* on capital improvement of the Bell System. AT&T top management has indicated that they intend this level of expenditure to continue. The annual revenue from telecommunications in the United States is around $35 billion and is growing at about $4 billion per year. Much of the capital expenditure in the telecommunications industry is going into the trunks and trunk switching that satellites and demand-assignment equipment could replace today.

Furthermore the cost of long-distance facilities will swing still more in favor of satellites as the design of the codecs and other digital equipment improves and as space costs drop with the space shuttle and associated vehicles in the 1980s.

Today's terrestrial facilities *exist,* however, and are rapidly being added to. The traffic is growing at a rapid rate, with long distance traffic growing much faster than local traffic. The volume of long distance traffic will probably double in the next five years, and double again in the following five years. A vast expenditure is therefore needed on *new* long-distance capacity. It makes sense that most of this new capacity should be in the form of satellite networks. It is one of the great ironies of the satellite story that the corporation which produced the first satellite to actively relay telephone signals, and the corporation most needing satellites, AT&T, was the one corporation prevented for many years from developing the satellites it needed, by government regulation of one form or another. AT&T has both the resources and the technological brilliance to develop satellite systems as advanced as the one just described. However, advanced satellite development may have to wait for smaller organizations who are willing to compete with AT&T.

Telecommunications in America would almost certainly have been better served in the space segment if AT&T had been allowed to go full steam ahead with satellites. However this would probably have wiped out any potential competition, and might have prevented (depending on the FCC) what now appear to be some of the most interesting satellite innovations—the developments in the ground segment permitting communications users to lease their own satellite antennas and build private corporate satellite networks.

NEW ENTRANTS

The tradeoffs in the design of satellite systems perceived by the traditional common carriers are likely to be different from those perceived by new entrants into the field. To a traditional common carrier a satellite is a means to *augment* its existing network. To a new entrant, a satellite is a means to *bypass* the established common carrier facilities.

AT&T would use satellites to dramatically increase the quantity of long-distance channels, to make Picturephone trunking more feasible, and to lower the investment cost of expanding the long-distance network. Western Union uses satellites to lower the cost of leasing long-distance channels from AT&T.

The Musak Corporation, on the other hand, sees a satellite as a means to avoid the cost of common carrier terrestrial distribution. The University of Hawaii sees a satellite as a means to interconnect computer terminal users everywhere to computers far afield. The military uses satellites to connect infantry units and field operations to distant command and control locations without the need of local telephone lines. SBS sees satellites as a means to give a large corporation antennas at or near its premises, to lower its telephone bill, and to provide new uses of telecommunications such as videoconferencing, interactive image transmission, and electronic mail.

A large common carrier can utilize an entire satellite for its traffic. To most non-common-carrier organizations satellite capacity is far in excess of their needs. Some organizations can use a whole transponder. To many, even one transponder has far too much capacity. The key to their using the satellite is techniques for *sharing* it.

Large common carriers are likely to build satellite systems with a relatively small number of earth stations. Western Union has five continental earth stations. A AT&T–GTE consortium using a Comsat General satellite has seven earth stations. Even Bell's spectacular 100-million-voice-channel proposal only used 50 earth stations. Satellites in such systems are thought of as pipelines in the sky to enhance a vast terrestrial network which has its own multiplexing, switching and distribution facilities.

New entrants, on the other hand, would like to build satellite systems with a large number of earth stations. If these are receive-only stations, as with Musak, the system architecture is relatively straight-forward. If the earth stations transmit and receive, then system control equipment is needed which may be complex, including demand-assignment multiple-access equipment, multiplexors, codecs, data concentrators, and switching equipment. Many of the common carrier earth stations do little other than receive and transmit the signals. They pass a television channel, mastergroup, or other multiplexed block on the land lines which have long been designed to carry it, and the traditional toll office equipment handles the multiplexing, routing and switching.

Box 26.1 summarizes these differences in viewpoint.

BOX 26.1 Differences in viewpoint of traditional common carriers and other organizations concerning the use of satellites

MOTIVATIONS OF INDEPENDENT ORGANIZATIONS IN ESTABLISHING SATELLITE CHANNELS	*MOTIVATIONS OF TRADITIONAL COMMON CARRIERS IN ESTABLISHING SATELLITE CHANNELS*
• Establish minimum-cost channels.	• Respond to rate-of-return regulation (in U.S.) which encourages high capital investment.
• Provide new business opportunities.	• Preserve the existing plant, much of which has a 40-year write-off.

continued

BOX 26.1 ***Continued***

• Introduce competition to established common carriers.	• Preserve their own control of the industry.
• Create facilities which carry all types of traffic.	• Augment the telephone network.
• Permit many organizations without very high traffic volumes to share the satellite.	• Use the satellite for their own traffic.
• Provide multiple-access demand-assignment of individual channels.	• Carry existing mastergroups and other groups.
• Provide burst multiplexing to give flexible demand-assignment of all types of signals.	• Extend terrestrial circuit-switched system.
• Bypass the local loop bottleneck so that wideband video and data signals can be carried.	• Preserve today's local distribution.
• Have satellite antennas at all major premises.	• Use a small number of earth stations in regions of highest traffic density.
• Use small, cheap antennas.	• Use large, expensive, earth stations which concentrate high traffic volumes.
• Minimize the total system cost.	• Maximize the transponder utilization.
• Use public telephone network to bypass the effects of storms (above 11 GHz).	• Use space diversity to bypass the effects of storms (above 11 GHz).
• *Bypass* today's terrestrial facilities.	• *Augment* today's terrestrial facilities.

MULTIPLE ACCESS

How can new entrants to the satellite arena achieve the objective of linking the satellite channels to users everywhere, at a low cost?

The first part of the answer to that question comes from the use of an appropriate multiple-access technique.

MULTIPLE SPOT BEAMS

As with simple multiplexing, the sharing of satellite capacity by geographically scattered users can be achieved by frequency-division, time-division, or space-division. Space-division, in this context, implies multiple spot beams from the satellite, as in Fig. 26.3, and some capability to switch between the beams. The phrase "switch in the sky" has been used to refer to a satellite which can switch spot beams on and off, and switch channels between the beams.

Space-division demand assignment is of limited value today. It takes an exceptionally large antenna to produce spot beams like those in Fig. 26.1. Even then a beam covers several hundred miles of earth so the technique relates to common carrier earth stations or television distribution covering a large geographical area. It is not of value for enabling many users with small earth stations to share a domestic satellite. Switching equipment on board the satellite adds to the risk of an unrepairable failure in space. Domestic satellites such as ANIK and WESTAR have only one major beam. Hence some other form of demand assignment is needed—frequency-division or time-division.

MULTIPLE TRANSPONDERS

The simplest way to divide satellite capacity by frequency is to give different users different transponders. A television organization uses one transponder, Musak uses another, a common carrier leases another, a large corporation uses a fourth, and so on. The trouble with this approach is that the transponders are of fixed capacity whereas many users want variable channel assignment. Furthermore the transponder capacity is much too big for most users. It may be a worthwhile tradeoff for future satellites to carry some smaller transponders. Small transponders, however, mean more weight on the satellite for a given channel capacity.

Even when a corporation leases a whole transponder, it still has a demand assignment problem in using that transponder. Some technique is needed for geographically dispersed users to share a transponder.

FDMA AND TDMA

When each of many earth stations has access to the same transponder, the bandwidth of that transponder may be shared by using *frequency-division multiple access* or *time-division*

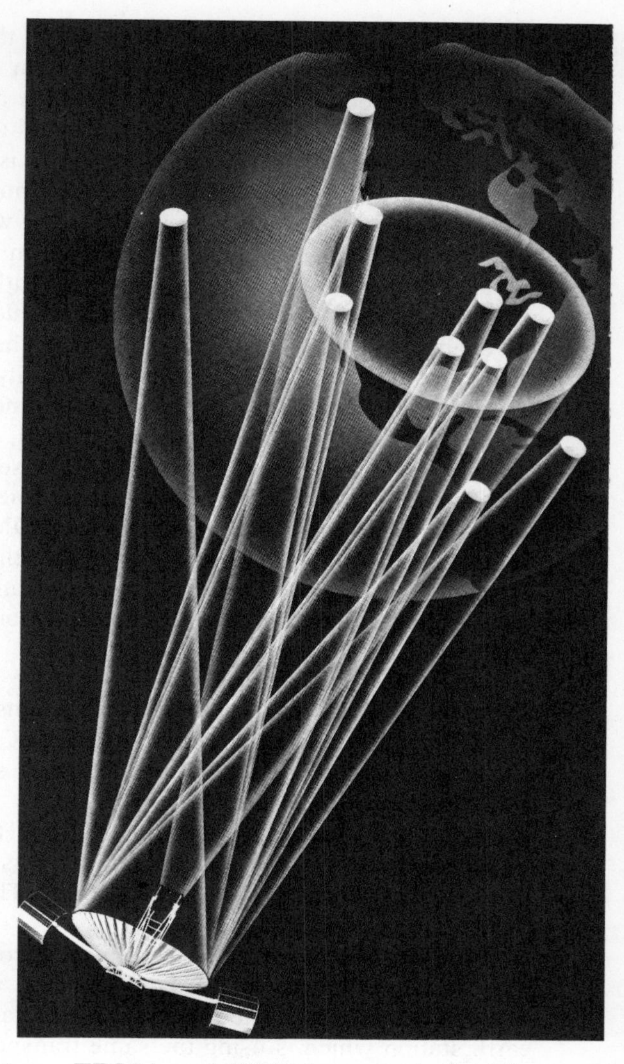

Figure 26.3 Multiple spot beams from NASA's ATS-6 satellite. To make full use of such beams it is desirable to have switching capability on the satellite.

multiple access. These are referred to as FDMA and TDMA respectively. Either can be employed with any existing satellite.

An FDMA system makes available a pool of frequencies and assigns these, on demand, to users. A TDMA system makes available a stream of time slots and assigns these, on demand, to users.

One channel, derived either by frequency division or time division, may be reserved to function as a control channel. The signals on this channel convey each station's requests for capacity, and inform each station about the channel assignments.

In theory, the satellite itself could be the controlling location. The reader might imagine a demand-assignment system operating with a little old lady sitting in the satellite talking to the users like a PBX operator and pushing plugs into a panel to connect them. In reality, if the satellite *were* the control point an on-board computer would be used. It is desirable, however, not to complicate the satellite too much because it must have extremely high reliability. Furthermore demand assignment is needed with *today's* satellites. Therefore, if a controlling location is used, it must be on earth and a transmission via satellite is needed to it and back, on a permanently assigned subchannel.

Such a scheme may be designed for allocating channels of equal capacity, for example telephone channels. On the other hand it may be designed so that earth stations can request different capacity channels. An earth station may request a low-speed data channel at one time, a high-speed data channel at another, a telephone channel, a video channel, or channels of other capacity.

Multiple-access techniques are discussed further in Chapter 31.

Demand assignment systems in commercial use with satellites in the first half of the 1970s were all FDMA, not TDMA. In the future the cost of high-speed digital equipment will drop and its reliability improve. Given appropriate cost and reliability, TDMA offers significant advantages over FDMA, giving higher satellite throughput and greater flexibility.

THE SBS SYSTEM

SBS (Satellite Business Systems), a partnership sponsored by IBM, Comsat General and the Aetna Insurance Company, plans to launch two satellites in 1979 which will use TDMA operation [2].

The satellites will operate at 12/14 GHz and will have eight transponders of 54 MHz bandwidth. Small earth stations using 5 or 7 meter dishes will be employed at or near customer premises. There will eventually be many hundreds of such earth stations.

All transmissions will be digitally encoded and transmitted in the form of high-speed bursts of bits. A burst modem will be used which operates at 41 mb/s. A burst transmitted by one earth station can be received by any other earth station which is using the same transponder. There is thus complete interconnectability between those stations which share a transponder. The bursts may be used to carry speech, data, images, or video signals. A station receiving a stream of bursts will extract those bursts which are addressed to it, and reject all others. Continuous signals such as speech will be reassembled from the bursts in a high-speed buffer. Tight security procedures will be used, where needed, to ensure that stations cannot use bursts which were not intended for them.

Figure 26.4 shows the planned contours of signal strength from an SBS satellite. The antenna pattern is designed so that the East Coast area of the United States receives a stronger signal. The majority of the customers will be

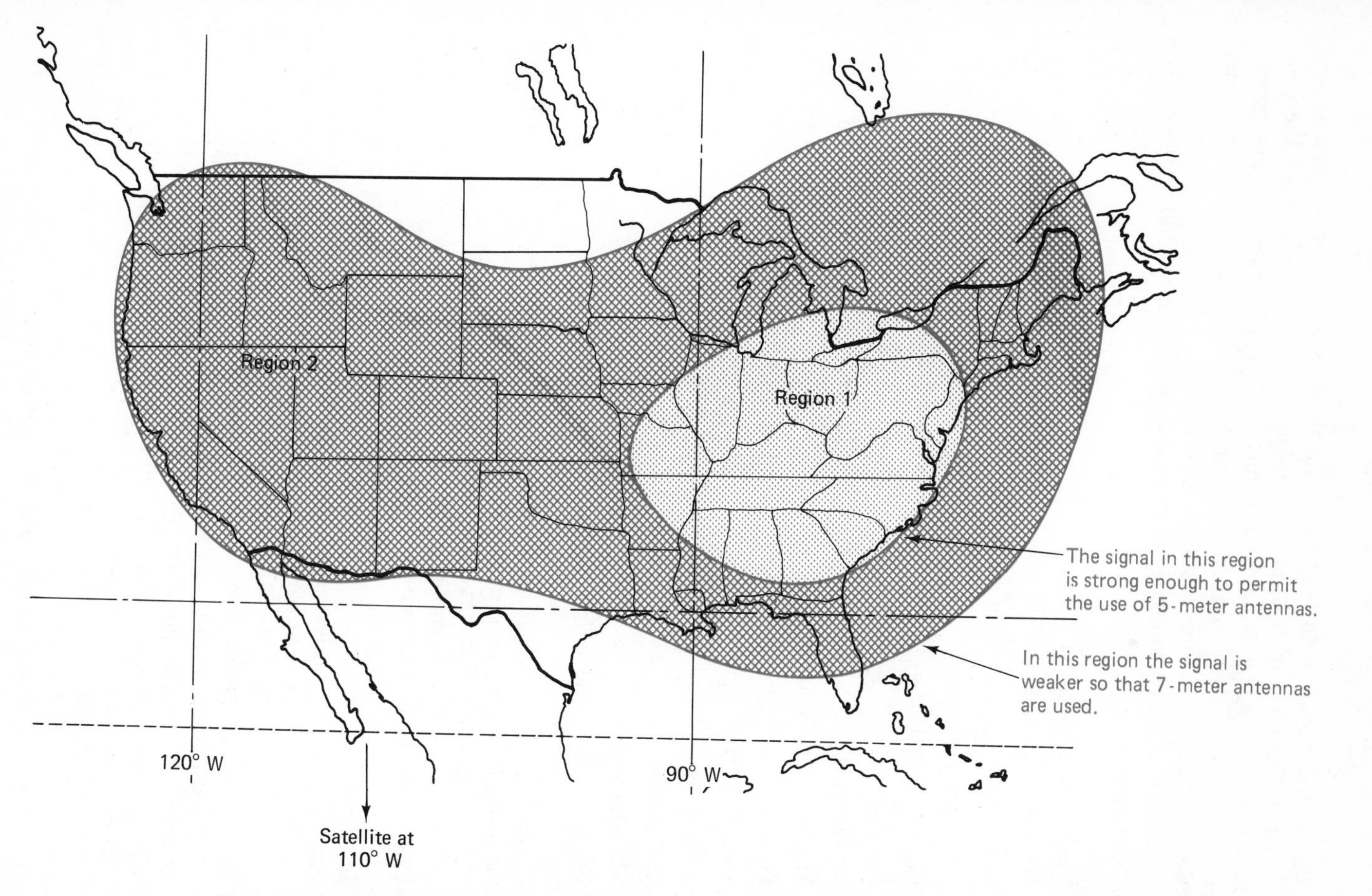

Figure 26.4 Contours of signal strength from a 12/14 GHz SBS satellite positioned at 110°W. The system is designed to permit 5-meter earth antennas on the East Coast where the traffic is denser and the weather attenuation is worse. Elsewhere in North America 7-meter earth antennas will be used.

in the area labeled Region 1 in Fig. 26.4, and these will receive a signal strong enough to permit the use of relatively inexpensive 5-meter antennas. The customers in Region 2 will use 7-meter antennas. A 7-meter antenna transmitting at 14 GHz is approximately equivalent in performance to a 15-meter antenna transmitting at today's satellite frequency of 6 GHz.

The communications controllers at customer locations will perform the following functions:

- Burst modem operation
- Formating, framing, and synchronizing the bursts
- Forward error correction in the bursts
- Echo suppression
- Voice activity compression (DSI, discussed in Chapter 28)
- Signalling
- Multiplexing
- Multiple access control
- Demand assignment control
- Security control
- Analog/digital conversion (the codec function)
- Call processing
- Circuit switching (PABX function)
- Customer/terrestrial system interfacing

In addition to providing better man-to-man communications in corporations with speech and image transmission, it is likely that the advent of systems such as SBS providing for bursts of multimegabit data transmission will substantially change the way computers are used. Data transfer by satellite can be equivalent in speed to data transfers *within* a computer system or its channels.

BROADCAST SATELLITES

Receive-only earth stations are a fraction of the cost of earth stations which transmit. It is possible to design systems with very low cost receiving stations as evidenced by the proposed Musak receivers and the satellite antennas in India which received television from ATS-6 at a cost of less than $100 each.

Existing commercial satellites such as WESTAR have enough power to broadcast television to regional centers with 10-meter dishes. They do not have enough power to broadcast television directly to subscribers. They could send low bandwidth signals directly to subscribers as with the Musak proposal by using a transponder at a small fraction of normal bandwidth.

The ideal broadcast satellite needs high on-board power, an antenna as large as ATS-6 or larger and operation at 12/14 GHz. The earth station transmitting to the satellite may be large and powerful, and the system designed to minimize the cost of the rooftop antennas. Canada's CTS satellite has enough power for TV to be received with 1-meter dishes (Fig. 26.6). One estimate of the costs of a broadcast system serving millions of subscribers is given in Fig. 26.5 [3]. In Japan, home satellite receivers have been designed which could be mass-produced for about $150 each [4].

Satellites in the near future will be used extensively for distributing signals to the heads of CATV systems in some countries rather than directly to subscribers. A CATV "Long Lines" network interlinking vast numbers of separate cable systems will probably grow out of today's CATV satellite transmission. Figure 26.7 shows a typical CATV earth station in North America.

LOW-COST EARTH STATION

The key to corporate use of satellite systems is the achievement of sufficiently low-cost earth stations. Among the factors that could permit lower transmit/receive earth station costs are the following:

1. *Small antenna.*
2. *Fixed (nontracking) antenna.*
3. *Uncooled receiver.*
4. *Cheap microwave circuit technology*—planar circuits [4].
5. *Demand-assignment* technique which is not overly sophisticated.
6. *Low bandwidth transponder* on satellite.
7. *Modulation technique* which can tolerate a substantial level of noise and interference.
8. *High modulation index.*
9. *Low information rate per channel* permitting unsophisticated modulation and concentration techniques.
10. *Frequency* which gives freedom from common carrier interference.
11. *Site location* which permits freedom from common carrier interference.
12. *Mass production,* i.e., a large community of similar earth station users.

In particular there is a tradeoff between maximizing satellite throughput and minimizing earth station cost. A given transponder may handle a large number of channels or a large bandwidth at high earth station cost, or a small number or small bandwidth at low earth station cost. The range of options is wide. The same WESTAR transponder, for example, has been used to handle 1200 voice channels totalling about 4 MHz in bandwidth with an earth station of about $1/2 million, or four hi-fi music channels totalling about 0.06 MHz with an earth station cost below $1000. A large common carrier may opt to have few earth stations and hence employ expensive earth station equipment to maximize the satellite throughput. An organization providing rooftop antennas

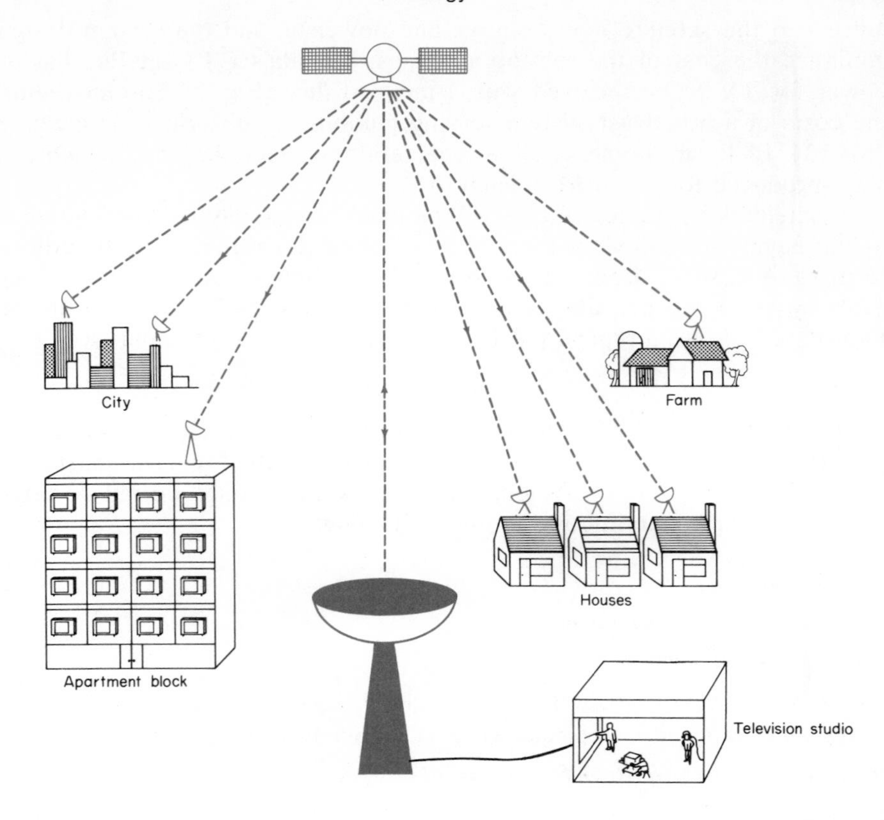

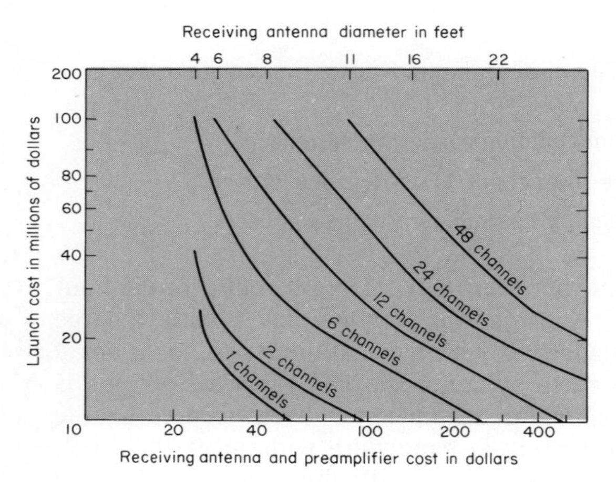

Figure 26.5 One cost estimate of a system designed to broadcast television directly to millions of viewers.

Figure 26.6 Canada's CTS Satellite. The large solar arrays of this satellite carry 27,000 individual solar cells and generate 1.2 kilowatts of power. This, with 12/14 GHz operation, permits good quality television to be received with a 1-meter earth antenna. The solar arrays expand like a concertina (below) when in orbit, giving a tip-to-tip span of 16½ meters.

Figure 26.7 Satellites such as Westar and RCA's Satcom can broadcast television to earth stations with 10-meter disks like this. Such earth stations are used to feed CATV systems. For inexpensive television reception at home with small antennas, the satellite needs more power, a larger on-board antenna, and operation at 12/14 GHz. (*Courtesy Scientific Atlanta.*)

will attempt to achieve a low earth station cost at the expense of having a lower satellite throughput.

There is a tradeoff between the use of power and the use of bandwidth. The signal at the bottom of Fig. 24.6 is of less power than the one at the top, but requires more bandwidth. On terrestrial transmission links, plenty of power is available and so the signal is packed tightly into a small bandwidth, thereby increasing the total information-carrying capacity. On satellite links power is a scarce resource, therefore FM modulation with a fairly high modulation index may be used.

On a terrestrial link, 900 voice channels occupy 3.72 MHz—a CCITT mastergroup. The voice channels are multiplexed together using amplitude modulation, with each voice channel occupying a slot of 4 KHz. On INTELSAT IV the 900 voice channels occupy an entire 36 MHz transponder. The mastergroup is used to frequency-modulate the 70 MHz IF carrier, using a high modulation index. Again, a 36 MHz transponder relays one television channel of 4.6 MHz.

FM transmission tends to withstand the effects of noise. The larger the modulation index the greater the protection from noise. If a portion of the frequencies at the bottom of Fig. 24.6 were removed, the information being sent could still be recovered. Rather like a hologram, you can destroy part of it and its information still exists. The worse the signal-to-noise ratio, the higher the modulation index used. In an extreme case, the Musak Corporation wants their customers to use 1-meter antennas with relatively inexpensive electronics. As the antenna size used with a 4/6 GHz satellite is dropped from 10 meters to 1 meter the noise-to-signal ratio drops from bad to horrifying. However music has a very low bandwidth compared to the transponder bandwidth, and by transmitting it with a sufficiently high modulation index excellent reproduction can be obtained. Four channels of Musak which would require less than 80 KHz on earth are transmitted over a 36 MHz bandwidth.

BAD WEATHER

The move to frequencies higher than the common carrier band of 4/6 GHz is important, first to avoid common carrier interference, and second because it permits smaller and cheaper earth station antennas.

12/14 GHz and 20/30 GHz links have excellent properties when the weather is good. Figure 26.8 shows the combined effects of loss, attenuation, and noise when the skies are clear. The high frequencies have slightly better properties than 4/6 GHz because the gain of antennas of the same diameter is higher at the higher frequencies. Figure 26.8 is drawn for an angle of elevation of the earth station antenna of 15°. For higher angles the curve favors the higher frequencies still more. Most domestic satellite systems can avoid earth station angles of elevation less than 20°.

The trouble with the higher frequencies is that their properties are much worse when the weather is very bad. Moderate rain and cloud cover does not do much harm, but the link deteriorates badly in very heavy storms. Figure 26.9 shows the composite effect of loss, attenuation, and noise during extremely heavy rain and cloud cover.

Figure 26.10 shows the effects of an intense rainstorm on transmission received from space at 16 and 30 GHz. The period illustrated in Fig. 26.9 was immediately after sunrise, so that angle of elevation is low.[1]

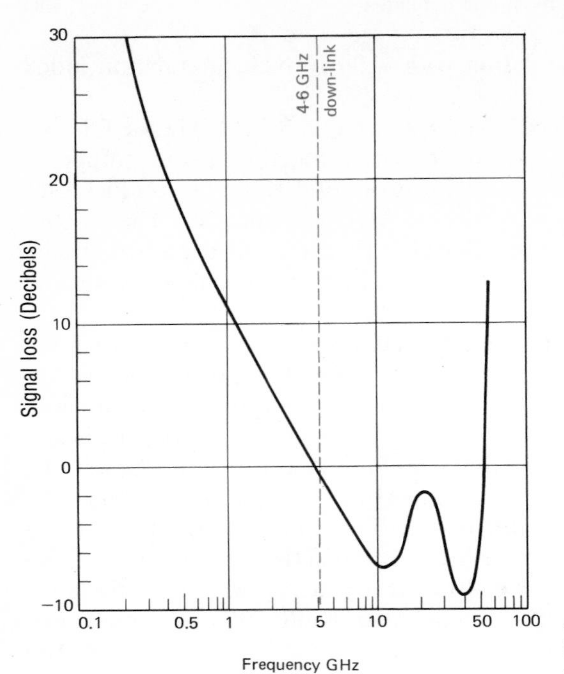

Figure 26.8 Composite effect of link loss, atmospheric attenuation, and atmospheric noise.

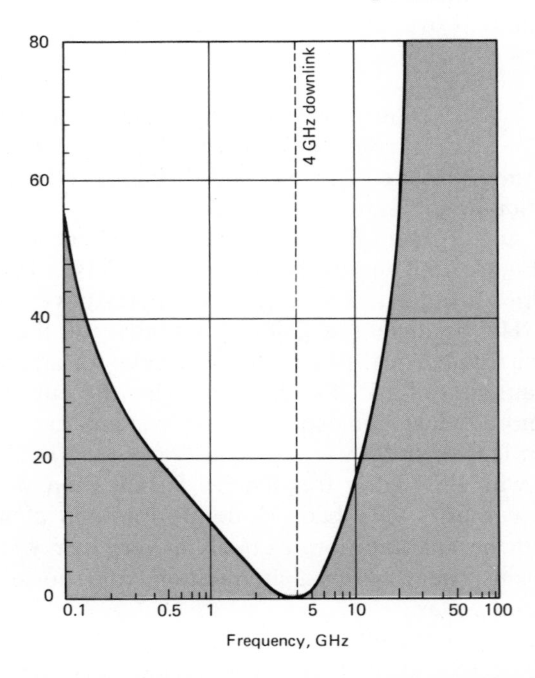

Figure 26.9 This chart shows how the loss illustrated in Fig. 26.8 is worsened by intense rain and severe fog or cloud cover.

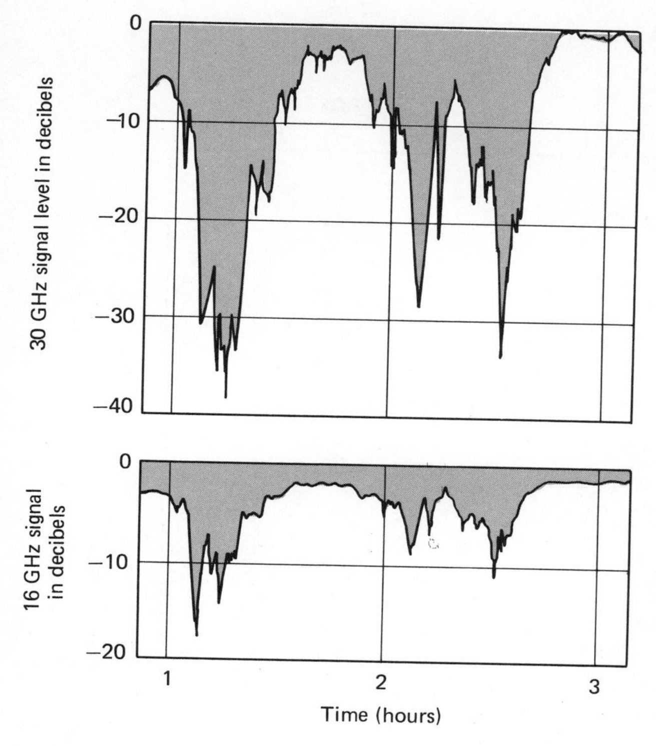

Figure 26.10 Attenuation of 16 GHz and 30 GHz transmission caused by an intense rainstorm (exceeding 100 millimeters per hour), measured at Crawford Hill, New Jersey. (*Redrawn from Reference* 5.)

Rain has far less effect at 12 and 14 GHz. Figure 26.11 summarizes statistics compiled over three years at Comsat Laboratories in Clarksburg, Maryland [2]. It will be seen that for about one hour per year the signal received is about 10 decibels worse than average at 12 GHz, and about 14 decibels worse than average at 14 GHz. The differences could be compensated for by using much larger antennas or more powerful amplifiers. On the other hand a low-cost earth station could be designed to have an availability of, say, 0.995. The high attenuation occurs only infrequently.

SPACE DIVERSITY

The most intense storms are in fact the most limited in area. Consequently, a trade-off called *space diversity* is discussed for common carrier systems. Two earth stations operating at 12/14 GHz would be separated by several miles. Because very intense storms are limited in diameter, when one station was suffering a loss of 16 decibels due to a storm the other would normally be experiencing less intense rain with less than 4 decibels loss. The same concept could be more valuable at 20/30 GHz when the station in a heavy storm could be suffering 40 decibels loss and its neighbor less than 10. Space diversity would require the ability to switch channels wherever possible to avoid the most intense storms. Such an ability

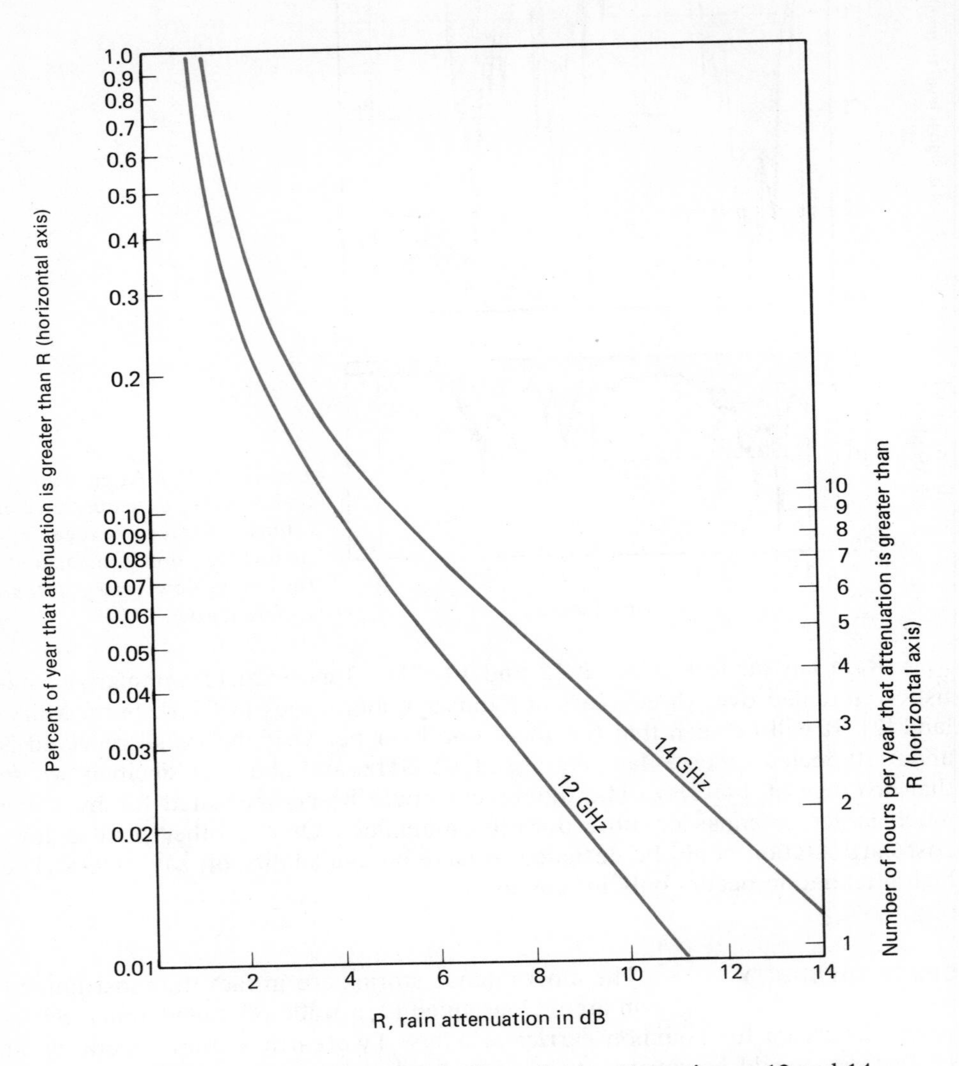

Figure 26.11 Statistics on the effect of rain attenuation at 12 and 14 GHz, measured at Clarksburg, Maryland. This attenuation occurs only a few hours per year.

could be made an integral part of systems such as the Bell System. Many common carrier earth stations are likely to have two antennas; *diversity-earth stations* could have the antennas several miles apart and linked to the control center by terrestrial microwave or cable. Eventually space diversity may help to make the high bandwidth of 20/30 GHz links appear attractive.

FAIR WEATHER ANTENNAS

Whereas space diversity could be cost-effective for common carriers, it is unlikely to be suitable for most corporations operating their own antennas. The cost-conscious corporation has two alternatives. First, during intense storms the control equipment can route certain traffic over the public telephone network. Second, some links may be engineered to have less than 100% availability. Most telecommunications engineers are horrified at the suggestion of links which do not work when the weather is bad. However, there are many situations in which they could be highly cost-effective.

Table 26.1 summarizes the frequency of bad weather effects by showing the percentile attenuations for rain, clouds, and air, and the sum of these [5]. The top section of the table relates to the 99.99 percentile. In other words, the attenuation is worse than the figures shown for 0.01% of the time—about one

Table 26.1 Percentile attenuations for rain, clouds, and air*

Percentile	Element	Attenuation Frequency (GHz) 4	16	35	90
99.99	Rain (db)	0.5	26.4	105	320
	Clouds (2 g/m³, 0° C) (db)	0.5	5.6	24	150
	Oxygen and water vapor (db)	—	0.3	0.8	3.5
	Total (db)	1.0	32.3	129.8	573.5
99.9	Rain (db)	0.1	6.6	26.4	80
	Clouds (1.5 g/m³, 0° C) (db)	0.3	4.0	17	120
	Oxygen and water vapor (db)	—	0.3	0.8	3.5
	Total (db)	0.4	10.9	44.2	203.5
97	Rain (db)	—	0.4	1.5	4.7
	Clouds (1.0 g/m³, 0° C) (db)	0.2	2.7	12	82
	Oxygen and water vapor (db)	—	0.3	0.8	3.5
	Total (db)	0.2	3.4	14.3	90.2

*Source: Adapted from reference 5.

hour per year. The other two sections give the 99.9 percentile and the 97 percentile figures. The figures relate to a site with fairly heavy rainfall, 100 cm per year.

We see that except for 10 hours in the year, the attenuation of 16-GHz waves is not greater than 10.0 db and that of 35-GHz waves is not more than 44.2 db. The attenuation of 90-GHz waves is high, however. These frequencies have been selected because they do not coincide with atmospheric absorption peaks. The attenuation levels shown suggest that with large and powerful satellites, 16- and 35-GHz waves could be used successfully.

Let us suppose that a 20/30 GHz satellite is used with 20-meter ground antennas, and that this provides good transmission 99.9% of the time when the attenuation in atmosphere, cloud, and rain is no worse than 30 decibels. 97% of the time it is no worse than 2 decibels, and so an antenna with 28 decibels less gain can be used. A 2-meter antenna would give good transmission 97% of the time.

Certain important locations, then, could be equipped with a large antenna. Other locations would have an inexpensive 2-meter unit. Let us assume that the effect of this would be that the locations with the large antenna are cut off from the satellite 0.1% of the time. This is 8.8 hours in total per year, but only 2 *working* hours. Terrestrial equipment failures would probably make it longer than this. Locations with the 2-meter antenna are cut off from the satellite 3% of the time, i.e., 262.8 hours in total but only 60 *working* hours per year.

What would be the effect of this?

A 99.9% system would be better than today's telephone system which experiences a low number of local loop and other failures. (In fact the availability of the earth station electronics is likely to be less than 99.9%.) The locations with the 2-meter antennas may be provided with a private branch exchange which automatically places calls between corporate locations on the satellite network if possible, but if not it places them on the toll telephone system. A priority system may be used so that only certain telephone extensions or computer terminals are permitted to place nonsatellite calls, in order to keep costs down.

If videoconferencing is used, there may be no alternative to satellite channels. There would rarely be problems with videoconferences at the important locations. The locations with 2-meter antennas would have good transmission 97% of the time. For the other 3%, video pictures with varying degrees of "snow" on the screen might be tolerated *providing that the sound quality is good* so that people's voices are clear. Good sound quality could be achieved by allocating several times the normal bit rate to the sound channel and using powerful error-correcting codes.

If music is transmitted to the 2-meter-antenna locations, it may be tolerable to do without it 3% of the time, and perhaps use local music tapes as back-up.

Great use will no doubt be made one day of satellites in education. Schools will have access to televised material and computer-assisted instruction, both of much higher quality than most that has been demonstrated to date. The antennas serving the schools will need to be inexpensive and a compromise of 3% unavailability is probably acceptable because alternate classroom material can be used.

NONREAL-TIME TRAFFIC

Perhaps the major appeal of antennas which only work in fair weather is for nonreal-time traffic. The sending of electronic mail, facsimile documents, monetary transactions, cables, or batches of computer data, can wait until a storm has passed. The potential volume of nonreal-time satellite traffic is gigantic if the earth station cost is kept low.

Cheap earth stations would be ideal for electronic mail. Perhaps one day post offices everywhere will have satellite antennas on their roofs.

SUMMARY

Box 26.2 summarizes the main tradeoffs in satellite system design. There are major economies of scale, as elsewhere in telecommunications. If the satellite can be designed for a high traffic volume it can be made heavy and powerful with a corresponding reduction in cost per channel and earth station cost. It is therefore desirable to design the technology to permit transponders to be shared by as many users as possible and to enable the satellite to carry all types of traffic.

REFERENCES

1. Leroy C. Tillotson, "A Model of a Domestic Satellite Communications System," *Bell System Tech J.*, December 1968.
2. Satellite Business Systems FCC Applications for a Domestic Satellite System, 1975. Available from the FCC, Washington, D.C.
3. Eugene V. Rostow, "A Survey of Telecommunications Technology," President's Task Force on Communications Policy, Staff Paper No. 1, Washington, D.C., 1969.

BOX 26.2 Major tradeoffs in satellite design

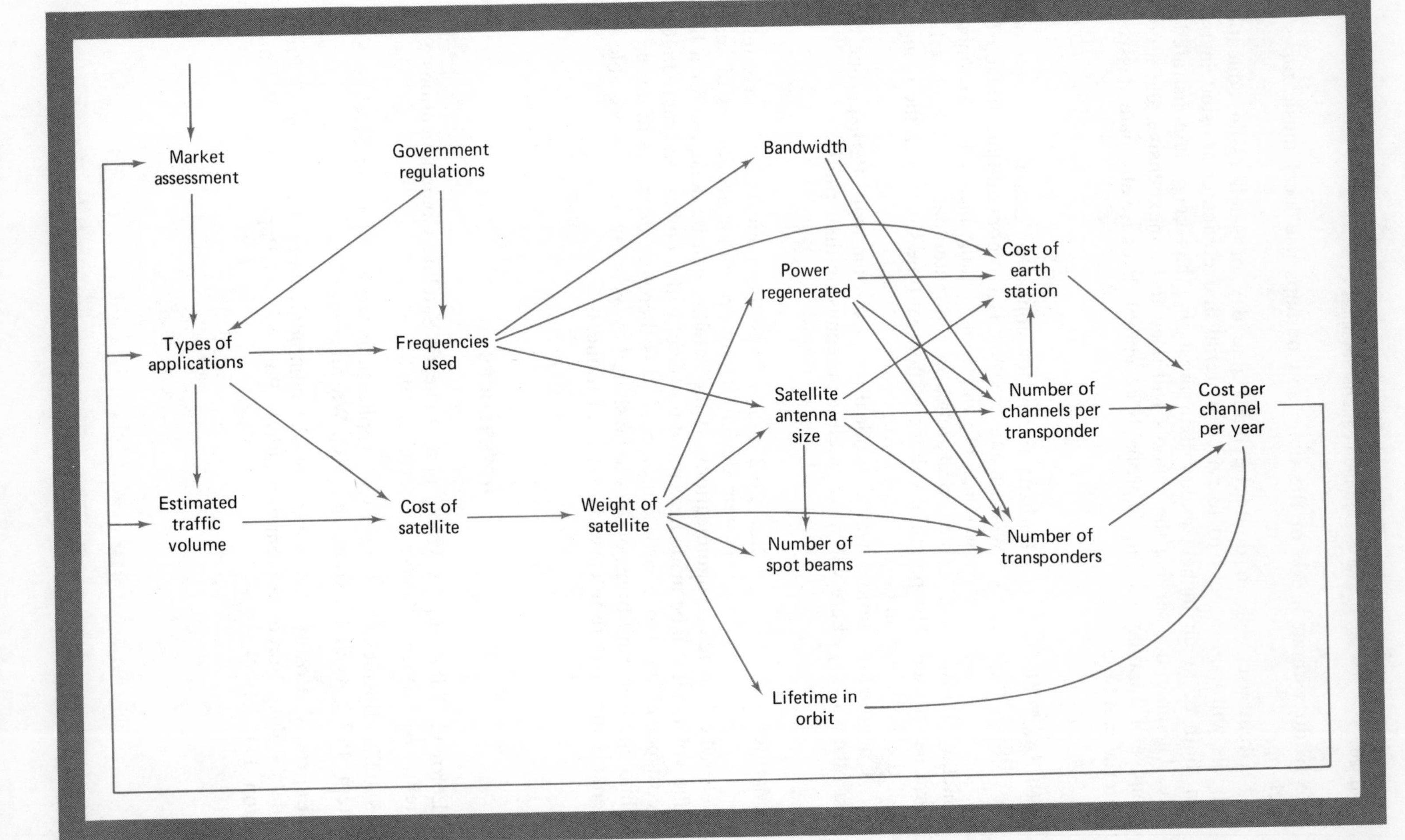

4. Yoshihiro Konishi, NHK Technical Research Lab., Japan, paper delivered at U.N. meeting on Satellite Broadcasting Systems for Education, Tokyo, 1974.

5. "Future Communications Systems via Satellites Utilizing Low-Cost Earth Stations" prepared by the Electronics Industries Association for the President's Task Force on Communications Policy, Washington, D.C., 1969.

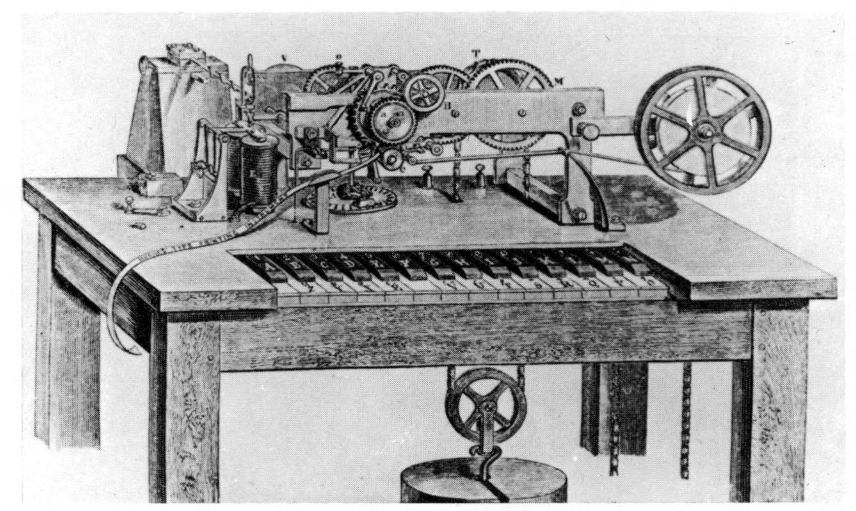

Hughes printing telegraph, designed 1855. (*Courtesy of The Post Office, England.*)

27 DIGITAL CHANNELS AND PCM

We have described several new telecommunications channels which operate digitally at 274 million bits per second. Let us summarize these:

1. AT&T's new coaxial cable system, the T4M system, contains 20 coaxial tubes giving 10 full duplex channels of 274 mb/s. A few segments of this are now in operation.
2. AT&T's new 18 GHz radio system, the DR18 system, transmits 14 channels of 274 mb/s, giving 7 channels in each direction. A few segments of this are now in operation.
3. AT&T's helical waveguide, the WT4 system, which modulates 120 sine wave carriers at 274 mb/s to give 60 channels of 274 mb/s in each direction. A few segments of this are now in operation.
4. A glass fiber system powered by a semiconductor laser. Experimental repeaters have been built in Bell Laboratories to relay 274 mb/s through fibers. The repeaters on such a system could be 10 kilometers apart. Many thousands of such fibers could occupy one cable.
5. Satellite transponders have been designed to operate at a wider bandwidth than most of today's transponders, and relay bit rates such as 274 mb/s.

In the AT&T (and hence the United States) telephone network there are four main levels of digital carrier rate of which 274 mb/s is the highest. A variety of different physical channels and multiplexors are built to conform to these four basic speeds. They are referred to as the T1, T2, T3, and T4 carriers:

T1 carrier: 1.544 mb/s

T2 carrier: 6.312 mb/s

T3 carrier: 45 mb/s

T4 carrier: 274 mb/s

The T1 carrier was designed in the 1960s to operate over twisted wire pair cables and carry 24 voice channels. The T2 carrier came into use in 1972 and can operate over good quality wire pair trunks to carry 96 voice channels or one Picturephone channel. The T3 carrier can carry one digitized mastergroup (600 voice choice channels). It is too fast for wire pair cables and too slow for coaxial. Its bit rate could be sent over one of today's 36 MHz satellite transponders, and experimental optical fibers have been designed for it. It is possible that no major T3 transmission system will come into use. It will serve merely as a bridge between the T1 or T2 and the T4 carriers.

The bit streams of the smaller digital carriers can be fed into the larger carriers by means of multiplexors. Thus four T1 streams can be fed into a T2 stream by means of an M12 multiplexor. Fig. 27.1 shows the digital levels and multiplexor links between them. The multiplexors are designated Mxy where x designates the lower level and y the higher level.

As indicated in Fig. 27.1 an additional level has been introduced above level 1 which doubled the capacity of T1 routes—level T1C. The T1C carrier transmits 3.2 mb/s over wire pair links and is designed to be similar to the T1 carrier for ease of upgrading T1 trunk routes.

LINKS IN OTHER NATIONS

Many other nations are installing digital transmission facilities. Few, however, use the same bit rates and digital structures as the United States. A proliferation of incompatible digital systems has been installed.

International recommendations for standards are emerging. CCITT, the international organization for agreement on telecommunications standards, has recommended detailed specifications for two basic PCM systems—an international equivalent of the Bell System Level 1. One uses 1.544 mb/s to transmit 24 voice channels and the other 2.048 mb/s to transmit 30 voice channels [1, 2]. The 1.544 mb/s recommendation is not identical to the Bell T1 carrier.

CEPT, the European authority (European Conference of Postal and Telecommunications Administrations) has recommended that the multiplexing should go up in steps of 4, giving a family of digital systems as follows:

Level	Millions of bits/second	Number of voice channels
1	2.048	30
2	8.448	120
3	34.304	480
4	139.264	1920
5	565.148	7680

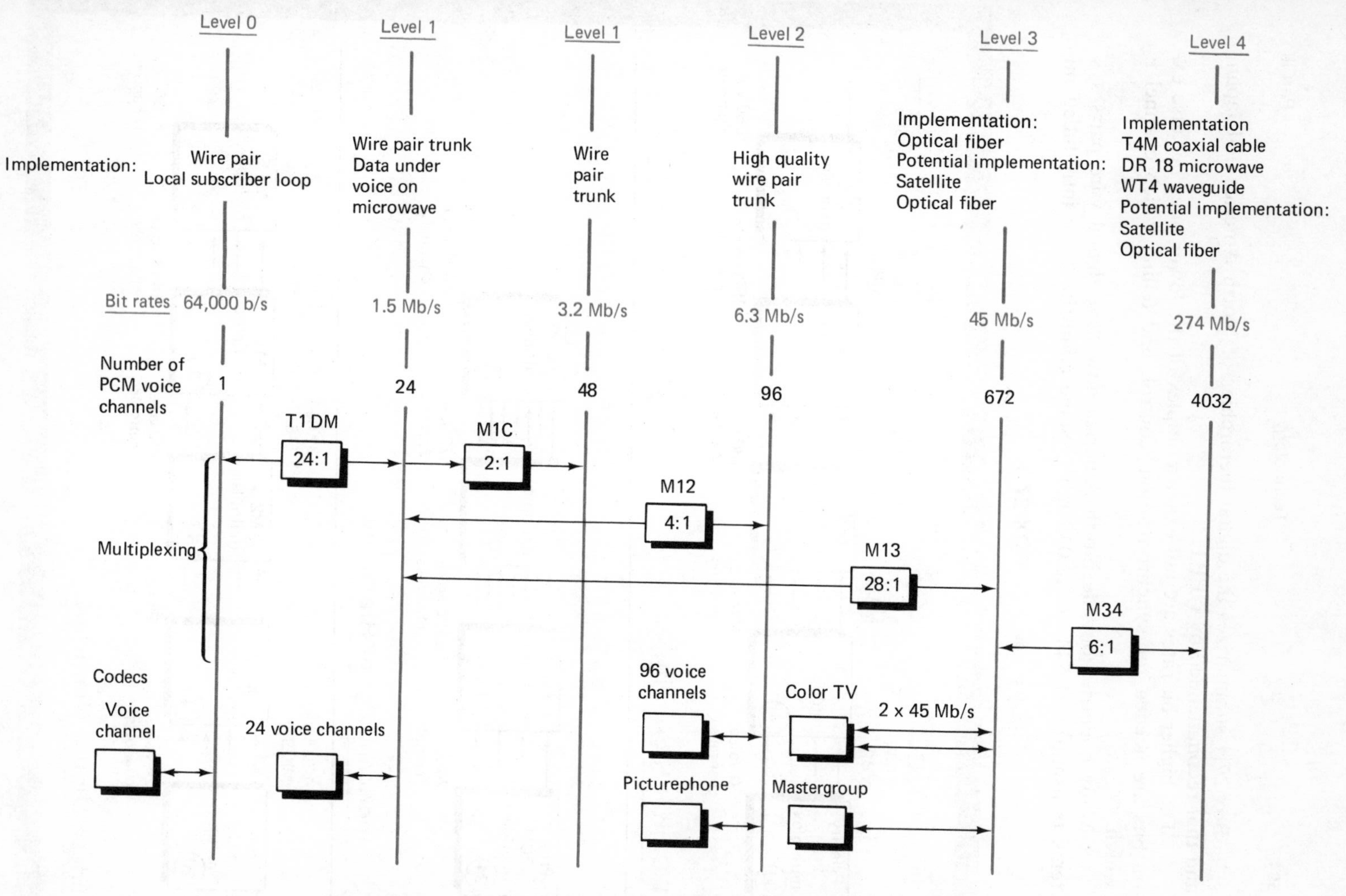

Figure 27.1 The hierarchy of digital channels on the Bell System.

Box 27.1 summarizes the digital hierarchies of North America and Japan, and that recommended by CEPT.

The higher bit rates are not exact multiples of the lower ones because extra bits are needed for synchronization, control, and telling which channel is which.

Many countries outside North America now have digital wire pairs systems transmitting 1.544 and 2.048 mb/s. Some countries have introduced mi-

BOX 27.1

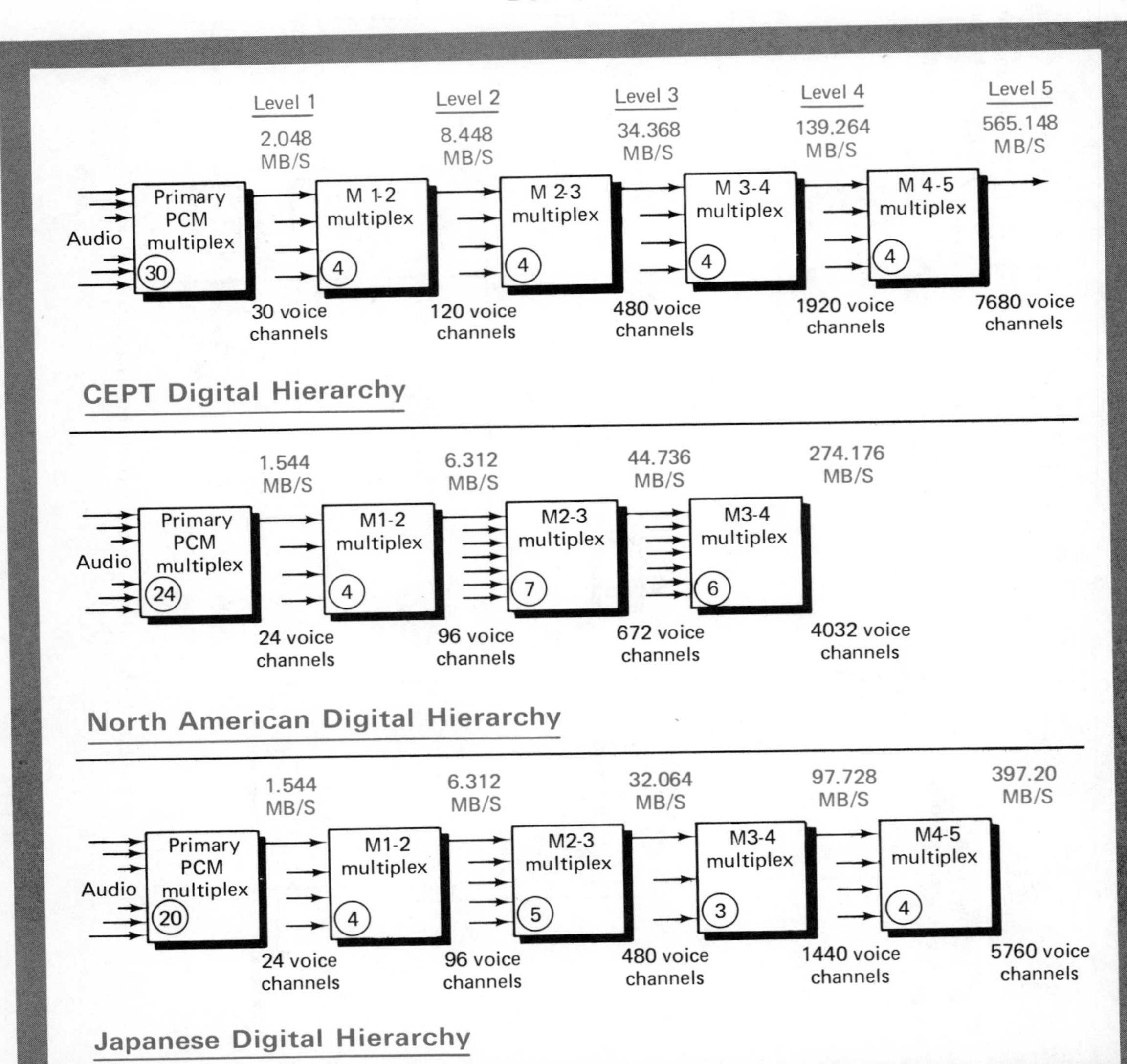

crowave systems operating at 13 GHz, with PSK modulation at 8.448, 34.304, and between these at 17.152 mb/s. Prototype coaxial cable systems operating at 34.304 and 139.264 mb/s have been developed. Italy uses a small coaxial cable carrying 8.448 mb/s, called microcoaxial.

THE BELL SYSTEM T1 CARRIER

Chapter 4 explained the essentials of time-division multiplexing and pulse code modulation and illustrated them in Figs. 6.2, 6.3, and 6.4. The most common use of pulse code modulation at the time of writing is in the Bell System T1 carrier (Fig. 27.2). As we commented the Bell T1 PCM System multiplexes together 24 voice channels. Seven bits are used for coding each sample. The system is designed to transmit voice frequencies up to 4 KHz, and therefore 8000 samples per second are needed. 8000 frames per second travel down the

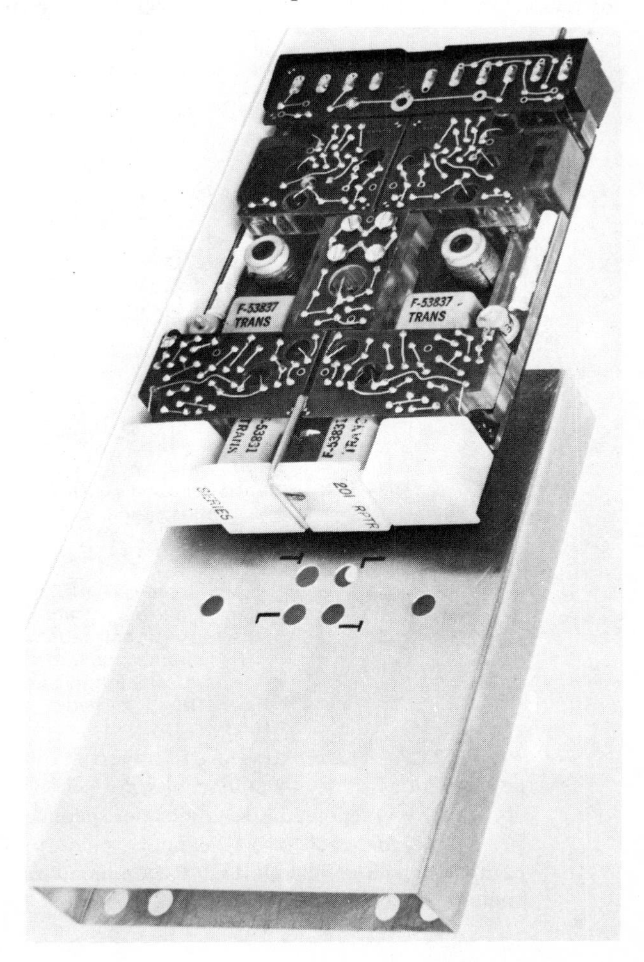

Figure 27.2 AT&T repeater for T1 carrier circuits.

line. Each frame, then, takes 125 microseconds. A frame is illustrated in Fig. 27.3. It contains eight bits for each channel. The eighth is used for supervisory reasons and signaling, for example, to establish a connection and to terminate a call. There are a total of 193 bits in each frame, and so the T1 line operates at $193 \times 8000 = 1{,}544{,}000$ bits per second.

The last bit in the frame, the 193rd bit, is used for establishing and maintaining synchronization. The sequence of these 193 bits from separate frames is established by the logic of the receiving terminal. If this sequence does not follow a given coded pattern, then the terminal detects that synchronization has been lost. If synchronization does slip, then the bits examined will in fact be bits from the channels—probably speech bits—and will not exhibit the required pattern. There is a chance that these bits will form a pattern similar to the pattern being sought. The synchronization pattern must therefore be chosen so that it is unlikely that it will occur by chance. If the 193rd bit were made to be always a one or always a zero, this could occur by chance in the voice signal.

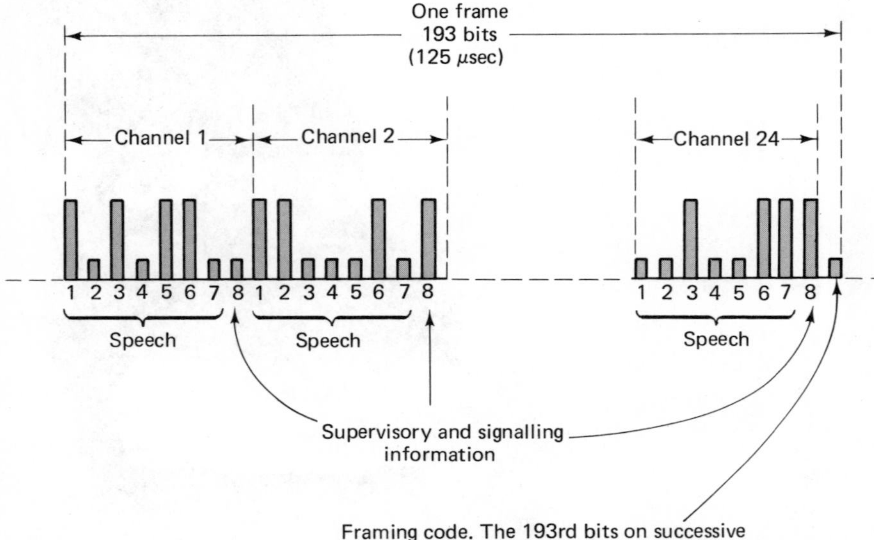

Figure 27.3 The bit structure of a North American PCM transmission link operating at 1.544 million bits per second (T1 carrier). The frame shown here is repeated 8000 times per second, thus giving 8000 samples per second on each of 24 channels, plus an 8000 bps bit stream for control signaling. The CCITT Recommendation for 1.544 bps PCM is slightly different (see Fig. 27.6) [2].

It was found that an alternating bit pattern, 0 1 0 1 0 1 . . . never occurs for long in any bit position. Such a pattern would imply a 4-KHz component in the signal, and the input filters used would not pass this. Therefore the 193rd bit transmitted is made alternately a one and a zero. The receiving terminal inspects it to ensure that this 1 0 1 0 1 0 . . . pattern is present. If it is not, then it examines the other bit positions that are 193 bits apart until a 1 0 1 0 1 0 . . . pattern is found. It then assumes that these are the framing pulses.

This scheme works very well with speech transmission. If synchronization is lost, the framing circuit takes 0.4 to 6 milliseconds to detect the fact. The time required to reframe will be about 50 milliseconds at worst if all the other 192 positions are examined; but normally the time will be much less, depending on how far out of synchronization it is. This is quite acceptable on a speech channel. It is more of a nuisance when data are sent over the channel and would necessitate the retransmission of blocks of data. Retransmission is required on most data transmission links, however, as a means of correcting errors that are caused by noise on the line and detected with error-detecting codes.

The permissible signal levels are not equally spaced in PCM encoding. The levels are bunched closer together at the lower signal amplitudes than at the higher ones. This gives better reproduction of low volume speech.

REGENERATIVE REPEATERS

The main reason why high bit rates can be achieved on wire-pair circuits using pulse code modulation is that repeaters are placed at frequent intervals to reconstruct the signal.

In most PCM systems working today the repeaters are placed at intervals of between 1 and 5 kilometers. The Bell T1 System, operational since 1962, uses repeaters at intervals of 1.8 kilometers, typically, which is the spacing of loading coils employed when the wires were used for analog transmission; the repeaters replace the loading coils. These repeaters reconstruct 1,544,000 pulses per second.

A regenerative repeater has to perform three functions, sometimes referred to as the 3 "Rs": reshaping, retiming, and regeneration. When a pulse arrives at the repeater, it is attenuated and distorted. It must first pass through a preamplifier and equalizer to *reshape* it for the detection process. A filter removes the DC component. A *timing* recovery circuit provides a signal to sample the pulse at the optimum point to decide whether it is a one or a zero bit. The timing circuit controls the *regeneration* of the outgoing pulse and ensures that it is sent at the correct time and is of the correct width.

As shown in Fig. 27.5 the pulses transmitted occupy half a time slot. A pulse represents a 1 bit, and absence of a pulse denotes a 0 bit. Each 1 transmitted has an opposite polarity to the previous 1. This concentrates the energy of the signal around 772 MHz rather than 1.544 MHz when a string of ones is

Figure 27.4 Repeaters for the Bell System T2 carrier which transmits 6.312 mb/s.

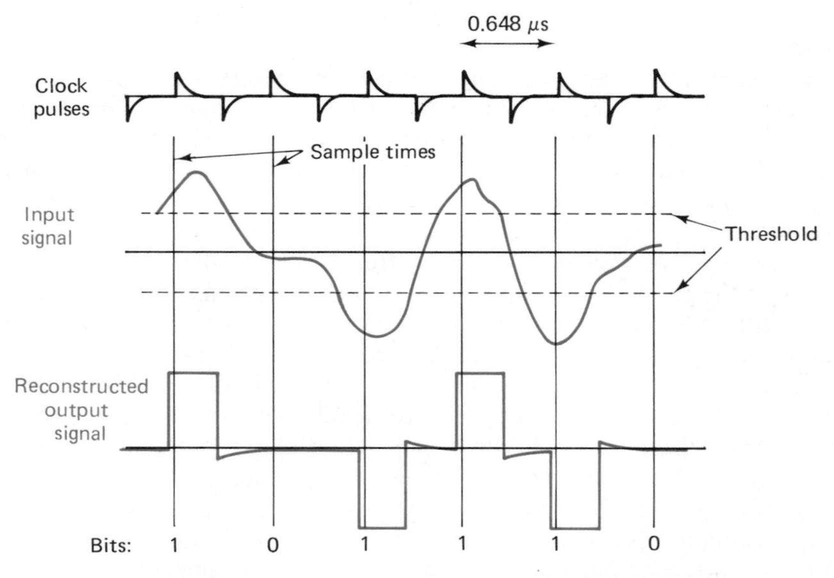

Figure 27.5 The repeaters, every 1.8 kilometers on a T1 circuit, reshape, retime, and regenerate the pulse stream.

transmitted. To avoid the DC component when a string of zeros is transmitted, a special code is substituted for each group of six zeros.

The T1 carrier uses a 3-level signal, as shown in Fig. 27.5, called a bipolar signal. Higher levels of digital carrier use a binary signal—a positive pulse for a 1 and no pulse for a 0. A binary signal is more efficient; a higher bit rate can be transmitted over a given bandwidth. The T1 carrier uses bipolar signaling to reduce margin against carrier in the decision circuit.

PCM cables usually transmit the power for the repeaters down the cable itself. A filter at each repeater separates the message signal from this DC current.

CROSSTALK

City and short-haul trunk cables often contain many hundreds of wire pairs. If all of these wire pairs transmit in the same direction they can all carry the 1.5 mb/s of the T1 carrier or the 3.2 mb/s of the T1C carrier, with the exception of a number of wire pairs reserved for fault location and order wires for communicating between manholes and central offices. A large wire-pair cable can thus be given a digital capacity of billions of bits per second. If however different wire pairs transmit in opposite directions there is a danger of crosstalk. The strong signal leaving a repeater may be next to a weak signal arriving, which has been attenuated by the length of the cable. The strong signal can interfere with the weak one. This problem is dealt with in one of three ways. First the cable may contain only a small proportion of digital wire-pairs and these are separated by normal analog telephone wire-pairs. Second the cable may be partitioned, an electrical shield separating the wires for each direction of transmission. Third, and best if it is possible, two cables may be used, one for each direction of transmission.

CCITT RECOMMENDATIONS

As mentioned, the CCITT has made recommendations for PCM systems for transmission at the T1 carrier speed of 1.544 mb/s, and for transmission at 2.045 mb/s [1, 2].

Figure 27.6 shows the CCITT recommendations for transmission at 1.544 mb/s. As is often the case, the CCITT recommendation is slightly different to the North American standard set by AT&T. Like AT&T, it employs a 193-bit frame with 8 bits per channel as in Fig. 27.3, but the frame alignment bit is the first bit, not the 193rd bit as in Fig. 27.3, and it carries a different synchronization pattern. Twelve such frames are grouped together to form one *multiframe*.

There are two versions of it. One has a common signaling channel associated with the block of 24 voice channels. The other has signaling associated with each voice channel. The common-channel signaling scheme is shown at the top of Fig. 27.6. The first bit of each frame serves two purposes. In odd frames it is used for maintaining synchronization, carrying a 1 0 1 0 1 0 . . .

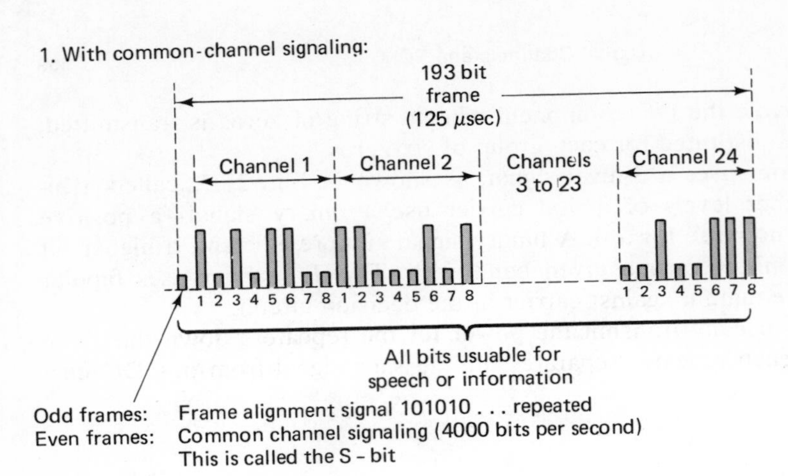

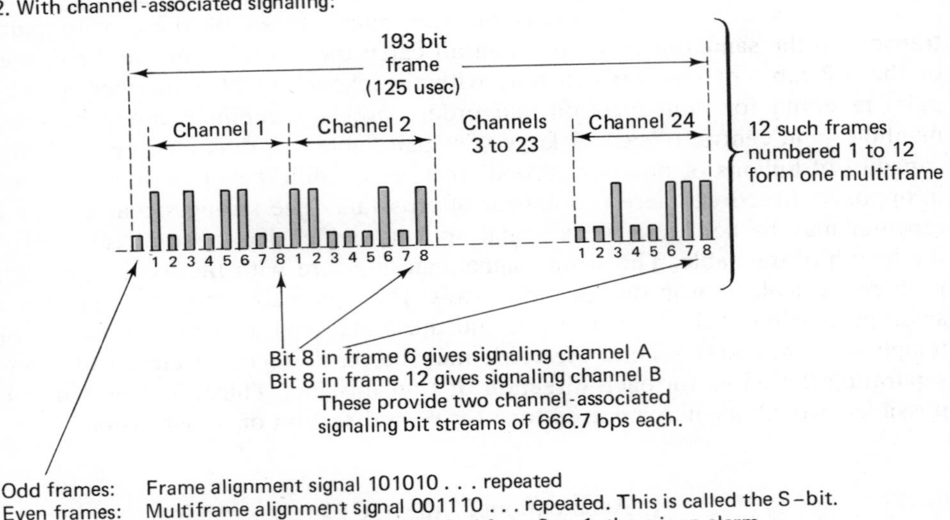

Figure 27.6 Signaling with PCM systems operating at 1.544 bps employing CCITT Recommendation No. G.733 [10].

pattern in successive frames (i.e., 1 in frame number 1, 0 in frame number 3, 1 in frame number 5 . . . and so on). In even frames it gives a bit stream (4000 bits per second because there are 8000 frames per second) which carries signaling information of the type necessary for controlling a telephone network, e.g., disconnect signals.

The version which allows channel associated signaling is shown at the bottom of Fig. 27.6. Here two signaling bit streams are associated with each channel. The frames are arranged into groups of 12, called a multiframe, and numbered 1 through 12. Bit 8 of each channel in frame 6 is reserved for signaling channel A. Bit 8 of each channel in frame 12 is reserved for signaling channel B. These bits each occur 666.7 times per second, hence each voice channel has two 666.7 bps signaling bit streams associated with it. The first bit of each frame is used for both frame and multiframe alignment.

With the first of these schemes the channels are composed of 8-bit words, as opposed to 7-bit words on the T1 carrier. They thus have 64,000 bits per second, as opposed to 56,000 bits per second on the T1 carrier. With the second scheme the words are also 8 bits, except that every sixth word in a channel has only 7 usable bits. The usable channel rate is therefore 62,666 bits per second.

Figure 27.7 shows the CCITT 2.048 mb/s recommendation, which most of the world outside North America and Japan is starting to use for PCM transmission. In this, 16 frames of 256 bits each form a multiframe. There are 32 8-bit time slots in each frame giving 30 speech channels of 64,000 bps each, plus one synchronization and alarm channel, and one signaling channel which is submultiplexed to give four 500 bps signaling channels for each speech channel.

The difference between the CCITT and North American standards will prevent the world becoming linked with digital channels on satellites and other media, without the need to convert from one system to another.

NATIONWIDE NETWORKS

The early PCM links were relatively short point-to-point connections between telephone offices, and were somewhat experimental in nature. The Bell T1 system grew into wide acceptance until much of the U.S.A.'s short-distance wire pairs trunks were converted to T1. It is now clear that the various digital facilities must link together to form a nationwide network in the United States and eventually a worldwide network.

When the T2 carrier came into use on the Bell System in 1972, it was designed for transmission up to 800 kilometers, but much attention had been paid in its design to eventually linking it into nationwide facilities. Except for very short distances, telephone calls are handled more economically by the T2 than by the T1 carrier. The T2 carrier, however, was designed to have an additional purpose—the trunking of Picturephone signals. Distortion on a Picturephone signal is much more harmful in its effect than on a telephone signal. For this reason Picturephone signals will not be sent long distances on analog trunks; digital trunks will be used. One Picturephone signal occupies one T2 bit stream. The T2 links were restricted to 800 kilometers. Therefore various techniques were designed to send high digital bit rates over existing analog channels—microwave and coaxial cable—to bridge the gap until nationwide digital trunking exists.

DATA TRANSMISSION

Whereas the need for nationwide Picturephone transmission is far from pressing as yet, nationwide data transmission over the digital channel is an ur-

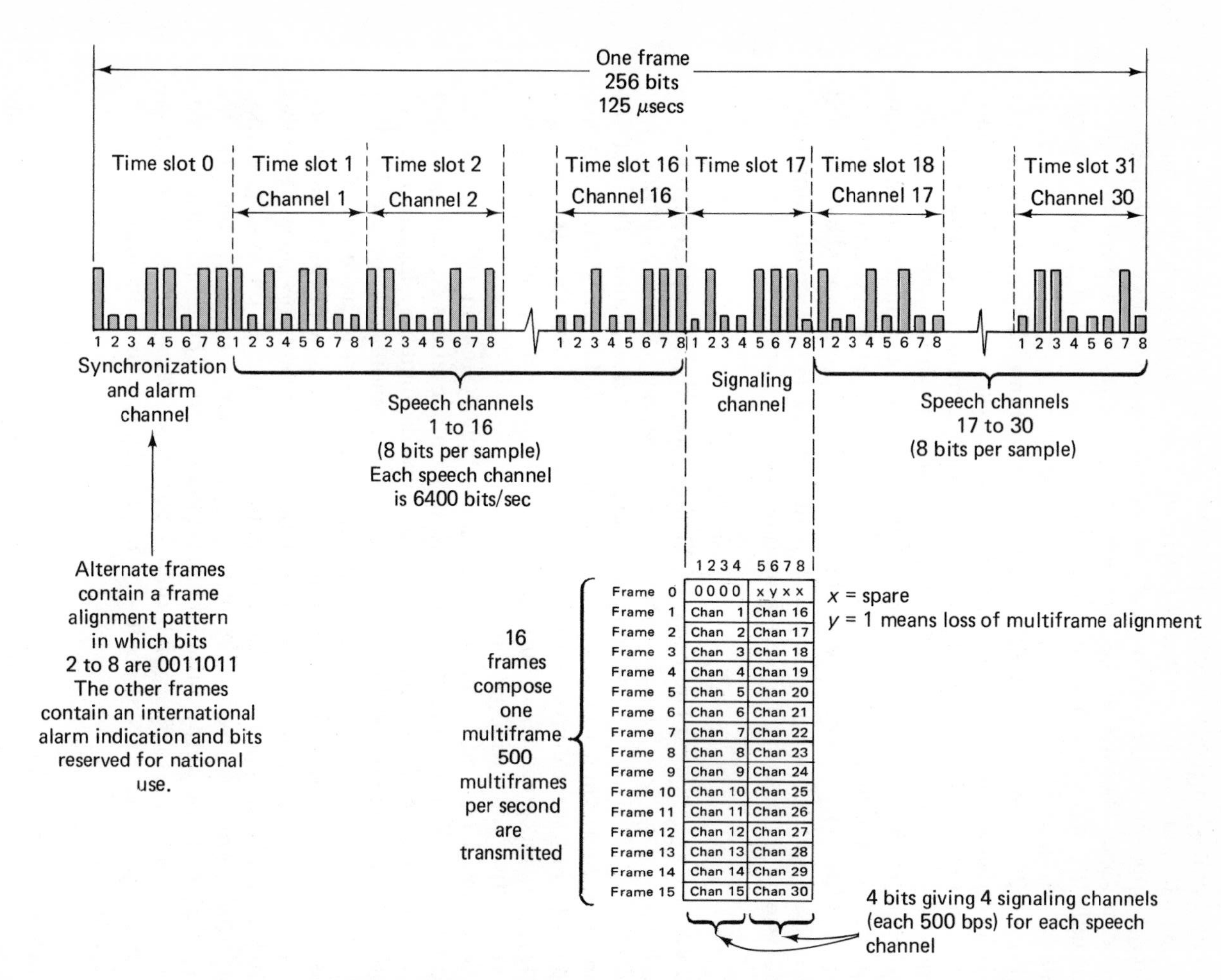

Figure 27.7 CCITT Recommendation for the structure of PCM channels for transmission at 2.048 million bits/seç [3]. 30 speech channels of 64,000 bps are derived, each with a signaling channel of 500 bps.

gent present-day requirement. It is far more efficient to send data over the PCM telephone channels than to convert it to analog form and send it over analog channels. AT&T has a service called DDS, the Dataphone Digital Service, in which data can be transmitted from subscriber to subscriber in direct digital form—i.e., no modems.

To provide an end-to-end data service several components are needed as well as the T1 and T2 trunks we have described. First, a data-carrying local loop must be established from the subscriber to his local central office. Second, if that central office does not yet have PCM trunks reaching it, the digital signal must be transmitted over other trunks. Third, to give a nationwide network the segments of data-carrying PCM system must be interconnected.

An economical means to transmit data over analog trunks has come into use on the Bell System, called *Data Under Voice, DUV*. As we have commented, analog voice channels are frequency-division multiplexed together to form *groups* which are commonly transmitted over a microwave link. A *mastergroup* for example consists of 600 voice channels, a *jumbogroup* of 3600. When these groups are sent over a microwave link there is a gap underneath them in the radio band which is transmitted. The gap is 564 KHz wide and nothing is transmitted in it except for a radio pilot which is used to indicate radio continuity (Fig. 27.8). The gap exists because the signal at the bottom edge of the band is too variable for good quality speech transmission. Data, however, can be transmitted in the gap with high accuracy.

When the T1 carrier bit stream is encoded as in Fig. 27.5, its spectrum is too wide to fit in the gap, as shown in Fig. 27.8. It is therefore recoded using a 7-level code. A data pilot is added, to monitor the signal, and the signal fits comfortably underneath the lowest group of telephone circuits. The radio continuity pilot has to be moved out of the way to a higher frequency.

Data Under Voice, DUV, made it possible to interconnect the digital carriers without building new physical links, and to build rapidly a nationwide data network. DUV will fill the gap until nationwide T4 links come into existence. Much of its potential value lies in the fact that microwave links go to most cities, including small ones.

DUV is designed to give a low error rate. A design objective is that on a 4000 mile connection better than 99.75% of all customer channel seconds shall be free of error. A 4000 mile connection will normally go over many different multihop radio systems. 16 radio systems would typically connect in tandem.

DUV and the T1 and T2 carriers made possible the AT&T Dataphone Digital Service (DDS) offering. Customers leased channels at speeds up to 56,000 b/s which are digital end-to-end, and hence require no modems. DDS could presumably handle data rates much higher than 56,000 b/s were it not for the limited capacity of the local telephone loops.

By putting digital repeaters on the local loop, somewhat like the T1 carrier, digital services could be expanded up to 1.544 mb/s. It is also possible to make digital services *switchable* in ways we discuss later.

Underneath the mastergroup transmitted by microwave are 564 KHz unused except for a pilot signal used to indicate radio continuity:

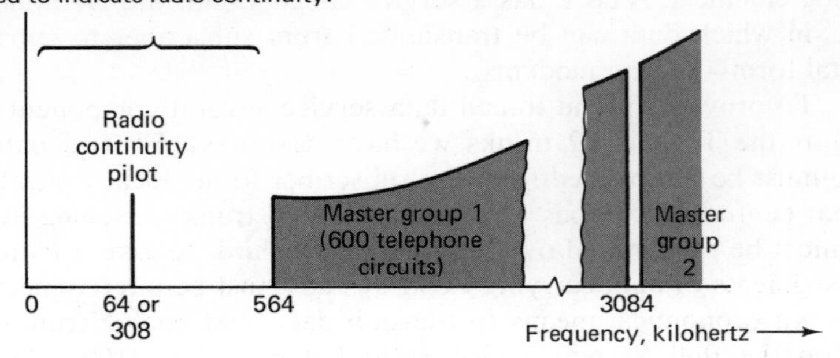

The baseband spectrum of data transmitted by the T1 carrier (1.544 mb/s) is as shown here:

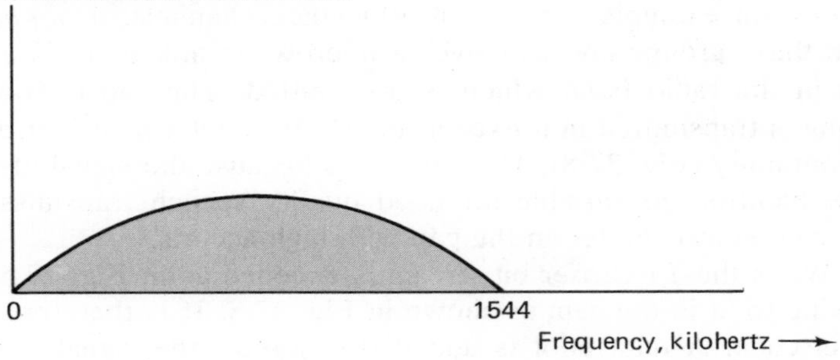

The data is compressed by converting it from the bipolar representation of Fig. 27.5 to a 7-level code. When this is done it fits underneath the mastergroup. The position of the radio continuity pilot has to be changed:

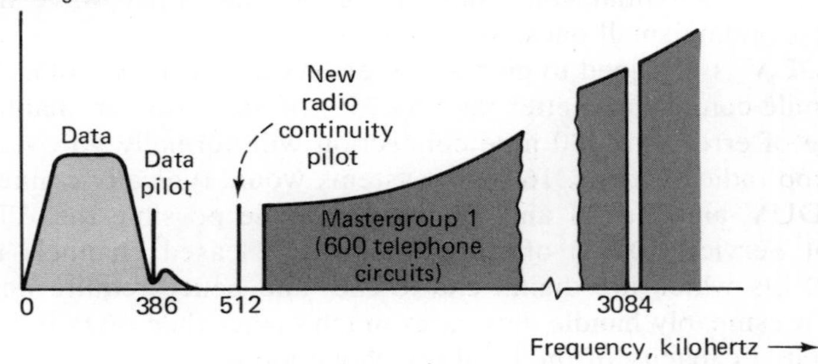

Figure 27.8 On the Bell System, segments of T1 carrier can be linked together nationwide by using DUV, Data Under Voice, on microwave channels.

Western Union, like AT&T, require nationwide digital facilities superimposing on an existing nationwide analog network. Substantially more than half of Western Union's traffic is digital in nature. Western Union are therefore modifying existing analog microwave routes to make them *hybrid microwave* carrying 6.3 mb/s in the lower part of their spectrum. On other microwave routes digital radio is being added, using the same towers and antennas, and operating initially at 20 mb/s. The latter approach is called *digital overbuild.* Western Union's WESTAR satellite transponders also transmit either analog groups or high-speed digital bit rates. Fig. 27.9 shows the Western Union digital hierarchy.

SYNCHRONIZATION

Synchronization is vital for digital transmission. It is essential for the receiving machine to know which bit is which. This is not too much of a problem on point-to-point lines. The Bell T1 carrier solved it by adding one extra bit per frame, thereby obtaining an 8000-bit-per-second signal that carries a distinctive pattern. If synchronization slips, then this pattern is searched for and synchronization can usually be restored in a few milliseconds.

This is fine so long as the channel bank remains intact. If, however, the bank were split up into its constituent channels and these channels were transmitted separately by pulse code modulation, then it would be advantageous to have a synchronization bit sequence for each channel rather than for the group of channels. The result would be one bit per character rather than one bit per frame, as in Fig. 27.3. One bit per character is already used for network control signaling, and the result is 8000 bits per second in this case. This rate seems far too much for any purposes that can be foreseen at the moment. Network signaling consists mainly of sending routing addresses (the number you dial) and disconnect signals (when you replace your receiver). Therefore it has been suggested that this bit position should be shared between the network control signaling function and the synchronization function.

Synchronization becomes a much more difficult problem when a large switched network is considered. Signals are transmitted long distances over different and variable media, and their time scale must inevitably differ slightly even if the transmitting locations are synchronized. This problem must be solved before a nationwide PCM network can be established. There are two types of solutions. The first is a *fully synchronous* approach in which an attempt is made to synchronize the clocks of the different switching offices and to compensate for any drift in synchronization of the information transmitted. The clocks of the different offices in the network must all operate at exactly the same speed. The second is a *quasi-synchronous* approach in which close but not perfect clock correlation is accepted, and the multiplexing operations must be designed to cope with the imperfections. For the time being the latter approach is the practicable one.

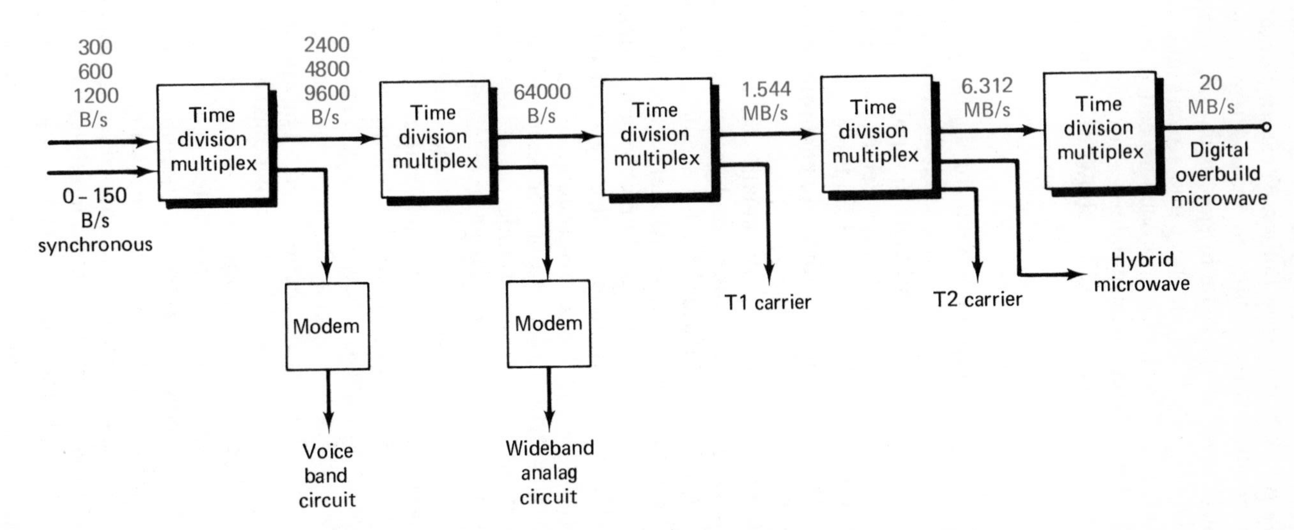

Figure 27.9 The Western Union hierarchy of digital channels.

Two bit streams, then, which are to be multiplexed together on a T1 or higher level channel, have very slightly different bit rates. To achieve the multiplexing, more pulses are available on the outgoing line than on the incoming lines. The excess time slots are filled with dummy pulses. The presence of the dummy pulses is signaled, and they are removed at the receiving end. This technique is referred to as *pulse stuffing*. The more accurate the clocks the less pulse stuffing is required. As the design of the digital hierarchy evolves the accuracy of the clocks employed may increase and the degree of pulse stuffing needed may decrease.

AT&T'S T4M SYSTEM

In most cities the ducts beneath the streets for telecommunications channels are almost full, and more capacity is needed. Digging up the streets to lay down larger ducts is extremely expensive. A more attractive option is to replace some of today's cables with digital cables which carry a higher traffic volume.

The Bell System T4M coaxial cable system comes in two versions, one designed to fit into existing 3½-inch ducts and the other to fit into the newer 4-inch ducts. The former has 18 coaxial tubes per cable, and the latter has 22 tubes per cable. The cables are the same as those used on today's analog long-haul systems, but electronics are entirely different. Each tube can transmit 274 mb/s, carrying 4032 voice channels. Two tubes in a cable are spare and can be automatically switched into operation if a failure occurs on another tube. Of the remaining tubes half transmit in one direction and half in the other. Thus the smaller T4M cable transmits a total of 2.192 billion b/s in each direction, carrying up to 32,256 two-way telephone conversations; the larger transmits a total of 2.740 billion b/s carrying up to 40,320 two-way telephone conversations.

The central conductor of each coaxial tube carries the DC current which is used to power the regenerators until the next maintenance office is reached. The span between such offices can be up to 180 kilometers.

In addition to occupying a dedicated cable the T4M bit streams are sometimes sent over certain coaxial tubes in cables which also carry other types of signals. Some composite cables, for example, have 8 coaxial tubes and 750 pairs of wires.

Fig. 27.10 illustrates the T4M carrier.

EVOLUTION

An enormous amount of capital is tied up in national analog telephone networks. So much money is involved that they cannot be converted to all-digital networks in a few years. The proportion of digital links will grow slowly. They will be installed first where they can be most profitable, or where the pressure for extra circuits is greatest, as in crowded urban and suburban areas. The T4M system was first installed in the New York to Newark area and was pressed into service early after a cata-

Figure 27.10 Bell System craftsmen connecting a section of the T4M digital coaxial system in New York. Each coaxial tube in the cable carries 274 million bits per second. The mass of wire pairs seen on the left of the photograph can each carry 1.544 million bits per second when they are employed as links of the T1 system. Such digital transmission systems will spread fast in the years ahead.(*Courtesy AT&T.*)

strophic fire in 1975. It had to be tailored to the available cable ducts and traffic requirements.

Figure 27.11 shows how PCM lines might be used in today's typical urban environment. It presents a highly simplified picture of part of the telephone network in London. The top half of the picture shows interconnections between local, tandem, and subtandem exchanges. The bottom half shows how the network could be simplified with the use of PCM links. The simplification shown by the illustration would have appeared far more striking if it had been drawn with the 200 or so local exchanges that exist rather than with the small number in the diagram.

The saving would be much greater, again, if PCM concentrators were used also, as will be described in the next chapter.

ADVANTAGES OF PCM

As noted earlier, PCM can give lower costs per telephone channel on short-haul lines and can greatly multiply the utilization of lines within city areas. Two trends will widen the range of economic application: the decrease in bandwidth cost due to the introduction of higher-capacity channels and the decrease in cost of logic circuitry, which is likely to be great when large-scale integration is fully developed.

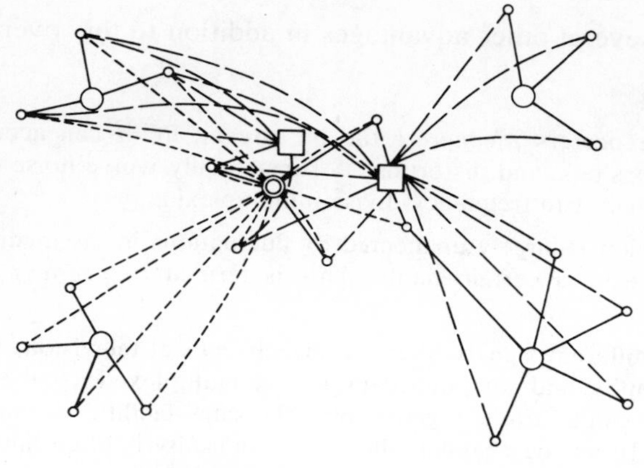

o Local exchange
◯ Subtandem exchange
◎ Tandem exchange
□ Toll/trunk outlet
——— Junction to subtandem
- - - - - Junction to tandem
— — — Junction to toll/trunk outlet

A highly simplified diagram of the present London step-by-step tandem network catering for subscriber trunk dialing (direct distance dialing in American parlance). Only a limited number of exchanges and routings are shown, in order to lessen the complexity of the figure. All direct junctions between local exchanges have been omitted.

With integrated PCM between exchanges, the routing might be simplified as follows:

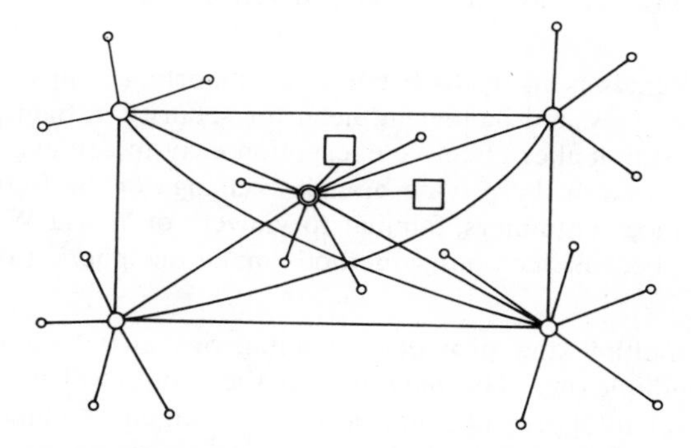

If the drawing had not been so highly simplified, the saving would have appeared much greater.

Figure 27.11 Redrawn with permission from *Techniques of Pulse Code Modulation in Modification Networks,* by C. C. Hartley, P. Morret, F. Ralph, and D. J. Tarran. Cambridge University Press, Cambridge, 1967.

PCM has several other advantages in addition to this overriding economic argument:

1. Because the repeaters *regenerate* the bit stream, PCM can accept high levels of line noise, cross talk, and distortion. A substantially worse noise-to-signal ratio can be accepted than with frequency-division multiplexing.
2. The transmission is largely unaffected by fluctuations in the medium, provided that they do not exceed certain limits. This is termed *ruggedness* in a transmission system.
3. All types of different signals, such as speech, music, television, Picturephone, facsimile, telegraphy, and computer data, will be multiplexed together in a digital form. These signals can all travel together over the same facilities without interfering with one another. In analog channels the system capacity is often limited by mutual interference between different types of signals.
4. Much higher data rates can be achieved than with analog transmission. This factor will become increasingly important economically as the use of computers and terminals increases. Data transmission is increasing much faster than other forms of transmission. Unless held back by unsuitable or expensive transmission facilities, it will continue to rise probably at an increasing rate. Most common-carrier predictions of this market seem gross underestimates. There will be little relationship between present-day telegraphy and the future uses of data transmission.
5. Some future transmission media, such as optical fibers transmitting laser pulses may be inherently digital in nature.
6. Some future satellite systems may be dominated by the use of demand-assigned multiple-access. Computerized earth stations handling time-division multiplexed bit streams will probably give the most economic use of satellites.
7. Many bits are available for network control signals. The possibilities for signaling and remote control of the network are therefore much greater than with today's analog plant.
8. Encryption of signals is likely to become an important subject with the increasing concern about privacy and increasing need for security in data processing systems. Digital transmission makes effective encryption easy to achieve. Encrypting devices may perhaps be used in the private branch exchange of the future, under computer control. The analog scramblers, familiar to viewers of World War II movies, are of little use today because computer methods make deciphering of the scrambled signals easy.
9. Time-division multiplexing provides advantageous *switching* methods as well as *transmission* multiplexing. The networks of the future will integrate switching and transmission technology, both, in part, using digital techniques. Time-division switching costs will be lower if the transmission also uses time-division multiplexing.
10. Concentrators can substantially lower the cost of the local distribution network. As will be discussed, digital techniques provide a way to build inexpensive concentrators. These devices could be used in large numbers in future telecommunication networks.

Figure 27.12 There have been cases of telephone eavesdropping and recording done on a large scale by intercepting microwave radio signals, and also by wiretapping. To make telephone transmission secure it is necessary to digitize the signals and use cryptography. (*Photo: AT&T.*)

Ironically it is likely that one of the most important reasons for installing PCM channels now is that PCM will not be used on these channels in the future. There are more efficient methods of transmitting calls digitally than PCM, which could *greatly* increase the call-carrying capacity of the digital channels which are being installed today. This is the subject of the following chapter.

REFERENCES

1. *CCITT Recommendation A.733* (on PCM multiplex equipment operating at 1544 KB/S), *Green Book,* Vol. III: Line Transmission, International Telecommunications Union, Geneva, 1973.
2. *CCITT Recommendation A.732* (on PCM multiplex equipment operating at 2048 KB/S), *Green Book,* Vol. III: Line Transmission, International Telecommunications Union, Geneva, 1973.
3. K. E. Fultz and D. B. Penick, "The T1 Carrier System," *Bell System Technical Journal,* 44 (Sept. 1965), pp. 1405–1452.
4. J. F. Travis and R. E. Yeager, "Wideband Data on T1 Carrier," *Bell System Technical Journal,* 44 (Oct. 1965), pp. 1567–1604.
5. J. H. Bobsin and L. E. Forman, "The T-2 Digital Line," *Bell Laboratories Record,* 51 (Sept. 1973).

Three-stage de Forest audion amplifier of the type first built in 1912 by the Federal Telegraph Company. This is the earliest known commercial cascade audio-frequency amplifier. It had a gain of 120. (*Courtesy ITT.*)

28 SIGNAL COMPRESSION

While bandwidths and bit rates increase in future, the signals we send over them will shrink. Complex techniques are being made to work for reducing the bandwidth or bit rates needed to transmit speech, television, and other signals. Such techniques will be translated into mass-producible microcircuits.

Digital channels such as those we have described are being introduced at a rapid rate into the telephone networks of many countries. By the early 1980s the Bell System will have more than a hundred million channel miles of digital trunks. Almost all of the world's telephone administrations are using PCM encoding of voice, so that a telephone conversation requires a channel of 64,000 b/s in each direction. There are, however, methods of encoding telephone conversations into a much smaller number of bits. These techniques offer the potential of increasing the capacity of the digital trunks by a factor of 4 or possibly much higher.

To encode speech into fewer bits than the PCM encoding now in such widespread use, elaborate encoding methods are needed. Such encoding would have been excessively expensive when the early PCM systems came into use. Now, however, complex coding methods lend themselves to LSI mass production. Video telephone, music, television, and facsimile signals can also be handled with fewer bits by employing more complex *codecs*.

A major thrust in the next five or ten years will be improvement of codec efficiency. From 1955 to 1965 the improvement of *modems* brought an order of magnitude increase in the data rate which could be sent over an analog telephone line. 1975 to 1985 will probably bring an order of magnitude increase in the number of telephone calls that could be sent over the world's digital channels.

PCM sampling 8000 times per second can reproduce any frequency up to 4000 Hz. However most of the energy in speech is at frequencies well below 4000 Hz; most of it is below 1000 Hz. The signal does not change too fast.

Consequently, rather than encoding the absolute value of each sample, the difference in value between a sample and the previous sample may be encoded. This technique is called *differential pulse code modulation, DPCM*. Fewer bits are needed per sample.

In practice, differential PCM encodes the difference between the amplitude of the current sample and a *predicted* amplitude, estimated from past samples. A DPCM circuit typically employs the last three speech samples to make a guess of what the next sample will be. The error in the guess can be encoded in about 5 bits rather than the 7 bits of conventional PCM. Hence $5 \times 8000 =$ 40,000 bits per second are needed instead of the 56,000 of conventional PCM.

A lower bit rate could be achieved if the guess were better. Elaborate schemes have been devised for improving the prediction based upon measured characteristics of the speech. Different voices, for example, have a different pitch—the fundamental frequency of vibration of the vocal chords. The shape of the voice waveform tends to repeat itself at the pitch-frequency interval. The waveform which tends to repeat at this interval depends upon the shape and position of the palate, tongue, and other speech articulators. A delay circuit with a delay equal to the above interval is used in some predictive circuits. The delay and other parameters which affect the prediction are updated continually This form of encoding is known as *adaptive predictive encoding,* or *differential PCM with an adaptive predictor*. Bit rates of 20,000 have been used with experimental adaptive circuits.

DELTA MODULATION

A technique called *delta modulation* also encodes signal *differences* but uses only one bit for each sample. The encoding indicates whether the waveform amplitude increases or decreases at the sampling instant. This binary sampling is illustrated in Fig. 28.1. The sample voltage is compared, each time, with a voltage obtained by integrating the previous samples. If it has gone up, a 1 bit is transmitted, if not a 0 bit.

The number of pulses needed for this form of encoding depends on the rate of change of the signal amplitude. If the peak amplitudes are of low frequency and if the high-frequency components are of low amplitude, fewer bits will be needed for the encoding than if all frequency components are of the same amplitude. This is the case with speech. Figure 28.2 shows a typical speech spectrum. It is also the case with television and Picturephone signals. In a 1-MHz Picturephone signal, most of the energy is concentrated below 50 KHz.

Overloading with this type of modulation will come not from too great a signal amplitude, but from too great a rate of change. The encoding on the right-hand side of Fig. 28.1 is barely keeping up with the signal change.

Another variation on this scheme uses two, three, or four bits per sample which permits four, eight, and sixteen gradations of signal change to be recorded each sample time, and then uses fewer samples. The levels are not linearly spaced.

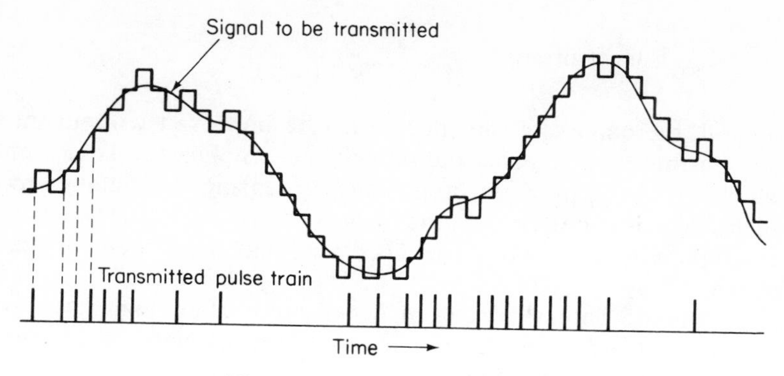

Figure 28.1 Delta modulation.

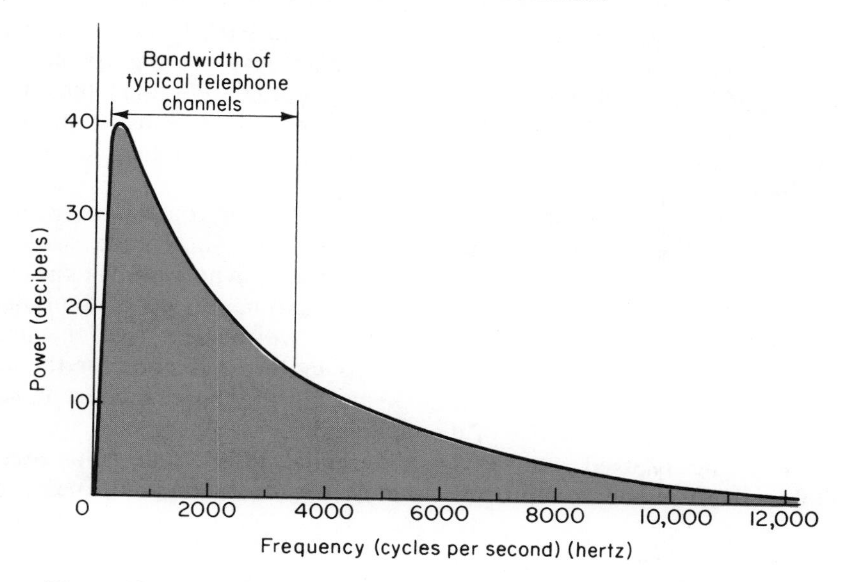

Figure 28.2 The spectrum of human voice. The high amplitudes are at low frequencies. This makes it particularly suitable for transmission by *delta modulation.*

Just as differential PCM can be made *adaptive,* so adaptive techniques can be used for delta modulation. Relatively inexpensive delta modulation circuits are in use which encode speech with good quality into 32,000 bits per second. This figure will soon drop to 24,000 and possibly to 16,000 or thereabouts.

COMPANDORS

Quantizing noise is heard as raspy quality in the speech received. The fewer the number of levels used in encoding a given signal, the worse will be the quantizing noise. If the sampling levels were fixed and equally spaced, a soft telephone voice would have worse quantizing noise than a loud voice. The quantizing noise would be greater relative to the signal strength. This can be overcome by using more closely spaced levels for a soft voice than for a loud voice.

In general, the quality of the speech can be improved without increasing the requisite bit rate by increasing the number of sampling levels for those amplitude values which occur most often, and decreasing the number for those values which occur least often. In other words, there is a variable spacing between the sampling levels, with the levels closer together where they are used the most often.

Small amplitude values of speech occur more often than large ones. A scheme called *companding* therefore increases the spacing of the quantizing levels for the stronger signal values.

A *compandor* is a device that, in effect, compresses the higher-amplitude parts of a signal before modulation and expands them back to normal again after demodulation. Preferential treatment is therefore given to the weaker parts of a signal. The weaker signals traverse more quantum steps than they would do otherwise, and hence the quantizing error is less. This is done at the expense of the higher-amplitude parts of the signal, for the latter cover fewer quantum steps.

The process is illustrated in Fig. 28.3. The effect of companding to move the possible sampling levels closer together at the lower-amplitude signal values, is sketched on the right-hand side of the figure, which shows the quantizing of a weak signal and a strong signal. The right-hand side of the diagram is with companding, the left-hand side without. It will be seen that on the left-hand side the ratio of signal strength to quantizing error is poor for the weak signal. The ratio is better on the right-hand side. Furthermore, the strong signal is not impaired greatly by the use of the compandor.

Companding is used with PCM, differential PCM, and delta modulation. Different telephone administrations must use the same companding rules

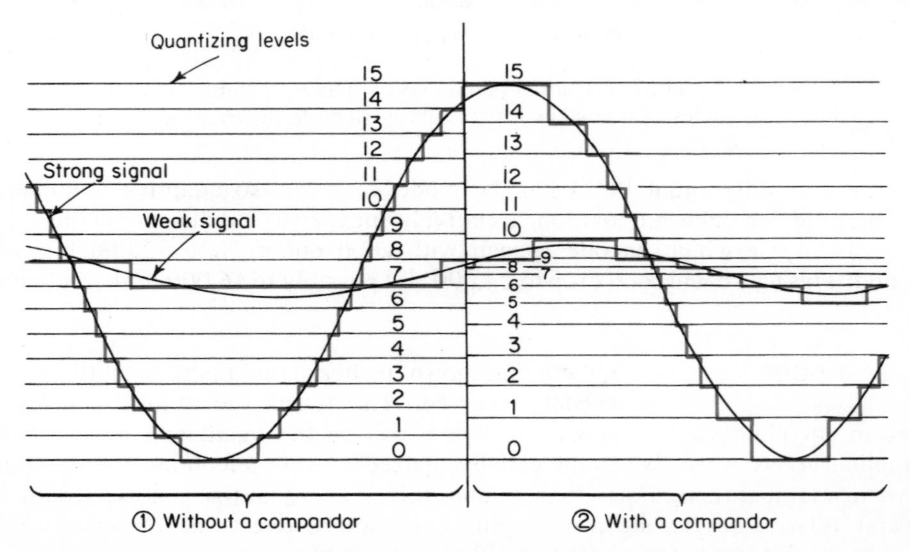

Figure 28.3 With a compandor the quantization of the weak signal gives more separate values, and therefore a better ratio of signal to quantizing noise. Standard sets of companding rules are specified in the CCITT recommendation G.711.

if they are to interchange signals. Two CCITT standards exist for the spacing of the quantizing levels of compandors, giving two alternative sets of companding rules known as A-law and μ-law companding [1].

A CHEAP 4800 B/S CODEC

Coding mechanisms which produce a high degree of speech compaction and little distortion are expensive. It is possible, however, to produce inexpensive codecs which achieve the same high compaction at the expense of speech quality.

One such device uses a simple mechanism for encoding speech into 4800 b/s [4]. It uses PCM encoding with only one bit per sample. If this technique were used alone the result would be terrible because no variation in loudness would occur. The pauses in speech would become noise as loud as the speech itself. Consequently a separate digital signal encodes the loudness. The loudness envelope of speech is the second illustration in Fig. 28.4.

The device samples of the speech at 4400 samples per second (meaning that frequencies above 2200 Hz are ignored). The samples are transmitted as a 4400 b/s bit stream. The loudness of the speech is encoded as a bit stream of 400 b/s. This is done by filtering the speech to obtain its envelope and encoding the envelope using companded delta modulation. The 4400 b/s and 400 b/s streams are multiplexed together to form a bit stream of 4800 b/s. Every 12th bit in this stream is the loudness encoding. No synchronization bits are transmitted. The receiving device can tell which is the loudness bit stream: it is the only bit stream of the twelve which does not approximately equal numbers of ones and zeros over a period of 100 milliseconds. Such a device could be built onto one LSI chip.

The speech quality is unpleasantly granular but it is a usable communications channel. The words of a clear speaker are intelligible. Such a device could form a useful addition to 4800 b/s modems. It could be used where spectrum space is very short as in mobile radio applications, or where the user wishes to derive cheap long-distance voice channels from a digital medium such as a satellite channel. More complex encoding, however, can achieve much better speech quality with the same bit rate.

VOCODERS

It is important to distinguish between speech encoding mechanisms which attempt to achieve a speech quality comparable with that we are familiar with on the telephone, and encoding which merely attempts to convey the meaning of the words spoken. In the latter case voice distortion is permissible provided that semantic clarity is preserved.

The above 4800 bps PCM codec is designed to convey intelligence but make no attempt at telephone-quality speech. Another important class of en-

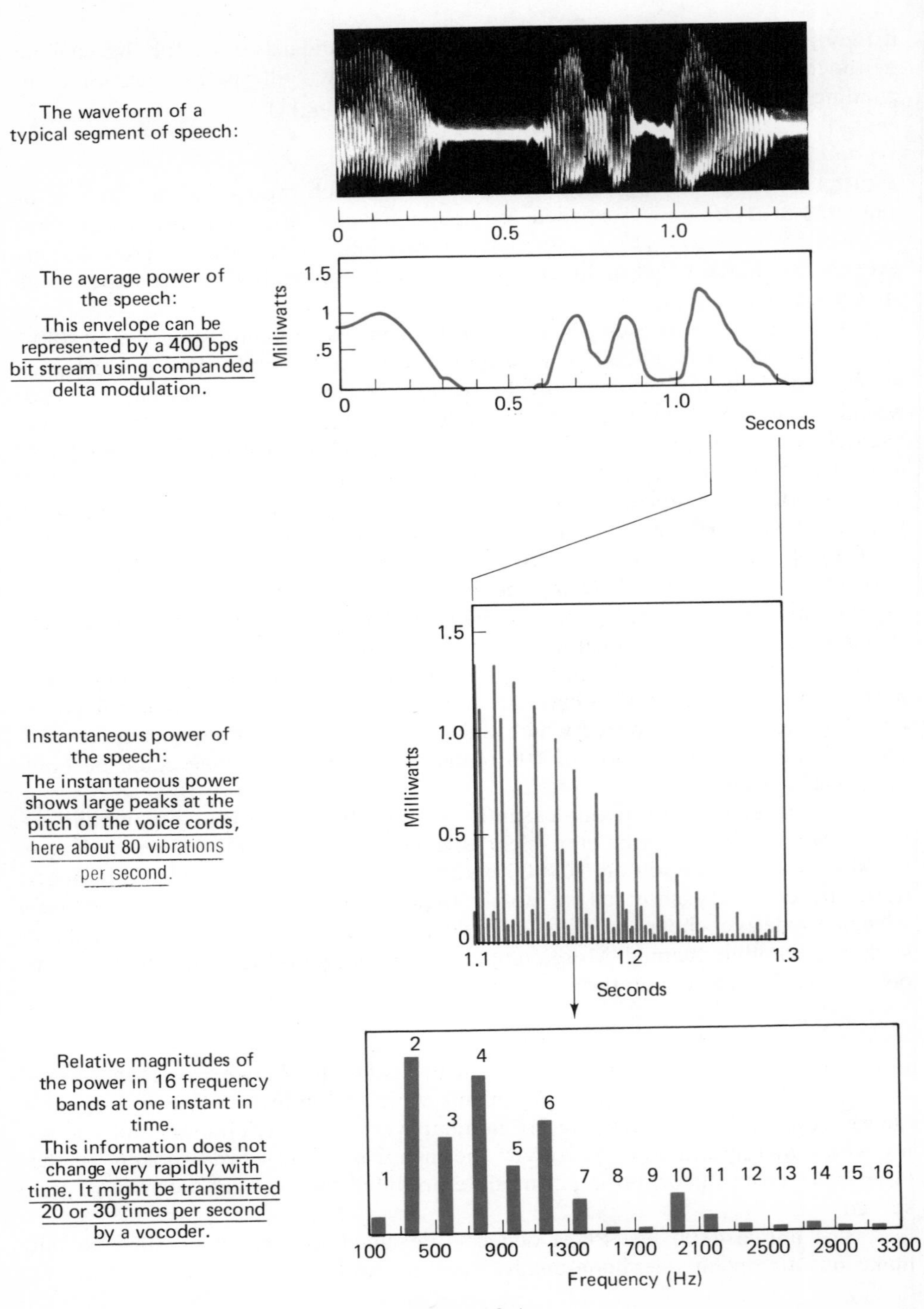

The waveform of a typical segment of speech:
0
0.5
1.0
The average power of the speech:
This envelope can be represented by a 400 bps bit stream using companded delta modulation.
Milliwatts
1.5
1
.5
0
0
0.5
1.0
Seconds
Instantaneous power of the speech:
The instantaneous power shows large peaks at the pitch of the voice cords, here about 80 vibrations per second.
Milliwatts
1.5
1.0
0.5
0
1.1
1.2
1.3
Seconds
Relative magnitudes of the power in 16 frequency bands at one instant in time.
This information does not change very rapidly with time. It might be transmitted 20 or 30 times per second by a vocoder.
1 2 3 4 5 6 7 8 9 10 11 12 13 14 15 16
100 500 900 1300 1700 2100 2500 2900 3300
Frequency (Hz)

Figure 28.4

coder which tries to produce semantic clarity but not copy the telephone voice precisely is referred to as the *vocoder*.

A vocoder transmits enough information for a voice to be *synthesized*, without attempting to preserve the original voice waveform. Enough significant parameters of the waveform are transmitted for artificial speech to be created by the receiving circuit. Vocoder designers are thus concerned with the human *perception* of speech rather than with recreating an exact waveform. In some cases vocoder speech does not sound much like the originating speaker; in some cases it sounds zombie-like—not quite human. But it conveys the words intelligibly. It is therefore unsuitable, at least so far, for the public telephone network, but has the possibility of saving money in the future on corporate or military networks, and could expand the use of the radio spectrum for mobile telephones.

There are a variety of forms of vocoder depending upon which parameters are used to synthesize speech. A common form is referred to as the *channel vocoder* and is illustrated in Fig. 28.5. A channel vocoder sends three types of information. First it sends information about the *pitch* of the vocal cords. This fundamental frequency can be seen in the top and third diagrams of Fig. 28.4 as the higher power lines which occur about every 0.0125 second—a frequency of about 80 Hz.

Second the vocoder indicates when the speech is "voiced" and "nonvoiced." "Voiced" means that the vocal cords are in operation. In a typical second of speech there are a few bursts of vocal cord operation, illustrated in the top diagram. Between these bursts there are sounds vital for speech intelligibility, which are lower in power and generally higher in frequency. During the voiced portions of speech the higher frequencies tend to be superimposed upon the vocal cord pitch—80 Hz in the illustration. During the nonvoiced portions the higher frequencies tend to be imposed upon a hiss, like Gaussian noise, with no harmonic structure. This sound is usually produced by a turbulence around a constriction in the mouth or throat, or created by the tongue. There is thus a major difference between the voiced and nonvoiced portions of the speech.

The third piece of information the channel vocoder transmits relates to the frequencies other than the fundamental pitch. The frequency range is divided in smaller bands of frequency and the energy in each of these bands is sampled. The illustration at the bottom of Fig. 28.4 shows 16 such bands each of 200 Hz. A vocoder may or may not use equally spaced bands. The information in this bottom diagram does not change rapidly. It may be transmitted by a vocoder 20 or 30 times per second. Using DPCM, this can utilize 2000 to 4000 bps.

The speech is synthesized from this transmitted information as shown in the right hand side of Fig. 28.5. Two initial sources of sound are used, a generator of hiss (Gaussian noise) and a generator of a buzzing sound consisting of pulses corresponding to the pitch of the vocal cords. At any instant one of the

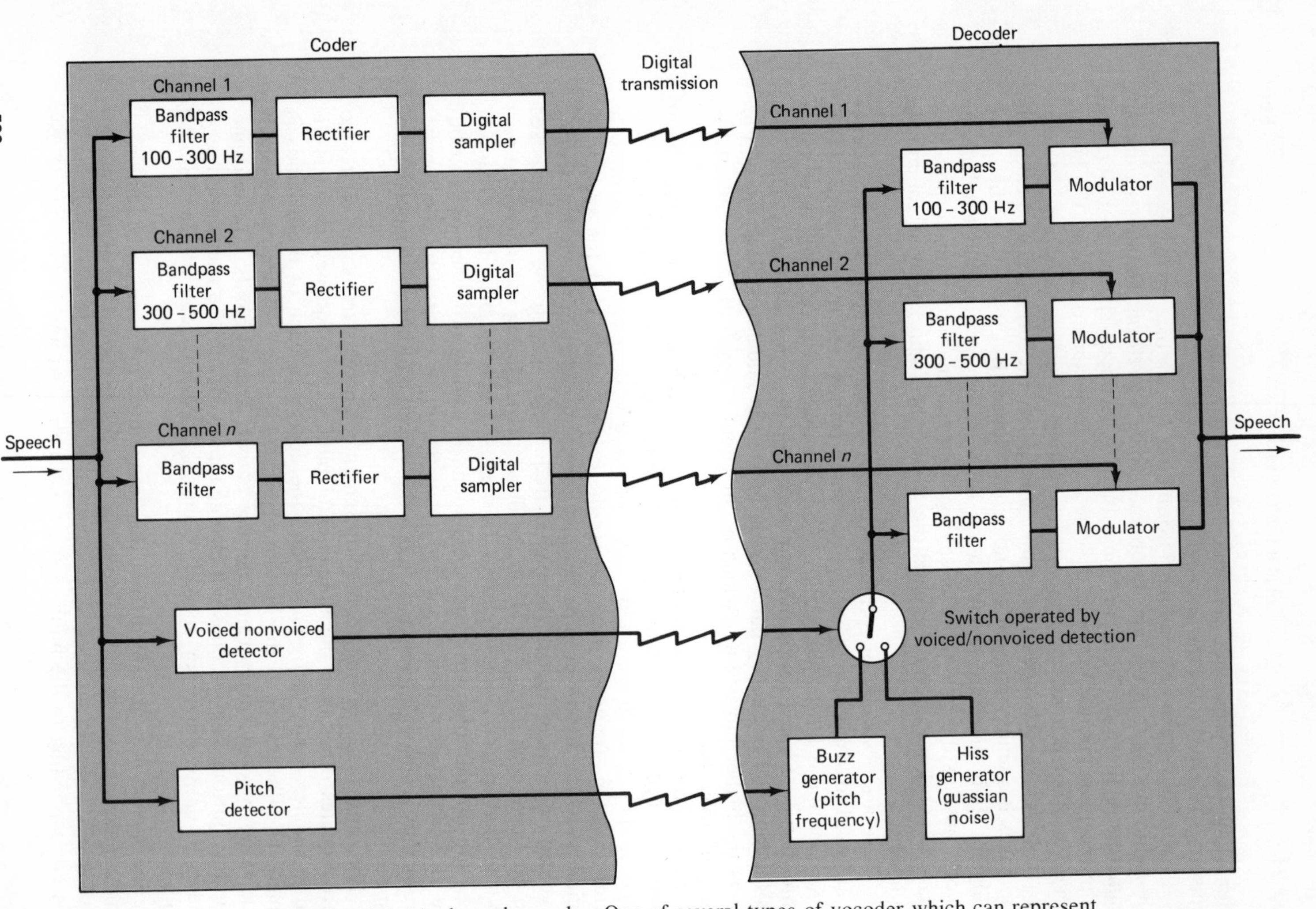

Figure 28.5 A channel vocoder. One of several types of vocoder which can represent speech in a few thousand bits per second.

sources is used. A switch switches on the buzz during *voiced* periods of speech, and the hiss during *nonvoiced* periods. The hiss or buzz is then split into 16 channels, each of which is modulated with the corresponding signals from the 16 frequency bands transmitted. The result is speech which has high intelligibility though the voice sounds somewhat unnatural. (*Conventional* telephone speech sounds somewhat unnatural but we have become used to that.)

This type of vocoder has been operated between 2400 and 4800 bps. Another form of vocoder transmits the frequency peaks in speech, known as *formants,* and this has been made to operate intelligibly at 600 bps.

MORE NATURAL VOCODERS

Because of the great potential value, a major effort is being made to design vocoders which produce more natural sound.

Some designers have used a hybrid of PCM transmission and channel vocoding. The low frequencies, say from 200 to 1000 Hz are transmitted complete in DPCM form. The higher frequencies are split into bands and are sampled 20 or 30 times per second as with a channel vocoder. The lower frequencies are then combined with the samples to synthesize the higher speech frequencies. This system, called a *voice-excited vocoder,* typically uses 7200 to 9600 bps and gives better quality than the various forms of *pitch-excited vocoder* such as the channel vocoder of Fig. 28.5.

LINEAR-PREDICTIVE CODING

The vocoders we have described extract information from the frequency spectrum of speech. A technique which appears very promising extracts information from speech in the form of a digital pulse stream, but unlike PCM or delta modulation does attempt to transmit the complete waveform.

This technique, called *linear-predictive coding,* produces an error signal giving the difference between an actual pulse value and a value predicted by using a number of previous samples. The weights used in the predictive coding are recalculated continuously as the speech statistics vary. Unlike schemes which attempt to transmit the complete waveform, linear-predictive coders extract only certain types of information from the error signal, including a gain factor, pitch information, and information about whether the speech is voiced or unvoiced. Far fewer bits are needed than for transmitting the entire error signal.

By using the error signal the speech tends to be encoded more accurately when it is changing fastest. This is not the case with frequency vocoders. Linear-predictive vocoders can therefore give more natural-sounding speech. Fairly natural-sounding speech may be transmitted in 10,000 bps or less, but fast and elaborate circuits are needed to achieve that.

Linear-predictive coders may become the means of multiplying the capacity of digital telephone lines by four or more. *The existence of such techniques greatly strengthens the arguments for building digital transmission facilities rather than analog.*

VIDEO SIGNAL COMPACTION

Differential PCM and delta modulation are valuable for the encoding of video signals as well as speech.

A 1-MHz Picturephone signal is carried by a bit stream of 6.3 million bits per second. The sampling rate must be 2 million samples per second (twice the maximum frequency, as discussed in Chapter 4). If standard PCM encoding is used, this gives 3 bits per sample. Hence eight discrete amplitude levels can be reconstructed. This scheme results in a grainy picture. Eight levels are not enough to avoid grain. Color television is transmitted using 10 bits per sample.

To overcome this problem differential encoding is used, as is illustrated in Fig. 28.6. A feedback loop is employed and a signal which is delayed by a small amount is subtracted from the incoming signal. This produces a difference signal, as seen at the bottom of Fig. 28.6. When the object in the camera view is still, the incoming signal is not changing, and the difference signal is zero. For slight movement in front of the camera, the difference signal is small. For rapid movement, the difference signal is large. Most of the time the Picturephone image changes slowly, and so the eight sampling levels are bunched together around zero amplitude, as shown. Fewer levels are used for larger difference signals.

Commercial codecs are available for compressing color television signals into less than half the number of bits needed with straight PCM. Tight differential encoding of television can be overloaded when the image is moving too fast. When the astronauts moved too fast on the television pictures from the moon, a blur followed them across the screen because of the digital encoding that was used.

Nevertheless it can be demonstrated that when the brain perceives motion in an image it does not require as much detail in the image as when the image is still. Accuracy of outlines loses its significance on moving objects and the mental processes of the viewer fill in the detail. We therefore have a situation which can be used in image compaction. When a portion of the image is still it has little difference from one frame to the next, so encoding only the difference will require relatively few bits. When a portion of the image is moving, the brain tends to fill in the details and will accept an undetailed image. One way or the other we can cut down on the number of bits needed.

The points which compose a video picture are referred to as *picture elements,* usually called *pels.* If a picture element is for a moving portion of the image it must represent the difference between frames, but not much accuracy or resolution is needed. If a picture element is for a nonmoving portion of the image it must represent the image accurately but can do so by making refer-

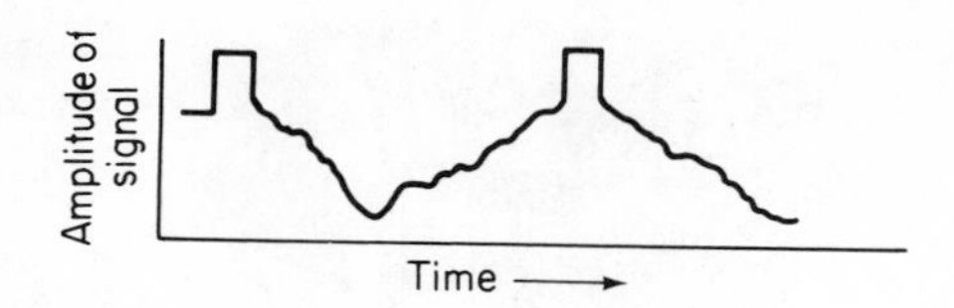

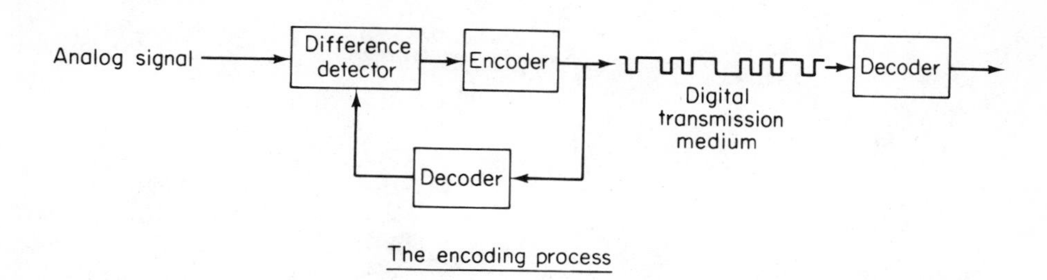

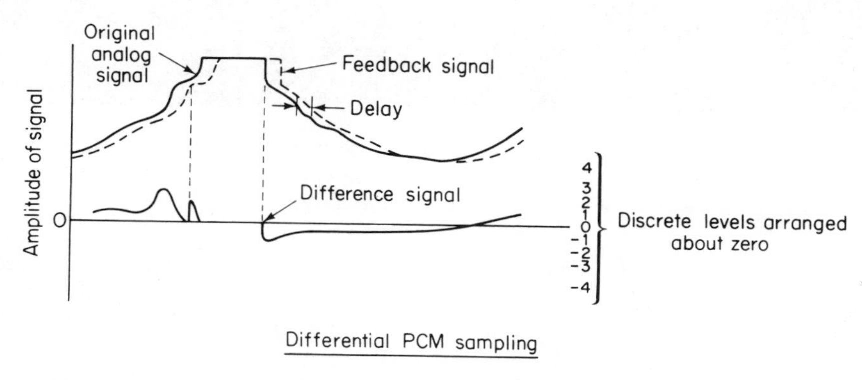

Figure 28.6 The differential encoding process in Picturephone transmission.

ence to the previous picture element in that position. In one implementation of this principle, only every other picture element is encoded, thus halving the required bit rate. These missing picture elements are replaced in one of two ways. If they are for a rapidly changing part of the picture they are replaced by the average of the adjacent picture elements in the same line. If they are for unchanging or slowly changing parts of the picture, they are replaced by the same picture element from the previous frame.

This procedure can be taken further. There is a variety of ways, some elaborate, of encoding either the frame differences or the image motion. Techniques have been demonstrated which require as little as 0.75 of a bit per pic-

Figure 28.7 Digging up the streets to repair or add additional cables is very expensive. The capacity of existing cables can be increased by using digital transmission and signal compression.

ture element, on average, whereas 9 or 10 bits per picture element are needed to straight PCM encoding.

It is thus clear that compaction techniques can reduce the number of bits needed for video encoding by an order of magnitude. However the circuits needed to do so are complex and in some cases need enough memory to hold an entire frame—3 or 4 million bits. As with speech there is a major trade-off between encoding expense and transmission expense.

The images used for video telephone and video conferencing have much less motion than the images used on television, especially those from movies with fast action and editing, panning and zooming. Therefore powerful forms of compaction can be used for video telephone and video conferencing which are not appropriate for television.

ANALOG TELEVISION COMPRESSION

Techniques exist for analog compression of television signals as well as digital. General Electric's *Sampledot* television system transmits color television over 600 KHz channels (or 900 KHz with an alternate model)—less than the analog bandwidth used by AT&T for their Model

II Picturephone. Two circuit cards 4 1/2″ × 4 1/2″ are used in the transmitter and receiver to achieve this, and there is slight degradation in picture quality.

It is thought by some that techniques will come into existence soon for sending television over a fraction of today's bandwidth with negligible image degradation. When this occurs it could further increase the capacity of CATV cables and will give regulatory authorities major problems with reallocating the radio spectrum.

The Sampledot system does not transmit a complete frame every thirtieth of a second. Instead it transmits one sixteenth of a frame with the picture elements selected in a pseudorandom fashion. The TV picture is divided up into segments. Each segment has 16 horizontal TV scan lines. The pseudorandom pattern in each segment has one picture element from each scan line. The same pattern is repeated in every segment. Then a new pattern used with again one picture element from each scan line in a segment. Every 1/30 of a second a new pattern is used and in 16/30 of a second every picture element on the screen is transmitted. In effect the frame scanning time has thus been extended from 1/30 second to 16/30 second. The trick lies in finding a pseudorandomness which is as undisturbing for the viewer as possible.

Such schemes are not yet perfected to the point where they are completely acceptable for all public broadcasting. They are certainly acceptable, however, for most TV applications in industry, and could greatly improve the economics of video telephones and video conferencing.

TASI

A technique has long been used on subocean cables for doubling their capacity by continually reassigning the channels to telephone speakers. It is called *TASI, Time Assignment Speech Interpolation.*

On a link carrying a conversation, both parties do not normally speak at once, and for a small proportion of the total connection time (usually about 10%) nobody is speaking. The long distance link is normally a "four-wire" circuit, meaning that there is a transmission path in both directions, so the path in one direction is in use on average only about 45% of the total time. In other words for 100 talkers, only about 45 on average will be speaking simultaneously. There is a spread about this average, and so it is necessary to provide more than 45 one-way channels. However if a large number of voice channels are carried together, statistics indicate that the ratio of talkers to one-way channels required becomes close to 1/0.45.

The TASI equipment is designed to detect a user's speech and assign him a channel in milliseconds after he begins to speak. An almost undetectable amount of his first syllable is lost. He retains the channel until he stops speaking. It is then taken away from him so that it can be allocated to another speaker if necessary.

DSI

With digitized speech, a digital form of TASI is used, called DSI (digital speech interpolation). DSI is faster and more efficient than its analog precursor.

DSI has been used on digital satellite circuits. As soon as a speaker pauses, the subchannel he used is made available for dynamic allocation to other speakers. The high-speed circuits detect speech almost immediately it begins and allocate a subchannel to it. If there is no free subchannel at that instant, the speaker will usually only have to wait milliseconds before one of the other speakers pauses and *his* subchannel is reallocated.

DSI more than doubles the capacity of channels it is used on.

INCREASED CAPACITY

A unit such as the AT&T D2 interface to the T2 carrier takes 96 telephone calls and digitizes them for transmission over a wire pair cable. It receives the bit streams coming in the opposite direction over an associated wire pair and converts them to analog form. At some time in the future the unit which interfaces with the T2 carrier may use predictive delta modulation and DSI. With circuits in experimental use today the two wire pairs could have a capacity of 500 telephone calls, rather than 96, without the user detecting significant impairment of the speech quality. More complex circuits could give a capacity of 1000 calls, but the codecs required need high speed logic which is expensive today.

The above figure of 500 telephone calls on the T2 carrier assumes that codecs digitize calls into 25,000 bits per second. This can be done today without significantly distorting the speech. If a degree of distortion is tolerable, much lower bit rates can be used. One device transmits four telephone conversations over a full duplex channel of 9600 b/s derived by using a modem on a leased telephone line. The distortion can be of such a nature that the words are almost always intelligible, but the speaker's voice is not necessarily recognizable. In business communication this can be acceptable if the cost is low. It may be particularly valuable for foreign telephone calls in corporations because the line cost is high and a caller has often never met the person he is calling.

Delta modulation and DSI are likely to be used on corporate networks before they are used on the public telephone network. Corporate networks may always employ higher levels of speech compression and higher levels of distortion than public networks.

MAIL AND MESSAGES

The high-speed digital channels, although designed for speech, are likely to have a major impact on the sending of messages of all types. All of message types in Box 15.1 can be sent digitally, and the nonreal-time ones can occupy channel space that would otherwise be idle in the telephone network (Fig.

15.1). It therefore seems highly desirable that the mail facilities of a nation should share the telephone facilities.

Mail and messages sent in computer code occupy an order of magnitude fewer bits than if sent in facsimile form. Nevertheless enormous quantities of items in facsimile form will be transmitted in the future. Digital facsimile will become very important. The digitization of documents is a different art to the digitization of telephone calls or moving video signals.

The bit stream for most documents needs to represent only black and white conditions. As a document is scanned, one scanning line at a time, the scanning encounters alternate runs of white and runs of black. A run of black may be very short—the scan crossing one thin line. A white run may be long—the entire width of the page. The encoding process can be designed to represent these *run lengths* in the minimum number of bits. In some facsimile schemes *predictive* encoding is used. The encoding process attempts to guess whether the next element to be encoded will be black or white, on the basis of what has been encoded so far (to the left and above). Only the incorrect guesses need to be transmitted.

Using today's digital facsimile encoding, the Bible could be transmitted in facsimile form over a T4M, DR18 or other T4-speed channel in about one second.

REFERENCES

1. CCITT Recommendation A.711, on compandors. CCITT Green Book Volume III, Line Transmission, International Telecommunications Union, Geneva, 1973.

2. CCITT Recommendation A.733 on PCM multiplex equipment operating at 1544 KB/S. CCITT Green Book Volume III, Line Transmission, International Telecommunications Union, Geneva, 1973.

3. CCITT Recommendation A.732 on PCM multiplex equipment operating at 2048 KB/S. CCITT Green Book Volume III, Line Transmission, International Telecommunications Union, Geneva, 1973.

4. R. M. Wilkinson, "A 4800 b/s adaptive PCM speech coder," Report No. 73031, Signals Research and Development Establishment, Christchurch, England, 1973.

An early automatic telephone exchange in Stockholm, using racks of 500-selector switches. (*Courtesy L. M. Ericsson.*)

29 PACKET SWITCHING

With digital channels proliferating and the use of computer terminals growing by leaps and bounds, a particularly important technology is the building of data transmission networks. As we discussed in Chapter 10, packet switching is one way to build a switched data network capable of handling the *burstiness* of interactive computer traffic. This chapter discusses more details of packet-switching networks.

The CCITT definition of a *packet* is as follows: *A group of binary digits including data and call control signals which is switched as a composite whole. The data, call control signals and possibly error control information are arranged in a specified format.* [1]

The associated CCITT definition of *packet switching* is: *The transmission of data by means of addressed packets whereby a transmission channel is occupied for the duration of transmission of the packet only. The channel is then available for use by packets being transferred between different data terminal equipment. Note—The data may be formatted into a packet or divided and then formatted into a number of packets for transmission and multiplexing purposes.* [1]

These definitions apply to the networks for which the terms were originally used, such as the ARPA network, the Telenet network, the Datapac service in Canada, and the British Post Office EPSS (Experimental Packet-Switched Service), but they apply also to networks with conventional data concentrators or to hierarchical network protocols such as IBM's SNA (System Network Architecture), or to conventional message switching networks. A major difference between such networks is the geographical layout. A network such as ARPA or Telenet serves many computer centers with a mesh-structured line layout such as that in Fig. 16.6. A typical data concentrator network or message switching network has a tree structure, taking traffic to or from one computer center, or possibly a few interlinked centers.

MESSAGE SWITCHING

Packet switching is a form of *store-and-forward* switching. Messages are stored at the switch nodes and then transmitted onwards to their destination. Store-and-forward switching has existed for decades in telegraphy where it is called *message switching* [2]. There are, however, major differences between packet switching and conventional message-switching.

Whereas message switching is intended primarily for nonreal-time people-to-people traffic, packet switching is intended primarily for real-time machine-to-machine traffic, including terminal-to-computer connections, and is employed to build computer networks. These differences in purpose are such that there are major differences in operation between message-switching and packet-switching networks. One important difference is in the speed of the network. A packet-switching network may be expected to deliver its packet in a fraction of a second, whereas a message-switching system typically delivers its message in a fraction of an hour. Each node passes the packet to the next node quickly like passing on a hot potato. Another important difference is that a message-switching system files a message for possible retrieval at some future time. A packet-switching system deletes the message from memory as soon as its correct receipt is acknowledged. Because a message-switching system files messages, usually at one location, it tends to use a *centralized* star-structured or tree-structured network. A packet-switching network usually has an amorphous structure with no particular location dominating the structure.

In many message-switching systems long messages are sent as a single transmission. In packet-switching systems long messages are chopped up into relatively small slices—1008 bits per packet on the U.S. Telenet system; 255 8-bit bytes on Canada's Datapac system. Because the packets are of limited size they can be queued in the main memory of the switching nodes, and passed on rapidly from node to node. At its destination the original message has to be reassembled from the slices.

The locations connected in the Telenet network shown in Fig. 16.8 are linked with wideband lines operating at 50,000 and 56,000 bits per second. Each of these terminates in a small network computer. The network computer has two main functions. First it acts as a link between the network and the data processing equipment which uses the network. Second, it carries out the switching operation, determining the route by which the data shall be sent, and transmitting it.

The customer's computers which the network serves are called *host* computers. When one host computer sends data to another, it passes the data with a destination address to its local network computer. The network computer formats the data into one or more packets. Each packet contains the control information needed to transmit the data correctly. The packets are transmitted from one network computer to another until they reach their destination. The final network computer strips the transmission control information from the packets, assembles the data and passes it to the requisite host computer.

A network computer receiving a packet places it in a queue to await attention. When it reaches the head of the queue, the computer examines its destination address, selects the next network computer on the route, and places the packet in an output queue for that destination. A packet-switched network is usually designed so that each network computer has a choice of routing. If the first-choice routing is poor because of equipment failure or congestion, it selects another routing. The packet thus zips through the network, finding the best way to go at each node of the network and avoiding congested or faulty portions of the network.

We might compare a telecommunications network with a railroad network. With circuit-switching there is an initial switch setting operation. It is like sending a vehicle down the track to set all of the switches into the desired position; the switches remain set and the entire train travels to its destination. With packet switching the cars of the train are each sent separately. When each car arrives at a switch, the decision is made where next to send it. If the network is lightly loaded the cars will travel to their destination by a route which is close to the optimum. If the network is heavily loaded, they may bounce around or take lengthy or zig-zag paths, possibly arriving in a different sequence to that in which they departed.

A train with only a single car can head off into the network with no initial set-up operation. However if the train has many cars it should not start its journey until it is sure that there is enough space for all of the cars at the destination. As we shall see, on some networks an engine has to be sent to the destination and return with a go-ahead message before the train can set off.

PACKETS

The packets might be thought of as envelopes into which data are placed. The envelope contains the destination address and various control information. The transmission network computers should not interfere in any way with the data inside the envelopes. In fact the system should be designed with security safeguards so that network computers cannot pry into the contents of the envelopes.

Figure 29.1 shows the structure of the ARPANET, or Telenet, packet. It has a maximum length of 1008 bits. The text (data being sent) is preceded by a start-of-message indicator and a 64-bit header. It is followed by an end-of-message indicator and 24 error-detection bits. The header contains the destination address, the source address, the link number and packet number which is used to ensure that no packets are lost and that packets in error are transmitted correctly, a message number with an indication of whether there is more of the message following in another packet, and some special-purpose control bits.

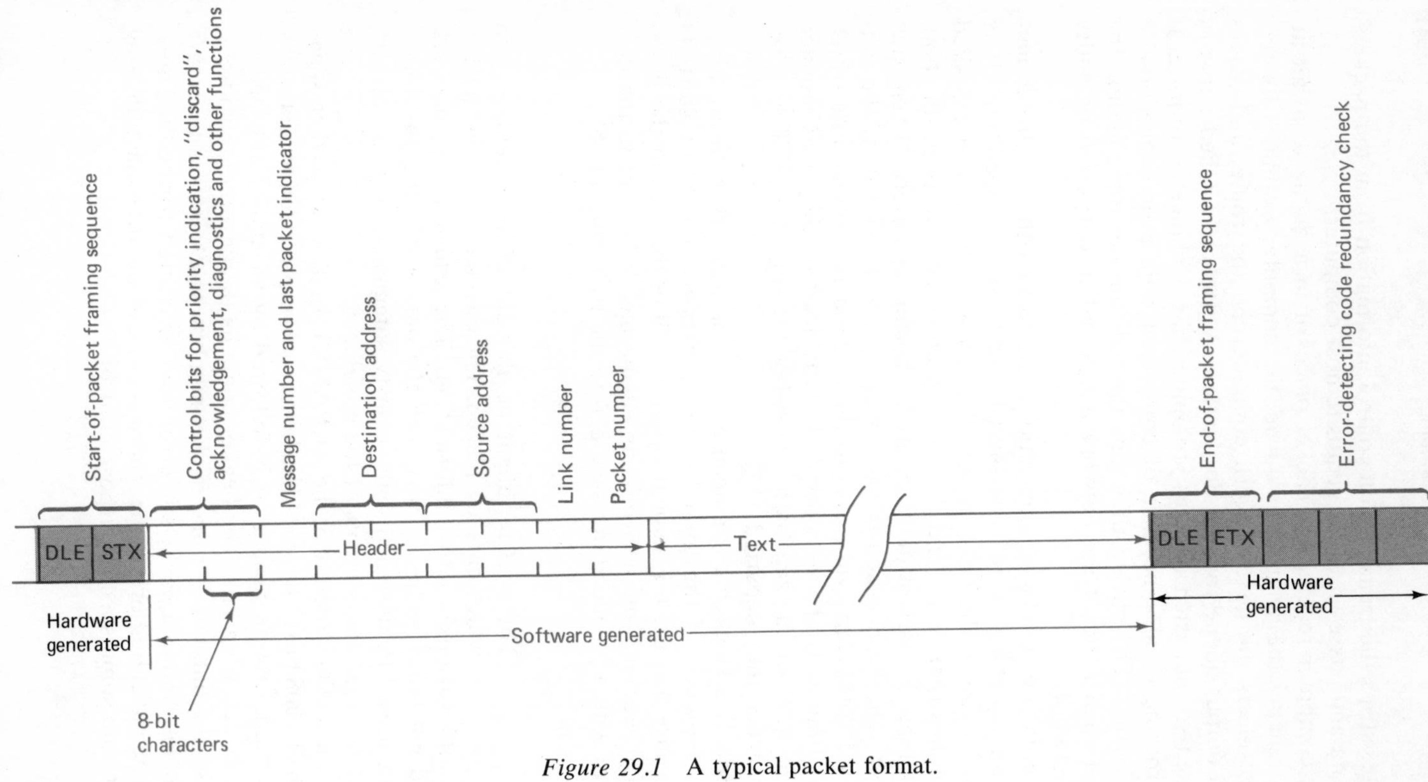

Figure 29.1 A typical packet format.

PACKET CONTROL PROCEDURES

The transmission of the packets through the network requires three types of control:

1. *Error control,* to deal with any transmission errors that occur.
2. *Routing control,* to determine the routes over which the packets are transmitted.
3. *Flow control,* to avoid congestion in the network which could cause lockouts or traffic jams.

ERROR CONTROL

Error control procedures are applied to each point-to-point link. When a node receives a packet, it checks its accuracy using the error-detecting code bits. Error-detecting codes can be made very powerful so that the probability of a transmission error being undetected is very low. It is estimated that the probability of an error bit being undetected on a link in the ARPA or Telenet system is 10^{-12}.

If a node finds the packet it receives to be correct it transmits an acknowledgment message to the sender. If it finds the packet to be incorrect or garbled, it ignores it. Whenever a node transmits a packet it retains it in storage until it receives the acknowledgment of correct transmission. If it does not receive such an acknowledgment within a specified time period (say 100 milliseconds) it automatically retransmits the packet. If repeated attempts to transmit the packet receive no acknowledgment, the sender tries to transmit by a different route.

In some packet-switching systems there exists this point-to-point error control between the switching nodes but there is no *end-to-end* error control between the destination machine and the originating machine. Consequently the malfunctioning of a switching node can, on rare occasions, result in the loss of a message.

ROUTING CONTROL

When a packet-switching computer receives a packet addressed to another location, it must determine which of the neighboring nodes of the network to send it to. The computer will have a programmed procedure for routing the packet. A variety of different routing strategies are possible:

1. Predetermined Routing

The route may be determined before the packet starts on its journey. The packet then carries routing information which tells network computers where to send it. The determination of the route may be done by the originating location, or it may be done by a "master" station controlling the entire network.

Most of the traffic consists not of lone wandering packets, but of "sessions" in which many packets pass back and forth between the same two locations.

Typical sessions are *dialogues* between a terminal user and a computer in which many messages and responses are sent, *data entry operations* in which an operator keys in many transactions, *file transfers* in which many packets of data are sent, and *monitoring operations* in which a remote computer monitors or controls a process. With predetermined routing, the routing decision can be made once for the entire session. Any packet which is part of that session contains the same routing instructions. If a line or network computer fails during the session, a recovery procedure will be necessary in which a new route is established.

The alternative to predetermined routing is that each network computer makes its own routing decision for each packet. ARPANET and most proposed packet-switching systems employ this nonpredetermined routing. With nonpredetermined routing the network computer has more processing to do but the packet envelope may be shorter because it contains a destination address not an entire route.

2. Calculated Routing

The address of the destination nodes in a network may be chosen in such a way that it is possible for any interim node to determine which way to send a packet by performing a simple calculation on its address. If a node has received information about a failure in that direction it may calculate a second-best routing.

Calculated routing is simple but in general too inflexible and hence unlikely to be used in practice.

3. Static Directory Routing

With directory routing, each node has a table telling it where to send a packet of a given destination. Figure 29.2 shows a possible form of such a table. The table shown gives a first choice and second choice path. If the first choice path is blocked or inoperative, a node will use the second choice path. (There is no need for the table in the illustration to give a third choice because no node has 3 lines going from it.) As a packet travels through the network, each node does a fast table look-up and sends it on its way.

4. Dynamic Directory Routing

The previous method uses a fixed table. A more versatile method is to use a table which can be automatically changed as conditions of the network change.

There are several possible criteria that could be used in selecting the entries for a table such as that in Fig. 29.3. They include:

1. Choosing a route with the minimum number of nodes.

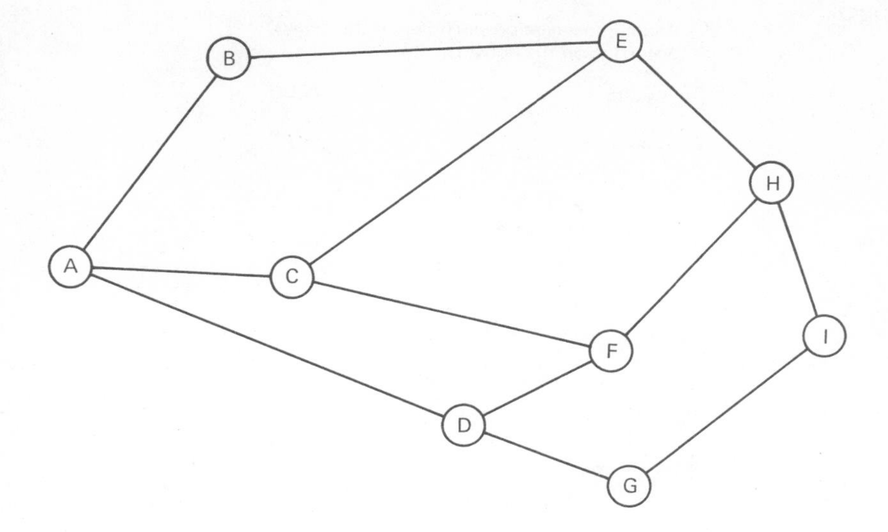

Packet destination	First choice node	Second choice node
A	A	F
B	A	E
D	A	F
E	E	A
F	F	A
G	F	A
H	E	F
I	E	F

Figure 29.2 A static routing table for node C in the above network, giving the node to which C should route a packet for a stated destination.

The numbers are proportional to the delays occurring on the network:

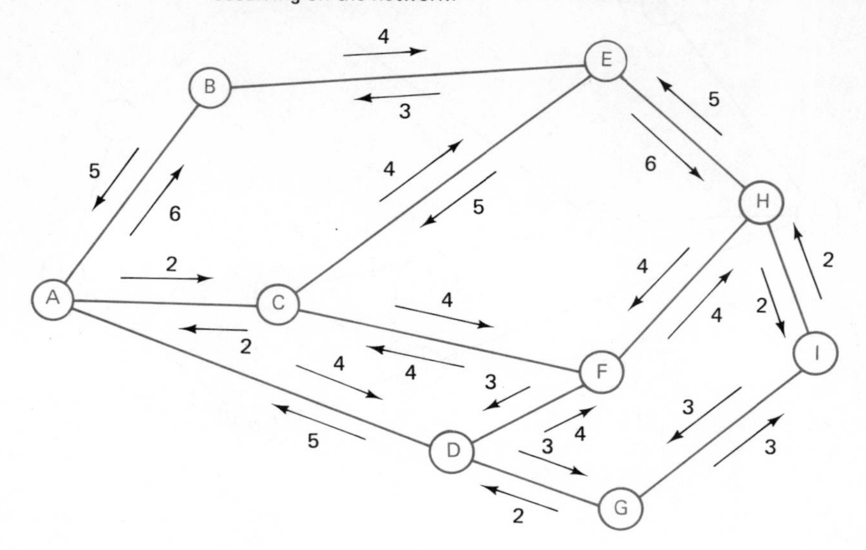

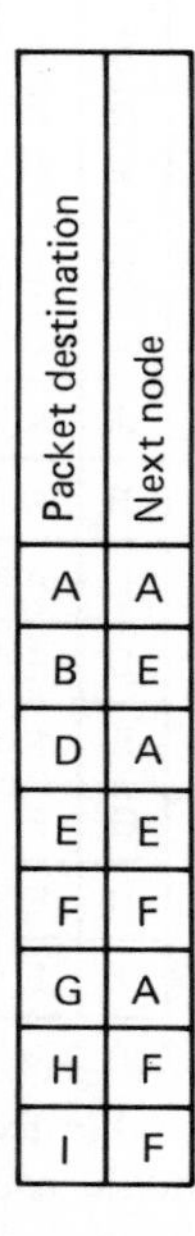

Packet destination	Next node
A	A
B	E
D	A
E	E
F	F
G	A
H	F
I	F

Figure 29.3 A dynamic routing table for node C intended to minimize transit delays under current network conditions.

2. Choosing routes which tend to spread the traffic to avoid uneven loading.
3. Choosing a route giving a minimum delay under *current* network conditions.

The last of these conditions implies that the table will be constantly modified to reflect the current delays on the network caused by congestion or failure. Figure 29.3 shows figures proportional to the delays on the network at one time, and a routing table for node C which takes these delays in consideration. As part of the time delay at each node is caused by queuing, the delays and the optimum routing table will change as the traffic patterns and volumes change.

The question now arises: how should a node be informed of what the network delays are? Several methods have been suggested. Paul Baran suggested that the delay on any route x → y is similar to the delay traveling in the opposite direction y → x. Consequently a node could obtain approximate information about network delays by finding out how long it had taken each packet reaching it to travel from its originating node. The time of departure would be recorded in each packet so that each node could determine this transit time. The method is rather like asking travelers on a rush-hour traffic system, "What is it like where you came from?"

Another system is for each node to send a service message at intervals to each of its neighbors. The message will contain the time it was originated, and also that node's knowledge of whether the delays have changed. The recipients will record how long the message took to reach them, and if this is substantially different to the previously recorded delay they will include this information in the next service message they send. Knowledge of changes in transit times will thus be disseminated throughout the network.

ADAPTIVE ROUTING

A scheme in which the routes selected vary with the conditions of the network is called *adaptive routing*. The ARPA and Telenet network use adaptive routing, each node sending a service message every half second. Most proposed public networks in other countries also use adaptive routing.

Adaptive routing sometimes results in oscillatory behavior, the routing pattern oscillating rapidly backwards and forwards under peak conditions. Minor changes in the routing algorithm can affect the routing behavior under heavy loading in ways which are difficult to predict without simulation of the network. Detailed study of routing algorithms is being made, and can substantially affect the peak network throughput.

It is possible that packets in such a network could fail to reach their destination because of temporary equipment failure or a data error in the address. Such packets might be passed indefinitely from one node to another if something did not stop them. To prevent this occurrence, a count field is used in each packet and the number of nodes that have relayed that packet is recorded in it. When the count exceeds a certain number, the packet is returned to its point of origin. This process protects the network from becoming clogged with roving, undeliverable messages.

FLOW CONTROL

Flow control is desirable to prevent too many packets converging on certain parts of the network so that traffic jams occur. The control messages which are passed between nodes to control the packet routing play a part in avoiding traffic jams. However if too many packets enter the network heading for a given destination the routing control alone will not prevent a traffic jam. Traffic congestion can be harmful because packets bounce around from node to node occupying an excessive share of the transmission capacity. The network performance degenerates out of all proportion to the increased load, like the roads out of a large city on a Friday evening.

The best way to prevent congestion is to control the *input* to the network. Control messages can warn all input nodes that congestion is beginning to build up. The most common cause of potential traffic jams is that one host computer suddenly sends a large volume of traffic to another. If the packets for this traffic follow each other at the speed of the input node, there may be a traffic jam on the route. The rate of input needs to be controlled rather than merely opening a sluice-gate wide.

The ARPA network controls such surges by permitting only one message at a time to be sent from an originating point to any one destination. The messages can consist of up to eight packets. When the destination node has completely received and assembled the message and delivered it to the machine for which it was intended, the destination node sends a control message back to the originating node saying that it is permitted to send another message. This is called a "Ready-for-next-message" signal, RFNM (pronounced "Rufnum"). The RFNM is formatted similarly to the other packets, and finds its way through the network using the same protocols as any other packet.

Figure 29.4 shows the passage of a message through a packet-switching system with protocols similar to the original ARPA network. The message goes from the host machine on the left of the diagrams to the host machine on the right. Three switching nodes, or IMPs, are shown. The first IMP is connected directly (i.e., not over a telephone company line) to the originating host machine. The host sends a message, which on the ARPA network can be up to 8063 bits, to its local IMP with a header saying where the message is to be sent. The IMP chops the message into slices, each slice becoming a packet with the format shown in Fig. 29.1. The IMP determines which line to transmit on, and sends the first packet. When it receives an acknowledgement from the next IMP saying that the packet was received correctly, it sends the next packet. The packets eventually arrive at the destination IMP, the third IMP in the figure. They may have come by different routes and hence could possibly arrive out of sequence. The destination IMP waits until it has received all of

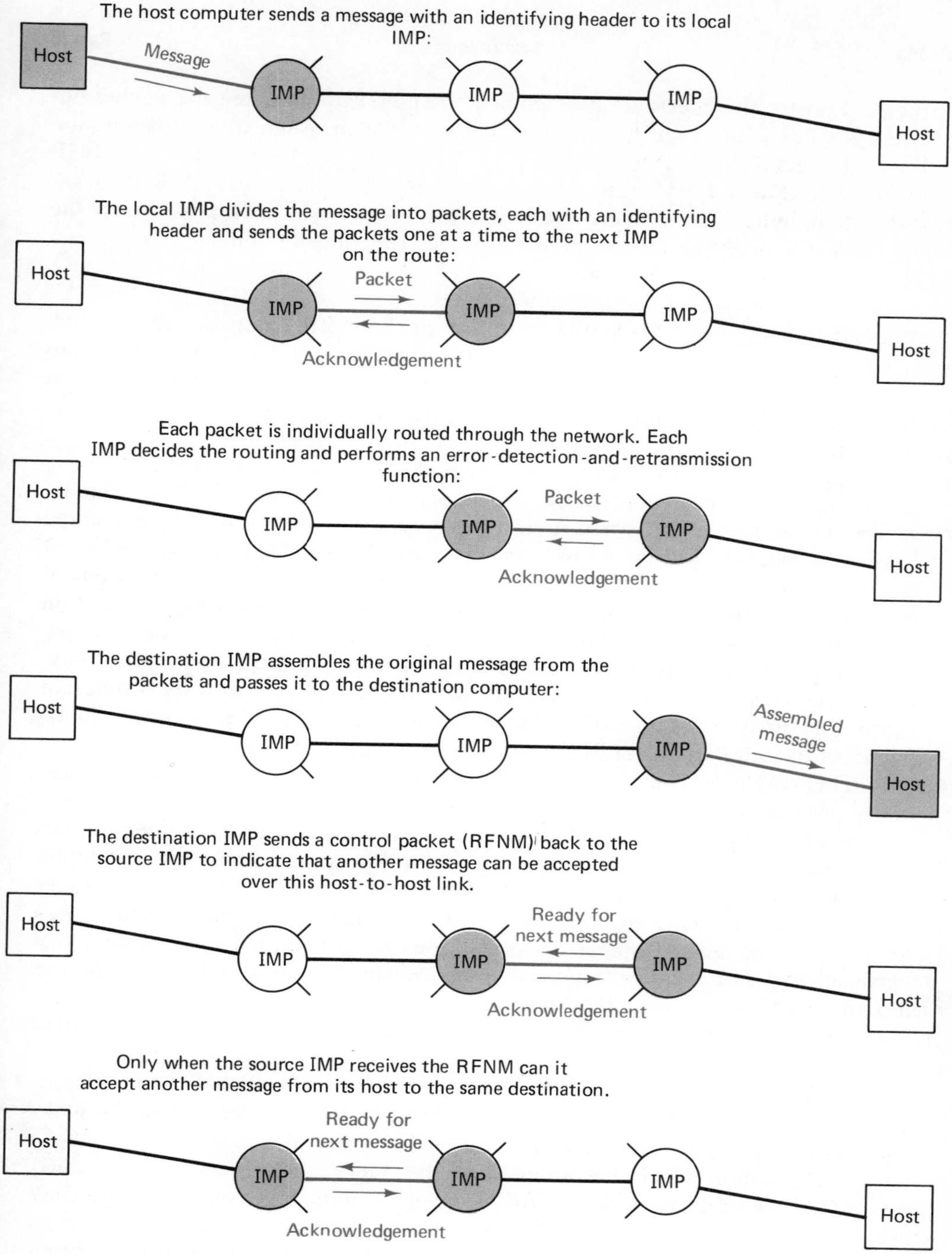

The host computer sends a message with an identifying header to its local IMP:

Host — Message → IMP — IMP — IMP — Host

The local IMP divides the message into packets, each with an identifying header and sends the packets one at a time to the next IMP on the route:

Host — IMP — Packet / Acknowledgement — IMP — IMP — Host

Each packet is individually routed through the network. Each IMP decides the routing and performs an error-detection-and-retransmission function:

Host — IMP — IMP — Packet / Acknowledgement — IMP — Host

The destination IMP assembles the original message from the packets and passes it to the destination computer:

Host — IMP — IMP — IMP — Assembled message → Host

The destination IMP sends a control packet (RFNM) back to the source IMP to indicate that another message can be accepted over this host-to-host link.

Host — IMP — IMP — Ready for next message / Acknowledgement — IMP — Host

Only when the source IMP receives the RFNM can it accept another message from its host to the same destination.

Host — IMP — Ready for next message / Acknowledgement — IMP — IMP — Host

Figure 29.4 The flow of data on the ARPA network. A problem exists with the original ARPA control mechanism shown here, in that shortage of memory in the destination IMP can cause a "reassembly lockout" in which messages partially received from different sources cannot be completely received and a traffic jam builds up. An additional reservation mechanism is necessary to prevent such lockouts.

the packets for this message; it assembles the message removing the packet envelopes, adds a message header, and delivers the message to the destination host. The receiving IMP then sends a RFNM addressed to the originating IMP on the left. The RFNM bounces like any other packet to the IMP it is addressed to. When the lefthand IMP receives the RFNM correctly it tells the host that it is ready to accept another message for transmission.

REASSEMBLY LOCKOUT

As the operators of the ARPA network discovered, the protocol shown in Fig. 29.4 was not enough to prevent congestion problems. A serious form of jam could occur when long messages from different sources converged on one destination.

The destination IMP would receive and acknowledge the first packets of several messages. It has only a certain amount of memory available for the assembly of messages, and when it starts to receive a message it does not know how many packets will arrive for that message. It may begin the reception of more messages than it can complete. It is then in serious trouble. It runs out of memory for reception of more packets, but cannot deliver to the destination host the incomplete messages which are clogging its memory. The neighboring nodes continue to transmit packets to it but it cannot accept them. After several attempts the neighboring nodes try to send the packets to it by a different route. But the destination IMP is locked solid. The traffic piles up in neighboring areas like city streets after an accident in the rush hour.

This condition is called a *reassembly lockout,* and was regarded as a bug in the protocol design of the ARPA network.

Reassembly lockout can be avoided in several ways. First, it does not occur if *only single-packet messages* are sent. The originating hosts could be made to slice up their own messages so that this is possible. Second, the lockout could be avoided if the destination IMPs could pass incomplete messages to the receiving host. Both of these solutions were regarded as unsatisfactory because they complicate the protocols needed in the hosts. The intent of the design was that the network should appear transparent to the hosts, and that they should not have to be concerned with slicing up or reassembling messages.

A third solution is to give the nodes a backing store so that they can temporarily move incomplete messages out of their main memory when the lockouts occur. This would increase the cost of the nodes substantially.

The solution which was adopted on the ARPA and similar networks was to prevent the originating nodes from sending a multipacket message until they were sure that enough space had been reserved in the destination node for its reception. To accomplish this, more control messages were needed in addition to those in Fig. 29.4. Before sending a multipacket message, the IMP on the left sends a reservation message to the destination IMP telling it how much space to reserve. The destination IMP sends an acknowledgment if it has the space, and then the originating IMP sends the packets. The overhead associated with the reservation process is not as severe as it may sound, because most of the messages sent are small enough to fit in a single packet.

It will be observed that the control of a mesh structured network is substantially more complex than that of the more conventional tree- or star-structured networks.

TWO TYPES OF PACKET SWITCHING

Because of the problems associated with message reassembly, two types of packet switching have evolved. The first handles multipacket messages and so has to have protocols which permit error-free message *assembly* without causing traffic jams. The second handles only single-packet messages, and hence avoids complex protocols which increase the network overhead. Canadian common carriers coined the term *datagram* to relate to the second kind of service. In a datagram service users can send messages up to but not exceeding the maximum capacity of one packet. This permits a network to be built with simple control procedures and switches, low overhead, and fast transit times.

A datagram network is of value for many applications including the vast future needs of electronic fund transfer. It could be designed to operate with very inexpensive terminals such as those in Fig. 3.2. Other applications need to send longer messages—including the vast future needs of facsimile mail services.

A CCITT proposal recommends a standard for packet switching, as discussed in Chapter 33.

NETWORK TRANSPARENCY

It is the intention of the designers of many packet-switching networks to make the communication techniques as unobtrusive as possible to the users. The network operation should be *independent* of the nature of the computing operations that employ the network. Many new types of computers and new types of operations could then employ it. The network should connect two computer processes, perhaps thousands of miles apart, as though they were directly interconnected via a precisely defined interface.

This illusion of direct interconnection is referred to as *network transparency*. To make the network appear transparent, the transmission must be fast and the software must hide the complexity of its operations from the process which uses it.

On the ARPANET and Telenet systems there are three layers of communication protocol illustrated in Fig. 29.5. On the outside of this diagram are two computer processes communicating by sending streams of bits to one another. In the innermost layer the IMPs are transmitting to one another. In between

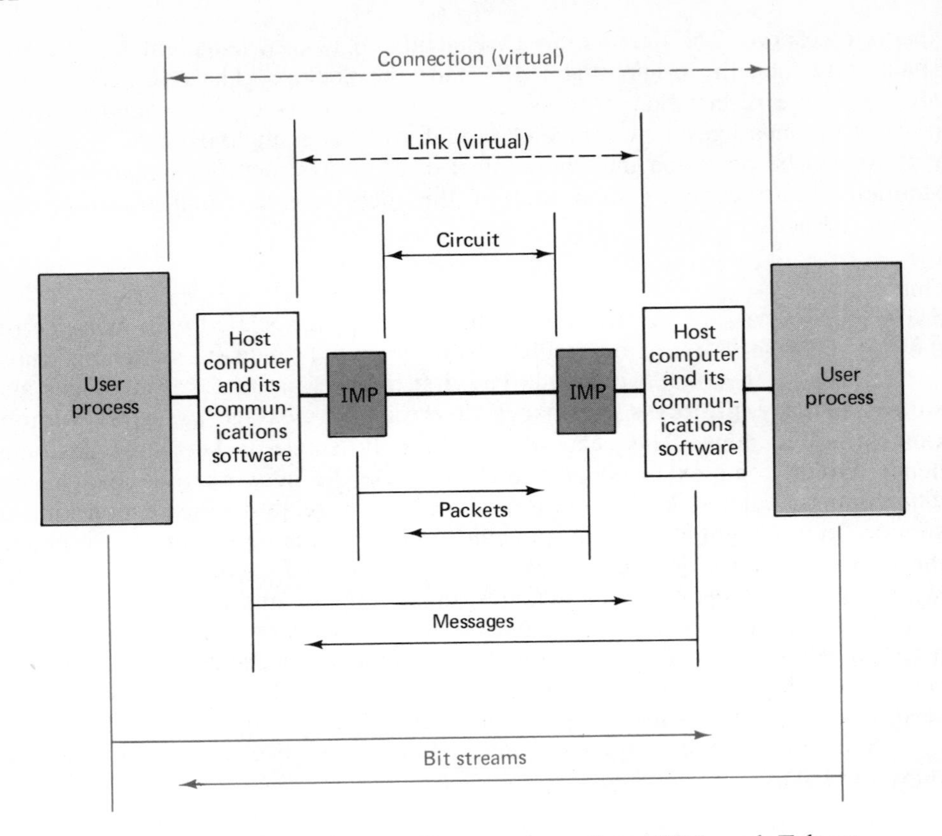

Figure 29.5 Three layers of protocol on the ARPA and Telenet networks.

the IMP network and the user process are protocols which reside in software and hardware in the host computers.

The connection between the IMPs is a physical circuit over which *packets* are transmitted. The connection between hosts is referred to as a "link," and conveys *messages*. The link is referred to as a "virtual" path, because in reality there is no physical path between hosts; it only appears that way. The connection between user processes, passing *bit streams,* is likewise referred to as a "virtual" path.

Three protocols are needed to permit communication between the user processes:

1. The IMP-to-IMP protocol which permits the packets to be routed between IMPs, free of errors.
2. The HOST-to-IMP protocol which permits the HOST and IMP to pass messages to one another. This requires appropriate hardware and software in the host.

3. The HOST-to-HOST protocol which provides rules permitting the user process to talk to one another.

TWO SEPARATE FUNCTIONS

There are two separate functions which must be carried out by nodes of a packet-switching network: communication with the users, and the switching of the packets. In the ARPA network both functions are carried out by the same machines—the IMPS or TIPS. Other proposals, including proposals for public switched networks in several countries, have employed separate machines for the two functions.

If the functions are separated, the switch can be a relatively small computer. The steps to be executed in switching one packet are relatively few, and so a high data throughput is possible with a fast machine. It has been estimated that more than a million bits per second could be switched using today's technology, and so the switch might be geared to PCM transmission channels of 1.544 or 2.048 million bits per second. Unlike a message-switching computer, the packet-switching computer does not have any accounting or filing functions to perform. It needs no peripheral storage. It merely receives packets, queues them, determines their routing, and retransmits them. During idle moments it will send diagnostic packets and update its routing table.

The subsystem which interfaces with the user has somewhat more complex functions than the switch. They may include:

1. Creating the Packets

Data from a terminal or host is placed into one or more envelopes and control information needed for transmission is written.

2. Reassembly of Data

After transmission data which are sent in more than one packet are reassembled. For some uses the entire message may be assembled prior to delivery; for others it may be better to deliver the data a block at a time as it arrives, provided that the blocks are in sequence.

3. Host-Network Protocol

The interface computer will observe a protocol for communicating with the host computers to ensure that the interchange functions correctly and that no data can be lost. Different protocols may be needed for different types of host.

4. Terminal Protocol

A protocol for communicating with user terminals will be observed. Some of the user terminals may be far away, connected to the interface computer by links incorporating multiplexor, concentrators, polling, public network dialing, or other procedures.

5. Special-Function Protocols

Special protocols may be used for functions such as the transfer of files, the use of graphics, "conference calls" in which more than one user participates in one dialogue, mail-box services in which the network passes messages between users and holds them until the users see them, facsimile transmission, transmission of exceptionally high security, and others.

6. Session Control

Many transmissions, as discussed earlier, are part of a "session" in which multiple messages are sent, as in a human telephone conversation. In this case the interface computer may control the session. It will store an envelope header for the session so that it does not have to be recreated for each message. It may use a session oriented protocol with the host computer or terminal. It may allocate a high priority to the packets of certain sessions, possibly after establishing whether this priority is sustainable with the current network load.

TWO LEVEL NETWORK

Various authorities, including D. W. Davies and his co-workers at the British National Physical Laboratory [3] and the designers of the PCI (Packet Communications Incorporated) network, the first of the U.S. value-added common carriers to receive FCC approval, have advocated that a packet-switched network should be constructed having two levels. The lower level would be the local area network, corresponding broadly to the central office and local loops of the telephone network. This would carry out the function of interfacing with the host computers and terminals. The higher level would be the long-distance packet-transmission and switching network, corresponding broadly to the trunking network of the telephone system.

A wide variety of different terminals can be attached to the lower level interface computer. Indeed, one of the chief advantages of this type of network is that entirely different types of terminals can intercommunicate. They can have widely different speeds and use different codes. They can be synchronous or start-stop. They can use polling, contention or be alone on a line. They can use different types of error control. If desired, they can be connected via multiplexers or concentrators. Above all, they can be inexpensive, for most of the costly terminal features like buffering and elaborate line control are not really needed. The interface computer maintains a list of the characteristics of all the terminals attached to it, their control mechanisms, speeds, and transmission codes. If the code differs from the network transmission code the interface computer converts it as the packet is being assembled. When the packet is received by the destination line-control computer, it is converted, if necessary, to the code of the receiving terminal. It is estimated that one interface computer could handle more than a thousand terminals in this way.

An additional function of the interface computer is to collect the information necessary for logging and billing subscribers.

The high level network can use links of different speeds for moving the packets. As satellite technology develops the high level network of such a system would probably incorporate satellite transmission. Where the high level network is installed by the same authority that operates the telephone network, the links can be the PCM links that carry telephone voice traffic. Packet transmission and voice transmission can thus share facilities.

THE FUTURE OF PACKET SWITCHING

Many countries have announced an intention to build or experiment with a packet-switching network. There will continue to be arguments about the relative merits of packet-switching and fast circuit-switching networks.

It seems likely that the existing packet-switching networks will steadily grow, acquiring more traffic and more nodes. If they become large and ubiquitous, economies of the scale may make some form of switched data network replace most of the private leased-line networks that corporations and government departments use today with techniques such as concentrators and polling.

There are several future directions in which packet-switching networks will probably evolve if their traffic grows sufficiently:

1. The high-speed PCM links we have discussed may become the links used by packet-switching networks. Transmission rates of millions of bits per second will permit very fast response-time systems to be built.
2. To fill such high speed links the networks will have to attract a high traffic volume (as will Datran to fill its 44 mbps trunks). Much of this traffic may come from relatively new uses of data links such as electronic mail, electronic fund transfer, and other forms of message delivery.
3. As networks grow very large it is economical for them to become multilevel networks with a hierarchy of switching offices. Just as the telephone network has five classes of office, so data networks may acquire two, three, and eventually more levels.
4. Several classes of traffic may be handled, differentiating perhaps *datagrams* and long messages.
5. Several classes of priority may be handled, including perhaps *immediate delivery, 2-second delivery for interactive computing, delivery in minutes,* and *overnight delivery.*
6. Some message traffic may be *filed* as on a message-switching system. Messages intended to be read on visual display units or spoken over the telephone may be filed until the recipient requests them. Distributed storage rather than centralized storage

may be used, especially for bulky data, depending upon the relative costs of storage and transmission. A hybrid between message-switching and packet-switching may thus emerge.

7. Fast-connect circuit switching has advantages over packet-switching for some types of traffic. The nodes of a large data network may be designed to select whether a circuit-switched or packet-switched path is used. A hybrid between circuit-switching and packet-switching may emerge.
8. The user interface computer may become separate from the switching computers and have an entirely different set of functions. It may be designed to convert the transmission of all terminals to a standard format, code, and protocol so that completely incompatible machines can be interconnected. A telex machine using Baudot code may transmit to a visual display terminal using the CCITT Alphabet No. 5.
9. Elaborate security procedures may become used.
10. The user interface devices may be designed to receive from and transmit to conventional facsimile machines or other analog devices. I.E. codecs may be built into the nodes.
11. The interface machines may be designed to compress messages before transmission, to increase the transmission efficiency. This is valuable with data, but especially valuable with facsimile messages.
12. The interface machines may be designed to handle packet radio terminals or controllers. Portable data terminals may be linked to the system.
13. One of the most cost effective data transmission facilities will be the satellite. Packets will probably be sent via satellite. To use future satellites in an optimal (broadcast) fashion will substantially change the topology and protocols of packet-switching networks.
14. Economies of scale and flexibility may require that telephone or continuous-channel traffic and burst traffic be intermixed. Networks, especially satellite networks, capable of handling both continuous-channel and packet-switched traffic may emerge.
15. An interlinking of separate national networks will occur. Satellites will interlink nodes in many countries, giving users of packet-switching the capability to use computers around the world and send messages worldwide.
16. It is extremely important that there should be internationally agreed standards for the interface to the networks which user machines employ. Such standardization is discussed in Chapter 33.
17. When vast numbers of computers are available on the networks, directory machines will be very important, for enabling users to find the facilities they need.

REFERENCES

1. CCITT Fifth Plenary Assembly, Green Book Volume VII, *Telegraph Technique,* published by the International Telecommunications Union, Geneva, 1973.

2. James Martin, *Telecommunications and the Computer, Second Edition,* Englewood Cliffs, N.J.: Prentice-Hall, Inc., 1976.

3. D. W. Davies, "The Principles of a Data Communication Network for Computers and Remote Peripherals," presented at IFIP Congress, Edinburgh, 1968.

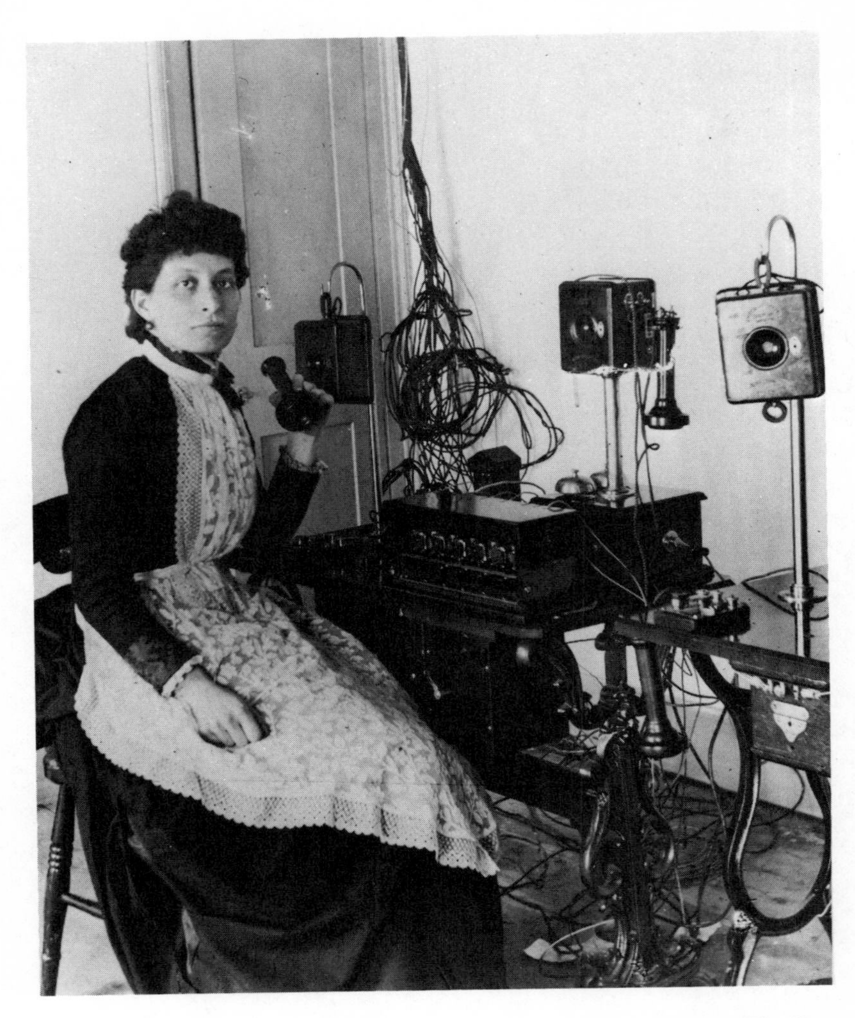

The telephone exchange at Croydon, London, 1884. (*Courtesy The Post Office, England.*)

30 TIME-DIVISION SWITCHING

For a century telecommunication authorities have tended to divide up the capacity of a channel by giving all users the same type of subchannel. Where switching offices existed they were designed to switch this type of subchannel. Thus the telephone network provided switched telephone channels. The telex network provided switched telex channels. Message switching networks provided low-speed message channels. Television networks provided television channels.

It cannot be stressed too strongly that we have now entered a different era in telecommunications when the capacity of a network needs to be switched in a much more flexible manner. The channels should be switchable between different types of users in a time-varying manner. One time the user at a given location may want a voice channel, at another a high-speed data channel, at another an interactive data channel with a high peak-to-average transmission ratio, in the future he may want a video channel, and so on. If the transmission network is to serve the users as well as it indeed can, the transmission capacity should be dynamically switchable between these bandwidths according to the users' instantaneous requirements.

There are several reasons why switching should swing from the allocation of fixed-bandwidth subchannels to the time-varying assignment of differing bandwidths. These reasons are:

1. The bandwidths available today are much higher than in previous decades and promise to become much higher still. It is desirable to use these high bandwidths to their full potential.
2. PCM transmission techniques permit all manner of signals, e.g., voice, video, facsimile, and data, to be freely intermixed.
3. High speed control equipment can be built to interleave the different signals economically in a time-varying manner. To have done so in earlier decades would have been prohibitively expensive.

4. The needs of machines talking to machines are fundamentally different to those of people talking to people. To force machine-to-machine transmission into the type of network needed for people-to-people transmission is to cripple the potential capabilities of today's machines.
5. In today's networks real-time and nonreal-time signals can be intermixed, and this is a key to network efficiency.
6. Satellite channels are not, in essence, point-to-point channels, but can interconnect users dispersed over one third of the earth's surface. To take advantage of such channels highly flexible allocation of their capacity between different users is needed.

MORE THAN PACKET SWITCHING

The designers of packet-switching networks recognized the need to build switched *data* networks in which the capacity is allocated in a rapidly time-varying fashion. One second a user may be receiving a high-speed burst of data; the next he is not receiving or transmitting anything. At one time he uses his link for fast-response man-computer dialogue; at another for slow batch transmission.

Today's packet-switching networks assume that all their users transmit *data*. The majority of the bandwidth of a corporation's telecommunication facilities, however, is employed for transmitting voice signals. Voice, video, and facsimile, will probably always outpace computer data in the demand for bandwidth, either in corporate networks or public networks. Data is only a small fraction of what is transmitted. To switch bandwidth in a time-varying manner between voice, data and other uses, needs a switching technique more flexible than those discussed in the previous chapters.

Time-division switching may provide the answer.

TIME-DIVISION TRANSMISSION AND SWITCHING

When time-division multiplexing is used, either of PCM signals or data, the type of switching called *time-division switching* becomes economical.

For decades, telecommunications organizations have completely separated the design of switching equipment from that of transmission equipment. With PCM transmission and time-division switching the two are intimately associated so that they cannot be designed in isolation from one another.

Figure 30.1 illustrates the principle. Here tomatoes are shown rolling down a chute and being sorted by gates that open and close at exactly the right times. There are four types of tomatoes, A, B, C, and D and they must be switched to four outgoing paths.

The tomatoes may be thought of as being samples of four signals traveling together in a time-division-multiplexed fashion. Signal C is to be switched to path 1, signal A to path 2, and so on. The gate to path 1 opens at exactly the

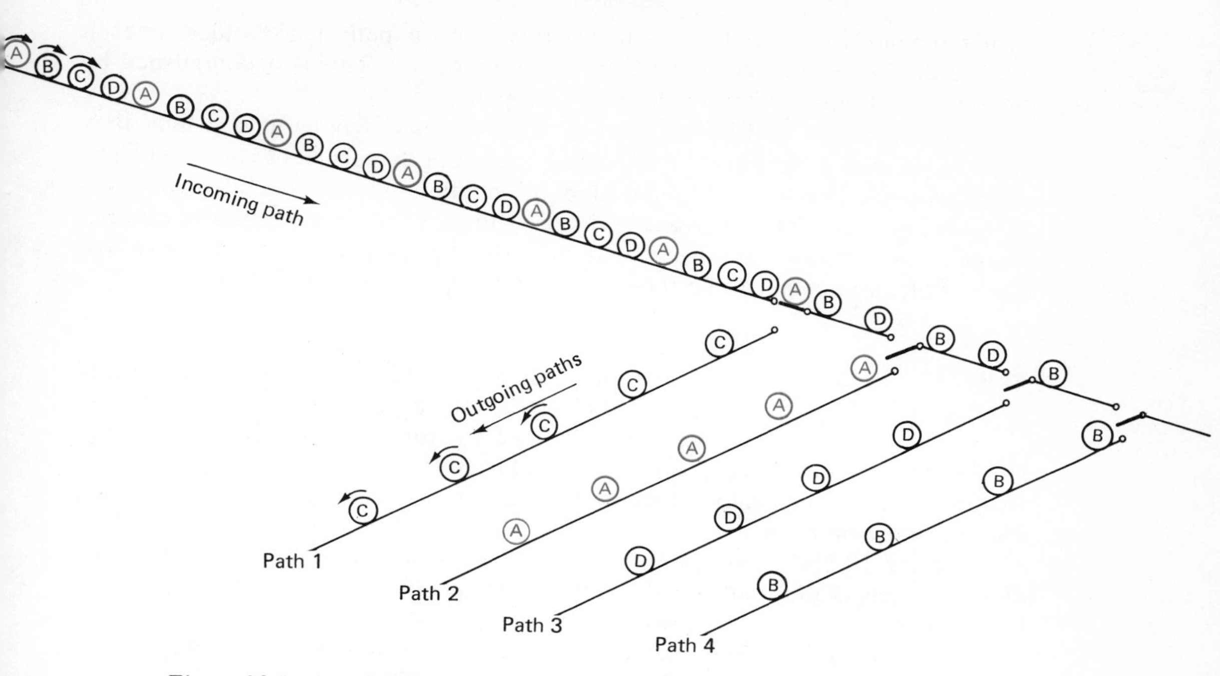

Figure 30.1 Synchronous time-division switching of a time-multiplexed stream.

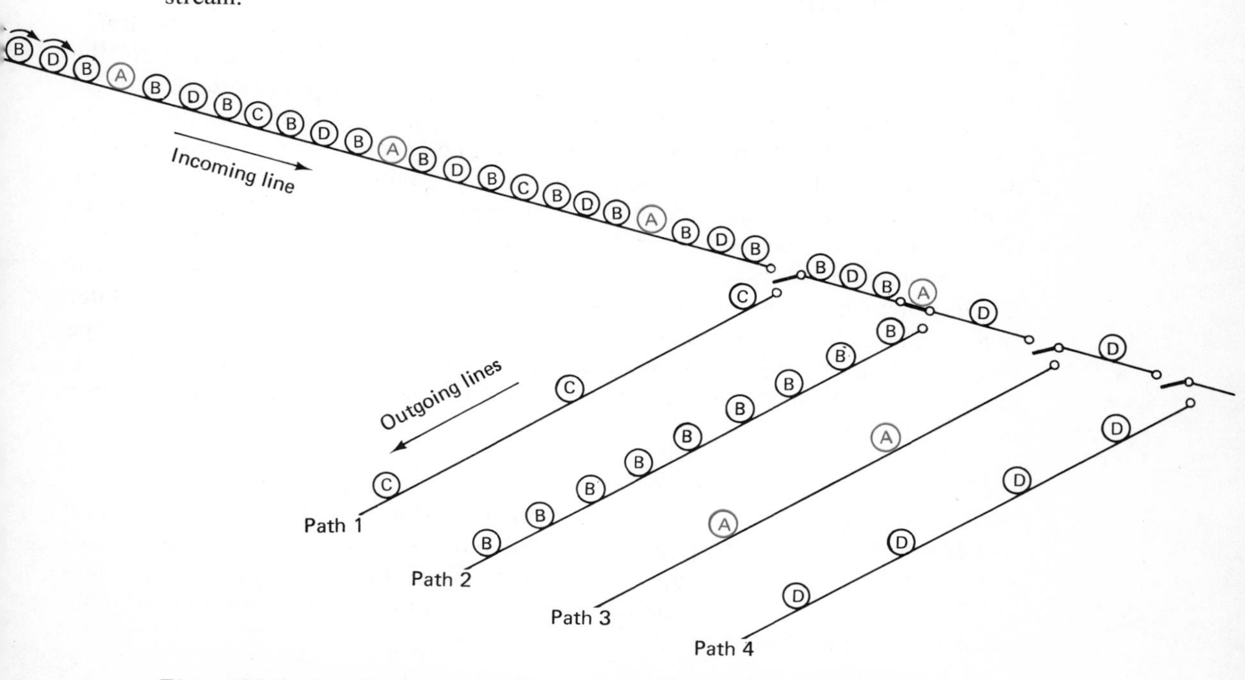

Figure 30.2 Synchronous time-division switching with different channel speeds.

right moments to make the C samples travel down path 1. At other times it may be desirable to make signal C travel down path 4. This is accomplished by changing the timings with which the gates open.

The tomatoes of Fig. 30.1 could represent *bits* arriving at a switch; they could represent *8-bit bytes;* they could represent larger *blocks* of data. The same principle applies: the separation of the time-multiplexed elements is a switching process. They may be arriving at the rate of several million bits per second. There may be hundreds or thousands of paths, and the gates open and close at electronic speed like the logic circuits in a computer.

VARYING-SPEED STREAMS

In Fig. 30.1 the signals being switched are of equal speed. One of the most attractive features of time-division switching for future networks is that it can handle streams of different speeds. The capacity of a high-speed channel and switch can be dynamically allocated between users needing high, medium, and low transmission rates.

Figure 30.2 illustrates the principle of a transmission line and switch handling channels of different speeds. A high data-rate channel is switched to path 2. Low data-rate channels are switched to paths 1 and 3.

When a user requires a channel in a time-division network employing different channel speeds, he will indicate the destination of the call, the channel speed he wants, and possibly the duration of the call. A network controller will then attempt to make a reservation for these requirements through the transmission channels and switches. If the operation is under computer control the reservation could be for a very short block of time. One user device might request a 200 bit per second channel for three minutes while another might request a 256,000 bit per second channel for one second. Both of these may be derived from a digital carrier operating at say 2.048 million bits per second. With time division of the capacity under computer control, subchannels of large or small capacity may be allocated to users for very brief periods of time.

While some users specify the duration of their call when the reservation is made, others may request a channel for the foreseeable future, and at a later time request the disconnection of that channel. A telephone caller (or the equipment serving him) may, for example, request a channel of 64,000 bps capacity to be returned to a pool of transmission capacity available for allocation to other callers.

TIME-DIVISION TELEPHONE SWITCHING

Most of the time-division switching equipment employed by telephone companies switches fixed-capacity channels, as in Fig. 30.1, but nevertheless promises major advantages over conventional telephone switching.

The incoming path of Fig. 30.1 may be a PCM or digital transmission line. There are clear economic benefits in this association of time-multiplexed switching with time-multiplexed transmission. Time-multiplexed switches are also used to switch channels which enter and leave the switch in a nonmultiplexed form. Figure 30.3 shows a number of lines entering a simple switch. The signals are time-multiplexed by the switch onto a fast bus, and the resulting stream is demultiplexed by accurately timed gates in accordance with the user's requirements.

When a switch such as that in Fig. 30.3 switches telephone voice signals, and not data, there is no need to completely digitize the voice. Instead it can be left in the form of pulses which represent the instantaneous value of the analog signal—the PAM (pulse amplitude modulation) pulses of Fig. 4.2. There

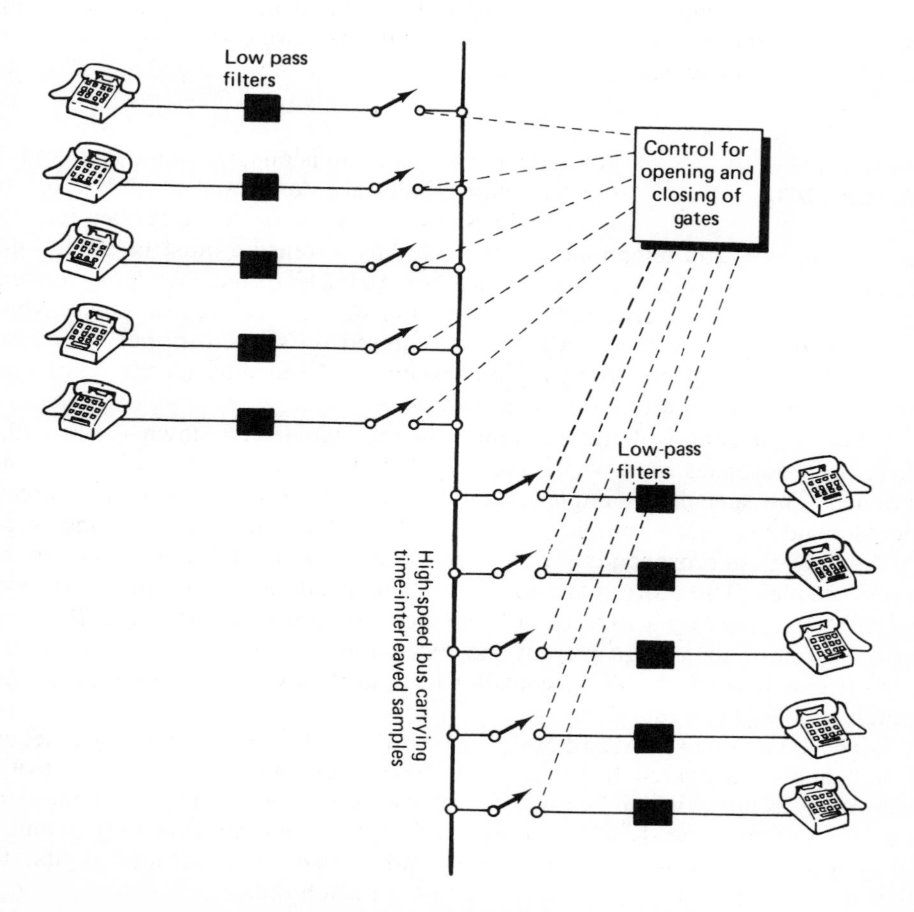

Figure 30.3 A simple time-division telephone switch.

are one eighth as many PAM pulses as there would be bits if the signal was completely digitized, therefore, a higher throughput can be handled by the gates.

It will be seen that time-division switching is quite different from conventional switching, in which a physical path is permanently connected (called space-division switching). The number of switch points is much lower. If there are N lines, then N switch points are needed. If, on the other hand, N lines are to be interconnected physically, then N^2 switch points are needed. The latter figure can be reduced by multistage switching with a limited number of simultaneous interconnections; if no more than one tenth of the lines are permitted to be interconnected at one time, 0.21 N^2 switch points could be used. Thus for a switch interconnecting 100 lines, time-division switching needs 100 switch points, whereas open space-division switching is likely to need at least 2100. Time-division switching is lower in cost today than space division for many applications. As the cost of fast logic circuitry drops lower, time-division switching will become increasingly economical.

ESS NO. 101 SWITCH UNITS

An example of time-division switching is found in the way subscribers are connected to Bell No. 101 Electronic Switching System. A subscriber location might have, say, 200 telephone extensions. Any extension must be able to dial numbers on the public network. In the No. 101 Electronic Switching System, the dialed digits are interpreted in a computer at the local central office, which then controls the switching. All that is required on the subscriber's premises is a small cabinet of electronics, which samples the signals on the voice lines and switches them in a time-division fashion.

The organization of the equipment in this cabinet is shown in Fig. 30.4. There are two buses onto which the samples of signals on the voice lines are switched. The speech is sampled 12,500 times per second. Thus there are 80 microseconds between sampling times. The duration of each sampling is approximately 2 microseconds, with a guard interval of 1.2 microseconds between samples. Therefore there can be 25 independent sets of samples, or 50 for both buses, giving a total of 50 time slots to be divided among 200 speech lines. No more than a quarter of the extensions can be in operation at one time—a higher traffic-handling capacity than normally encountered on private branch exchanges of this size.

The buses thus carry $12{,}500 \times 25 = 312{,}500$ PAM samples per second each. Samples are gated to the appropriate speech lines under the control of electronic circuitry, which is directed by the computer at the local central office. For this purpose, two data lines go from the time-division switch unit to the central office computer. One carries the dialed or Touchtone digits; the other carries the switching instructions to the switch unit.

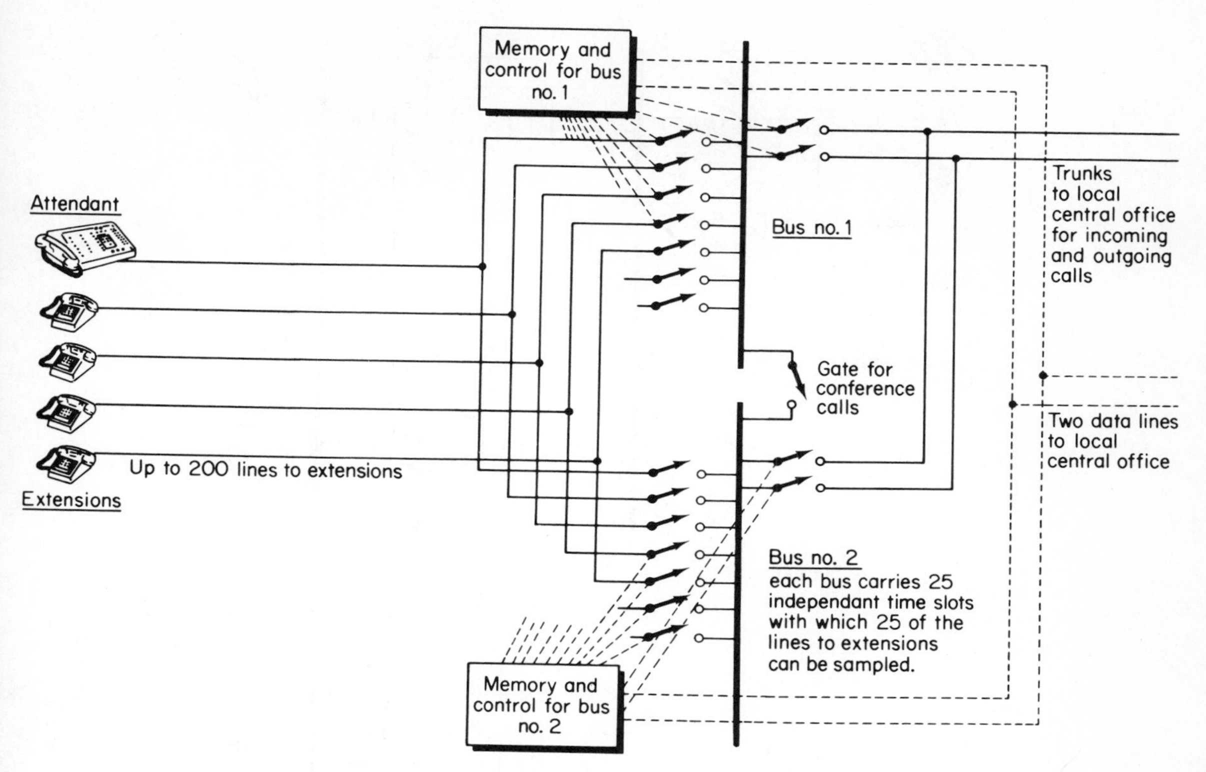

Figure 30.4 The organization of the time-division switching units that are connected to the Bell System No. 101 electronic switching system.

The use of two buses doubles the number of time slots available to the customer, but perhaps more important, it greatly enhances reliability. If one bus or its controls fail, the other can handle any of the extensions. In this way, 25 time slots instead of 50 are then available. The gate shown between two buses also gives a convenient means for establishing conference calls. Figure 30.5 shows the bus arrangement of a larger switch unit. Here there are up to four buses. Each carries 60 independent time slots, giving a total of 240, which means 3 million samples per second. Subgroups of 32 speech lines are connected through the time-division switch to subgroup buses. The subgroup buses in turn are connected through more time-division switches to the 60-time-slot buses.

The upper limit of this technology today is several million PAM samples per second.

PAM samples are not used for long-distance transmission purposes be-

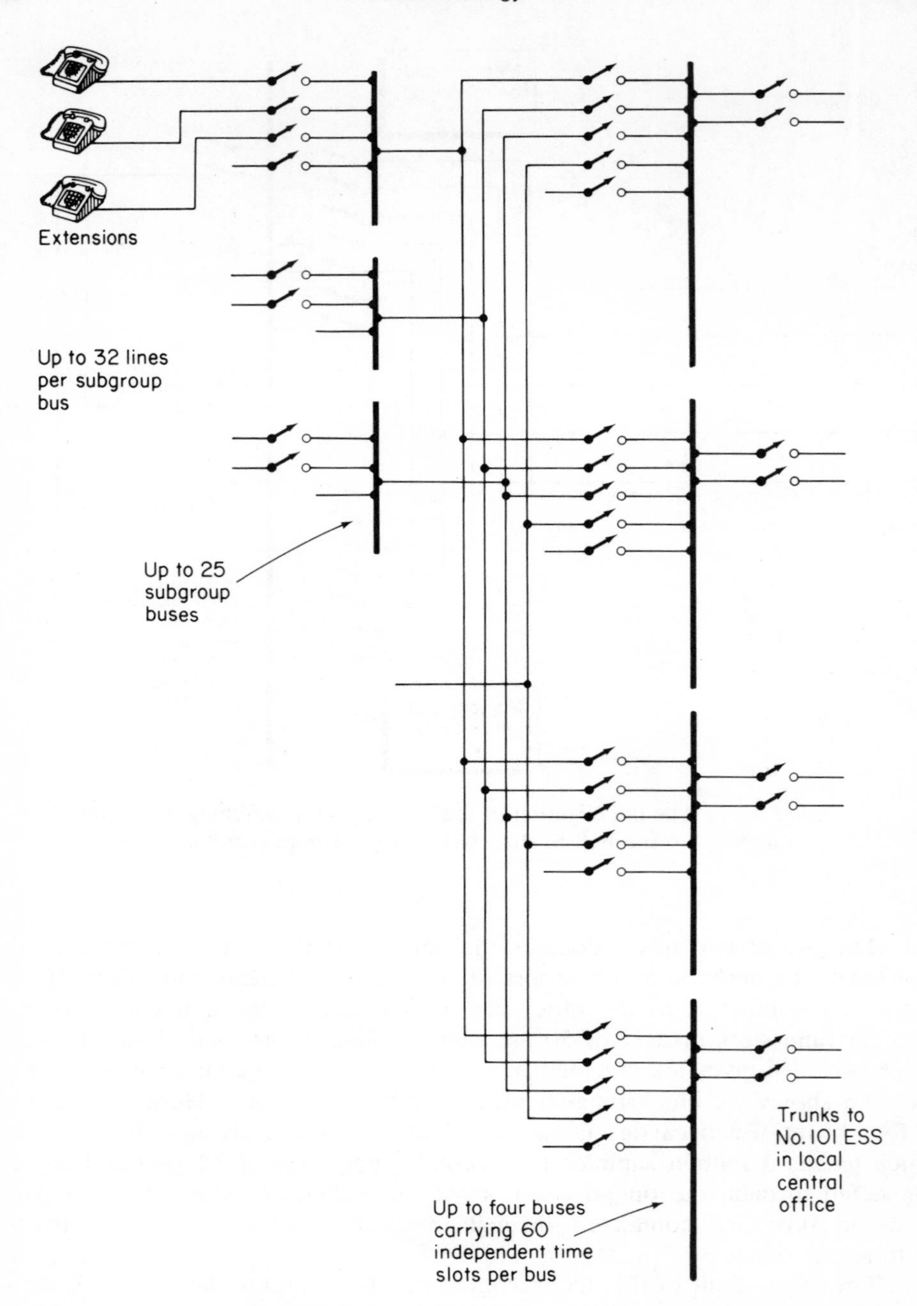

Figure 30.5 A time-division switching unit organization like that in Fig. 30.4 but capable of handling 800 telephone extensions.

cause, being susceptible to noise and distortion, they do not have the advantages of binary encoding (PCM as described in Chapter 4). In trunk switching, therefore, PCM samples can flow through a time-division switch.

CONCENTRATORS

An economically attractive configuration for future telephone networks is to have subscribers connected to time-division concentrators that are linked to the nearest telephone exchange by PCM lines. The concentrator unit may also have the function of digitizing the calls—the codec function.

100 subscriber lines might be connected to a T1 carrier link by means of such a concentrator. Probability calculations based on today's telephone traffic indicate that the 24 channels of the T1 carrier could handle the peak calls from more than 100 typical subscribers with a good grade of service. That is, there would be a very low probability of a subscriber obtaining a "busy" signal because no T1 channel was free. CCITT 2.048 mbps lines could have more than 125 subscribers concentrated onto them.

Figure 30.5 shows concentrators and PCM facilities coexisting with old telephone cables and exchanges.

ELASTIC BUFFERS

When digital transmissions are switched, as with PCM trunk switching, it is common to employ a buffer storage. Figure 30.6 shows such a buffer with 120 cells each of which can hold eight bits (one PCM sample). 120 telephone calls are time-division multiplexed. A time slot from incoming call 1 is read into cell 1; a time slot from incoming call 2 is read into cell 2, and so on. The contents of the cells are read out in a different sequence, depending upon the call switching instructions that have been set up, and are time-multiplexed onto the output channel. In this way, time-slots are interchanged between the incoming and outgoing lines. The storage must adjust to the time differences of the incoming samples and because it does so it is sometimes referred to as an *elastic store.*

To handle 120 calls of 8000 samples per second each, a switching time of $\frac{1}{(120 \times 8000)}$ seconds, or approximately 1 microsecond, is needed. The switching speed of reasonable-cost circuitry sets a limit to the number of calls that can be handled by a single time-slot interchange unit. To switch a large number of calls, many such units are employed and calls going between them must themselves be switched. Fast circuit-switching units are employed for this purpose (referred to as *space-division switching*). A larger electronic switch thus consists of a combination of time-division and space-division units.

ESS NO. 4

This is the manner of operation of the Bell System trunk switch, ESS No. 4.

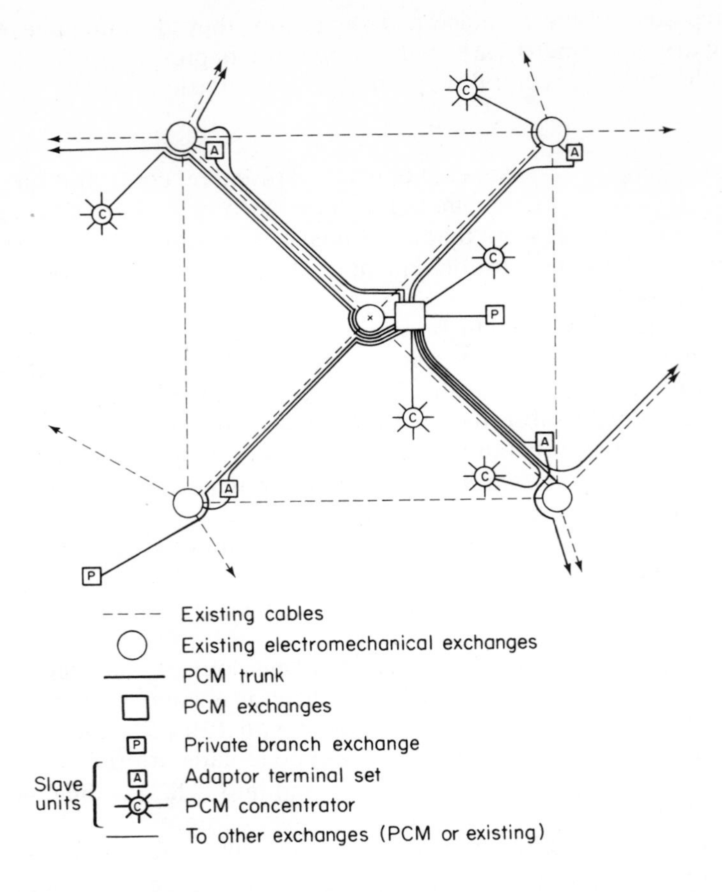

Figure 30.6 A new PCM network with PCM exchanges and concentrators, superimposing upon today's existing network. (Reproduced with permission from *Techniques of Pulse Code Modulation in Communication Networks* by G. C. Hartley, P. Morret, F. Ralph, and D. J. Tarran. Cambridge University Press, Cambridge, 1967.)

In ESS No. 4 the telephone voice sampling interval of 1/8000 second is further subdivided into 128 time slots of .977 microsecond each. This is the basic cycle time of the switch. In each .977 microsecond time slot an 8-bit PCM sample is transferred from an incoming trunk to an outgoing trunk, and another sample is transferred in the opposite direction to give two-way conversation. One *time-slot interchange* unit, like that in Fig. 30.6, handles 120 voice channels. The 8-bit samples from these 120 voice channels are read into the buffer storage and then read out onto the appropriate outgoing lines.

Many of these time-slot interchange units are employed in parallel, and are interconnected by a space-division switch as shown in Fig. 30.7. Other designs of time-division switches have employed different combinations of space-division and time-division.

When a call is received by ESS No. 4, a *trunk-hunt* program is used first to search for a free trunk to the required destination. Then a *path-hunt* program searches for a free path out of the many possible paths (in space and time) that could interconnect the two trunks. The commands for each switching stage are then written in a cyclically repeating memory and read out 8000 times per second to control the time-slot interchanges.

ESS No. 4 can interconnect more than 100,000 trunks, and can handle more than three times the traffic of the largest electromechanical trunk switch. Many of the incoming trunks are analog, not digital. Before these can be switched their calls must be converted to PCM form, and then after switching converted back again, if necessary. Even with these conversion steps ESS No.

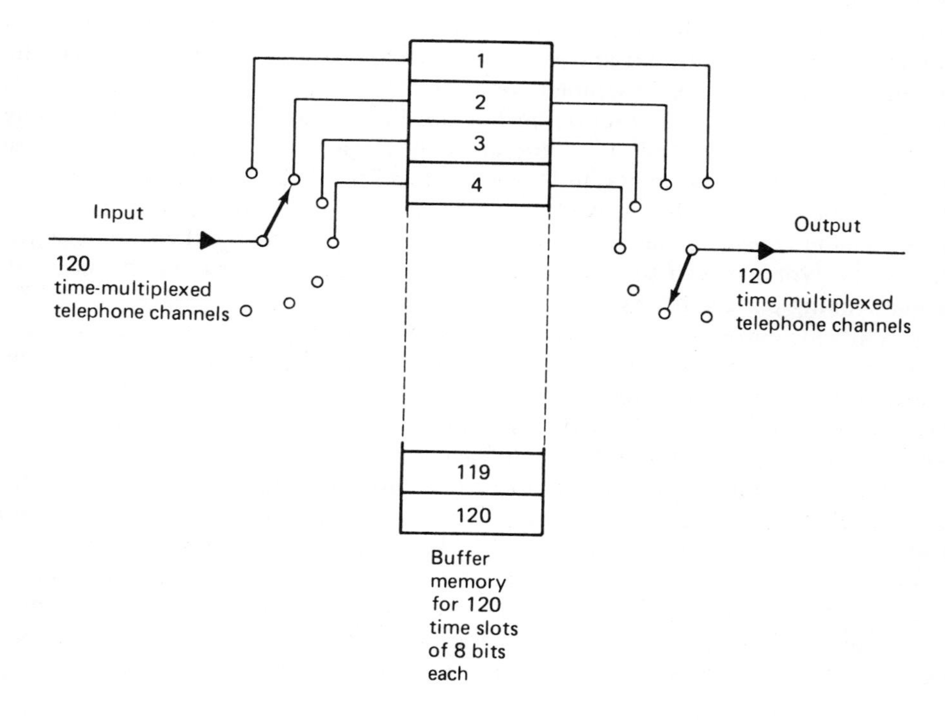

Figure 30.7 Use of a buffer memory for switching 120 telephone channels. 120 telephone channels require a switching time of about 1 microsecond. If a larger number of telephone channels are to be switched then time-division stages are intermixed with space-division switching stages as in Fig. 30.8.

4 is more economical than electromechanical toll switches. With digital trunks, however, it is much more economical.

RESERVED AND NONRESERVED CHANNELS

In the time-division schemes discussed so far, the subchannels are *reserved* before the transmission takes place. A request for a subchannel is made and the necessary time-slots are allocated to it. The allocated stream of time-slots cannot then be used for any other purpose until the subchannel is "disconnected." With burst data transmission the reservation may be for a very short period of time, but nevertheless the transmission cannot commence until the reservation is made.

Another form of time-division is that in which no reservation is made. This form of operating is suitable when the transmission does not need to be *continuous,* but can be broken up in time in a nonsynchronous fashion. Telephone and television, for example, require continuous transmission. Data transmission does not usually need to be continuous but can proceed in bursts providing that the transmitting and receiving machines have buffers in which the stream of data can be assembled and disassembled. When a burst travels through a network and a reservation has not been made for it at the switching points, it may be delayed. It is therefore necessary to be able to store the bursts at the switching points until time-slots are free which the bursts can use.

Time-division with reserved channels may be referred to as *synchronous* time-division, because the fragments of the calls fit into regular pre-assigned time-slots. Nonreserved time-division may be called *asynchronous* because no regular timing pattern is observed.

With synchronous time-division the time-slots which a call occupied determine where that call will be switched. The switch has to be given a routing instruction once only, at the start of the call. With asynchronous time-division each fragment of a call must carry the address of its destination, and the switch is given a routing instruction for each fragment.

Packet-switching is a form of asynchronous time-division switching. In today's packet-switching network the entire leased channel carries packets, and nothing else. The control signals are themselves packets. Another possibility is to intermix reserved time-slots and nonreserved time-slots such as packets. If this is done then the advantages of packet-switching, enumerated in the previous chapter, can be combined with the advantage of synchronous time-division switching that it can carry continuous transmissions such as voice.

COMBINED SYNCHRONOUS AND ASYNCHRONOUS TIME-DIVISION

Most data transmission is subject to peak rate requirements which are much higher than the average rate. As indicated in Table 10.1, the ratio of peak rate to average rate on interactive systems often exceeds 1000. Hence an asynchronous operation such

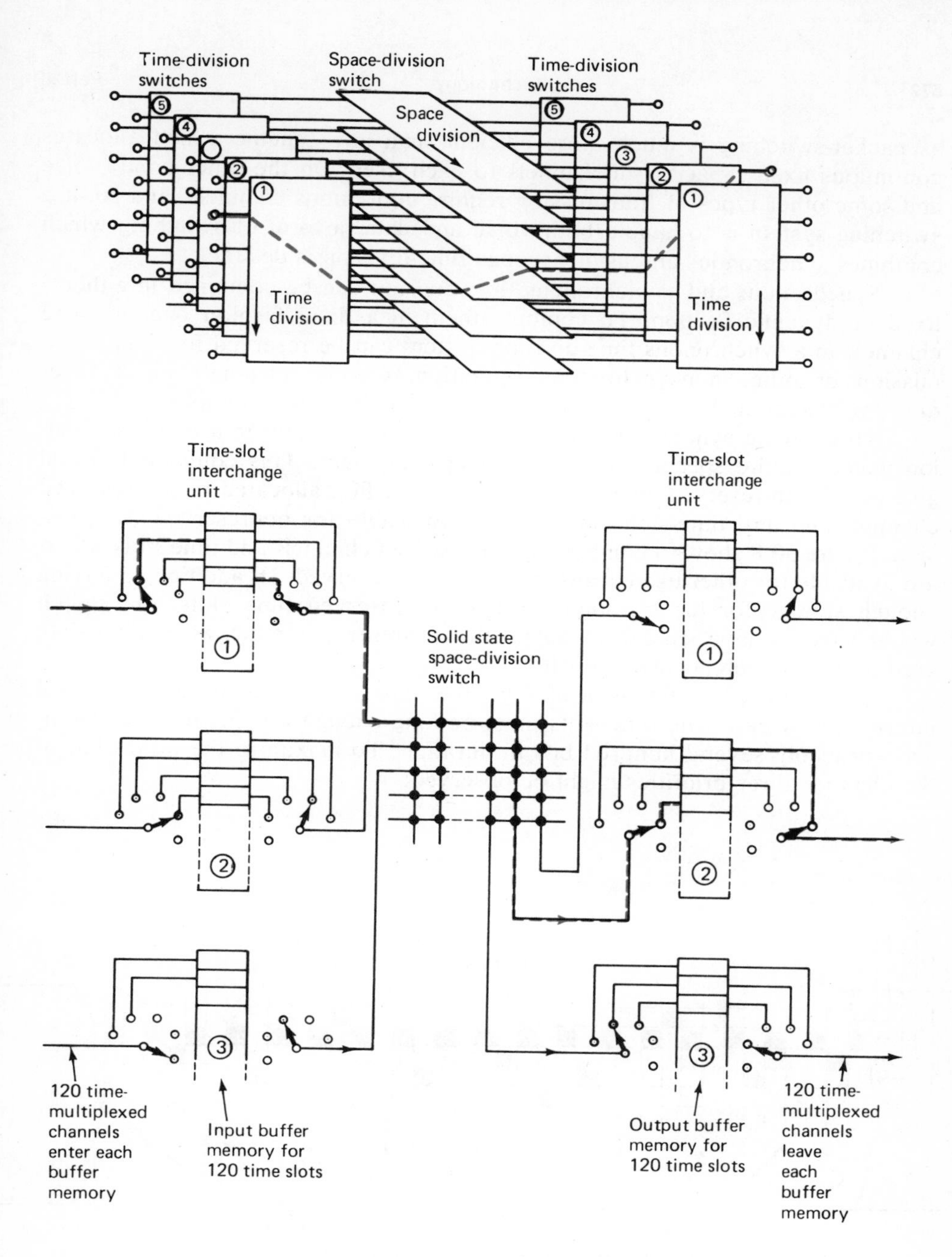

The ESS No. 4 trunk switch can switch calls between more than 100,000 trunks by a set of time-division switches interconnected by a space-division switch. The dotted line shows the path of a typical call through the exchange.

Figure 30.8 The ESS No. 4 trunk switch can switch calls between more than 100,000 trunks by a set of time-division switches interconnected by a space-division switch. The dotted line shows the path of a typical call through the exchange.

as packet-switching is much more efficient than any scheme which allocates continuous fixed-capacity subchannels to each user. On the other hand, voice and some other types of transmission require continuous channels, and so, if a switching system is to handle both voice and data, a form of switching which combines synchronous and asynchronous time-division is desirable.

Synchronous and asynchronous time-division can be combined in either a fixed or dynamic fashion. To combine them in a fixed fashion one or more channels in a synchronous time-division system can be reserved for data transmission or other nonsynchronous operation. Certain channels on a time-division system could be reserved for packet switching, for example.

To combine synchronous and asynchronous operation in a dynamic fashion would require specially built switching equipment. This equipment would give priority to reserved channels. Time-slots are first allocated to the reserved channels and any time-slots left over are available for nonreserved transmission. Figure 30.8 shows a combination of reserved channels and time slots which are available for other use because they are not reserved. In addition to having enough storage for the maximum number of reserved time slots, the switch would have to have storage to queue the nonreserved traffic when time-slots are not immediately available for it.

Each unreserved time-slot of Fig. 30.8 must carry addressing and control information with it. The time-slots must be long enough for this to be economical—probably several hundred bits in duration. To maximize the utilization of the channel, three priorities might be observed:

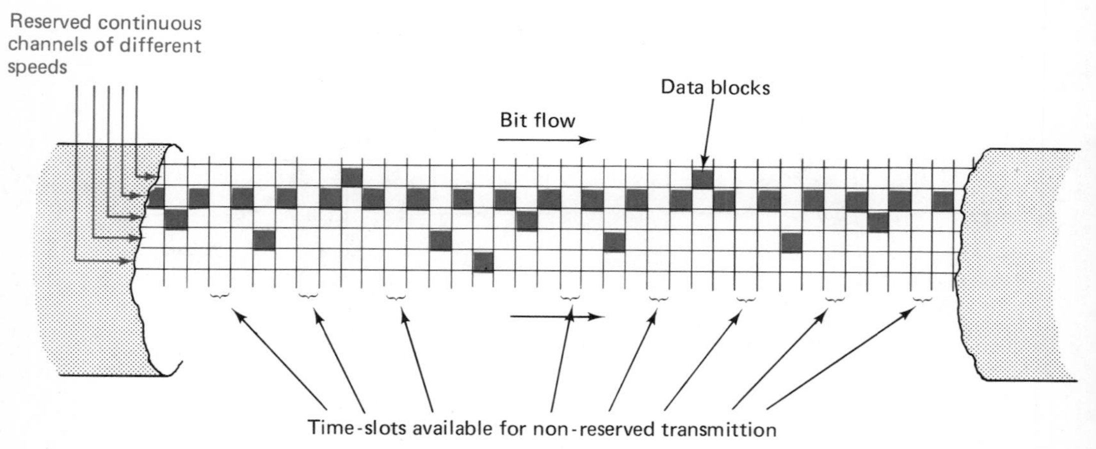

Figure 30.9 Time-division multiplexing or switching in which synchronous and asynchronous channels (continuous and burst channels) are interleaved. The bursts must fit into the gaps left in the reserved channels. The squares or time-slots in this diagram represent appropriately-sized blocks of bits.

Priority 1: Continuous channels (e.g., voice).

Priority 2: Interactive or real-time data.

Priority 3: Batch or non-time-dependent data.

Also to maximize channel utilization, time-slots of different sizes may be employed. It is uneconomic to divide batch data up into small time-slots each with addressing and control information, and it is uneconomic to put brief interactive responses into large time slots.

There are many possible variations in the details of the scheme expressed in principle in Fig. 30.8.

As the merging of computer technology and telecommunications pursues its inevitable course we can expect to find more uses of flexible demand-assigned time-division transmission and switching.

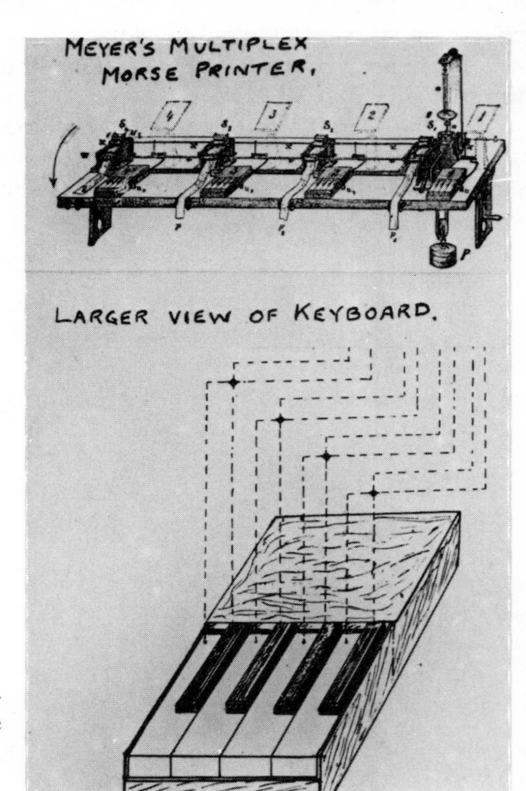

One of the earliest users of multiplexing—Meyer's Multiplex Morse Printer. (*Courtesy The Post Office, England.*)

31 TDMA

When a large number of locations are to be interconnected, switching offices are used to lower the number of links that are required. When high capacity channels are available at reasonable cost there is an alternative to switching offices. Many geographically-dispersed access points could share the same channel, with a control mechanism used to enable them to intercommunicate as they need.

High capacity channels are now becoming available—the PCM links, digital radio, CATV cables, waveguides and satellites. If they are analog channels they can be shared by frequency-division. If they are used in a digital fashion they can be shared by time-division. Most telephone trunks are shared by point-to-point multiplexing. If there are many access points rather than two, the technique is referred to as *multiple-access*. If it is done in a digital time-division fashion it is called *time-division multiple-access,* generally referred to as *TDMA*.

Frequency-division multiple-access, *FDMA,* is also used. Frequency-division implies that separate frequencies are allocated to different users, as in radio broadcasting; a transmitter or receiver must be tuned to the frequency which is assigned to it. Time division implies that different time slots are allocated to different users, and each user must know at what times he must transmit and receive.

DEMAND ASSIGNMENT

Multiple-access schemes can operate with either fixed channel assignments to the various users, or with channels being assigned in a varying fashion according to demand. The latter is called *demand-assignment*. Demand assignment implies that when a user wants a channel he makes a request, and after a brief time-lag a channel is allocated to him. The acronym *DAMA* is used for *demand-assigned multiple-access.* There are various types of DAMA control mechanisms.

Demand-assigned TDMA has been used experimentally on cable television channels. Because a television cable interconnects many locations there could be communication *among* the locations if the right control mechanisms were used. A digital stream of many millions of bits per second is derived from one television channel downstream, and a similar bit rate is used upstream. The bit stream can be subdivided for voice, message, or still-image communication. At some future time when optical fiber CATV cables are used there will be a high enough bit rate to allow video communication among homes on the cable or between homes and facilities at the cable head.

A wide variety of transmission facilities could employ TDMA. Figure 31.1 illustrates some of them. They include radio systems, satellite transponders, and multipoint terrestrial cables.

Multiple-access is the key to efficient use of satellites with many user locations, as discussed in Chapter 13. Satellite earth stations are coming into use with control mechanisms which permit many locations to share a transponder on a demand basis. In the past such mechanisms have used frequency-division multiple-access. In the future high-speed burst modems for satellites will permit use of TDMA which can give more flexible channel allocation. A somewhat similar TDMA mechanism could permit the sharing of a radio channel with geographically scattered users, a CATV cable, a tree structured data link, or a telephone company PCM channel such as the T1, T2 or T4 carrier links, or CCITT equivalents. TDMA would be of value with digital telephone links when the demand for channels fluctuates greatly.

Some computer systems have employed a wideband ring channel with many users sharing it, and multiple-access control. Multiple-access has been used for decades on lower bandwidth lines interconnecting telegraph machines or data terminals, with the purpose of lowering the total line mileage and hence the cost. Such lines are referred to as *multidrop* lines and are usually controlled today with some form of *polling* discipline in which each location in turn is asked whether it has anything to send.

RESERVATIONS

Most TDMA systems employ some form of reservation mechanism rather like an airline seat booking system. The time slots during which a user can transmit may be thought of as being rather like the seats on airplanes. Many users in many geographical locations may want to book seats. It is therefore necessary to link these users to a central reservation computer.

A TDMA system may employ a reservation computer. A time-derived subchannel is used as a control channel which connects all users to the reservation computer. When a user station wants to transmit it sends a request on the control channel. The reservation computer receives the request and deter-

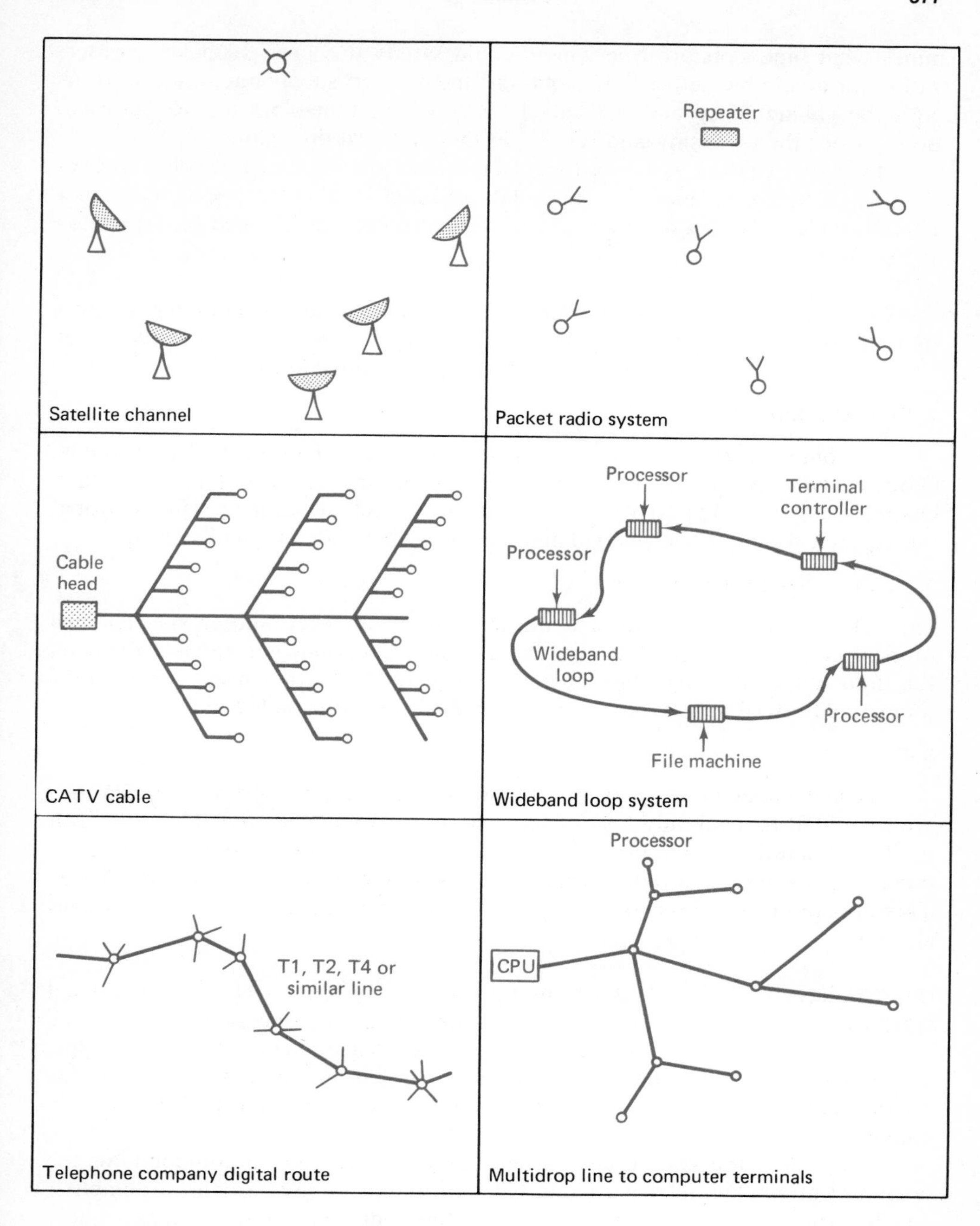

Figure 31.1 Types of channel needing multiple-access control. Box 33.2 summarizes the techniques used for multiple-access control.

mines what time-slots are free which would satisfy this user. It makes a reservation for a suitable set of time-slots—a time-derived subchannel—and informs both the calling party and the called party which time-slots are to be used. Both parties then transmit and receive in the allocated time-slots.

In some TDMA systems the reservations are all for the same type of channel—a voice channel, say, or a data channel of a given speed. On other systems widely varying channels are derived from the high speed bit stream as in Fig. 30.8.

THREE TYPES OF CONTROL

A mechanism is necessary for maintaining control over channel allocations. There are three types of ways of controlling the transmission:

1. Central Control

A computer at a central control point can accept requests for channels, allocate channels to the requestors, and inform the interested parties which channels are allocated to whom. The requests and allocations will be transmitted on a common control-signaling channel which all stations listen to.

2. Decentralized Control

There may be no central control location, but each station has its own form of control. Any station requiring channel space makes a request for it on the common control-signaling channel. Every other station hears the request, and some form of joint protocol determines how the channels are allocated.

3. Contention

A high-capacity channel may be shared in a free-for-all fashion. Stations are permitted to send only a short burst of information at a time. They do so at random. Sometimes the bursts from different stations collide and damage each other. Each station can detect when this happens and so retransmits its bursts. This apparently reckless form of operation has advantages in certain circumstances.

CENTRALIZED CONTROL

Many multiple-access systems will use centralized control. A computer at one location receives the requests for channel space, makes the allocations, and notifies the calling and called parties. Each station communicates with this central controlling computer continuously on a permanently derived subchannel.

To protect the system against the failure of the control computer, or the communications link to it, a second location may also have a control computer on "hot standby," i.e., ready to take over the controlling function immediately if necessary. On some ultra-reliable designs, *any* station could take over the control function.

On a satellite system, a station wanting to transmit has to communicate with the satellite via the subchannel which is permanently used for control. In Fig. 31.2 the centralized control is performed by a computer at earth station 2.

If earth station 1 wants to transmit to earth station 10, it sends a request on the common control channel. The request received by earth station 2, where a controlling computer examines a list of free channels and allocates one to satisfy earth station 1's request. Earth station 2 sends a message on the common control channel to earth stations 1 and 10 saying that a certain channel has been allocated for transmission between them. When the transmission ends, either earth station 1 or 10 signals earth station 2, indicating that the transmission is over. The controlling computer updates its list of free channels so that the channel which was used can now be allocated to another user.

Figure 31.2 Centralized control of demand-assignment performed by a computer at earth station 2.

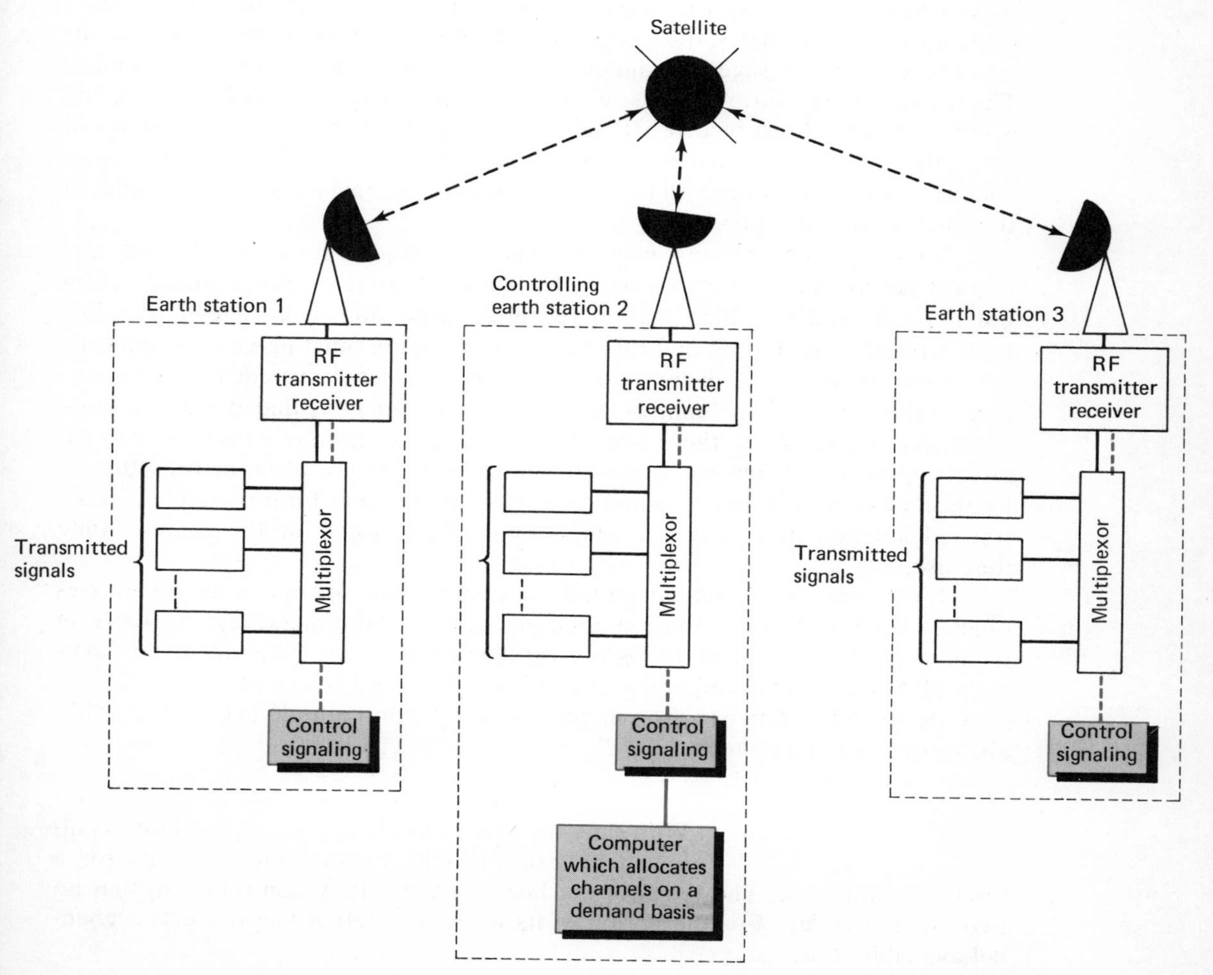

On a satellite system, when a station requests a channel, it cannot have it immediately. It usually has to wait at least 540 milliseconds for its request to reach the controlling location and the response to be returned. For much transmission this delay does not matter. It is trivial compared with the time it takes a telephone user to obtain a dialed connection.

If a transmission is very short, for example, a few characters traveling between a computer and a terminal, then the overhead of reserving a channel for it and terminating the reservation may be excessively high. Different techniques have therefore been used for computer terminal users employing satellite channels.

DECENTRALIZED CONTROL

With decentralized control of channel reservation, the list of available channels must be maintained at *every* station rather than at a controlling location. A station wishing to transmit to another station selects a free channel and sends a control message to that station requesting permission to transmit on that channel. The control message is again sent on a common control-signaling channel. The recipient station responds if the selected channel is still free saying whether it is ready to receive. If the originating station receives this go-ahead correctly, it transmits. When the channel is finished with (the transmission is ended), a control message will be sent to that effect so that all stations can update their list of available channels.

Two separate stations may, by chance, select the same channel and request permission to transmit on it. In that case some preprogrammed protocol must determine which station receives permission to go ahead. Possibly both requests are negated and the stations in question must make new requests.

One advantage of decentralized control is that the system is not vulnerable to the failure of one control station. A disadvantage is that it may be more expensive especially if there are many stations. Centralized control may be more appropriate when elaborate allocation schemes are used such as the allocation of many different channel capacities. With either form of control a centralized location may have the function of billing users for the channel time they use.

Both centralized and decentralized control can operate with either frequency-division or time-division multiple access. Multiple-access systems in major use prior to the late 1970s used frequency-division. They included Comsat's SPADE system [3,6] and General Electric's ES 144 system.

As digital techniques drop in cost, time-division multiple access will probably move into dominance.

BURSTS

With time-division multiple access, each station is allowed to transmit a high speed burst of bits for a brief period of time. The times of the bursts are carefully controlled so that no two bursts overlap. For the period of its burst the station has the entire channel available to it.

In order to make TDMA operate, *burst modems* are employed to transmit high-speed bursts of data. Each burst carries its own means of synchronization, so can be transmitted and received in isolation. Each burst starts with a synchronization pattern which permits the receiving modem to recover the carrier and demodulate it correctly, and to recover the clock timing and know which is the first of the data bits. The burst contains a powerful error-detecting code which permits the receiving device to know whether the data have been received correctly. It also contains control information saying where it originated and where it is addressed to. A small time gap is left between bursts to make sure that they do not interfere. Each burst contains synchronization bits and a preamble containing control information. The stations thus transmit their bursts, one station after another, with accurately controlled timing.

In the simplest form of TDMA, each station in turn is allocated an equal-length burst (in a round-robin fashion). To be efficient, however, the stations must be able to vary their transmission rate, and so either the bursts will be of variable length, or else the scheme must permit some stations to transmit more often than others. A TDMA system with *fixed* rather than *variable* channel assignment needs somewhat less expensive control mechanisms. A *fixed-assignment* station would always transmit the same length burst at the same time in a cycle. A *demand-assignment* station would vary the timing and the duration of its bursts according to the instantaneous traffic volume.

A burst can carry voice, video, data or anything which is digitally encodable. A station may be transmitting many voice channels or possibly a video channel, in which case it is allocated frequent bursts or large bursts. On the other hand it may be allocated infrequent bursts or small bursts because it has relatively few bits to send.

The control channel which is used to convey the requests for channel assignment, and inform all stations what assignments have been made, is sometimes called an *order wire*—a term dating from the early days of manual telephone switching.

A burst from a station may contain information to many other stations. Each station receives every burst. A station extracts only those items from the bursts which are addressed to it. It stores the items in a buffer and assembles, burst after burst, the telephone signals and other traffic that were sent to it. The control of the bursts, and allocation of bursts to users, is done by a mechanism which resembles parts of a computer operating system. Because the bursts can be of widely varying length, and the bits can carry all types of signals, the technique is extremely flexible.

FUNCTIONS OF A SATELLITE TDMA SYSTEM

Figure 31.3 shows the essential functions of a TDMA system which handles voice and data traffic.

The control units which encode the signal into the digital form in which it is transmitted and decode it after transmission, could be remote from the earth station. They relay a bit stream to the multiple-access units, possibly over a terrestrial T1 or other PCM

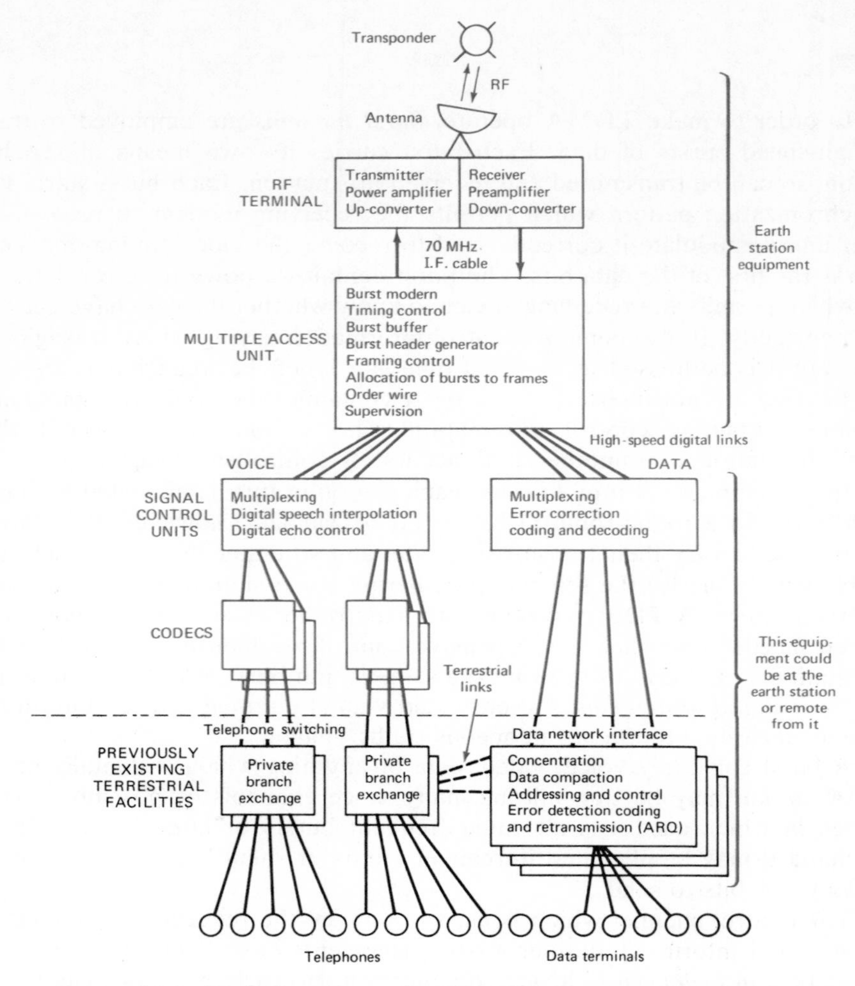

Figure 31.3 Basic functions of a TDMA system used for telephone and data traffic.

carrier link, possibly over a private microwave or millimeterwave link or other wideband facility.

The unit which processes speech ready for transmission is separate from that which prepares data. Data require an error-correcting code, whereas speech may not. Digital speech interpolation is shown being used to condense the telephone traffic. Data, therefore, cannot be sent over the voice channels because of the clipping of speech that occurs. Digital *echo* control is of major importance in the speech processing.

High speed bit streams from both the speech and data access units are sent to the multiple-access facility which buffers this traffic and transmits it in appropriately timed bursts.

The speech and data access units are designed to interconnect to existing terrestrial facilities such as private branch exchanges, corporate tie-line networks, data network concentrators, and so on.

The U.S. domestic satellite system proposed by SBS for 1979 launch uses demand-assigned TDMA. Users can request a variety of voice, date, and image channels, any of which can be derived from one transponder using burst modems operating at 41 mb/s. Separate corporate networks, east coast or west coast, can share transponders. Particularly interesting, the system will provide a *pool* of capacity which is not permanently allocated, but which can be shared by all stations which require a short period of data transmission at a rate in excess of 448 Kb/s. Such demand-assignment of high-speed data channels could lead to new types of uses of computers. The FCC application for the SBS system comments as follows [7].

Today, a customer often cannot connect his remote locations to his central computer facilities because of the large communications costs involved. With the proposed SBS system, the distinction between central and remote computing is virtually eliminated. Central computers will be able to communicate with remote computers at virtually the same high data rates at which they process data internally. A company's data base can, in effect, be moved out to remote processors at all traffic concentration points, and thus be much closer to the company's most remote operations. This new accessibility will reduce the customer's terrestrial communications expense and improve the data processing service that his remote locations receive.

The wideband capability of the proposed SBS system is also expected to be a catalyst for development of an emerging series of facsimile and video applications and products that will help companies communicate more efficiently.

REFERENCES

1. These calculations are discussed in the author's *Systems Analysis for Data Transmission* (Englewood Cliffs, N.J.: Prentice-Hall Inc., 1972).
2. Concentrators are discussed in the above book.
3. E. R. Cacciamani, "The SPADE System as Applied to Data Communications and Small Earth Station Operation," COMSAT Tech. Rev. Vol. 1, No. 1, Fall 1971.
4. B. I. Edelson & A. M. Werth, "SPADE System, Progress and Application," COMSAT Tech. Rev. Vol. 2 No. 1, Spring 1972.
5. N. Abramson, "Packet Switching with Satellites," AFIPS Conference Proc., National Computer Conference, Vol. 42, 1973.
6. L. Kleinrock, and S. Lam, "Packet Switching in a Slotted Satellite Channel," AFIPS Conference Proc., National Computer Conference, Vol. 42, 1973.
7. Satellite Business Systems FCC application for a Domestic Communications Satellite systems, FCC, Washington, D.C., 1976.

In 1897, Marconi succeeded in communicating by wireless eight miles across the Bristol channel with this apparatus—the first wireless transmission across water. The upper two pieces are a Righi spark gap (left) and an induction coil. Below is a Morse key. (*Courtesy The Marconi Co., Ltd.*)

32 PACKET RADIO

Packet radio, discussed in Chapter 11, is one alternative to the local telephone loops for interactive computer terminals. Because of this, it could be very valuable in countries with underdeveloped telephone systems. In highly developed countries, it could provide mobile or portable terminals. Its enthusiasts imagine a large market of pocket-calculator-sized machines containing microcomputers and a radio link to the nearest packet-switching node.

Packet radio needs a form of multiple-access control which permits many scattered devices to transmit on the same radio channel without harming each other's transmission. There is no control problem in broadcasting from the control station *to the devices.* The problem lies in transmitting *from the devices* to the central station—a central computer, concentrator, or packet-switching node.

Unlike a satellite channel, the transmitting devices need to be cheap and simple and channel bandwidth is not great. The messages sent from the radio terminals are mostly very short. An elaborate form of multiple-access control with messages going back and forth to a control center presents too much overhead. Yet demand-assignment of channel capacity is definitely needed.

A simple form of control referred to as ALOHA control is used on such systems. It was first employed by Abramson at the University of Hawaii, where he applied it to satellite channels as well as to terrestrial radio links [1].

RANDOM PACKETS

ALOHA control is a very simple, but effective, *contention* scheme. Each terminal is attached to a transmission control unit with a radio transmitter and receiver. The data to be transmitted are collected in the transmission control unit which forms them into a "packet." The packet contains the addresses of the receiving location and the originating terminal, and some control bits, in a header. The Hawaii packet contains up to 640 bits of data following a 32-bit header, and is protected by a powerful error-detecting code which uses 32 redundant bits in each packet. The transmission control units cost $3000 to build in 1973. Abramson estimates that they could cost less than $500 to produce in quantity [2]. They could be cheaper than modems.

When the packet is complete, the transmission control unit transmits it at the maximum speed of the link. These packet transmissions take place *at random*. All devices receiving in that frequency receive the packet. They all ignore it with the exception of the device (or devices) which the packet is addressed to. A device which receives a packet addressed to it transmits an acknowledgement if the packet appears to be free from error.

The sending device waits for the acknowledgement confirming correct receipt. It waits for a given period and then if it has not received an acknowledgement, it transmits the packet again.

If one central station is transmitting to many outlying stations—for example, a computer transmitting to its terminals—the central station has control over the times at which the packets are sent. When the outlying stations transmit to the central station, the packets are transmitted at random times. Consequently, they occasionally collide. If two packets are transmitted at overlapping times, both will be damaged, and their error-detecting codes will indicate the damage. The transmission control units which sent the packets will therefore receive no acknowledgement. After a given time, each transmission control unit retransmits its packet. It is important that they should not retransmit simultaneously or the packets will again collide. Each transmission control unit should wait a different time before retransmitting.

The different waiting times could be achieved by giving each transmission control unit a different built-in delay. However, there are many such control units and the one with the longest delay would be at a disadvantage. Therefore, the ALOHA scheme gives each control unit a randomizing circuit so that the time it waits before retransmission is a random variable. The retransmitted packets are unlikely to collide a second time but there is a very low probability that this will happen and a third retransmission will be made. A second, or even third collision does little harm if the delays before reattempting transmission are low compared to the desired terminal response time. If a packet is damaged by other causes, such as radio noise, it will similarly be retransmitted. The ALOHA protocol is illustrated in Fig. 32.1.

Packet radio is a little like an unruly auction in which participants all shout their bids and the auctioneer acknowledges them. If a bid is not acknowledged, the person shouts it again. Occasionally, two persons shout a bid at once and then both have to repeat their bid.

FREQUENCIES

A packet radio system can use one frequency or two. The Hawaii system uses two 100 KHz channels in the UHF band at 407.350 MHz and 413.475 MHz, one for transmitting *to*

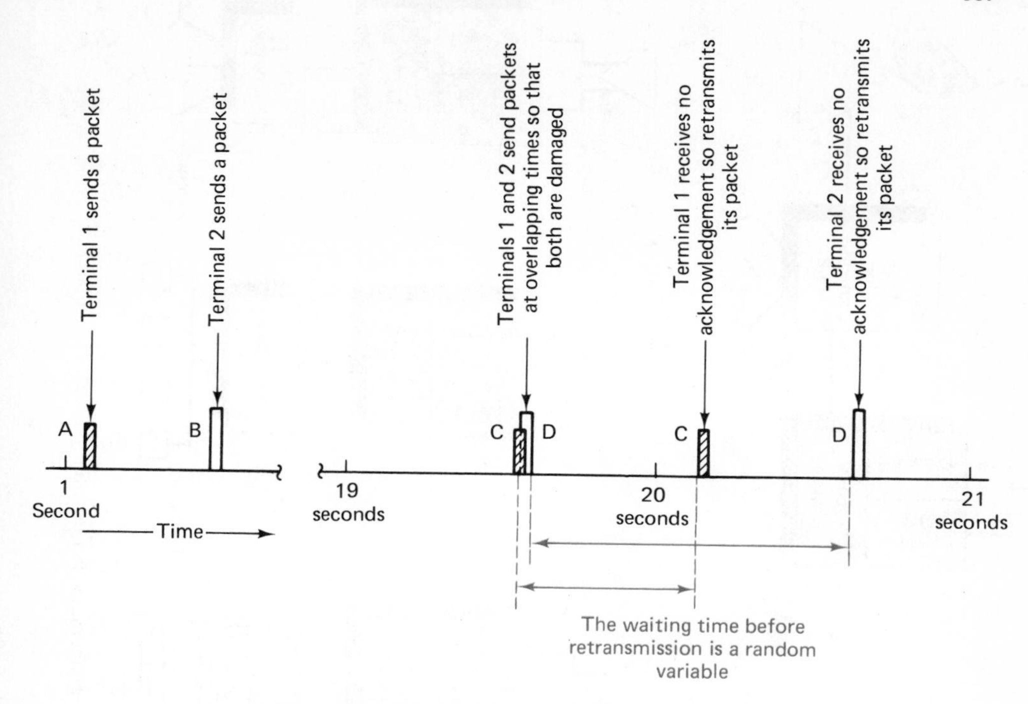

Figure 32.1 The "classical" ALOHA protocol.

the terminals and one for transmitting *from* them. If a separate channel is used for transmitting from the computer to the terminals, there is no problem with interference on this channel. (The bidders can hear everything the auctioneer says.) In most computer dialogues, the terminal says much less than the computer, so the channel from the terminals, on which packets can collide, will be utilized much less than the channel *to* the terminals.

On the original Hawaii system, the 100 KHz bandwidth channels transmitted 24,000 bits per second. Fixed-length packets of 704 bits were used. Each packet thus occupied the channel for approximately 29 milliseconds. Figure 32.1 is drawn approximately to scale using these figures.

Many different types of data-processing devices may be interconnected using ALOHA broadcasting (Fig. 32.2). The transmission control unit may be attached to a terminal, or concentrator or controller handling many terminals, a computer in a central data processing location, and IMP of a packet-switching network, a satellite earth station controller, and so on.

CHANNEL UTILIZATION

It is clear that the ALOHA protocol will work well if the number of transmissions is small compared with the total available channel time. A more interesting

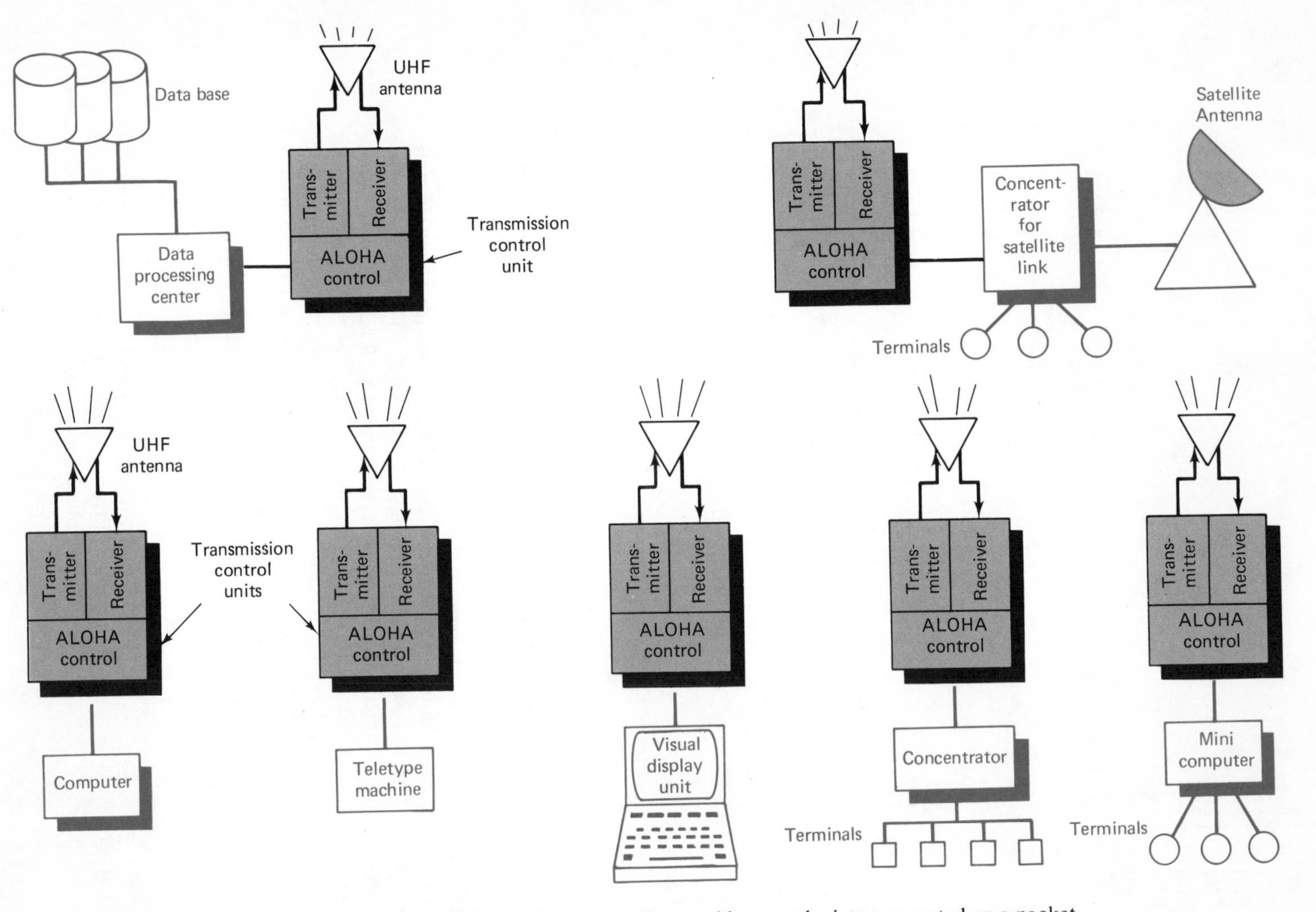

Figure 32.2 Many different data processing machines can be interconnected on a packet broadcasting network.

question is: what happens when the channel becomes busy? Abramson's analysis of the ALOHA channel proceeds as follows:

Suppose that all of the users of an ALOHA channel originate λ packets per second on average.

Suppose that the packets are of fixed length of duration T seconds.

The utilization of the channel as perceived by the users, which in conventional queuing theory is referred to as p, is then:

$$\rho = \lambda T$$

$$\rho \text{ is } \frac{\textit{the total time the channel is in use for sending original packets}}{\textit{the total time}}$$

In reality, there will be more than λ packets using the channels because some packets have to be transmitted more than once because of collisions. Let the total number of packets per second, including retransmitted packets, be λ'.

Because of the retransmitted packets, the actual channel utilization will be greater than that perceived by the users—greater than ρ. To understand how the channel behaves when it becomes busy, we wish to know the relationship between ρ and λ'.

The assumption normally made in the analysis of interactive systems is that the users originate messages *independently* and *at random*. In other words, the probability of a message starting in a small time Δt is proportional to Δt. If this assumption is valid, then there is a Poisson distribution of the number of messages originating per second. In this calculation, we assume a Poisson distribution of the number of packets per second.

The probability of n packets originating in a second is:

$$\text{Prob } (n) = \frac{\lambda'^n \, e^{-\lambda'}}{n!}$$

The probability of no packets originating in a time of duration t is:

$$[\text{Prob } (n=0)]^t = e^{-\lambda' t}$$

Suppose that one particular packet originates at a time, t_o. There will be a collision if any other packet originates between the times $t_o - T$ and $t_o + T$. In other words, there is a time period of duration $2T$ in which no other packet must originate if a collision is to be avoided.

The probability of no packet originating in the time, $2T$, is $e^{-\lambda' . 2T}$. This is the probability a packet will have no collision.

Let R be the fraction of packets that have to be retransmitted.

Then:

$$R = 1 - e^{-\lambda' . 2T} \qquad [32.1]$$

The relation between the number of user-originated packets and the actual number of packets on the channel is therefore:

$$\lambda = \lambda' \ [1 - R] = \lambda' \ e^{-\lambda'.2T}$$

Substituting into $\rho = \lambda \ T$:

$$\rho = \lambda' \ T \ e^{-\lambda.2T}$$

Sometimes a retransmitted packet will collide a second time and have to be retransmitted again. The mean number of times a given data packet is retransmitted is:

$$N = 1 + R + R^2 + R^3 \ldots$$

$$= \frac{1}{1 - R}$$

$$= e^{\lambda'.2T} \quad [32.2]$$

Substituting $\lambda' \ T$ in equation 32.1 we have

$$\rho = \frac{\log_e N}{2N} \quad [32.3]$$

In Fig. 32.3, ρ is plotted against N.

It will be seen that as the channel utilization (as perceived by the user) increases, the traffic which attempts to use the channel builds up at an increasing rate. A chain reaction develops, with the retransmitted packets themselves causing retransmissions so that above a certain throughput the channel becomes unstable.

Differentiating equation 32.2 shows that the maximum value of ρ is $1/2e = 0.184$.

The maximum utilization of a classical ALOHA channel is 18.4%.

It is clearly desirable to stop the users of an ALOHA channel transmitting when they reach the unstable part of the curve in Fig. 32.3. Batch transmission devices might be made to pause for a while when any packet is retransmitted *twice*. Interactive users might be slowed down either by giving them an artificially long response time, or by giving them a visible warning. (Reference 9 discusses the stability of heavily-loaded packet radio systems.)

PERFORMANCE IMPROVEMENTS

The teleprocessing expert used to optimizing landline systems to give high line utilization will be horrified by the figure of 18.4% maximum channel utiliza-

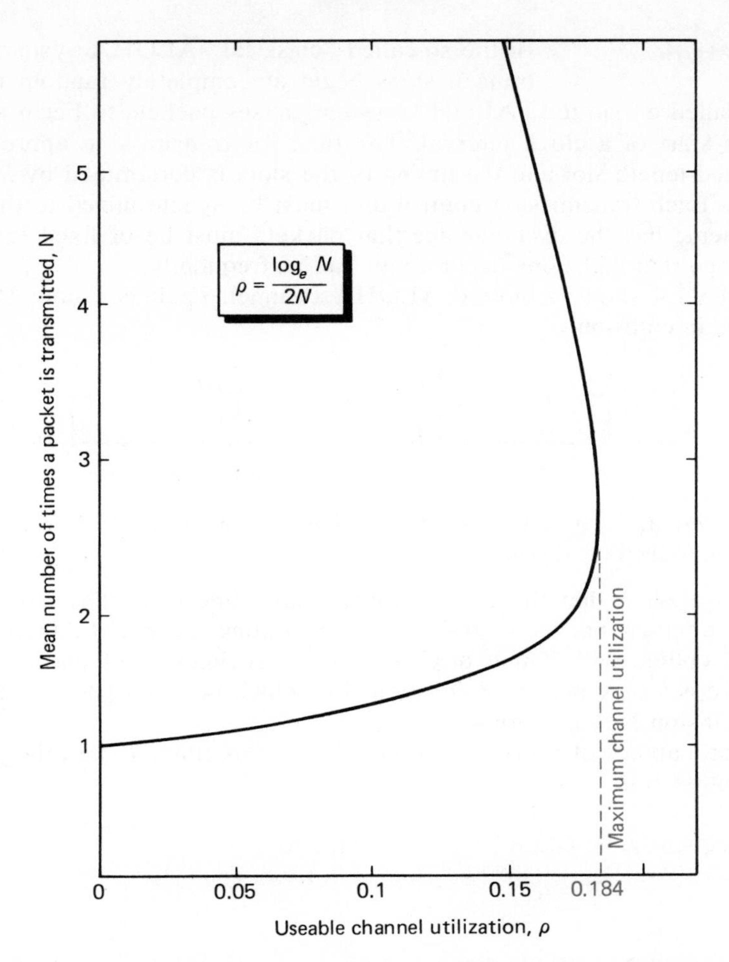

Figure 32.3 Traffic on a "classical" ALOHA channel. The maximum useable channel utilization is $\frac{1}{2e} = 0.184$.

tion. In reality an ALOHA system is unlikely to run at more than about 16% channel utilization because of need to avoid the unstable part of the curve in Fig. 32.3.

There are, however, several ways to improve the channel utilization. We will discuss three techniques which have been used:

1. Slotted ALOHA
2. Carrier sense
3. FM discrimination

With a combination of these three techniques, channel utilization becomes as high as that on typical land-line systems.

SLOTTED ALOHA

In the so-called "classical" ALOHA system packet, transmissions begin at completely random times. A variation called a "slotted" ALOHA system causes packets to begin transmission at the start of a clock interval. The time for transmission approximately fills one fixed-length slot and the timing of the slots is determined by a system-wide clock. Each transmission control unit must be synchronized to this clock. Such a scheme has the disadvantage that packets must be of fixed length, but the advantage that collisions occur about half as frequently.

Figure 32.4 shows a slotted ALOHA channel. Packets C and D on this diagram are in collision.

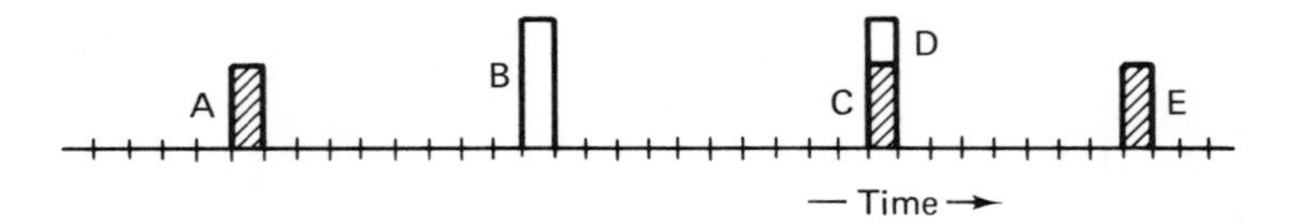

Figure 32.4 A slotted ALOHA channel. Each packet must begin on one of the time divisions.

Suppose again that the packet transmission time is T. The slot width is also T. If a given packet is transmitted beginning at time t_0, then another packet will collide with it if it originates between times $t_0 - T$ and t_0. In other words, there is a time period of duration T in which no other packet must originate if a collision is to be avoided.

The probability of no packet originating in this time, T, i.e., the probability of no collision is $e^{-\lambda' T}$.

Consequently, equation 32.1 is modified to:

$$\rho = \lambda' T\ e^{-\lambda' T} \quad [32.4]$$

Equation 32.2 is modified to:

$$N = e^{\lambda' T} \quad [32.5]$$

Substituting, we then have:

$$\rho = \frac{\log_e N}{N} \quad [32.6]$$

This is plotted in Fig. 32.5. The channel utilizations for a given retransmission rate are double those for classical ALOHA channel.

The maximum utilization of a slotted ALOHA channel is 36.8%.

The transmission control unit is slightly more complicated because it must maintain the slot timing. A transmitter will send out periodic timing pulses for the entire system and each control unit must synchronize its activity to these pulses.

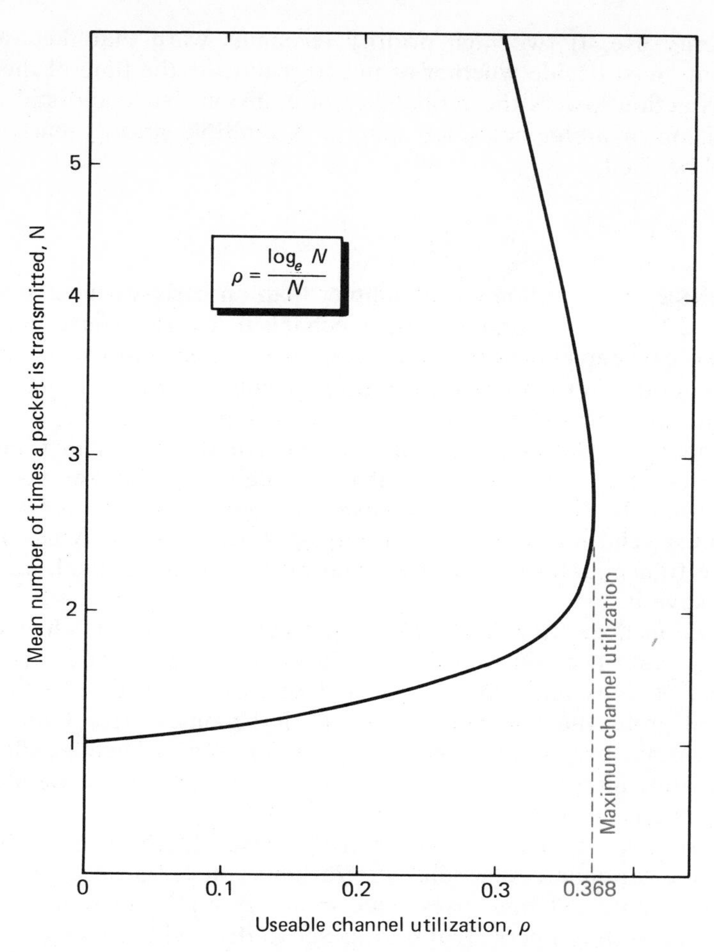

Figure 32.5 Traffic on a slotted ALOHA channel (Fig. 32.4). The channel utilization is twice that of the classical ALOHA channel of Fig. 32.3, but only fixed-length packets are sent.

EARLY WARNING Any technique which prevents the separate transmission control units from transmitting at the same instant will improve the utilization of a packet radio channel. One possibility is to use a reservation protocol which permits the control units to reserve certain time slots. Another technique is to use an early-warning system.

An early-warning system may divide the terminals up by priority. If there are two priority classes, the high priority terminals send out a signal at a fixed interval before they transmit. All transmission control units hear this warning signal. The low priority ones avoid the time slot in question leaving it free for the high priority use. If two high priority terminals warn that they want the same slot, both must decide whether or not to reallocate the time of their transmission. This action lowers the probability of collision, but could still result in either a collision or a slot being left empty. A multiple priority early-warning system could be used.

CARRIER SENSE

Somewhat simpler than an early-warning system is a *carrier sense* mechanism. Carrier sense implies that early terminal can detect when another terminal is transmitting—it can detect its carrier. If another terminal is transmitting, it will avoid starting its transmission at that instant because it would cause a collision.

Unfortunately, while all terminals can receive the powerful transmission from the center, they may not all be able to receive the weaker transmission from other terminals. In this case, to make a carrier sense mechanism operate, the center must send out a *busy signal* saying when it is receiving. The busy signal will be transmitted on a narrow bandwidth control channel, and all stations will receive it.

What action does a terminal take if it wants to transmit when the busy signal is being sent? It could wait until the busy signal ceases and then immediately transmit. This is referred to as *persistent* carrier sense. The terminal is persistent and grabs the channel as soon as it becomes free. Unfortunately, there may be other persistent terminals that do the same. Then a collision may occur immediately after the busy signal ceases as more than one persistent terminal grabs the channel.

An alternative is *nonpersistent* carrier sense. Here, if the channel is sensed busy the terminal reschedules its transmission for a later time, using a random delay again, and then tries once more. A higher channel throughput can be achieved with nonpersistent carrier sense than with persistent. The delays are longer, at least at low channel utilizations, but usually not long enough to matter.

A third type of carrier sense gives maximum channel utilizations almost as high as nonpersistent carrier sense, but gives lower delays when the traffic is heavy. This is called *p-persistent* carrier sense. Here, if the channel is sensed idle, the terminal transmits the packet with a probability p. With a probability $1 - p$, it waits briefly and then repeats the process. p can be set at whatever figure gives the best performance.

These three types of carrier sense can be used with slotted or nonslotted ALOHA systems. They are summarized in Table 32.1, applying to slotted systems.

Figure 32.6 shows Fig. 32.5 redrawn to show the effects of carrier sense [4] Using nonpersistent carrier sense or p-persistent where the value of p is low, channel utilization can be as high as 80%.

The effectiveness of carrier sense depends upon the propagation delay

Table 32.1 Three types of carrier sense mechanism

	Nonpersistent Carrier Sense	Persistent Carrier Sense	$p-$Persistent Carrier Sense
IF THE CHANNEL IS SENSED IDLE:	The terminal transmits the packet.	The terminal transmits the packet.	With probability p, the terminal transmits the packet. With probability 1−p, the terminal delays for one time slot and starts again.
IF THE CHANNEL IS SENSED BUSY:	The terminal reschedules the transmission to a later time slot according to a retransmission delay distribution. At this new time it repeats the algorithm.	The terminal waits until the channel goes idle and then immediately transmits the packet.	The terminal waits until the channel goes idle and then repeats the above algorithm.
		NOTE: This is a special case of $p-$persistent carrier sense	

Figure 32.6 The effect of carrier sense techniques on packet radio channel utilization [4].

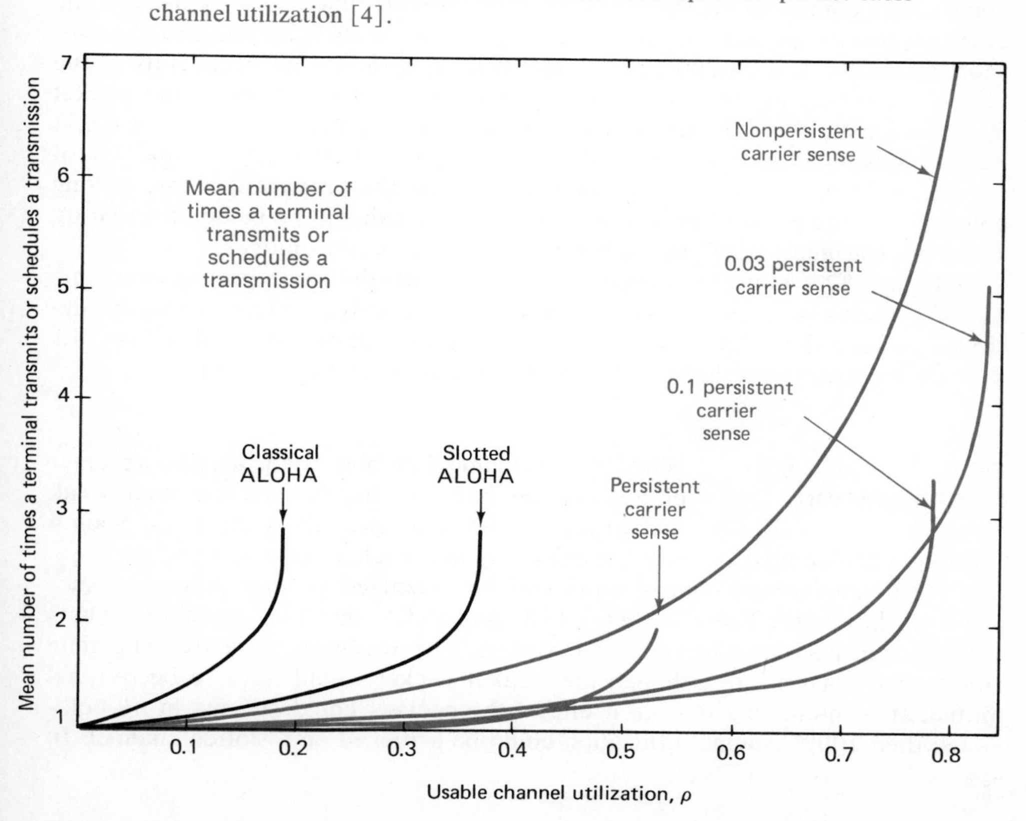

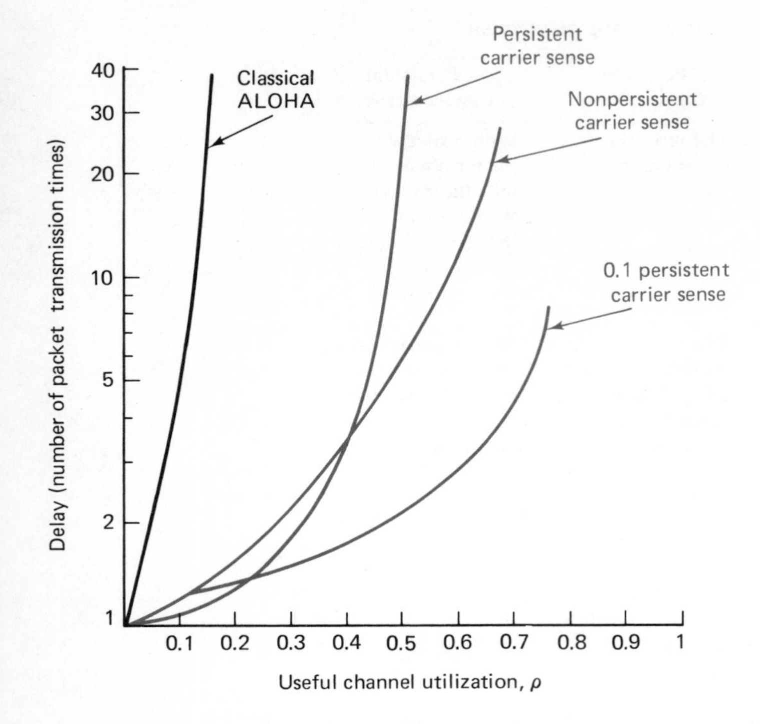

Figure 32.7 Packet delay with different carrier sense techniques. (*Redrawn from Reference 4.*)

being short compared to the message transmission time. This is usually the case because the packets are transmitted at the speed of light, and usually over short distances. If a 24,000 bps channel is used with 704 bit packets as in the Hawaii system, and the average distance transmitted is 10 miles, the packet transmission time is 29 milliseconds and the radio propagation time is 0.054 milliseconds. The curves in Fig. 32.6 are drawn for a propagation time 1% of the packet transmission time. If for some reason the propagation time is long compared to the packet transit time, carrier sense can still be made effective by sending a warning "busy" signal before the packet is transmitted.

Figure 32.7 plots the delays that occur (with the same propagation time ratio). To achieve a fast response time on an interactive system, it may be desirable to keep the delay less than 10 packet transit times, say. In this case, 0.1 persistent carrier sense appears the best of the three in Fig. 32.7 [4].

FM DISCRIMINATION

When the FM modulation is used, a radio receiver circuit can be built to discriminate between weak and strong signals, and reject the weak ones. Such a technique can greatly increase the efficiency of a packet radio system.

The transmission control units can be organized so that they vary substantially in power. There is then a high probability that two colliding packets will differ in power sufficiently that one will be received correctly. The time slot is then not wasted, although the weaker packet would have to be retransmitted. It is interesting to note if one of the packets could survive in *all* collisions, then 100% channel utilization could be achieved on a slotted channel. In practice, the discrimination could never be this good, but it might be a reasonable objective that one packet should survive in half of the collisions. The maximum channel utilization could then be high, especially if carrier sense is also used.

Unless a transmission control unit had the ability to change its transmitting power, discrimination would have the effect that some terminals always had low priority. This may or may not be a disadvantage. It would be advantageous, for example, to give batch terminals a lower priority than interactive ones. Interactive terminals with a dialogue structure needing a fast response should be given priority over those for which a longer response time is acceptable.

REPEATERS

In a packet radio system the terminals are likely to transmit to some controlling location. This might be a computer center. It might be an ARPANET IMP or a Telenet Central Office. It might be a satellite earth station.

Some of the terminals may be too far away from the controlling location for an inexpensive transmitter to reach it. In such a case, radio repeaters are used. The transmitters used in the Hawaii system have a range of many miles, but repeaters are used for interisland transmission and to reach terminals in the shadow of mountains.

Conventional radio repeaters cannot receive and transmit on the same frequency at the same time because the strong transmitted signal blots out the weak incoming signal. They receive on one frequency band, translate the signal to a different frequency band, and then transmit it. A packet repeater can, however, operate on a single frequency. It switches off its receiver momentarily while it transmits a packet. Some packets will be lost when this happens, and, as with a collision, they will have to be retransmitted. Operating at a single frequency saves the expense of the frequency translation equipment.

The Hawaii repeaters use a single frequency for relaying packets to the control location, and a different frequency for relaying packets back to the terminals. If two frequencies are used in this way, the repeater antennas pointing towards the central station can be highly directional.

When the terminals, perhaps hand-held terminals, are distributed over a wide geographic area, many repeater stations may be used (Fig. 32.8). Such may be the case in areas of low population density and large distances, where the alternatives to radio would be expensive. In areas of high population density, such as a city, repeaters may be used to avoid problems with high-rise buildings.

When large numbers of terminals are used, it is desirable to select the design options such that the central location is complex, the repeaters less expensive, and the terminals as inexpensive as possible.

When multiple repeaters are used, a problem that must be solved is the cascading of packets. A packet from certain terminals may reach more than one repeater, or reach the central station as well as a repeater. Unless the repeaters have directional antennas, packets from one repeater may reach other

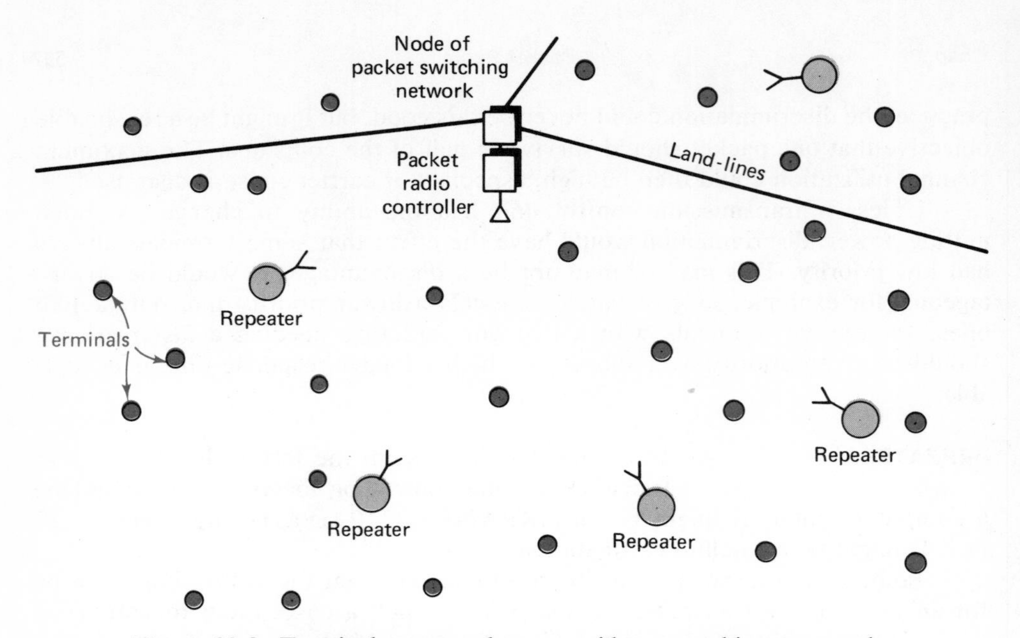

Figure 32.8 Terminals scattered over a wide geographic area may be served by multiple packet radio repeaters relaying transmissions to a central control station.

repeaters. Some packets might travel endlessly in loops. The unnecessary cascading of packets could substantially lower the effective channel capacity. It is necessary to devise a protocol that prevents packets from multiplying themselves.

HAND-HELD TERMINALS AGAIN

Let us return to the pocket terminal. Such devices might be used on a 50,000 bps UHF channel. In a man-terminal dialogue using the 256 character screen, the *average* transmission rate to the terminal from a distant computer over a long period of dialogue might typically be 10 characters per second. If these are transmitted as 7-bit characters, the channel might accommodate 500 active terminals. Many messages *to the computer* would be one character responses and some would be longer data entries. The transmission rate to the computer might typically average 1 character per second. The channel requirement *to the computer* would not exceed that *from the computer* even with the low channel utilization of a classical ALOHA system. A single system might therefore have thousands of users of whom not more than 500 would be likely to be active at once.

A typical input packet would be a few milliseconds in duration. A full-screen output message would be about 50 milliseconds. The response time of the device could therefore be very fast.

Its potential applications are as diverse as the computer industry itself.

REFERENCES

1. The ALOHA System is a research system at the University of Hawaii, supported by ARPA under NASA Contract No. NAS2-6700 and by the National Science Foundation under NSF Grant No. AJ-33220. N. Abramson, "The ALOHA System—Another Alternative for Computer Communication," Fall Joint Computer Conference, AFIPS Conf. Proc., Vol. 37, 1970.
2. N. Abramson, "Packet Switching with Satellites," National Computer Conference, AFIPS Conf. Proc., Vol. 82, 1973.
3. L. Kleinrock, and S. Lam, "Packet Switching in a Slotted Satellite Channel," National Computer Conference, AFIPS Conf. Proc., Vol. 42, 1973.
4. L. Kleinrock, and F. A. Tobagi, "Carrier Sense Multiple Access for Packet Switched Radio Channels," Proc. International Conference on Communications, IEEE, Minneapolis, 1974.
5. S. Fralick and J. Garrett, "A Technology for Packet Radio," *AFIPS Conference Proceedings,* Volume 44, 1975, AFIPS Press, Montvale, N.J.
6. H. Frank, I. Gitman, and R. VanSlyke, "Packet Radio Network Design—System Considerations," *AFIPS Conference Proceedings,* Volume 44, 1975, AFIPS Press, Montvale, N.J.
7. S. Fralick, D. Brandin, F. Kuo, and C. Harrison, "Digital Terminals for Packet Broadcasting," *AFIPS Conference Proceedings,* Volume 44, 1975, AFIPS Press, Montvale, N.J.
8. J. Burchfiel, R. Tomlinson, and M. Beeler, "Functions and Structure of a Packet Radio Station," *AFIPS Conference Proceedings,* Volume 44, 1975, AFIPS Press, Montvale, N.J.
9. L. Kleinrock and F. Tobagi, "Random Access Techniques for Data Transmission over Packet Switched Radio Channels," *AFIPS Conference Proceedings,* Volume 44, 1975, AFIPS Press, Montvale, N.J.
10. R. Binder, et al., "Aloha Packet Broadcasting—A Retrospect," *AFIPS Conference Proceedings,* Volume 44, 1975, AFIPS Press, Montvale, N.J.
11. L. Roberts, "Extension of Packet Switching to a Hand Held Personal Terminal," *AFIPS Conference Proceedings,* SJCC72, pp. 295–298.
12. V. Cerf and R. Kahn, "A Protocol for Packet Network Intercommunication," *IEEE Transactions on Communications,* May 1974, pp. 637–648.
13. R. E. Kahn, "The Organization of Computer Resources into a Packet Radio Network," *NCC 1975 Proceedings.*

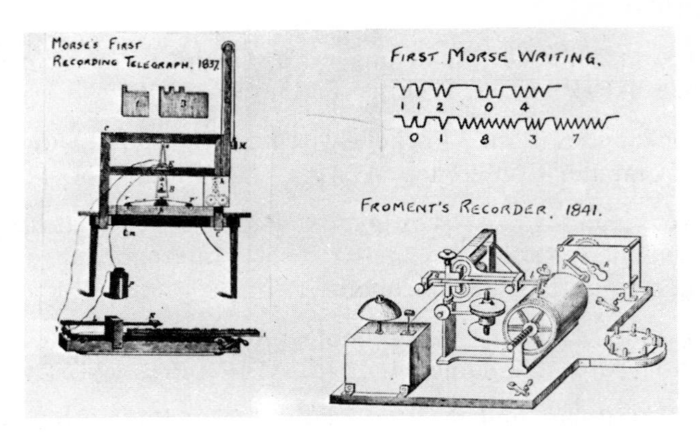

Opening the first London-Paris telegraph circuit November 1852. (*Courtesy The Post Office, England.*)

33 OPTIONS IN NETWORK DESIGN

It can now be seen that there are many options in the design of telecommunication networks. *This chapter summarizes them.*

The traditional telephone networks have been built with an architecture employing frequency-division multiplexing and continuous-channel electromechanical switching. Various services other than telephone have been accommodated into this basic architecture. The time has now come for the architecture which has served so well for decades to give way to new techniques using digital circuits, computers, memory, and satellites.

Digital trunks are becoming available carrying extremely high bit rates. Most such trunks are designed primarily for voice transmission, but clearly offer enormous potential for the burgeoning fields of data transmission, electronic mail, and other services. The digital channels are being created both by the conversion of earlier analog channels, as with Data Under Voice and the digitization of wire-pair trunks, and also by entirely new channels including satellites, waveguides, digital coax, digital microwave, and in the future optical fibers.

Computer designers have been content with data transmission channels of a few thousand bits per second. Now telecommunication highways transmitting many millions of bits per second are being built. If 56,000 bits per second are used for voice transmission, as is common today, then data transmission appears cheap by comparison with voice. In the mid-1960s a switched voice channel was equivalent to 600 bits per second—the speed of the modems in use. In the early 1970s a switched voice channel was equivalent to 2400 or 4800 bps. When PCM voice is used it is equivalent to 56,000 or 64,000 bps. (CCITT recommended speeds for voice transmission.)

To make the digital channels as valuable as possible to future computer users it is desirable that appropriate switching facilities should be connected to them. Fast burst switching is required rather than the slow continuous-channel switching of conventional telephony (the vertical bars in Fig. 10.2 rather than the horizontal bars). New switching facilities can be provided by:

1. First-tier common carriers—the established telephone companies or specialized carriers.
2. Second-tier or value-added common carriers, such as Telenet or Graphnet.
3. Computer manufacturers, providing a range of equipment for interactive computing.
4. Independent manufacturers of data transmission equipment, including manufacturers of satellite earth stations.

In the first two cases the switched channels are a public tariffed offering from common carriers. Public data networks are being built in many countries. In the second two cases the network would be built by a private organization using leased channels. Some of the new types of switched networks will be public networks; some will be private networks built for internal corporate use; some will be private networks built to serve multiple corporations, such as networks transmitting funds between banks or reservations between airlines.

MULTIPLEXING TECHNIQUES

Some new networks will be designed with point-to-point transmission links between the switching or concentration nodes. Others will have multi-point or broadcast links interconnecting many nodes. In the latter case *multiple-access* techniques will be necessary to enable many separate and independent users to share the channel.

Box 33.1 summarizes the multiplexing techniques used on point-to-point links. The channel capacity may be subdivided using *continuous-channel* multiplexing (horizontal bar in Fig. 10.2) or *burst multiplexing* (vertical bar in Fig. 10.2). Continuous-channel multiplexing can be done with a *fixed assignment* of channels, or *dynamic assignment*—the channels being allocated to users in a varying fashion according to their needs.

In traditional telephony a full duplex channel is allocated to the conversation. In future networks *speech interpolation* will be more common—the voice channel being snatched away from a user when he pauses and made available for other speakers. This has been done with frequency-division multiplexing on trans-ocean cables (TASI). In the future it will be done using time-division multiplexing with digital speech in a faster and more effective fashion (DSI, Digital Speech Interpolation).

With traffic of a burst nature, *burst multiplexing* is more efficient than continuous-channel multiplexing, and a variety of burst-multiplexing techniques exist. Burst multiplexing is also referred to as *asynchronous multiplexing*. Packet-switching networks use a form of burst multiplexing. The bursts are likely to vary widely in length, so a control procedure designed for variable-length bursts is more efficient than one for fixed-length bursts.

Burst multiplexing commonly employs storage at the nodes so that the bursts can be queued while they wait for transmission. This would describe the multiplexing that is used with packet-switching networks, and that used with data concentrators. When fast-connect *circuit-switching* is used there is not necessarily any storage at the nodes. Rather than queuing the data itself at the nodes, reservations for transmission time slots may be queued and the circuits are set up instantaneously at the requisite times.

When the multiplexing technique permits *reservations* of time-slots, the reservations could be made well ahead of the time when the information is actually ready for transmission. Look-ahead reservations would be of value when there are severe time constraints on the transmission. For example if a fast real-time response is essential, or if relatively slow continuous signals are interleaved among the burst traffic, such as speech or batch data from an input machine which runs nonstop.

In some transmission it is desirable to multiplex a mixture of burst and continuous-channel traffic. Any large-capacity channel may be subdivided in such a way that it can handle both. This hybrid multiplexing is perhaps particularly important with the high bit rates of satellite channels. Hybrid multiplexing is commonly accomplished by splitting a high-capacity channel into continuous subchannels, and then subdividing these subchannels when necessary with burst multiplexing. The 50 Kb/s lines which are burst-multiplexed in networks like ARPANET are themselves continuous subchannels of higher capacity links.

With many time-division multiplexing schemes, all of the channels derived are identical. With more complex control mechanisms, channels of different speeds can be interleaved and, if desirable, burst channels can be interleaved with continuous channels. Figure 30.8 illustrated a bit stream carrying both continuous and burst traffic. The squares or time-slots in this diagram represent appropriately-sized data blocks. The blocks allocated to continuous channels are reserved. The others are unreserved and can be used asynchronously for burst traffic.

MULTIPLE-ACCESS Box 33.1 was drawn for multiplexing techniques used on channels connecting two points. When a channel connects multiple points the requisite control mechanisms are more complex. When more than two geographically separate locations share the same channel this is referred to as *multiple access*. Multiple-access control mechanisms are summarized in Box 33.2.

Multiple-access is the key to efficient use of satellites, as we discussed in Chapter 13. Satellite earth stations are coming into use with control mechanisms which permit many users to share a broadcast channel. Similar multiple-access control can permit many users to share a CATV cable interactively, or a radio or packet radio channel. Some systems have employed a wideband ring channel with many users sharing it, and multiple-access control. Multiple-access

BOX 33.1 A SUMMARY OF MULTIPLEXING TECHNIQUES

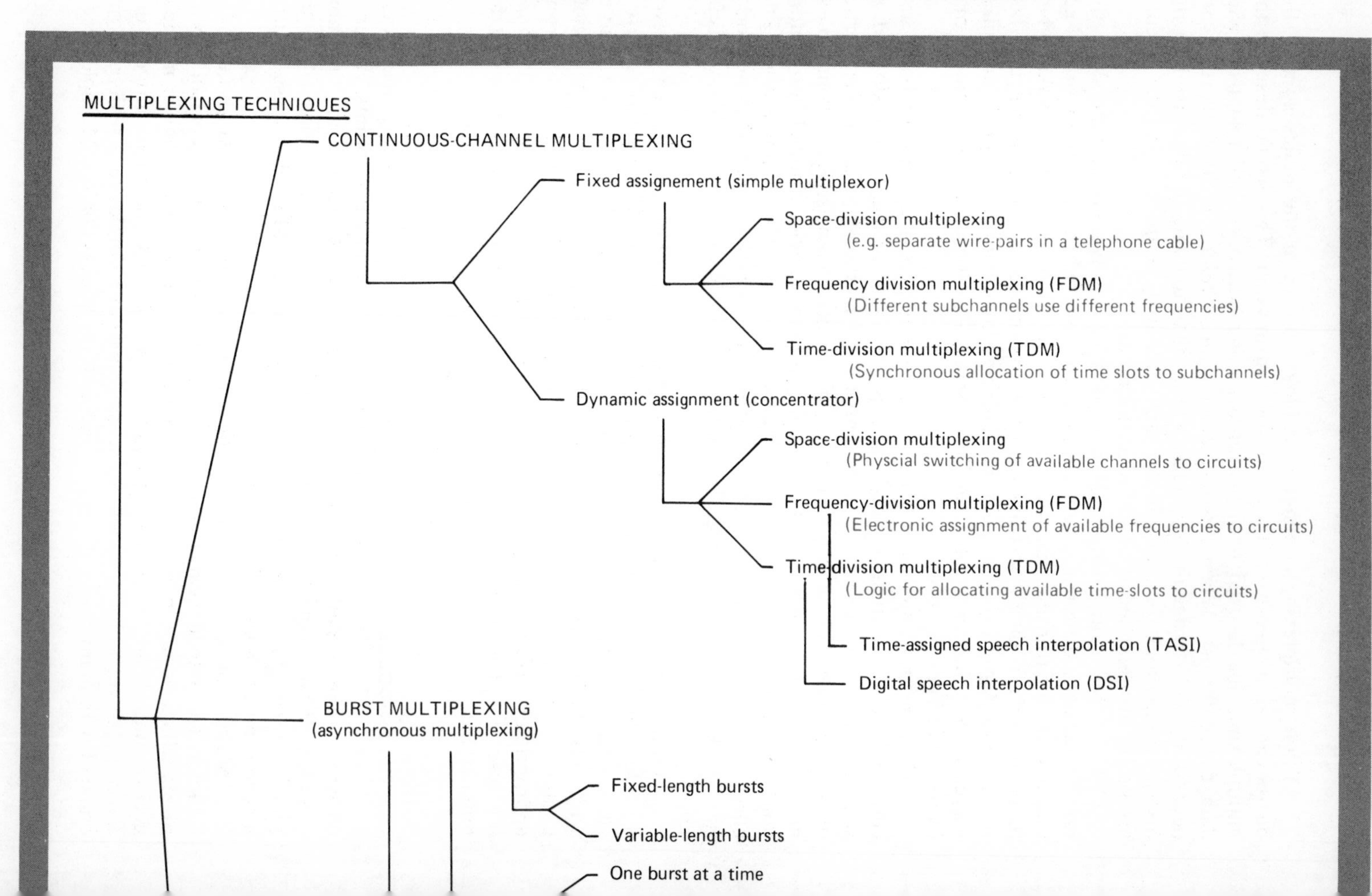

MULTIPLEXING TECHNIQUES
CONTINUOUS-CHANNEL MULTIPLEXING
Fixed assignement (simple multiplexor)
Space-division multiplexing
(e.g. separate wire-pairs in a telephone cable)
Frequency division multiplexing (FDM)
(Different subchannels use different frequencies)
Time-division multiplexing (TDM)
(Synchronous allocation of time slots to subchannels)
Dynamic assignment (concentrator)
Space-division multiplexing
(Physcial switching of available channels to circuits)
Frequency-division multiplexing (FDM)
(Electronic assignment of available frequencies to circuits)
Time-division multiplexing (TDM)
(Logic for allocating available time-slots to circuits)
Time-assigned speech interpolation (TASI)
Digital speech interpolation (DSI)
BURST MULTIPLEXING
(asynchronous multiplexing)
Fixed-length bursts
Variable-length bursts
One burst at a time

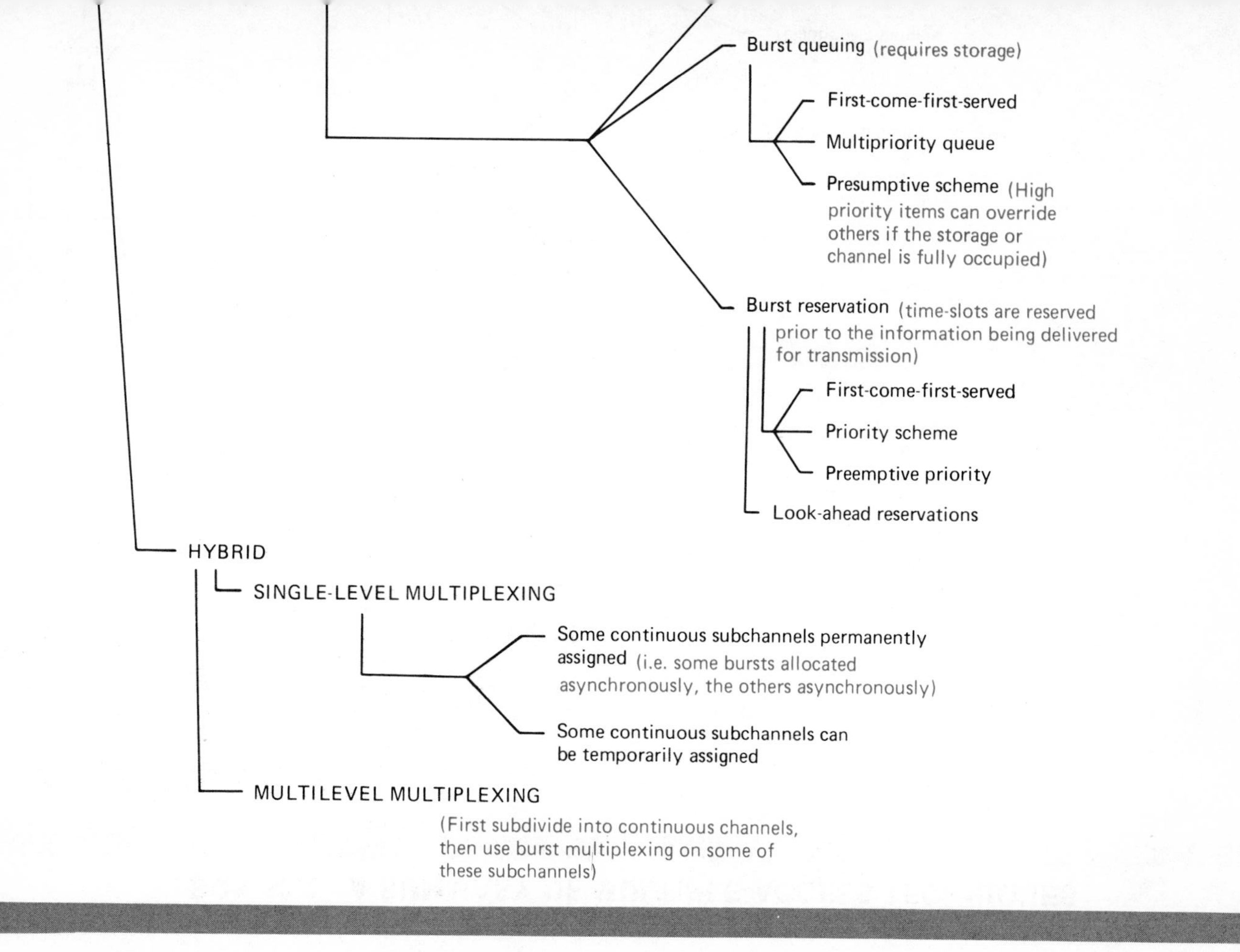

Burst queuing (requires storage)
First-come-first-served
Multipriority queue
Presumptive scheme (High priority items can override others if the storage or channel is fully occupied)
Burst reservation (time-slots are reserved prior to the information being delivered for transmission)
First-come-first-served
Priority scheme
Preemptive priority
Look-ahead reservations
HYBRID
SINGLE-LEVEL MULTIPLEXING
Some continuous subchannels permanently assigned (i.e. some bursts allocated asynchronously, the others asynchronously)
Some continuous subchannels can be temporarily assigned
MULTILEVEL MULTIPLEXING
(First subdivide into continuous channels, then use burst multiplexing on some of these subchannels)

BOX 33.2 A SUMMARY OF MULTIPLE-ACCESS TECHNIQUES

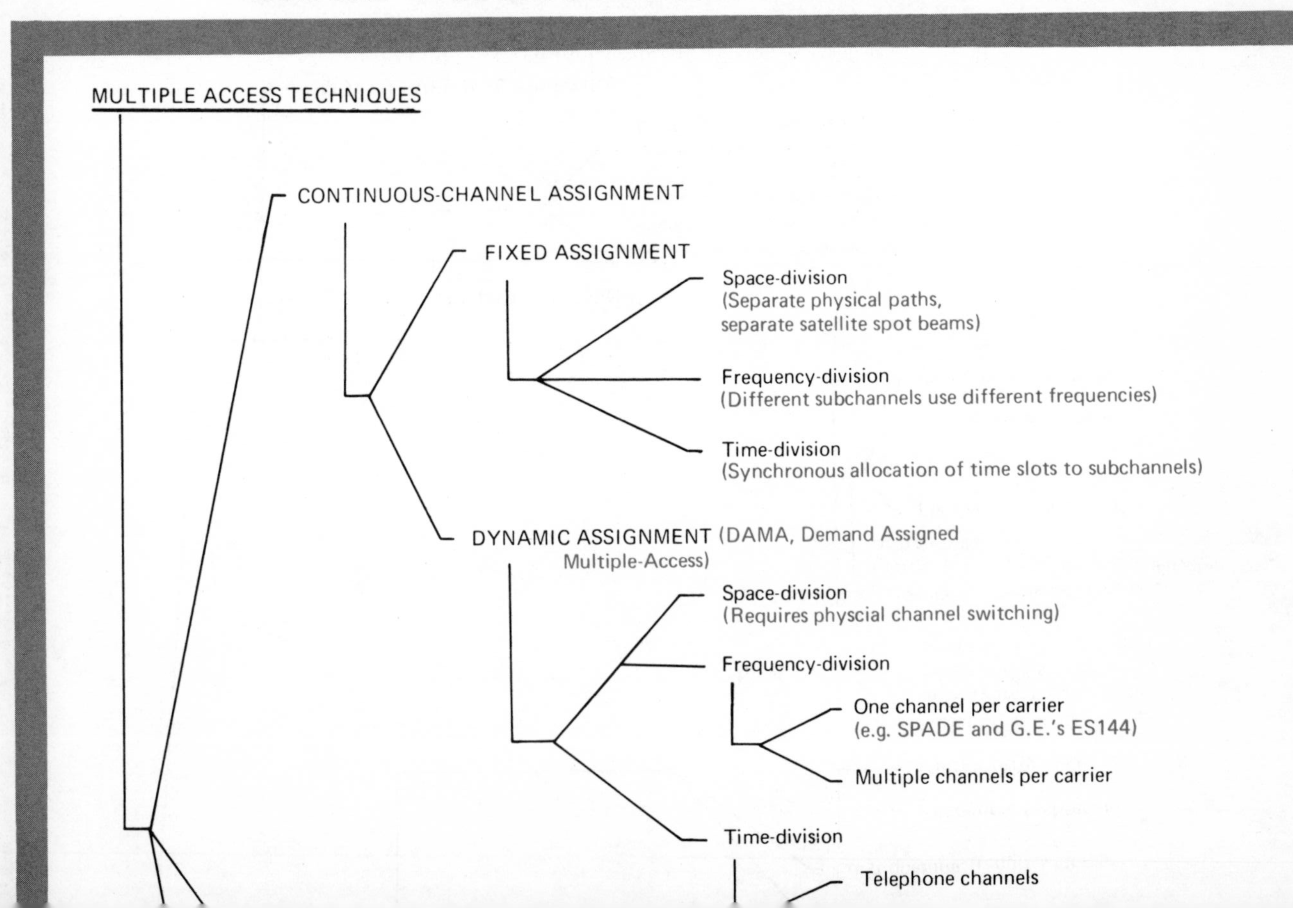

MULTIPLE ACCESS TECHNIQUES
CONTINUOUS-CHANNEL ASSIGNMENT
FIXED ASSIGNMENT
Space-division
(Separate physical paths,
separate satellite spot beams)
Frequency-division
(Different subchannels use different frequencies)
Time-division
(Synchronous allocation of time slots to subchannels)
DYNAMIC ASSIGNMENT (DAMA, Demand Assigned
Multiple-Access)
Space-division
(Requires physcial channel switching)
Frequency-division
One channel per carrier
(e.g. SPADE and G.E.'s ES144)
Multiple channels per carrier
Time-division
Telephone channels

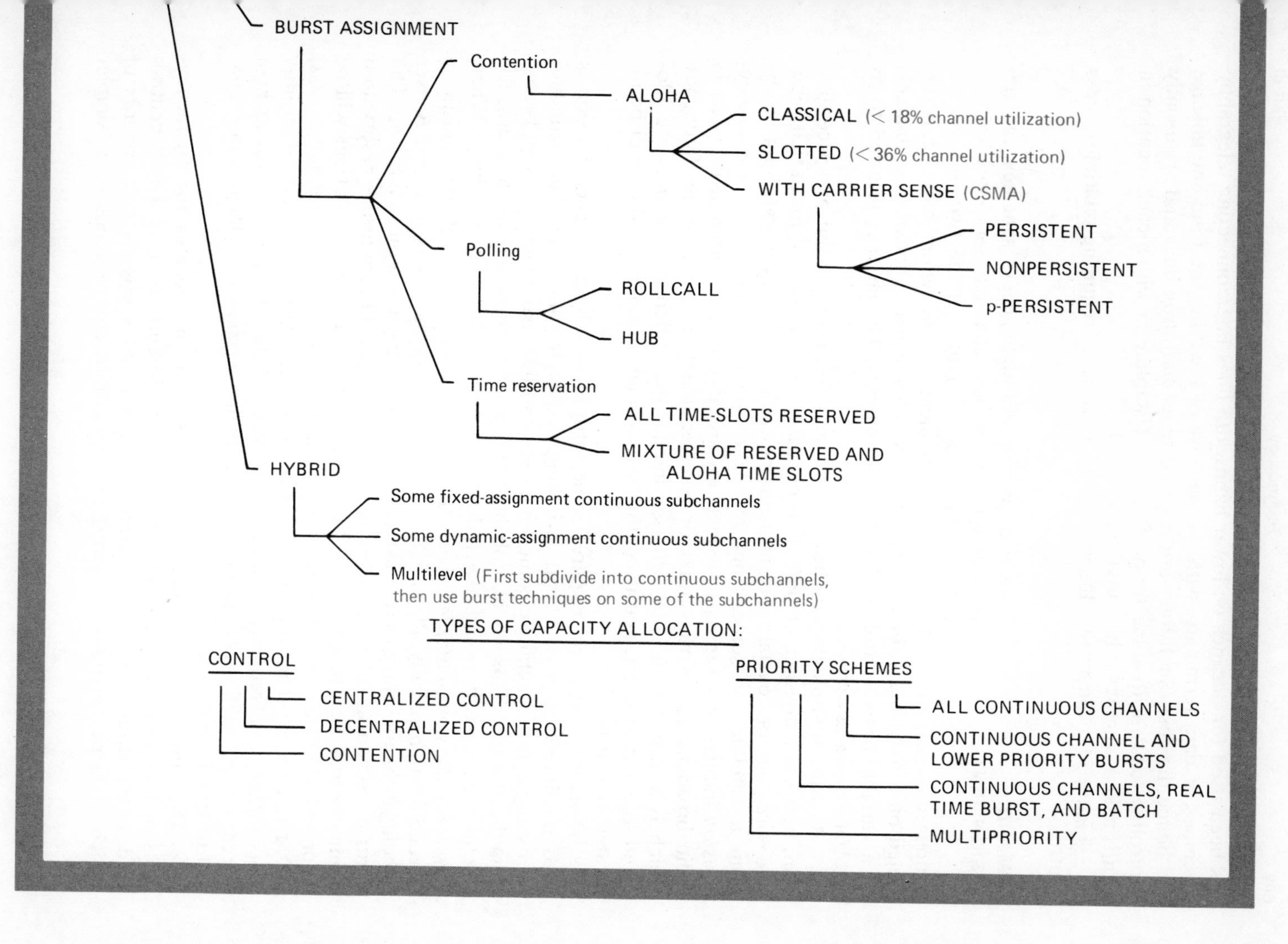
BURST ASSIGNMENT
Contention
ALOHA
CLASSICAL (< 18% channel utilization)
SLOTTED (< 36% channel utilization)
WITH CARRIER SENSE (CSMA)
PERSISTENT
NONPERSISTENT
p-PERSISTENT
Polling
ROLLCALL
HUB
Time reservation
ALL TIME-SLOTS RESERVED
MIXTURE OF RESERVED AND ALOHA TIME SLOTS
HYBRID
Some fixed-assignment continuous subchannels
Some dynamic-assignment continuous subchannels
Multilevel (First subdivide into continuous subchannels, then use burst techniques on some of the subchannels)
TYPES OF CAPACITY ALLOCATION:
CONTROL
CENTRALIZED CONTROL
DECENTRALIZED CONTROL
CONTENTION
PRIORITY SCHEMES
ALL CONTINUOUS CHANNELS
CONTINUOUS CHANNEL AND LOWER PRIORITY BURSTS
CONTINUOUS CHANNELS, REAL TIME BURST, AND BATCH
MULTIPRIORITY

has been used for decades on lower bandwidth lines interconnecting telegraphy machines or data terminals, with the purpose of lowering the total line mileage and hence the cost. Such lines are referred to as *multidrop* lines and are usually controlled today with some form of *polling* discipline in which each location in turn is asked whether it has anything to send.

Figure 31.1 illustrated different types of systems requiring multiple-access control.

DEMAND ASSIGNMENT

Like multiplexing, multiple-access schemes can derive either continuous channels or asynchronous burst channels, or a mixture of the two. When continuous channels are derived, who communicates with whom may be determined on a fixed basis, or a more complex scheme may allocate the channels on a variable basis according to user requirements. The latter is referred to as *DAMA—demand-assigned multiple-access.*

As with multiplexing, multiple-access may be accomplished by space division, frequency division, or time division. *Space-division* would mean that separate wire pairs or coaxial tubes in the same cable are used, separate spot-beams from a satellite, or some other means of physical separation. *Frequency-division* implies that separate frequencies are allocated to different users as in radio broadcasting; a transmitter or receiver must be turned to the frequency which is assigned to it. *Time-division* implies that different time slots are allocated to different users, and if continuous channels are derived they will be allocated synchronously.

Demand-assignment requires some form of switching. If space-division is used, it will require physical path switching. A switch could be used in a satellite, for example, for switching channels between multiple searchlight-like spot beams from the satellite. (This would probably not be the best form of multiple-access for satellites because it would lower the overall reliability of the satellite.) When *frequency-division multiple-access, FDMA,* is used a control mechanism is needed for dynamically allocating the available frequency slots to users. A user needing a channel makes a request for it, like requesting a seat on an airplane. The control mechanism allocates a frequency if one is free, like an airline reservation computer making a seat booking. The transmitting and receiving stations will be informed what frequency has been allocated. With *time-division multiple-access, TDMA,* the request and allocation process is similar in principle except that time-slots rather than frequency-slots are allocated. If the channel operates at high speeds—many millions of bits per second—then the time-slots allocated may consist of lengthy bursts of bits.

The control, or reservation, mechanism which allocates the frequency-slots or time-slots can be either centralized or decentralized. If it is decentralized, every station listens to a control channel and keeps its own records of which frequency or time-slots are free. Centralized control tends to be simpler and is less expensive than if every station has its own reservation mechanism. However there is a danger of failure in the controlling station, so one or more other stations should act as an immediate (hot) standby.

BURST MULTIPLE-ACCESS

When the information sent is in the form of bursts rather than continuous signals, multiple-access schemes can have different control mechanisms designed for interleaving bursts of transmission *asynchronously*. This would apply to interactive data transmission, telegrams, voicegrams, or any of the message types listed in Box 15.1.

There are three categories of burst control used on multiple-access channels: polling, contention, and time reservation. Polling has been the main form of control used on multidrop telephone and telegraph lines since the 1950s. A master station "polls" the other stations in turn telling them to transmit if they have anything to send. Polling can achieve fairly high line utilization if the following conditions apply:

1. There are not too many terminals sharing the channel.
2. The propagation delay, including the time taken to turn around the direction of transmission, is not too great compared with the message transmission time.
3. The messages are not too short compared with the polling overhead.

When many terminals share a high capacity channel polling is a poor control mechanism. This condition may apply to high speed loop channels, CATV channels, or packet radio channels. When, in addition, the propagation delay is high, polling is even worse. This is the case with satellite channels.

With a contention scheme a station does not wait to be asked; it makes its own decision when to transmit. Contention used to be used on early telegraph multidrop lines because a basic contention scheme is very simple. Today it is valuable for packet radio, interactive satellite channels and possibly other forms of multiple-access channel with too many stations or too great a propagation delay for polling to be efficient.

The simplest form of contention used today is an *ALOHA channel,* Somehwat more efficient is a *slotted* ALOHA *channel,* and better still is an *ALOHA channel with carrier sense*—all described in Chapter 32.

TIME-SLOT RESERVATIONS

As discussed previously, simple ALOHA cannot achieve a very high channel utilization. Better utilization can be achieved if reservations for time slots are made ahead of the time when the transmission occurs. Making reservations prevents the transmitted bursts from destroying one another as some of them do with the ALOHA protocols (e.g., the first transmissions of C and D in Fig. 32.1). On the other hand there is some overhead involved in making time-slot reservations, and the transmissions will be delayed slightly while the reservation is made. The mean overall transmission time will be greater than with a lightly loaded ALOHA channel, although occasional transmissions suffer a delay several times the mean on an ALOHA channel.

There are many possible ways in which a time-reservation protocol can be organized. On some systems it may be desirable that reserved-time messages are mixed with non-reserved-time ALOHA messages.

Figure 33.1 illustrates one example of a demand-assigned multiple-access time-reservation protocol.

L. G. Roberts [1] proposed a reservation system which could be used as an extension of terrestrial packet-switching networks such as ARPANET, using satellite channels of 50,000 bits per second or more. In this system all *protocol* messages are handled in an ALOHA fashion. The system is always in either a *reservation state* or an *ALOHA state*. In the reservation state reserved blocks of data are sent; in the ALOHA state, protocol or short data messages are sent.

Figure 33.1 could relate to a 50,000 bit-per-second channel. The time allocated for reserved bursts is 25 milliseconds. Each such burst begins with a bit pattern which establishes synchronization, has a header giving the addresses of its destination and source, and ends with a comprehensive error detection pattern. The burst would contain more than 1000 data bits.

At the left of the diagram, earth station A makes a request to transmit three blocks of data. The request travels on the ALOHA subchannel. 270 milliseconds later it listens to its own request and hears that it has not been relayed correctly. Another earth station had made a request at the same instant, causing an ALOHA collision. Station A makes its bid again, selecting a new ALOHA slot at random. This time it succeeds. All earth stations hear the request, know that station A is going to transmit 3 blocks, and calculate, knowing the present number of items in the queue, which time slots these blocks will occur. No other earth station will then attempt to use those positions in the queue. Station A does not wait for reply to its request. It assumes that every station has acted upon it, and so transmits the data.

Every station receives the transmitted blocks and examines their destination addresses. Stations C and Z will recognize their addresses and accept the data. All other stations will ignore the data. Stations C and Z check the error-detection bits in the blocks, conclude that the data were received correctly, and transmit acknowledgment message. The acknowledgment messages are sent, like the reservation request, in the period the channel is in ALOHA made. Earth station A retains the data until correct acknowledgments have been received. If no such acknowledgments arrive, station A will transmit the data again.

At the right hand side of the diagram, station A transmits a short message. This message is short enough to fit into one of the ALOHA slots and so no reservation is made for it. Again, station A retains it until a correct acknowledgment is received.

It is not known how much of the traffic will be short and how much will be long messages. The scheme is therefore designed to vary the mix it can handle automatically. If the long messages predominate, the channel is in the ALOHA state one sixth of the time as on the left hand side of the diagram. This is enough to handle the necessary protocol messages without excessive ALOHA collisions. If there are no long messages the channel reverts to continuous ALOHA state. This can be seen happening in the center of the diagram.

If the channel is in continuous ALOHA state, the first reservation request will cause it to allocate a time slot in reservation mode. As soon as the reservation queue goes to zero the system reverts to continuous ALOHA mode.

SWITCHING TECHNIQUES

In addition to the multiplexing, concentration, and multiple-access mechanisms, switching mechanisms may be required in a network. There are several possible types of switching and these are summarized in Box 33.3. The choice of switching from Box 33.3 is related to the choice of multiplexing or multiple-access from Boxes 33.1 and 33.2. Some networks do not have separate switching mechanisms because the routing is inherent in the concentration or multiple-access techniques. An ALOHA radio or satellite channel, for example, in effect does its own switching.

There are four main categories of switching in use, and several variations on each of these. They are:

1. Slow circuit-switching such as conventional telephone switching.
2. Fast circuit-switching capable of connecting and disconnecting paths in a small fraction of a second.
3. Conventional message-switching in which messages are filed and routed.
4. Packet-switching in which messages are chopped into slices (packets) and these are routed at high speed.

Box 33.4 summarizes the differences between these types of switching.

Switching, like multiplexing, can be categorized into continuous-channel switching and burst switching, illustrated by horizontal and vertical bars respectively in Fig. 10.2. Telephone switching, by far the most commonly used because of its universality, is continuous-channel switching, and as such is wasteful of both switching and transmission capacity when the traffic consists of short or high-speed bursts.

BOX 33.3 A SUMMARY OF SWITCHING TECHNIQUES

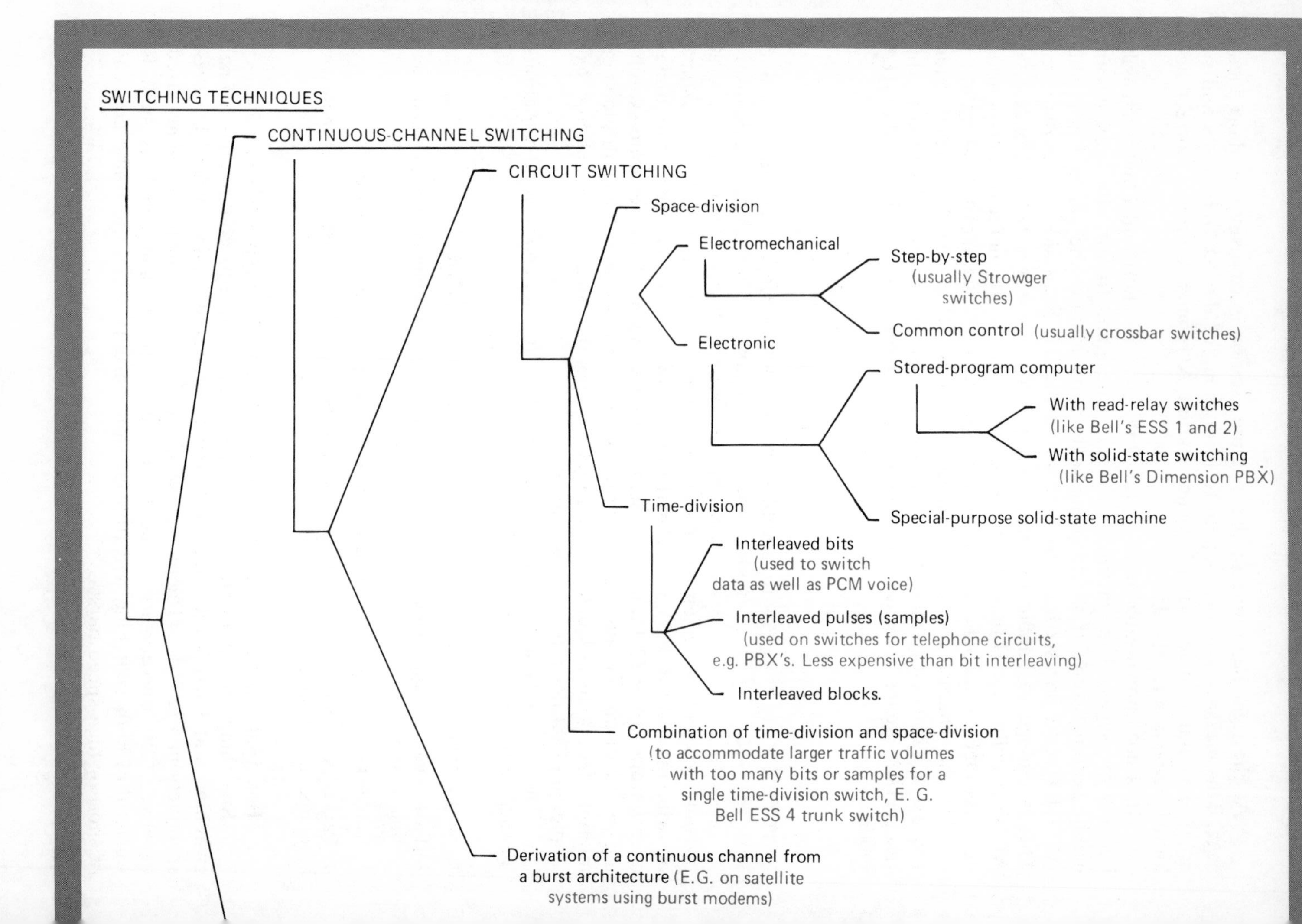

SWITCHING TECHNIQUES
CONTINUOUS-CHANNEL SWITCHING
CIRCUIT SWITCHING
Space-division
Electromechanical
Step-by-step (usually Strowger switches)
Common control (usually crossbar switches)
Electronic
Stored-program computer
With read-relay switches (like Bell's ESS 1 and 2)
With solid-state switching (like Bell's Dimension PBX)
Special-purpose solid-state machine
Time-division
Interleaved bits (used to switch data as well as PCM voice)
Interleaved pulses (samples) (used on switches for telephone circuits, e.g. PBX's. Less expensive than bit interleaving)
Interleaved blocks.
Combination of time-division and space-division (to accommodate larger traffic volumes with too many bits or samples for a single time-division switch, E. G. Bell ESS 4 trunk switch)
Derivation of a continuous channel from a burst architecture (E.G. on satellite systems using burst modems)

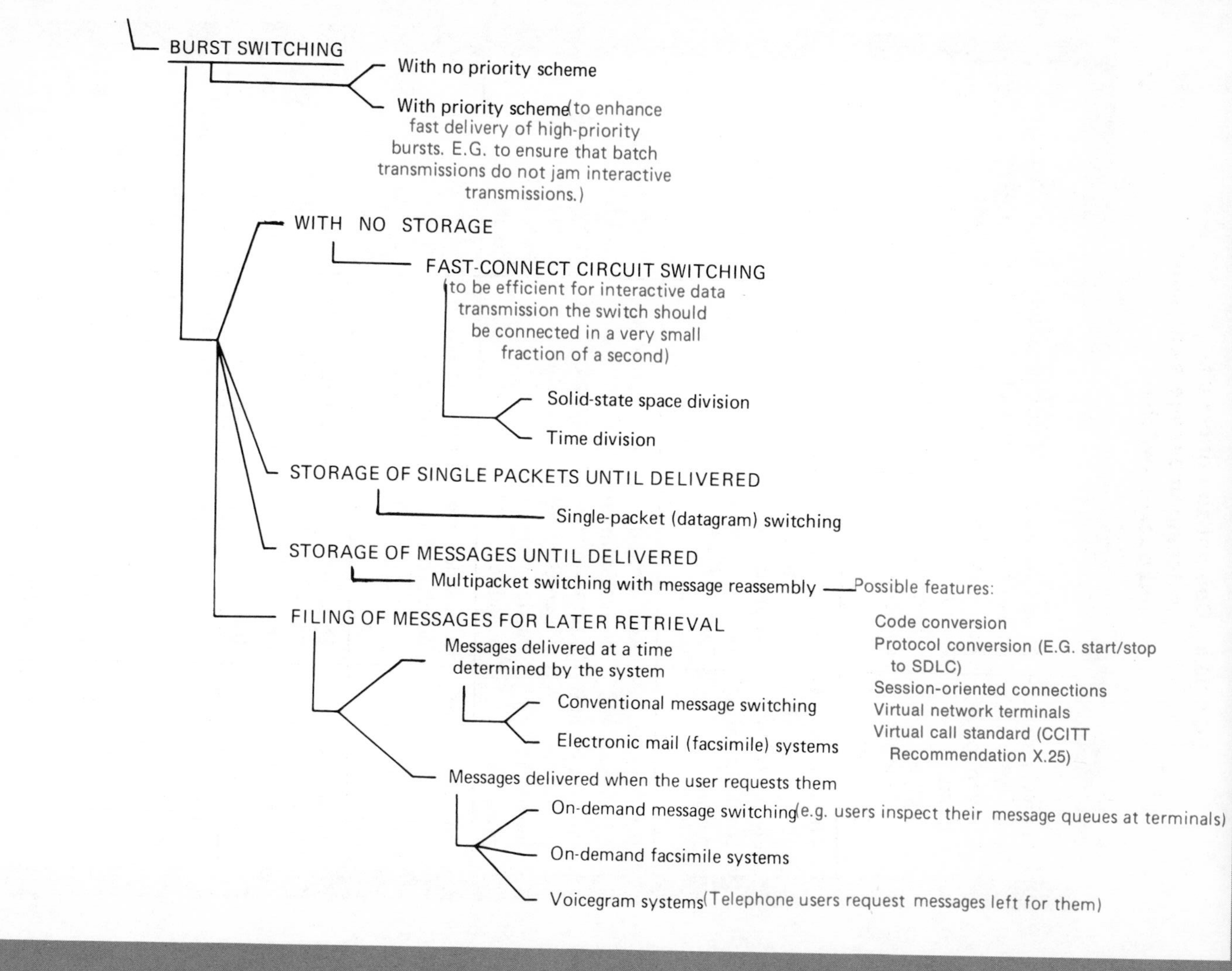
BURST SWITCHING
With no priority scheme
With priority scheme (to enhance fast delivery of high-priority bursts. E.G. to ensure that batch transmissions do not jam interactive transmissions.)
WITH NO STORAGE
FAST-CONNECT CIRCUIT SWITCHING
(to be efficient for interactive data transmission the switch should be connected in a very small fraction of a second)
Solid-state space division
Time division
STORAGE OF SINGLE PACKETS UNTIL DELIVERED
Single-packet (datagram) switching
STORAGE OF MESSAGES UNTIL DELIVERED
Multipacket switching with message reassembly
Possible features:
Code conversion
Protocol conversion (E.G. start/stop to SDLC)
Session-oriented connections
Virtual network terminals
Virtual call standard (CCITT Recommendation X.25)
FILING OF MESSAGES FOR LATER RETRIEVAL
Messages delivered at a time determined by the system
Conventional message switching
Electronic mail (facsimile) systems
Messages delivered when the user requests them
On-demand message switching (e.g. users inspect their message queues at terminals)
On-demand facsimile systems
Voicegram systems (Telephone users request messages left for them)

BOX 33.4 Comparison of circuit-switching, message-switching, and packet-switching

Main Characteristics of Conventional Circuit-switching (e.g., telephone switching)	Main Characteristics of Fast-connect Circuit-switching systems	Main Characteristics of Conventional Message-switching	Main Characteristics of Packet-switching
The equivalent of a wire circuit connects the communicating parties	The equivalent of a wire circuit is connected between the end buffers for brief periods.	No direct electrical connection.	No direct electrical connection.
Real-time or conversational interaction between the parties is possible.	Real-time or conversational interaction between the parties is possible.	Too slow for real-time or conversational interaction.	Fast enough for real-time or conversational interaction between data machines.
Messages are not stored.	Messages are not stored.	Messages are filed for later retrieval.	Messages are stored until delivered, but not filed.
Designed to handle long continuous transmissions.	Designed to handle short sporadic transmissions.	Designed to relay messages.	Designed to handle bursts of data.
The switched path is established for the entire conversation.	The switched path is repeatedly connected and disconnected during a lengthy interaction.	The route is established for each individual message.	The route is established dynamically for each packet.

continued

BOX 33.4 Continued

There is time delay in setting up a call and then neglible transmission delay.	A delay which ought to be less than one second, associated with setting up the call and delivering the message.	Substantial delay in message delivery.	Negligible delay in setting up the call. Delay of usually less than one second in packet delivery.
Busy signal if called party is occupied.	Delay, or busy signal, if called party is occupied.	No busy signal if called party is occupied.	Packet returned to sender if undeliverable.
Effect of overload: Increased probability of blocking, causing a network busy signal. No effect on transmission once the connection is made.	*Effect of overload:* Increased delay and/or increased probability of a busy signal.	*Effect of overload:* Increased delivery delay.	*Effect of overload:* Increased delivery delay (but delivery time is still short). Blocking when saturation is reached.
Electromechanic or computerized switching offices are used.	Computerized switching offices are used.	Fairly complex message-switching center is needed, with facilities.	Small switching computers are used with no filing facilities.
Protection against loss of messages is the responsibility of of the end users.	The network may be designed to protect the users against loss of messages.	Elaborate procedures are employed to prevent loss of messages. The responsibility of the network for the message is emphasized.	Some protection against loss of packets. End user protocols can be employed in message protection because of the conversational interaction.

continued

BOX 33.4 *Continued*

Relatively expensive to a user whose transmissions are very short.	Charges for short transmissions can be lower than over the telephone network.	Charges for message delivery lower than over the telephone network.	Charges for short transmissions can be lower than over the telephone network.
Any length of transmission is permitted.	Any length of transmission *may* be permitted.	Lengthy messages can be transmitted directly.	Lengthy transmissions are chopped into short packets. Very long messages must be divided by the users.
Economical with low traffic volumes if the public telephone network is employed.	High traffic volumes needed for justification.	Economical with moderate traffic volumes.	High traffic volumes needed for economic justification.
The network cannot perform speed or code conversion.	*May* provide speed or code conversion.	The network can perform speed or code conversion.	The network can perform speed or code conversion.
Does not permit delayed delivery.	*May* permit delayed delivery if the delay is short.	Delayed delivery if the recipient is not available.	Does not permit delayed delivery (without a special network facility).
Point-to-point transmission	Point-to-point transmission.	Permits broadcast and multi-address messages.	Does not permit broadcast and multi-address messages (without a special network facility).

continued

BOX 33.4 *Continued*

Fixed band-width trans-mission.	Users effec-tively employ small or large bandwidth ac-cording to need.		Users effec-tively employ small or large bandwidth ac-cording to need.

Telephone switching (and continuous channel switching in general) is evolving through several forms, and the early forms are still in use in most locations:

1. Manual.
2. Strowger switch (and step-by-step electromechanical exchanges in general).
3. Crossbar (and common-control electromechanical exchanges in general).
4. Computer-controlled read-relay switches (e.g., ESS 1 and 2).
5. Computer-controlled solid-state space-division.
6. Time-division switching of PAM samples (e.g., AT&T's dimension PBX).
7. Time-division switching of bits streams (to handle digitized voice and data).
8. Mixed time-division and space-division to give a large switching capacity (e.g., ESS 4).
9. Continuous-channels derived from a burst-switching mechanism handling blocks of data.

BURST SWITCHING Burst switching, efficient with sporadic rather than continuous traffic, has several variants. Box 33.3 categorizes them according to how data are stored in the switching system. Fast circuit-switching—fast enough to handle high-speed bursts of traffic, ideally switching in less than a millisecond—may have no storage at the switch. If time-division circuit-switching is used it will have enough storage to hold the bits or words passing through the switch, but no data will be queued.

Packet-switching machines store and queue the packets they handle. There are two options which we discussed in Chapter 29. A packet-switching network can take the easy way out and handle only single packets, or it can accept and deliver messages of widely varying length, chopping the longer ones into many packets. In the latter case the network must have enough storage to queue entire messages, and to reassemble them after transmission.

A packet-switching network (as presently conceived) does not retain the packets or messages once they have been delivered correctly. It acts simply as a mail service. A message switching network on the other hand *files* them for possible retrieval at a later time. In some cases the filing is a legal requirement of common-carrier message switching. In some cases it is a system requirement designed to ensure that messages are accountable and recoverable even when people or machines malfunction. Traditional message-switching systems have delivered their messages at times determined by *the system*. New forms of message-switching systems will hold the messages for some users until the users request them. The user sits at his terminal, and asks if there are any messages for him. He may use a screen unit so there is no paper generated. He may use a telephone asking if there are any telephone messages stored for him. When the messages are stored in facsimile or voice form the storage required will be much larger than when they are stored in teletype or other character-coded forms. The future will bring message-switching systems which can handle many of the forms of traffic listed in Box 15.1.

PRIORITY

In many conventional switching mechanisms there is no priority scheme; traffic is handled on a first-come-first-served basis. Priority schemes are easier to implement when the switching is computer-controlled, and can make many systems more valuable.

On a continuous-channel system, busy signals are given when too many callers request connections. A priority scheme can categorize the callers or the calling stations into grades of priority. High-priority callers can then be given a lower probability of receiving a network busy signal. A queuing mechanism can be used instructing high-priority callers to wait for a short time until a channel can be allocated to them, and meanwhile no channels are allocated to low-priority callers. A more extreme action would be to allow a rarely-used highest priority call to *interrupt* lower priority calls if necessary.

With packet-switching or message-switching systems high priority packets can go to the head of the queues so that their delivery time is as short as possible. Priority schemes are valuable when there is a mix of traffic of widely differing time constraints. In data networks, such as the public packet switching networks, some of the traffic is from interactive terminals requiring a fast response time, and some is batch traffic or message traffic having no critical time constraint. Priority should be given to the interactive traffic. Both traffic occasionally surges into such a network in great quantity, and if no priority scheme is used a temporary flood of such traffic could severely degrade the performance of interactive systems.

Future burst-switching networks may be designed to handle both burst traffic and some continuous traffic, the continuous traffic occupying a repetitive (synchronous) stream of bursts among which other traffic is interleaved. In such a system the bursts for the continuous traffic need priority over other bursts. There may be a requirement for four priorities inherent in the nature of the traffic:

Priority 1: Continuous traffic (e.g., telephone speech).

Priority 2: Fast-response computer traffic (e.g., from interactive terminals).

Priority 3: Batch or one-way message traffic to be delivered as quickly as is convenient.

Priority 4: Traffic that can be delivered by the following morning.

In addition to such priorities inherent in the nature of the traffic, other priorities may give certain users or certain terminals preferential treatment.

INCOMPATIBILITY

It will be seen that there are a variety of fundamentally different ways to build communication networks. For each of the fundamentally different techniques there are many possible, and attractive variations. Different countries, and often different organizations within the same country, are building (or have built) differently structured data networks. Some are burst networks; some are continuous-channel networks. Some are switched; some not. Some are fast; some slow. Some use satellites; some not. Some burst networks use packet-switching; some use fast-connect circuit switching. Some use packet-switching with message reassembly; some have proposed packet-switching without message reassembly. Different packet sizes have been used.

The users, and the manufacturers of terminals, especially those who operate internally are faced with a growing incompatibility in the transmission facilities available to them. This incompatibility causes problems in system design and is making terminals and teleprocessing software more expensive. Software designed for one type of network will often not operate on another.

It is true in other aspects of computer system design also that there are a variety of attractive but incompatible mechanisms for handling data. We are in an era of great invention, both in network design and in other aspects of system design. The user, the programmer, and the software designer, need to be protected from the heady proliferation of ingenious new techniques. In data base design the necessary protection is provided by creating "data independence." This means that the programmer does not know, in detail, how the data are stored or physically organized. He has his own simple view of the data and if the system stores data in a different form it must convert it to and from the view which the programmer sees. As networking techniques proliferate we need "network independence." The user should be able to give data to the network in standard, simple, form, and not worry about whether it travels by packet-switching, circuit-switching, satellite, slotted ALOHA, wideband loops or whatever.

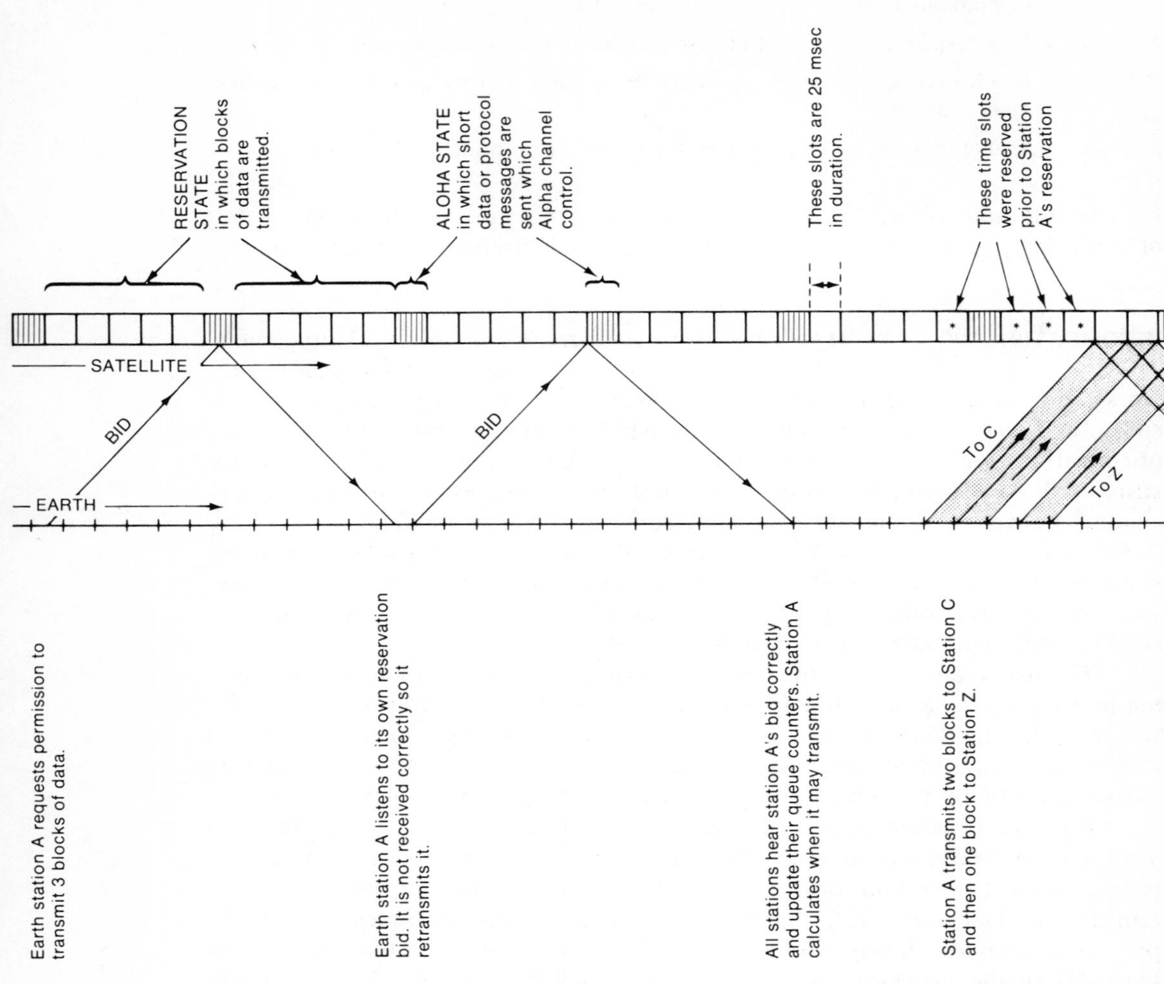
RESERVATION STATE in which blocks of data are transmitted.
ALOHA STATE in which short data or protocol messages are sent which Alpha channel control.
These slots are 25 msec in duration.
These time slots were reserved prior to Station A's reservation
SATELLITE
BID
BID
EARTH
To C
To Z
Earth station A requests permission to transmit 3 blocks of data.
Earth station A listens to its own reservation bid. It is not received correctly so it retransmits it.
All stations hear station A's bid correctly and update their queue counters. Station A calculates when it may transmit.
Station A transmits two blocks to Station C and then one block to Station Z.
Station A requests permission to transmit

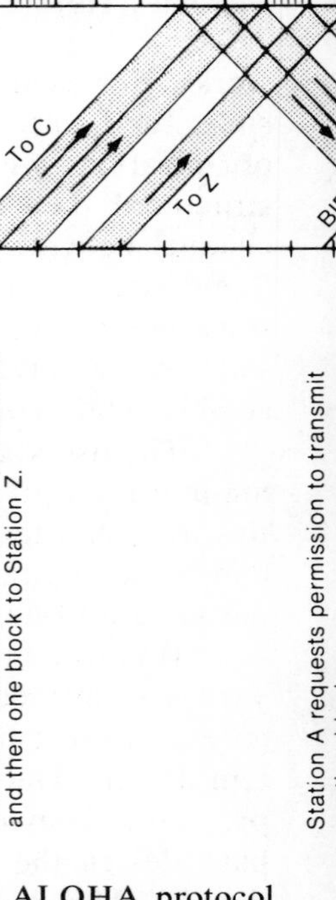

Figure 33.1 A burst reservation system with an ALOHA protocol mechanism (1).

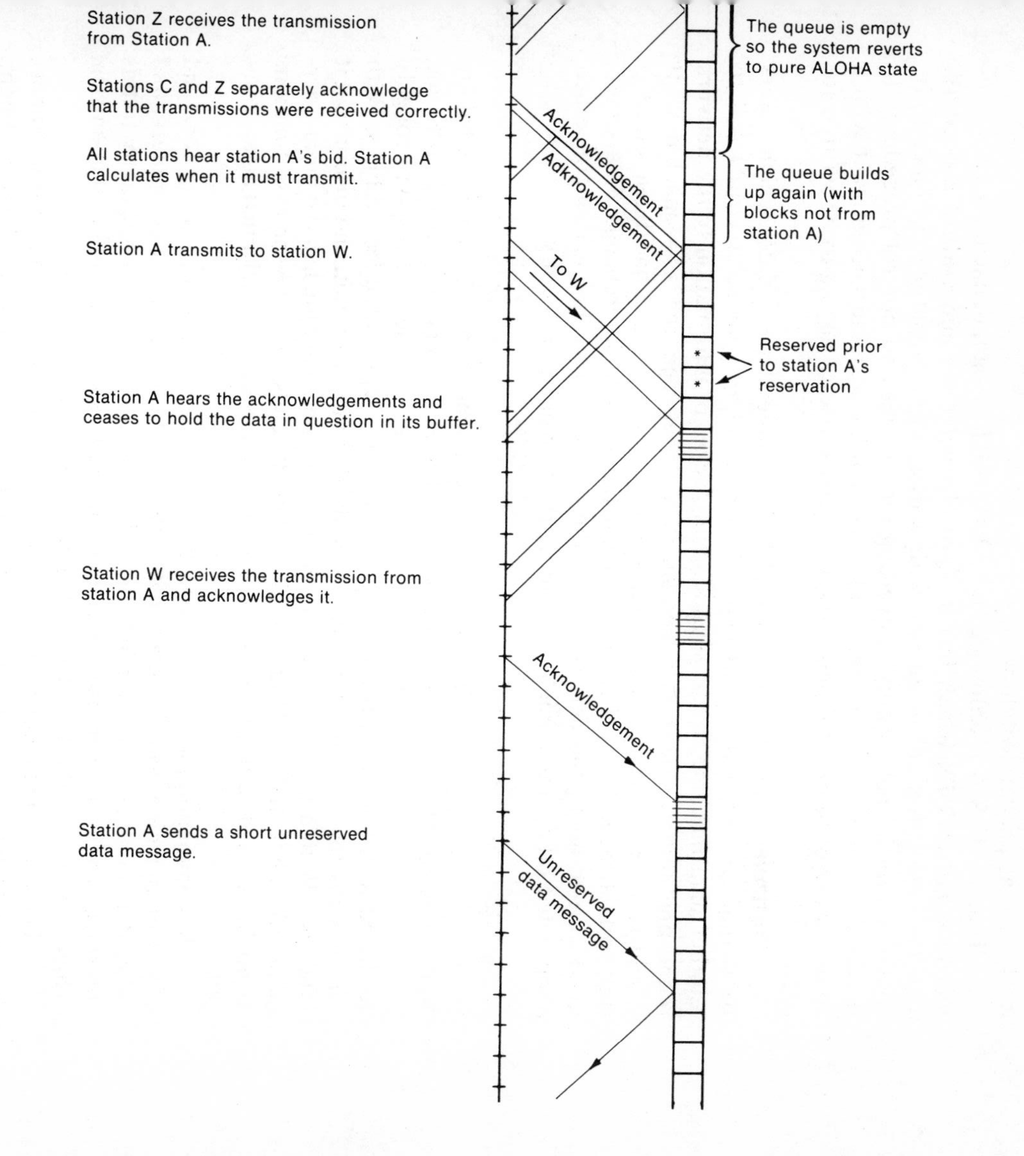

Station Z receives the transmission from Station A.
Stations C and Z separately acknowledge that the transmissions were received correctly.
All stations hear station A's bid. Station A calculates when it must transmit.
Station A transmits to station W.
Station A hears the acknowledgements and ceases to hold the data in question in its buffer.
Station W receives the transmission from station A and acknowledges it.
Station A sends a short unreserved data message.
Acknowledgement
Adknowledgement
To W
Acknowledgement
Unreserved data message
The queue is empty so the system reverts to pure ALOHA state
The queue builds up again (with blocks not from station A)
Reserved prior to station A's reservation

TRANSPARENCY

There have been several proposals for how the diverse networks of the world should be made "transparent" to the users. Sometimes, however, the proposals relate to one class of network such as packet switching.

Telenet (U.S.A.), following the ARPANET design, defined a *network virtual terminal* (NVT), which has its own character set and control procedure. The Telenet central offices allow a wide variety of incompatible machines to be connected to the network by converting their codes and control procedures to those of the network virtual terminal. This enables customers to interact with a large variety of terminal types without special software having to be written for each.

Particularly important is CCITT Recommendation X.25 which refers to *virtual circuits* and *virtual calls*. A data machine making a virtual call to another data machine would send a small packet of information to the network requesting that a connection be established between the two machines. It would receive a packet back saying whether or not the call had been successfully established. If so, then the two machines could interchange data freely, unconcerned about the details of how the data were traveling and unconcerned about what other traffic might be sharing the transmission facility.

CCITT RECOMMENDATIONS

The CCITT Recommendation X.25 states that there should be two classes of service: *virtual calls* and *permanent virtual circuits*. They correspond roughly to dialed telephone connections and permanently connected leased lines. With *virtual calls* the using machine has to establish a virtual circuit, and disconnect it after use. With a *permanent virtual circuit* the using machine has the circuit permanently connected, in effect, and all packets it sends go to the same destination. It might be used to form a permanent connection between a terminal and a computer.

Recommendation X.25 defines the interface between a data network and any user machine connected to that network. Many terminals, computers, and other data communications products, will be designed to conform exactly to this interface. They can then use a variety of different networks in different countries, which conform to the CCITT Recommendation.

The user equipment is referred to as the DTE (Data Terminal Equipment). It is connected by a multi-pin plug to the Data Circuit-terminating Equipment, DCE (which in most cases is a modem). The CCITT V and X series of recommendations define the nature of this interface and what is sent across it. The V series relate to data transmission over telephone and telex networks. The X series relate to transmission over public networks, and will have a major effect on the evolution of packet-switching and other data networks. Box 33.5 lists the V recommendations which give modem standards. Box 33.6 lists the X recommendations which will effect future data networks.

BOX 33.5 CCITT recommendations for modem standards [1]

The interface between the modem and the data-processing machine

CCITT	Recommendation
V.24	Definition of interchange circuits between modems and data terminal equipment
V.25	Automatic calling and/or answering equipment and disabling of echo suppressors

Modem standards for use on the switched telephone network

CCITT	Recommendation
V.21	200-baud modem
V.23	600/1200-baud modem
V.26b	2400/1200-bps modem
V.30	Parallel data transmission modems
V.22	Standardization of data-signaling rates
V.25	Automatic calling and/or answering equipment and disabling echo suppressors
V.15	Use of acoustical coupling

Modem standards for use on leased telephone lines

CCITT	Recommendation
V.26	2400-bps modem
V.27	4800-bps modem
V.22b	Standardization of data-signaling rates
V.35	Data transmission at 48,000 bps using 60-108-kHz (channel group) circuits

BOX 33.6 CCITT Recommendations for data transmission over public data networks [1]

V 41	Code-independent error control system
X 1	International user classes of service in public data networks
X 2	International user facilities in public data networks
X 20	Interface between Data Terminal Equipment (DTE) and Data Circuit-Terminating Equipment (DCE) for start-stop transmission services on public data networks
X 20 bis	V21-compatible interface between Data Terminal Equipment (DTE) and Data Circuit-Terminating Equipment (DCE) for start-stop transmission services on public data networks
X 21	The general purpose interface between Data Terminal Equipment and Data Circuit-Terminating Equipment for synchronous operation on public data networks
X 21 bis	The use on public data networks of DTE's which are designed for interfacing to synchronous V-series modems
X 24	List of definitions for interchange circuits between Data Terminal Equipment (DTE) and Data Circuit-Terminating Equipment (DCE) on public data networks
X 25	Interface between Data Terminal Equipment and Data Circuit-Terminating Equipment for terminals operating in the packet mode on public data networks
X 50	Fundamental parameters of a multiplexing scheme for the international interface between synchronous data networks
X 51	Fundamental parameters of a multiplexing scheme for the international interface between synchronous data networks using 10-bit envelope structure
X 60	Common channel signalling for synchronous data applications—data user part
X 70	Terminal and transit control signalling system for start-stop services on international circuits between anisochronous data networks

continued

BOX 33.6 *Continued*

X 71	Decentralized terminal and transit control signalling system on international circuits between synchronous data networks
X 92	Hypothetical reference connections for public synchronous data networks
X 95	Network parameters in public data networks
X 96	Call progress signals in public data networks

There are three levels at which the interface between the DTE and DCE is defined:

LEVEL 1 defines the physical, electrical, functional, and procedural characteristics to establish, maintain, and disconnect the physical link between the DTE and DCE. It is not concerned with what bits are sent, but with the electrical connections between the DTE and modem, or other DCE. It is defined in Recommendations X.21 and X.21 *bis*.

LEVEL 2 defines the link control procedure for data interchange across the link between the DTE and DCE. In other words what bits form the start and end of a message—the envelope in which information is sent—the start-of-message indicator, the address, the control bits, the end-of-message indicator, and the error-detecting code. Recommendation X.25 contains a link control procedure similar to HDLC (Higher-level Data Link Control) specified by ISO (the International Standards Organization) in ISO documents DIS 3309.2 and DIS 4335.

LEVEL 3 defines the packet format and control procedures for the exchange of packets containing control information and user data, between the DTE and DCE. It defines the formats of packets which set up and disconnect a virtual call, which carry information on a virtual call or permanent virtual circuit, and which are used to help recover from the various types of malfunction.

There is little argument about levels 1 and 2. Level 1 is the modem interface which is already widely available and accepted. Level 2, HDLC, is a full duplex line control, similar to IBM's SDLC but with some improvements; it is generally accepted as being more efficient than earlier line control procedures. It meets the needs of most terrestrial applications, though with satellites some modifications may be desirable. Level 3, the *virtual circuit* protocol, has raised some controversy, especially from authorities who believe that a *datagram* protocol is desirable, and that virtual calls and permanent virtual circuits, if standardized at all, should be a higher level protocol which *employs* datagrams.

DATAGRAMS

Employing the proposed *datagram* service the user sends and receives single packets, called *datagrams,* having a standard format. These packets cannot contain more than 255 bytes (2040 bits) of data. With datagram service the message reassembly problems described in Chapter 29 are avoided.

Datagram operation would be simple, fast, and cheap. A datagram machine would be much less complex than that employing the X.25 protocol. It is thought that hundreds of thousands of cheap datagram terminals, for example for electronic fund transfer, would soon be employed if datagram service were available. The terminals could be an inexpensive addition to a telephone set, such as those in Figures 3.2 and 3.3.

A datagram does not need a call to be established before it is sent, and disconnected after it is sent. It is transmitted to a packet-switching network entirely independently of any packet previously transmitted or subsequently transmitted. It is self-contained.

In a *virtual call* service, a datagram would be used for setting up the call and disconnecting it. A long message would be composed of multiple datagrams. Datagrams would be sent for purposes of error control, flow control, recovery from failures, and general control of the virtual circuits. The wording of a CCITT proposal relating to datagrams is given in Box 33.7.

It would be unfortunate if the concept of *datagrams* were lost in the standardization process, and only the more complex *virtual call* and *permanent virtual circuit* protocols were employed. All three are of value.

PROTECTING THE USERS

We have characterized the years ahead as an *era of great invention* in telecommunications. In an era of great invention the users need to be protected from the proliferation of new mechanisms. Protection can come from appropriate standards and "virtual" techniques which make any form of network appear as though it were a simple standard network, or any form of file or data organization appear as though it were a simple standard organization. Users ought to be demanding such protection, both nationally and internationally. Standards come into existence both from government-supported organizations such as CCITT.

BOX 33.7 CCITT proposal for virtual call service (CCITT Study Group VII, Question 1/VII Point C, April 1974)

1. INTRODUCTION

A number of European administrations are currently implementing or studying packet switched services. This contribution has taken into account results from consultations between those administrations.

2. INTERNATIONAL SERVICES

Two basic types of services are proposed, a datagram service and a virtual call service. Both types of services may offer a number of internationally agreed upon additional special facilities, as well as national or bilaterally agreed extra facilities.

The two basic service types proposed are:

2.1 Datagram Service

This service, as seen by the customers, has the following basic characteristics:

(a) Customers may deliver and receive short messages, "datagrams", in a standard format to and from the network, observing a simple transmission control procedure with (customer line) acknowledgement per datagram.

(b) The datagrams will be handled independently of each other by the network and with the minimum of delay. It is possible that datagrams may not be delivered to the addressee in the order in which they are input to the network.

(c) The chance of loss of a datagram in transit through the network will be kept small, after it has been acknowledged to the sending customer.

(d) The sending customer will be informed by the network if the addressee cannot be reached by a datagram, due to situations like out-of-order, non-existent address, or barred access conditions.

(e) The internal network management system will include the possibility to identify and block those datagrams from a customer which would otherwise have to be discarded.

continued

BOX 33.7 *Continued*

2.2 Extra-facilities of the Datagram Service

Some possible extra facilities in the datagram service could be:

(a) datagram delivery confirmation; i.e., the sending customer will be informed that the addressee has acknowledged the datagram, or that the datagram has been discarded by the network;
(b) closed user group; i.e., datagrams to or from unauthorized customers will be discarded by the network.
(c) retained datagram order; i.e., a sequence of datagrams delivered to the network for one addressee will be delivered in the same sequence order.
(d) priority; i.e., these datagrams will always be handled before lower priority datagrams by the network.
(e) fixed address; i.e., datagrams in abbreviated format from a customer will have a preassigned address and other handling information added by the network.
(f) multiaddress; i.e., the network will deliver the same datagram to several different addressees.
(g) datagram assembling/disassembling; i.e., the network will accept/deliver the data content of a datagram from/to a customer on a character-by-character basis and deal with the appropriate formatting within the network.

2.3 The Virtual Call Service

Basically, this service should enable two customers to initiate and carry out the transfer or exchange of strings of messages according to a standard communication procedure, with network end-to-end preservation of sequence order. The following service implementation is proposed:

(a) The calling customer will initiate a call by delivering a standard "call request + address" packet.
(b) The calling customer will receive a response from the network that the addressee is accepting the call from the calling party. Message transmission may commence before this response (with the risk of data loss if the response is negative).

continued

BOX 33.7 *Continued*

(c) Each message may include several packets. The network may split packets and/or regroup consecutive packets.
(d) The rate at which customers transmit packets is controlled by the network in a manner acceptable to both network and receiving customer.
(e) Either party may clear the call by sending a special type packet, which will also be conveyed to the opposite party.

2.4 Extra-facilities of the Virtual Call Service

Some possible extra-facilities in the virtual call service could be:

Message delivery confirmation.
Closed user group.
Priority.
Direct Call.
Abbreviated address calling.
Packet assembling and disassembling.
Preservation of packet integrity, i.e., the splitting and regrouping mentioned in point 2.3 (c) will not apply.

3. PACKET SIZE AND FORMAT

3.1 Datagrams

The following basic principles are proposed for the format employed in the datagram service:

(a) A packet should consist of a header, a customer's data field and an error check field.
(b) Both packet header field and customer's data field may have a variable length in increments of 8-bit bytes.
(c) The maximum length of the customer's data field should be 255 bytes (2040 bits).
(d) Indicators will be provided for both header and data field lengths.
(e) Preliminarily, it is envisaged that the error check field will consist of two bytes, using the polynomial defined in CCITT Recommendation V41 over the header and data fields.

continued

BOX 33.7 *Continued*

3.2 Packets for Virtual Calls

In the virtual call service, paragraphs 3.1 (a), (b), (d) and (e) apply.

The subject of packet size requires further study.

4. CUSTOMER LOGICAL INTERFACE

Packet formats on customer's access lines should be in accordance with paragraph 3.

Error correction should be performed by retransmission of those packets detected to be received in error.

Manufacturers or designers, given appropriate virtual standards, can then be free to invent all manner of ingenious new mechanisms and organizations.

The virtual call and virtual circuit recommendations relate to packet-switching over terrestrial telephone circuits. This may be a dominant technology for a period, but, as we have seen there are many other techniques and uses of telecommunications. If satellites, CATV, fast-connect circuit-switching, packet radio, teletext, TDMA, T4 carrier, and switches like Bell's ESS 4 play a major role in some countries, then virtual operation is desirable which makes links using these facilities appear the same as other types of links. As the transmission of facsimile documents, voicegrams, video signals, music, electronic mail, etc., assume major importance, standards and forms of virtual circuit will be needed for them also.

REFERENCES

1. The CCITT Recommendations relating to data transmission are contained proceedings of the Fifth Plenary Assembly, 1972, Vol. VIII, and the Sixth Plenary Assembly, 1976.
2. ISO/TC 97/SC6, *High-level data link control procedures*. Proposed draft international standard on elements of procedures. Doc. DP 4335 (Oct. 1975).
3. L. Pouzin, *Virtual call issues in network architectures*. Eurocomp, Brunel University (Sept. 1975), 603–618.
4. Minutes of meeting of Rapporteur's group on packet switching. CCITT–COM VII–No. 237-E (Oct. 1975), 142 p.

GLOSSARY

There is little point in redefining the wheel, and where useful the definitions in this glossary have been taken from other recognized glossaries.

A suffix "2" after a definition below indicates that it is the CCITT definition, published in *List of Definitions of Essential Telecommunication Terms,* International Telecommunication Union, Geneva.

A suffix "1" after a definition below indicates that the definition is taken from the *Data Communications Glossary,* International Business Machines Corporation, Poughkeepsie, 1967 (Manual number C20–1666).

Address. A coded representation of the destination of data, or of their originating terminal. Multiple terminals on one communication line, for example, must have unique addresses. Telegraph messages reaching a switching center carry an address before their text to indicate the destination of the message.

Alphabet (telegraph or data). A table of correspondence between an agreed set of characters and the signals which represent them. (2).

Alternate routing. An alternative communications path used if the normal one is not available. There may be one or more possible alternative paths.

Amplitude modulation. One of three ways of modifying a sine wave signal in order to make it "carry" information. The sine wave, or "carrier," has its amplitude modified in accordance with the information to be transmitted.

Analog data. Data in the form of *continuously variable* physical quantities. (Compare with **Digital data.**) (1).

Analog transmission. Transmission of a continuously variable signal as opposed to a discretely variable signal. Physical quantities such as temperature are continuously variable and so are described as "analog." Data characters, on the other hand, are coded in discrete separate pulses or signal levels, and are referred to as "digital." The normal way of transmitting a telephone, or voice, signal has been analog; but now digital encoding (using PCM) is coming into use over trunks.

Application program. The working programs in a system may be classed as *application programs* and *supervisory programs.* The application programs are the main data-processing programs. They contain no input-output coding except in the form of macroinstructions that transfer control to the supervisory programs. They are usually unique to one type of application, whereas the supervisory programs could be used for a variety of different application types. A number of different terms are used for these two classes of program.

ARQ (Automatic Request for Repetition). A system employing an error-detecting code and so conceived that any false signal initiates a repetition of the transmission of the character incorrectly received. (2).

ASCII (American Standard Code for Information Interchange). Usually pronounced "ask′-ee." An eight-level code for data transfer adopted by the American Standards Association to achieve compatibility between data devices. (1).

Asynchronous transmission. Transmission in which each information character, or sometimes each word or small block, is individually synchronized, usually by the use of start and stop elements. The gap between each character (or word) is not of a necessarily fixed length. (Compare with **Synchronous transmission.**) Asynchronous transmission is also called *start-stop transmission.*

Attended operation. In data set applications, individuals are required at both stations to establish the connection and transfer the data sets from talk (voice) mode to data mode. (Compare **Unattended operation.**) (1).

Attenuation. Decrease in magnitude of current, voltage, or power of a signal in transmission between points. May be expressed in decibels. (1).

Attenuation equalizer. (*See* **Equalizer.**)

Audio frequencies. Frequencies that can be heard by the human ear (usually 30 to 20,000 cycles per second). (1).

Automatic calling unit (ACU). A dialing device supplied by the communications common carrier, which permits a business machine to automatically dial calls over the communication networks. (1).

Automatic dialing unit (ADU). A device capable of automatically generating dialing digits. (Compare with **Automatic calling unit.**) (1).

Bandwidth. The range of frequencies available for signaling. The difference expressed in cycles per second (hertz) between the highest and lowest frequencies of a band.

Baseband signaling. Transmission of a signal at its original frequencies, i.e., a signal not changed by modulation.

Baud. Unit of signaling speed. The speed in bauds is the number of discrete conditions or signal events per second. (This is applied only to the actual signals on a communication line.) If each signal event represents only one bit condition, baud is the same as bits per second. When each signal event represents other than one bit (e.g., see **Dibit**), baud does not equal bits per second. (1).

Baudot code. A code for the transmission of data in which five equal-length bits represent one character. This code is used in most DC teletypewriter machines where 1 start element and 1.42 stop elements are added. (1).

Bel. Ten decibels, q.v.

BEX. Broadband exchange, q.v.

Bias distortion. In teletypewriter applications, the uniform shifting of the beginning of all marking pulses from their proper positions in relation to the beginning of the start pulse. (1).

Bias distortion, asymmetrical distortion. Distortion affecting a two-condition (or binary) modulation (or restitution) in which all the significant conditions have longer or shorter durations than the corresponding theoretical durations. (2).

Bit. Contraction of "binary digit," the smallest unit of information in a binary system. A bit represents the choice between a mark or space (one or zero) condition.

Bit rate. The speed at which bits are transmitted, usually expressed in bits per second. (Compare with **Baud.**)

Broadband. Communication channel having a bandwidth greater than a voice-grade channel, and therefore capable of higher-speed data transmission. (1).

Broadband exchange (BEX). Public switched communication system of Western Union, featuring various bandwidth FDX connections. (1).

Buffer. A storage device used to compensate for a difference in rate of data flow, or time of occurrence of events, when transmitting data from one device to another. (1).

Cable. Assembly of one or more conductors within an enveloping protective sheath, so constructed as to permit the use of conductors separately or in groups. (1).

Carrier. A continuous frequency capable of being modulated, or impressed with a second (information carrying) signal. (1).

Carrier, communications common. A company which furnishes communications services to the general public, and which is regulated by appropriate local, state, or federal agencies. The term strictly includes truckers and movers, bus lines, and airlines, but is usually used to refer to telecommunication companies.

Carrier system. A means of obtaining a number of channels over a single path by modulating each channel on a different carrier frequency and demodulating at the receiving point to restore the signals to their original form.

Carrier telegraphy, carrier current telegraphy. A method of transmission in which the signals from a telegraph transmitter modulate an alternating current. (2).

Central office. The place where communications common carriers terminate customer lines and locate the switching equipment which interconnects those lines. (Also referred to as an *exchange, end office,* and *local central office.*)

Chad. The material removed when forming a hole or notch in a storage medium such as punched tape or punched cards.

Chadless tape. Perforated tape with the chad partially attached, to facilitate interpretive printing on the tape.

Channel. 1. (CCITT and ASA standard). A means of one-way transmission. (Compare with **Circuit.**)
2. (Tariff and common usage). As used in the tariffs, a path for electrical transmission between two or more points without common-carrier-provided terminal equipment. Also called *circuit, line, link, path,* or *facility.* (1).

Channel, analog. A channel on which the information transmitted can take any value between the limits defined by the channel. Most voice channels are analog channels.

Channel, voice-grade. A channel suitable for transmission of speech, digital or analog data, or facsimile, generally with a frequency range of about 300 to 3400 cycles per second.

12-channel group (of carrier current system). The assembly of 12 telephone channels, in a carrier system, occupying adjacent bands in the spectrum, for the purpose of simultaneous modulation or demodulation. (2).

Character. Letter, figure, number, punctuation or other sign contained in a message. Besides such characters, there may be characters for special symbols and some control functions. (1).

Characteristic distortion. Distortion caused by transients which, as a result of the modulation, are present in the transmission channel and depend on its transmission qualities.

Circuit. A means of both-way communication between two points, comprising associated "go" and "return" channels. (1).

Circuit, four-wire. A communication path in which four wires (two for each direction of transmission) are presented to the station equipment. (1).

Circuit, two-wire. A metallic circuit formed by two conductors insulated from each other. It is possible to use the two conductors as either a one-way transmission path, a half-duplex path, or a duplex path. (1).

Common carrier. (*See* **Carrier, communications common.**)

Compandor. A compandor is a combination of a compressor at one point in a communication path for reducing the volume *range* of signals, followed by an expandor at another point for restoring the original volume range. Usually its purpose is to improve the ratio of the signal to the interference entering in the path between the compressor and expandor. (2).

Compressor. Electronic device which compresses the volume range of a signal, used in a compandor (q.v.). An "expandor" restores the original volume range after transmission.

Conditioning. The addition of equipment to a leased voice-grade channel to provide minimum values of line characteristics required for data transmission. (1).

Contention. This is a method of line control in which the terminals request to transmit. If the channel in question is free, transmission goes ahead; if it is not free, the terminal will have to wait until it becomes free. The queue of contention requests may be built up by the computer, and this can either be in a prearranged sequence or in the sequence in which the requests are made.

Control character. A character whose occurrence in a particular context initiates, modifies, or stops a control operation—e.g., a character to control carriage return. (1).

Control mode. The state that all terminals on a line must be in to allow line control actions, or terminal selection to occur. When all terminals on a line are in the control mode, characters on the line are viewed as control characters performing line discipline, that is, polling or addressing. (1).

Cross-bar switch. A switch having a plurality of vertical paths, a plurality of horizontal paths, and electromagnetically operated mechanical means for interconnecting any one of the vertical paths with any of the horizontal paths. (2).

Cross-bar system. A type of line-switching system which uses cross-bar switches.

Cross talk. The unwanted transfer of energy from one circuit, called the *disturbing* circuit, to another circuit, called the *disturbed* circuit. (2).

Cross talk, far-end. Cross talk which travels along the disturbed circuit in the same direction as the signals in that circuit. To determine the far-end cross talk between two pairs, 1 and 2, signals are transmitted on pair 1 at station A, and the level of cross talk is measured on pair 2 at station B. (1).

Cross talk, near-end. Cross talk which is propagated in a disturbed channel in the direction opposite to the direction of propagation of the current in the disturbing channel. Ordinarily, the terminal of the disturbed channel at which the near-end cross talk is present is near or coincides with the energized terminal of the disturbing channel. (1).

Dataphone. Both a service mark and a trademark of AT&T and the Bell System. As a service mark it indicates the transmission of data over the telephone network. As a trademark it identifies the communications equipment furnished by the Bell System for data communications services. (1).

Data set. A device which performs the modulation/demodulation and control functions necessary to provide compatibility between business machines and communications facilities. (*See also* **Modem** *and* **Subset.**) (1).

Data-signaling rate. It is given by $\sum_{i=1}^{m} \frac{1}{T_i} \log_2 n_i$, where m is the number of parallel channels, T is the minimum interval for the ith channel, expressed in seconds, n is the number of significant conditions of the modulation in the ith channel. Data-signaling rate is expressed in bits per second. (2).

Dataspeed. An AT&T marketing term for a family of medium-speed paper tape transmitting and receiving units. Similar equipment is also marketed by Western Union. (1).

DDD. (*See* **Direct distance dialing,** q.v.)

Decibel (db). A tenth of a bel. A unit for measuring relative strength of a signal parameter such as power, voltage, etc. The number of decibels is ten times the logarithm (base 10) of the ratio of the measured quantity to the reference level. The reference level must always be indicated, such as 1 milliwatt for power ratio. (1). See Fig. 9.5.

Delay distortion. Distortion occurring when the envelope delay of a circuit or system is not constant over the frequency range required for transmission.

Delay equalizer. A corrective network which is designed to make the phase delay or envelope delay of a circuit or system substantially constant over a desired frequency range. (*See* **Equalizer.**). (1).

Demodulation. The process of retrieving intelligence (data) from a modulated carrier wave; the reverse of modulation. (1).

Diagnostic programs. These are used to check equipment malfunctions and to pinpoint faulty components. They may be used by the computer engineer or may be called in by the supervisory programs automatically.

Diagnostics, system. Rather than checking one individual component, system diagnostics utilize the whole system in a manner similar to its operational running. Programs resembling the operational programs will be used rather than systematic programs that run logical patterns. These will normally detect overall system malfunctions but will not isolate faulty components.

Diagnostics, unit. These are used on a conventional computer to detect faults in the various units. Separate unit diagnostics will check such items as arithmetic circuitry, transfer instructions, each input-output unit, and so on.

Dial pulse. A current interruption in the DC loop of a calling telephone. It is produced by the breaking and making of the dial pulse contacts of a calling telephone when a digit is dialed. The loop current is interrupted once for each unit of value of the digit. (1).

Dial-up. The use of a dial or pushbutton telephone to initiate a station-to-station telephone call.

Dibit. A group of two bits. In four-phase modulation, each possible dibit is encoded as one of four unique carrier phase shifts. The four possible states for a dibit are 00, 01, 10, 11.

Differential modulations. A type of modulation in which the choice of the significant condition for any signal element is dependent on the choice for the previous signal element. (2).

Digital data. Information represented by a code consisting of a sequence of discrete elements. (Compare with **Analog data.**) (1).

Digital signal. A discrete or discontinuous signal; one whose various states are discrete intervals apart. (Compare with **Analog transmission.**) (1).

Direct distance dialing (DDD). A telephone exchange service which enables the telephone user to call other subscribers outside his local area without operator assistance. In the United Kingdom and some other countries, this is called *Subscriber Trunk Dialing* (STD).

Disconnect signal. A signal transmitted from one end of a subscriber line or trunk to indicate at the other end that the established connection should be disconnected. (1).

Distortion. The unwanted change in waveform that occurs between two points in a transmission system. (1).

Distributing frame. A structure for terminating permanent wires of a telephone central office, private branch exchange, or private exchange, and for permitting the easy change of connections between them by means of cross-connecting wires. (1).

Double-current transmission, polar direct-current system. A form of binary telegraph transmission in which positive and negative direct currents denote the significant conditions. (2).

Drop, subscriber's. The line from a telephone cable to a subscriber's building. (1).

Duplex transmission. Simultaneous two-way independent transmission in both directions. (Compare with **Half-duplex transmission.** Also called *full-duplex transmission.*) (1).

Duplexing. The use of duplicate computers, files or circuitry, so that in the event of one component failing an alternative one can enable the system to carry on its work.

Echo. An echo is a wave which has been reflected or otherwise returned with sufficient magnitude and delay for it to be perceptible in some manner as a wave distinct from that directly transmitted.

Echo check. A method of checking data transmission accuracy whereby the received data are returned to the sending end for comparison with the original data.

Echo suppressor. A line device used to prevent energy from being reflected back (echoed) to the transmitter. It attenuates the transmission path in one direction while signals are being passed in the other direction. (1).

End distortion. End distortion of start-stop teletypewriter signals is the shifting of the end of all marking pulses from their proper positions in relation to the beginning of the start pulse.

End office. (*See* **Central office.**)

Equalization. Compensation for the attenuation (signal loss) increase with frequency. Its purpose is to produce a flat frequency response while the temperature remains constant. (1).

Equalizer. Any combination (usually adjustable) of coils, capacitors, and/or resistors inserted in transmission line or amplifier circuit to improve its frequency response. (1).

Equivalent four-wire system. A transmission system using frequency division to obtain full-duplex operation over only one pair of wires. (1).

Error-correcting telegraph code. An error-detecting code incorporating sufficient additional signaling elements to enable the nature of some or all of the errors to be indicated and corrected entirely at the receiving end.

Error-detecting and feedback system, decision feedback system, request repeat system, ARQ system. A system employing an error-detecting code and so arranged that a signal detected as being in error automatically initiates a request for retransmission of the signal detected as being in error. (2).

Error-detecting telegraph code. A telegraph code in which each telegraph signal conforms to specific rules of construction, so that departures from this construction in the received signals can be automatically detected. Such codes necessarily require more signaling elements than are required to convey the basic information.

ESS. (Electronic Switching System). Bell System term for computerized telephone exchange. ESS 1 is a central office. ESS 101 gives private branch exchange (PBX) switching controlled from the local central office. (*See* Chapter 19.)

Even parity check (odd parity check). This is a check which tests whether the number of digits in a group of binary digits is even (even parity check) or odd (odd parity check). (2).

Exchange. A unit established by a communications common carrier for the administration of communication service in a specified area which usually embraces a city, town, or village and its environs. It consists of one or more central offices together with the associated equipment used in furnishing communication service. (This term is often used as a synonym for "central office," q.v.)

Exchange, classes of. Class 1 (*see* **Regional center**); class 2 (*see* **Sectional center**); class 3 (*see* **Primary center**); class 4 (*see* **Toll center**); class 5 (*see* **End office**).

Exchange, private automatic (PAX). A dial telephone exchange that provides private telephone service to an organization and that does *not* allow calls to be transmitted to or from the public telephone network.

Exchange, private automatic branch (PABX). A private automatic telephone exchange that provides for the transmission of calls to and from the public telephone network.

Exchange, private branch (PBX). A manual exchange connected to the public telephone network on the user's premises and operated by an attendant supplied by the user. PBX is today commonly used to refer also to an automatic exchange.

Exchange, trunk. An exchange devoted primarily to interconnecting trunks.

Exchange service. A service permitting interconnection of any two customers' stations through the use of the exchange system.

Expandor. A transducer which for a given amplitude range or input voltages produces a larger range of output voltages. One important type of expandor employs the information from the envelope of speech signals to expand their volume range. (Compare **Compandor.**) (1).

Facsimile (FAX). A system for the transmission of images. The image is scanned at the transmitter, reconstructed at the receiving station, and duplicated on some form of paper. (1).

Fail softly. When a piece of equipment fails, the programs let the system fall back to a degraded mode of operation rather than let it fail catastrophically and give no response to its users.

Fall-back, double. Fall-back in which two separate equipment failures have to be contended with.

Fall-back procedures. When the equipment develops a fault the programs operate in such a way as to circumvent this fault. This may or may not give a degraded service. Procedures necessary for fall-back may include those to switch over to an alternative computer or file, to change file addresses, to send output to a typewriter instead of a printer, to use different communication lines or bypass a faulty terminal, etc.

FCC. Federal Communications Commission, q.v.

FD or **FDX.** Full duplex. (*See* **Duplex.**)

FDM. Frequency-division multiplex, q.v.

Federal Communications Commission (FCC). A board of seven commissioners appointed by the President under the Communication Act of 1934, having the power to regulate all interstate and foreign electrical communication systems originating in the United States. (1).

Figures shift. A physical shift in a teletypewriter which enables the printing of numbers, symbols, upper-case characters, etc. (Compare with **Letters shift.**) (1).

Filter. A network designed to transmit currents of frequencies within one or more frequency bands and to attenuate currents of other frequencies. (2).

Foreign exchange service. A service which connects a customer's telephone to a telephone company central office normally not serving the customer's location. (Also applies to TWX service.) (1).

Fortuitous distortion. Distortion resulting from causes generally subject to random laws (accidental irregularities in the operation of the apparatus and of the moving parts, disturbances affecting the transmission channel, etc.). (2).

Four-wire circuit. A circuit using two pairs of conductors, one pair for the "go" channel and the other pair for the "return" channel. (2).

Four-wire equivalent circuit. A circuit using the same pair of conductors to give "go" and "return" channels by means of different carrier frequencies for the two channels. (2).

Four-wire terminating set. Hybrid arrangement by which four-wire circuits are terminated on a two-wire basis for interconnection with two-wire circuits.

Frequency-derived channel. Any of the channels obtained from multiplexing a channel by frequency division. (2).

Frequency-division multiplex. A multiplex system in which the available transmission frequency range is divided into narrower bands, each used for a separate channel. (2).

Frequency modulation. One of three ways of modifying a sine wave signal to make it "carry" information. The sine wave or "carrier" has its frequency modified in accordance with the information to be transmitted. The frequency function of the modulated wave may be continuous or discontinuous. In the latter case, two or more particular frequencies may correspond each to one significant condition.

Frequency-shift signaling, frequency-shift keying (FSK). Frequency modulation method in which the frequency is made to vary at the significant instants. 1. By smooth transitions: the modulated wave and the change in frequency are continuous at the significant instants. 2. By abrupt transitions: the modulated wave is continuous but the frequency is discontinuous at the significant instants. (2).

FSK. Frequency-shift keying, q.v.

FTS. Federal Telecommunications System.

Full-duplex (FD or FDX) **transmission.** (*See* **Duplex transmission.**)

Half-duplex (HD or HDX) **circuit.**

1. CCITT definition: A circuit designed for duplex operation, but which, on account of the nature of the terminal equipments, can be operated alternately only.
2. Definition in common usage (the normal meaning in computer literature): A circuit designed for transmission in either direction but not both directions simultaneously.

Half-duplex transmission. (*See* **Half-duplex circuit.**)

Handshaking. Exchange of predetermined signals for purposes of control when a connection is established between two data sets.

Harmonic distortion. The resultant presence of harmonic frequencies (due to nonlinear characteristics of a transmission line) in the response when a sinusoidal stimulus is applied. (1).

HD or **HDX.** Half duplex. (*See* **Half-duplex circuit.**)

Hertz (Hz). A measure of frequency or bandwidth. The same as cycles per second.

Home loop. An operation involving only those input and output units associated with the local terminal. (1).

In-house. (*See* **In-plant system.**)

In-plant system. A system whose parts, including remote terminals, are all situated in one building or localized area. The term is also used for communication systems spanning several buildings and sometimes covering a large distance, but in which no common carrier facilities are used.

International Telecommunication Union (ITU). The telecommunications agency of the United Nations, established to provide standardized communications procedures and practices including frequency allocation and radio regulations on a world-wide basis.

Interoffice trunk. A direct trunk between local central offices.

Intertoll trunk. A trunk between toll offices in different telephone exchanges. (1).

ITU. International Telecommunication Union, q.v.

Keyboard perforator. A perforator provided with a bank of keys, the manual depression of any one of which will cause the code of the corresponding character or function to be punched in a tape. (2).

Keyboard send/receive. A combination teletypewriter transmitter and receiver with transmission capability from keyboard only.

KSR. Keyboard send/receive, q.v.

Leased facility. A facility reserved for sole use of a single leasing customer. (*See also* **Private line.**) (1).

Letters shift. A physical shift in a teletypewriter which enables the printing of alphabetic characters. Also, the name of the character which causes this shift. (*Compare* with **Figures shift.**) (1).

Line switching. Switching in which a circuit path is set up between the incoming and outgoing lines. Contrast with message switching (q.v.) in which no such physical path is established.

Link communication. The physical means of connecting one location to another for the purpose of transmitting and receiving information. (1).

Loading. Adding inductance (load coils) to a transmission line to minimize amplitude distortion. (1).

Local exchange, local central office. An exchange in which subscribers' lines terminate. (Also referred to as *end office.*)

Local line, local loop. A channel connecting the subscriber's equipment to the line terminating equipment in the central office exchange. Usually metallic circuit (either two-wire or four-wire). (1).

Longitudinal redundancy check (LRC). A system of error control based on the formation of a block check following preset rules. The check formation rule is applied in the same manner to each character. In a simple case, the LRC is created by forming a parity check on each bit position of all the characters in the block (e.g., the first bit of the LRC character creates odd parity among the one-bit positions of the characters in the block).

Loop checking, message feedback, information feedback. A method of checking the accuracy of transmission of data in which the received data are returned to the sending end for comparison with the original data, which are stored there for this purpose. (2).

LRC. Longitudinal redundancy check.

LTRS. Letters shift, q.v. (*See* **Letters shift.**)

Mark. Presence of signal. In telegraph communications a mark represents the closed condition or current flowing. A mark impulse is equivalent to a binary 1.

Mark-hold. The normal no-traffic line condition whereby a steady mark is transmitted. This may be a customer-selectable option. (Compare with **Space-hold.**) (1).

Mark-to-space transition. The transition, or switching from a marking impulse to a spacing impulse.

Master station. A unit having control of all other terminals on a multipoint circuit for purposes of polling and/or selection. (1).

Mean time to failure. The average length of time for which the system, or a component of the system, works without fault.

Mean time to repair. When the system, or a component of the system, develops a fault, this is the average time taken to correct the fault.

Message reference block. When more than one message in the system is being processed in parallel, an area of storage is allocated to each message and remains uniquely associated with that message for the duration of its stay in the computer. This is called the *message reference block* in this book. It will normally contain the message and data associated with it that are required for its processing. In most systems, it contains an area of working storage uniquely reserved for that message.

Message switching. The technique of receiving a message, storing it until the proper outgoing line is available, and then retransmitting. No direct connection between the incoming and outgoing lines is set up as in line switching (q.v.).

Microwave. Any electromagnetic wave in the radio-frequency spectrum above 890 megacycles per second. (1).

Modem. A contraction of "modulator-demodulator." The term may be used when the modulator and the demodulator are associated in the same signal-conversion equipment. (*See* **Modulation** *and* **Data set.**) (1).

Modulation. The process by which some characteristic of one wave is varied in accordance with another wave or signal. This technique is used in data sets and modems to make business machine signals compatible with communications facilities. (1).

Modulation with a fixed reference. A type of modulation in which the choice of the significant condition for any signal element is based on a fixed reference. (2).

Multidrop line. Line or circuit interconnecting several stations. (Also called *multipoint line.*) (1).

Multiplex, multichannel. Use of a common channel in order to make two or more channels, either by splitting of the frequency band transmitted by the common channel into narrower bands, each of which is used to constitute a distinct channel (frequency-division multiplex), or by allotting this common channel in turn, to constitute different intermittent channels (time-division multiplex). (2).

Multiplexing. The division of a transmission facility into two or more channels either by splitting the frequency band transmitted by the channel into narrower bands, each of which is used to constitute a distinct channel (frequency-division multiplex), or by allotting this common channel to several different information channels, one at a time (time-division multiplexing). (2).

Multiplexor. A device which uses several communication channels at the same time, and transmits and receives messages and controls the communication lines. This device itself may or may not be a stored-program computer.

Multipoint line. (*See* **Multidrop line.**)

Neutral transmission. Method of transmitting teletypewriter signals, whereby a mark is represented by current on the line and a space is represented by the absence of current. By extension to tone signaling, neutral transmission is a method of signaling employing two signaling states, one of the states representing both a space condition and also the absence of any signaling. (Also called *unipolar*. Compare with **Polar transmission.**) (1).

Noise. Random electrical signals, introduced by circuit components or natural disturbances, which tend to degrade the performance of a communications channel. (1).

Off hook. Activated (in regard to a telephone set). By extension, a data set automatically answering on a public switched system is said to go "off hook." (Compare with **On hook.**) (1).

Off line. Not in the line loop. In telegraph usage, paper tapes frequently are punched "off line" and then transmitted using a paper tape transmitter.

On hook. Deactivated (in regard to a telephone set). A telephone not in use is "on hook." (1).

On line. Directly in the line loop. In telegraph usage, transmitting directly onto the line rather than, for example, perforating a tape for later transmission. (*See also* **On-line computer system.**)

On-line computer system. An on-line system may be defined as one in which the input data enter the computer directly from their point of origin and/or output data are transmitted directly to where they are used. The intermediate stages such as punching data into cards or paper tape, writing magnetic tape, or off-line printing, are largely avoided.

Open wire. A conductor separately supported above the surface of the ground—i.e., supported on insulators.

Open-wire line. A pole line whose conductors are principally in the form of open wire.

PABX. Private automatic branch exchange. (*See* **Exchange, private automatic branch.**)

Parallel transmission. Simultaneous transmission of the bits making up a character or byte, either over separate channels or on different carrier frequencies on the channel. (1). The simultaneous transmission of a certain number of signal elements constituting the same telegraph or data signal. For example, use of a code according to which each signal is characterized by a combination of 3 out of 12 frequencies simultaneously transmitted over the channel. (2).

Parity check. Addition of noninformation bits to data, making the number of ones in a grouping of bits either always even or always odd. This permits detection of bit groupings that contain single errors. It may be applied to characters, blocks, or any convenient bit grouping. (1).

Parity check, horizontal. A parity check applied to the group of certain bits from every character in a block. (*See also* **Longitudinal redundancy check.**)

Parity check, vertical. A parity check applied to the group which is all bits in one character. (Also called *vertical redundancy check.*) (1).

PAX. Private automatic exchange. (*See* **Exchange, private automatic.**)

PBX. Private branch exchange. (*See* **Exchange, private branch.**)

PCM. (*See* **Pulse code modulation.**)

PDM. (*See* **Pulse duration modulation.**)

Perforator. An instrument for the manual preparation of a perforated tape, in which telegraph signals are represented by holes punched in accordance with a predetermined code. Paper tape is prepared off line with this. (Compare with **Reperforator.**) (2).

Phantom telegraph circuit. Telegraph circuit superimposed on two physical circuits reserved for telephony. (2).

Phase distortion. (*See* **Delay distortion.**)

Phase equalizer, delay equalizer. A delay equalizer is a corrective network which is designed to make the phase delay or envelope delay of a circuit or system substantially constant over a desired frequency range. (2).

Phase-inversion modulation. A method of phase modulation in which the two significant conditions differ in phase by π radians. (2).

Phase modulation. One of three ways of modifying a sine wave signal to make it "carry" information. The sine wave or "carrier," has its phase changed in accordance with the information to be transmitted.

Pilot model. This is a model of the system used for program testing purposes which is less complex than the complete model, e.g., the files used on a pilot model may contain a much smaller number of records than the operational files; there may be few lines and fewer terminals per line.

Polar transmission. A method for transmitting teletypewriter signals, whereby the marking signal is represented by direct current flowing in one direction and the spacing signal is represented by an equal current flowing in the opposite direction. By extension to tone signaling, polar transmission is a method of transmission employing three distinct states, two to represent a mark and a space and one to represent the absence of a signal. (Also called *bipolar*. Compare with **Neutral transmission.**)

Polling. This is a means of controlling communication lines. The communication control device will send signals to a terminal saying, "Terminal A. Have you anything to send?" if not, "Terminal B. Have you anything to send?" and so on. Polling is an alternative to contention. It makes sure that no terminal is kept waiting for a long time.

Polling list. The polling signal will usually be sent under program control. The program will have in core a list for each channel which tells the sequence in which the terminals are to be polled.

PPM. (*See* **Pulse position modulation.**)

Primary center. A control center connecting toll centers; a class 3 office. It can also serve as a toll center for its local end offices.

Private automatic branch exchange. (*See* **Exchange, private automatic branch.**)

Private automatic exchange. (*See* **Exchange, private automatic.**)

Private branch exchange (PBX). A telephone exchange serving an individual organization and having connections to a public telephone exchange. (2).

Private line. Denotes the channel and channel equipment furnished to a customer as a unit for his exclusive use, without interexchange switching arrangements. (1).

Processing, batch. A method of computer operation in which a number of similar input items are accumulated and grouped for processing.

Processing, in line. The processing of transactions as they occur, with no preliminary editing or sorting of them before they enter the system. (1).

Propagation delay. The time necessary for a signal to travel from one point on a circuit to another.

Public. Provided by a common carrier for use by many customers.

Public switched network. Any switching system that provides circuit switching to many customers. In the U.S.A. there are four such networks: Telex, TWX, telephone, and Broadband Exchange. (1).

Pulse-code modulation (PCM). Modulation of a pulse train in accordance with a code. (2).

Pulse-duration modulation (PDM) **(pulse-width modulation) (pulse-length modulation).** A form of pulse modulation in which the durations of pulses are varied. (2).

Pulse modulation. Transmission of information by modulation of a pulsed or intermittent, carrier. Pulse width, count, position, phase, and/or amplitude may be the varied characteristic.

Pulse-position modulation (PPM). A form of pulse modulation in which the positions in time of pulses are varied, without modifying their duration. (2).

Pushbutton dialing. The use of keys or pushbuttons instead of a rotary dial to generate a sequence of digits to establish a circuit connection. The signal form is usually multiple tones. (Also called *tone dialing, Touch-call, Touch-tone.*) (1).

Real time. A real-time computer system may be defined as one that controls an environment by receiving data, processing them, and returning the results sufficiently quickly to affect the functioning of the environment at that time.

Reasonableness checks. Tests made on information reaching a real-time system or being transmitted from it to ensure that the data in question lie within a given range. It is one of the means of protecting a system from data transmission errors.

Recovery from fall-back. When the system has switched to a fall-back mode of operation and the cause of the fall-back has been removed, the system must be restored to its former condition. This is referred to as *recovery from fall-back.* The recovery process may involve updating information in the files to produce two duplicate copies of the file.

Redundancy check. An automatic or programmed check based on the systematic insertion of components or characters used especially for checking purposes. (1).

Redundant code. A code using more signal elements than necessary to represent the intrinsic information. For example, five-unit code using all the characters of International Telegraph Alphabet No. 2 is not redundant; five-unit code using only the figures in International Telegraph Alphabet No. 2 is redundant; seven-unit code using only signals made of four "space" and three "mark" elements is redundant. (2).

Reference pilot. A reference pilot is a different wave from those which transmit the telecommunication signals (telegraphy, telephony). It is used in carrier systems to facilitate the maintenance and adjustment of the carrier transmission system. (For example, automatic level regulation, synchronization of oscillators, etc.) (2).

Regenerative repeater. (*See* **Repeater, regenerative.**)

Regional center. A control center (class 1 office) connecting sectional centers of the telephone system together. Every pair of regional centers in the United States has a direct circuit group running from one center to the other. (1).

Repeater.

1. A device whereby currents received over one circuit are automatically repeated in another circuit or circuits, generally in an amplified and/or reshaped form.
2. A device used to restore signals, which have been distorted because of attenuation, to their original shape and transmission level.

Repeater, regenerative. Normally, a repeater utilized in telegraph applications. Its function is to retime and retransmit the received signal impulses restored to their original strength. These repeaters are speed- and code-sensitive and are intended for use with standard telegraph speeds and codes. (Also called *regen.*) (1).

Repeater, telegraph. A device which receives telegraph signals and automatically retransmits corresponding signals. (2).

Reperforator (receiving perforator). A telegraph instrument in which the received signals cause the code of the corresponding characters or functions to be punched in a tape. (1).

Reperforator/transmitter (RT). A teletypewriter unit consisting of a reperforator and a tape transmitter, each independent of the other. It is used as a relaying device and is especially suitable for transforming the incoming speed to a different outgoing speed, and for temporary queuing.

Residual error rate, undetected error rate. The ratio of the number of bits, unit elements, characters or blocks incorrectly received but undetected or uncorrected by the error-control equipment, to the total number of bits, unit elements, characters or blocks sent. (2).

Response time. This is the time the system takes to react to a given input. If a message is keyed into a terminal by an operator and the reply from the computer, when it comes, is typed at the same terminal, response time may be defined as the time interval between the operator pressing the last key and the terminal typing the first letter of the reply. For different types of terminal, response time may be defined similarly. It is the interval between an event and the system's response to the event.

Ringdown. A method of signaling subscribers and operators using either a 20-cycle AC signal, a 135-cycle AC signal, or a 1000-cycle signal interrupted 20 times per second. (1).

Routing. The assignment of the communications path by which a message or telephone call will reach its destination. (1).

Routing, alternate. Assignment of a secondary communications path to a destination when the primary path is unavailable. (1).

Routing indicator. An address, or group of characters, in the heading of a message defining the final circuit or terminal to which the message has to be delivered. (1).

RT. Reperforator/transmitter. q.v.

Saturation testing. Program testing with a large bulk of messages intended to bring to light those errors which will only occur very infrequently and which may be triggered by rare coincidences such as two different messages arriving at the same time.

Sectional center. A control center connecting primary centers; a class 2 office. (1).

Seek. A mechanical movement involved in locating a record in a random-access file. This may, for example, be the movement of an arm and head mechanism that is necessary before a read instruction can be given to read data in a certain location on the file.

Selection. Addressing a terminal and/or a component on a selective calling circuit. (1).

Selective calling. The ability of the transmitting station to specify which of several stations on the same line is to receive a message. (1).

Self-checking numbers. Numbers which contain redundant information so that an error in them, caused, for example, by noise on a transmission line, may be detected.

Serial transmission. Used to identify a system wherein the bits of a character occur serially in time. Implies only a single transmission channel. (Also called *serial-by-bit.*) (1). Transmission at successive intervals of signal elements constituting the same telegraph or data signal. For example, transmission of signal elements by a standard teleprinter, in accordance with International Telegraph Alphabet No. 2; telegraph transmission by a time-divided channel. (2).

Sideband. The frequency band on either the upper or lower side of the carrier frequency within which fall the frequencies produced by the process of modulation. (2).

Signal-to-noise ratio (S/N). Relative power of the signal to the noise in a channel. (1).

Simplex circuit.

1. CCITT definition: A circuit permitting the transmission of signals in either direction, but not in both simultaneously.
2. Definition in common usage (the normal meaning in computer literature): A circuit permitting transmission in one specific direction only.

Simplex mode. Operation of a communication channel in one direction only, with no capability for reversing. (1).

Simulation. This is a word which is sometimes confusing as it has three entirely different meanings, namely:

Simulation for design and monitoring. This is a technique whereby a model of the working system can be built in the form of a computer program. Special computer languages are available for producing this model. A complete system may be described by a succession of different models. These models can then be adjusted easily and endlessly, and the system that is being designed or monitored can be experimented with to test the effect of any proposed changes. The simulation model is a program that is run on a computer separate from the system that is being designed.

Simulation of input devices. This is a program testing aid. For various reasons it is undesirable to use actual lines and terminals for some of the program testing. Therefore, magnetic tape or other media may be used and read in by a special program which makes the data appear as if they came from actual lines and terminals. Simulation in this sense is the replacement of one set of equipment by another set of equipment and programs, so that the behavior is similar.

Simulation of supervisory programs. This is used for program testing purposes when the actual supervisory programs are not yet available. A comparatively simple program to bridge the gap is used instead. This type of simulation is the replacement of one set of programs by another set which imitates it.

Single-current transmission (inverse), **neutral direct-current system.** A form of telegraph transmission effected by means of unidirectional currents. (2).

Space. 1. An impulse which, in a neutral circuit, causes the loop to open or causes absence of signal, while in a polar circuit it causes the loop current to flow in a direction opposite to that for a mark impulse. A space impulse is equivalent to a binary 0. 2. In some codes, a character which causes a printer to leave a character width with no printed symbol. (1).

Space-hold. The normal no-traffic line condition whereby a steady space is transmitted. (Compare with Mark-hold.) (1).

Space-to-mark transition. The transition, or switching, from a spacing impulse to a marking impulse. (1).

Spacing bias. (*See* **Bias distortion.**)

Spectrum. 1. A continuous range of frequencies, usually wide in extent, within which waves have some specific common characteristic. 2. A graphical representation of the distribution of the amplitude (and sometimes phase) of the components of a wave as a function of frequency. A spectrum may be continuous or, on the contrary, contain only points corresponding to certain discrete values. (2).

Start element. The first element of a character in certain serial transmissions, used to permit synchronization. In Baudot teletypewriter operation, it is one space bit. (1).

Start-stop system. A system in which each group of code elements corresponding to an alphabetical signal is preceded by a start signal which serves to prepare the receiving mechanism for the reception and registration of a character, and is followed by a stop signal which serves to bring the receiving mechanism to rest in preparation for the reception of the next character. (Contrast with **Synchronous system.**) (Start-stop transmission is also referred to as *asynchronous transmission,* q.v.)

Station. One of the input or output points of a communications system—e.g., the telephone set in the telephone system or the point where the business machine interfaces the channel on a leased private line. (1).

Status maps. Tables which give the status of various programs, devices, input-output operations, or the status of the communication lines.

Step-by-step switch. A switch that moves in synchronism with a pulse device such as a rotary telephone dial. Each digit dialed causes the movement of successive selector switches to carry the connection forward until the desired line is reached. (Also called *stepper switch.* Compare with **Line switching** and **Cross-bar system.**) (1).

Step-by-step system. A type of line-switching system which uses step-by-step switches. (1).

Stop bit. (*See* **Stop element.**)

Stop element. The last element of a character is asynchronous serial (start-stop) transmissions, used to ensure recognition of the next start element. In Baudot teletypewriter operation it is 1.42 mark bits. (1).

Store and forward. The interruption of data flow from the originating terminal to the designated receiver by storing the information enroute and forwarding it at a later time. (*See* **Message switching.**)

Stunt box. A device to 1. control the nonprinting functions of a teletypewriter terminal, such as carriage return and line feed; and 2. a device to recognize line control characters (e.g., DCC, TSC, etc.). (1).

Subscriber trunk dialing. (*See* **Direct distance dialing.**)

Subscriber's line. The telephone line connecting the exchange to the subscriber's station. (2).

Subset. A subscriber set of equipment, such as a telephone. A modulation and demodulation device. (Also called *data set,* which is a more precise term.) (1).

Subscriber's loop. (*See* **Local line.**)

Subvoice-grade channel. A channel of bandwidth narrower than that of voice-grade channels. Such channels are usually subchannels of a voice-grade line. (1).

Supergroup. The assembly of five 12-channel groups, occupying adjacent bands in the spectrum, for the purpose of simultaneous modulation or demodulation. (2).

Supervisory programs. Those computer programs designed to coordinate service and augment the machine components of the system, and coordinate and service application programs. They handle work scheduling, input-output operations, error actions, and other functions.

Supervisory signals. Signals used to indicate the various operating states of circuit combinations. (1).

Supervisory system. The complete set of supervisory programs used on a given system.

Support programs. The ultimate operational system consists of supervisory programs and application programs. However, a third set of programs are needed to install the system, including diagnostics, testing aids, data generator programs, terminal simulators, etc. These are referred to as *support programs.*

Suppressed carrier transmission. That method of communication in which the carrier frequency is suppressed either partially or to the maximum degree possible. One or both of the sidebands may be transmitted. (1).

Switch hook. A switch on a telephone set, associated with the structure supporting the receiver or handset. It is operated by the removal or replacement of the receiver or handset on the support. (*See also* **Off hook** *and* **On hook.**) (1).

Switching center. A location which terminates multiple circuits and is capable of interconnecting circuits or transferring traffic between circuits; may be automatic, semiautomatic, or torn-tape. (The latter is a location where operators tear off the incoming printed and punched paper tape and transfer it manually to the proper outgoing circuit.) (1).

Switching message. (*See* **Message switching.**)

Switchover. When a failure occurs in the equipment a switch may occur to an alternative component. This may be, for example, an alternative file unit, an alternative communication line or an alternative computer. The switchover process may be automatic under program control or it may be manual.

Synchronous. Having a constant time interval between successive bits, characters, or vents. The term implies that all equipment in the system is in step.

Synchronous system. A system in which the sending and receiving instruments are operating continuously at substantially the same frequency and are maintained, by means of correction, if necessary, in a desired phase relationship. (Contrast with **Start-stop system.**) (2).

Synchronous transmission. A transmission process such that between any two significant instances there is always an integral number of unit intervals. (Contrast with **Asynchronous** or **Start-stop system.**) (1).

Tandem office. An office that is used to interconnect the local end offices over tandem trunks in a densely settled exchange area where it is uneconomical for a telephone company to provide direct interconnection between all end offices. The tandem office completes all calls between the end offices but is not directly connected to subscribers. (1).

Tandem office, tandem central office. A central office used primarily as a switching point for traffic between other central offices. (2).

Tariff. The published rate for a specific unit of equipment, facility, or type of service provided by a communications common carrier. Also the vehicle by which the regulating agencies approve or disapprove such facilities or services. Thus the tariff becomes a contract between customer and common carrier.

TD. Transmitter-distributor, q.v.

Teleprocessing. A form of information handling in which a data-processing system utilizes communication facilities. (Originally, but no longer, an IBM trademark.) (1).

Teletype. Trademark of Teletype Corporation, usually referring to a series of different types of teleprinter equipment such as tape punches, reperforators, page printers, etc., utilized for communications systems.

Teletypewriter exchange service (TWX). An AT&T public switched teletypewriter service in which suitably arranged teletypewriter stations are provided with lines to a central office for access to other such stations throughout the U.S.A. and Canada. Both Baudot- and ASCII-coded machines are used. Business machines may also be used, with certain restrictions. (1).

Telex service. A dial-up telegraph service enabling its subscribers to communicate directly and temporarily among themselves by means of start-stop apparatus and of circuits of the public telegraph network. The service operates world wide. Baudot equipment is used. Computers can be connected to the Telex network.

Terminal. Any device capable of sending and/or receiving information over a communication channel. The means by which data are entered into a computer system and by which the decisions of the system are communicated to the environment it affects. A wide variety of terminal devices have been built, including teleprinters, special keyboards, light displays, cathode tubes, thermocouples, pressure gauges and other instrumentation, radar units, telephones, etc.

TEX. (*See* **Telex service.**)

Tie line. A private-line communications channel of the type provided by communications common carriers for linking two or more points together.

Time-derived channel. Any of the channels obtained from multiplexing a channel by time division.

Time-division multiplex. A system in which a channel is established in connecting intermittently, generally at regular intervals and by means of an automatic distribution, its terminal equipment to a common channel. At times when these connections are not established, the section of the common channel between the distributors can be utilized in order to establish other similar channels, in turn.

Toll center. Basic toll switching entity; a central office where channels and toll message circuits terminate. While this is usually one particular central office in a city, larger cities may have several central offices where toll message circuits terminate. A class 4 office. (Also called "toll office" and "toll point.") (1).

Toll circuit (American). *See* **Trunk circuit** (British).

Toll switching trunk. (American). *See* **Trunk junction** (British).

Tone dialing. (*See* **Pushbutton dialing.**)

Touch-call. Proprietary term of GT&E. (*See* **Pushbutton dialing.**)

Touch-tone. AT&T term for pushbutton dialing, q.v.

Transceiver. A terminal that can transmit and receive traffic.

Translator. A device that converts information from one system of representation into equivalent information in another system of representation. In telephone equipment, it is the device that converts dialed digits into call-routing information. (1).

Transmitter-distributor (TD). The device in a teletypewriter terminal which makes and breaks the line in timed sequence. Modern usage of the term refers to a paper tape transmitter.

Transreceiver. A terminal that can transmit and receive traffic. (1).

Trunk circuit (British), **toll circuit** (American). A circuit connecting two exchanges in different localities. *Note:* In Great Britain, a trunk circuit is approximately 15 miles long or more. A circuit connecting two exchanges less than 15 miles apart is called a *junction circuit.*

Trunk exchange (British), **toll office** (American). An exchange with the function of controlling the switching of trunk (British) [toll (American)] traffic.

Trunk group. Those trunks between two points both of which are switching centers and/or individual message distribution points, and which employ the same multiplex terminal equipment.

Trunk junction (British), **toll switching trunk** (American). A line connecting a trunk exchange to a local exchange and permitting a trunk operator to call a subscriber to establish a trunk call.

Unattended operations. The automatic features of a station's operation permit the transmission and reception of messages on an unattended basis. (1).

Vertical parity (redundancy) check. (*See* **Parity check, vertical.**)

VOGAD (Voice-Operated Gain-Adjusting Device). A device somewhat similar to a compandor and used on some radio systems; a voice-operated device which removes fluctuation from input speech and sends it out at a constant level. No restoring device is needed at the receiving end. (1).

Voice-frequency, telephone-frequency. Any frequency within that part of the audio-frequency range essential for the transmission of speech of commercial quality, i.e., 300–3400 c/s. (2).

Voice-frequency carrier telegraphy. That form of carrier telegraphy in which the carrier currents have frequencies such that the modulated currents may be transmitted over a voice-frequency telephone channel. (1).

Voice-frequency multichannel telegraphy. Telegraphy using two or more carrier currents the frequencies of which are within the voice-frequency range. Voice-frequency telegraph systems permit the transmission of up to 24 channels over a single circuit by use of frequency-division multiplexing.

Voice-grade channel. (*See* **Channel, voice-grade.**)

Voice-operated device. A device used on a telephone circuit to permit the presence of telephone currents to effect a desired control. Such a device is used in most echo suppressors. (1).

VRC. Vertical redundancy check. (*See also* **Parity check.**)

Watchdog timer. This is a timer which is set by the program. It interrupts the program after a given period of time, e.g., one second. This will prevent the system from going into an endless loop due to a program error, or becoming idle because of an equipment fault. The Watchdog timer may sound a horn or cause a computer interrupt if such a fault is detected.

WATS (Wide Area Telephone Service). A service provided by telephone companies in the United States which permits a customer by use of an access line to make calls to telephones in a specific zone in a dial basis for a flat monthly charge. Monthly charges are based on the size of the area in which the calls are placed, not on the number or length of calls. Under the WATS arrangement, the U.S. is divided into six zones to be called on a full-time or measured-time basis. (1).

Word. 1. In telegraphy, six operations or characters (five characters plus one space). ("Group" is also used in place of "word.") 2. In computing, a sequence of bits or characters treated as a unit and capable of being stored in one computer location. (1).

WPM (Words per minute). A common measure of speed in telegraph systems.

INDEX

A

Absorption, 413, 415, 491–501
Action Communications, Inc.: WATS Box, 90–1
ALOHA protocol, 191, 197, 360, 585–98
Analog signal, 29
  convert to digital, 46, 49–55
Analog transmission, 45–55 *passim,* 154, 233
  converted to digital transmission, 46, 49–55
  and multiplexing (*see* Multiplexing)
  and switching (*see* Switching)
ANIK Canadian Satellite, 211–12, 227, 482
Answerback, voice (*see* Voice answerback)
Antennae:
  multiple, for mobile radio, 198
  satellite, 4, 211–36 *passim,* 264, 473–99
    low cost, 4, 216, 473, 486–87
    roof top, 13, 166, 216, 264, 382, 479–82
ARPANET, 77, 282–83, 371, 385, 539–56 *passim*
ATS-1, 178
ATS-6, 225, 302, 340–45, 360–61, 382
AT&T, 18, 33, 350–56, 450–51, 480 (*see also* Bell System)
  DAA (Direct Access), 352, 363–65
  DDS (Dataphone Digital Service), 59, 62, 162–63, 164, 272, 279, 350, 368, 513–15
    offering speech, 279
  Digital local loops tariff, 273
  Dimension PABX, 73, 350–57, 383
  DR-18 (Data Radio), 416–18, 501, 537
  DUV, 513–14
  Helical waveguide, 450–51
  L5 carrier, 27, 448

AT&T *(cont.)*
LD4 (*see* Bell System T4 carrier)
WT4, 64, 383, 454, 501
Attenuation, 415, 453, 491
of waveguides, 453
AUTOVON, 305–7, 355

B

Bandwidth, 28–9, 31–2, 46, 48, 57, 405–7
and cost, 31
for digital transmission, 48, 52 (Table 4.1)
increased, on local loops, 32, 47, 54–5
and information capacity, 29, 405
and speed of channels, 57–68
and TV, 57
Baseband digital signal, 165
Bell, Alexander Graham, 11–2, 26, 44, 56, 116
Bell Laboratories, 21, 132, 271, 467
satellite study, 475–76
Bell System, 18, 31, 76, 83
components cost breakdown, 31, 474
DR 18 (Digital Radio 18 System), 416–19
Bell System Centrex, 92–3
digital radio system, 416–19
ESS, 92–3, 383, 565–70
PABX functions, 76–83 (Box 6.1)
Bell System T1 carrier, 53–5, 63, 121–22, 124, 165, 383, 501–2, 505–21
on premises, 383
T1C, 502, 509
Bell System T2 carrier, 53–5, 63, 121–22, 124, 165, 383, 501–2, 505
TD2, 121, 502, 536
Bell System T3 carrier, 501–2, 505
Bell System T4 carrier, 64, 121–22, 383, 388, 418, 477, 501–2, 505, 537
T4M, 517, 537
Bell System TV distribution, 150
Bits and bandwidth calculations, 65
Bit stream, digital, 5
conversion, from analog, 46
Broadband signals, 122, 216
Broadcast data (radio), 172, 175, 284, 385
Broadcasting, 35, 284, 385
Broadcast signals, 21, 67, 219
Buffering, 567–71
burst, and data transmission, 160
Burst modem, 234, 576, 581
Burst multiplexing, 156
Burst transmission, 39, 155–58, 386, 474, 486, 580–81

C

Cable, marine, 210–11, 218, 224, 361, 383, 535
Cable TV, 5, 17–8, 31, 57, 64, 133–51, 166, 273, 322–24, 350–53, 371–72, 382, 384, 469, 576
  to bypass local loops, 273
  capacity of, 27–8, 138–41, 328–29, 330–37, 469
  combined with telephone cables, 332, 371–72
  compared with telephone cables, 134, 329–37, 371–72
  and facsimile transmission, 138–39, 148
  and optical fiber, 388
  public access channel, 138, 372
  and satellite transmission, 51, 139, 145, 225, 227, 382, 388, 487, 490
    tradeoffs, 487–89
  switched, 329, 333–35
    two-way, 141–45, 147–51, 328–29, 371–72
    and wired city, 328–29
Carterfone decision, 351, 362
CATV (*see* Cable TV)
CB radio, 18, 182–85, 324, 385
CCITT, 59–60, 63, 67, 95, 110, 165, 259, 381, 383, 502, 509–13, 539
  digital standards, 502, 506, 509–13, 539
CCSA, 81
CEEFAX, 172–74, 175–77, 207, 284, 350, 358, 384
Cellular Radio Telephone System organization, 191, 193–203, 385
Centrex, 77, 92–3
Channel capacities:
  and bandwidth, 28–9
  table of, 27
Channels:
  speeds of, 57–68
  utilization of, 160, 260–61
    and grade of service, 37, 40, 68
Circuit switching networks, 168–69, 279, 486 (*see also* Switching)
Coaxial cable, 31, 32, 52, 64, 121, 133–51 *passim,* 165, 259, 383, 445–48, 453, 501 (*see also* Bell System T1 carrier, *et al.*)
CODEC, 5, 48, 383, 477–79, 486, 523, 527, 532, 536
Code Conversion, computer, 161
Cognitronics, 97
Common carriers, 153–54, 368 (*see also* Bell System)
  and cream-skimming, 368
  new, 271–86
  satellite, 271, 280
  specialized, 162, 271, 273–80, 368
  value added, 162, 280–86, 368
Communications extender, 91
Communications links, categories of, 32 (esp. Box 3.1), 38–9
Communications satellites (*see* Satellites)

Community Antenna Television (*see* Cable TV)
Compaction, 161, 261–62, 323
Compandor, compandoring, 525–27
Compression, of signals, 323–37
Computer Assisted Instruction, 299, 327, 376, 399–400
Computer-generated mail, 261, 289–301 *passim*
Computerized networks, 304–11
Computerized switching (*see* Switching)
Computer utilities, 14, 368
Comsat, 213–14, 219–20, 280, 354–55 (*see also* Satellites)
Concentrator:
  and local loops, 32, 67, 124, 165, 279, 567
  and PCM, 518–19, 520
  and time-division switching (*see* Time-division switching)
  (*see also* Switching)
Conference calls, 297–98
Confra Vision, 128
Contention, 578, 585–86
Corporate networks, 162, 289–302, 305–11
Cost:
  breakdown, of Bell System components, 31, 474
  of cable television, 388–89
  of capital outlays, 355–56, 347–77 *passim*, 477–79
  of coaxial voice channel, 448
  of corporate communications, 289–301
  differences, of old and new common carriers, 285, 327–37
  of digital transmission, 47–9, 386
    vs. analog, 47–9
  of Frequency Division Multiplexing, 53, 55
  of frequency-division switching (*see* Switching)
  of LSI, 379–80
  of microprocessors, 380
  of microwave, 337
  minimizing techniques, 289–301 *passim*
  of optical fibers, 337, 454
  of packet radio, 586 (*see also* Packet radio)
  of PCM transmission, 517–18
  of Picturephone, 119–22, 388
  of satellite transmission, 219, 221–23, 336–37, 343, 360–61, 392, 396, 473–99
    of earth (ground) segment, 219–23, 479, 486–87
    of satellite "long lines," 475
    of space segment, 219–23, 479
    and tradeoffs, 487–91, 498
  of stored speech memos, 295
  of subocean (marine) cables, 361
  of switching, 31, 474
  of telecommunications networks in general, 289–301, 327–37, 395, 445, 477–79
  and telecommunications revenues, 478

Cost *(cont.)*
of terminals in the home, 328–37, 395
of Time Division Multiplexing, 53, 55
of waveguide, 452, 454
of "wired city" links, 328–29, 331–37
Cream-skimming, 274, 368
Credit cards, magnetic stripe, 244–49
Crime prevention, 41, 153, 208, 252, 255, 313, 381
Cross talk, 446, 509
Cryptography, 161, 255, 307

D

DAMA, 233–34, 474, 486, 575
Data banks, 5, 300
Data broadcasting, 5, 385
DATA PAC, 381, 539
Data radio, 171–79, 342
Data transmission, 47–8, 55, 511–21
broadcast, 385
cable television (*see* Cable TV)
and crime, 41, 48
digital channels and PCM, 501–21
switched networks for, 153–69
user requirements, 155–58, 159 (Box 10.1), 160–63
optimized, least cost, 293–94
by satellites, 217–25 (*see also* Satellites)
and TDMA, 575–83
and voice, 45–7, 293–94
waveguide, 454
Datran (now Southern Pacific), 162, 164, 274–79, 365
Delta modulation, 67, 261, 524–25, 532
Demand Assignment Multiple Access (*see* DAMA)
Dial-a-bus, 181
Dialogue, man-computer, 159
Digital Speech Interpolation (DSI), 35–6, 536
Digital transmission, 39, 44–55, 124, 233–34
and bandwidth, 48
by cables, 273
conversion to analog, economics of, 47–9, 511, 521
encoding for Picturephone, 124
for local loops, 32, 273
and mobile transceivers (*see* Mobile radio, transceivers)
and PAM (*see* PAM)
by satellite, 233–34
and specialized common carriers, 273, 279, 501–21
and switching (*see* Switching)

Digital transmission *(cont.)*
and synchronization of, 515–17
and TDMA (Time Division Multiple Access), 575–83
Digitizing conventional (analog) transmission, 28, 261–62, 501, 521
and bandwidth, 28–9, 48
of data, 48
of voice, 45–9
Dispersion, in optical fiber transmission, 457–60, 464
Distributed data-base systems, 380
Distributed intelligence, 68, 380, 386
Distributed processing, 380
Distributed storage, 67–8, 380
Distributed switching, 71–99 (*see also* Intelligent terminals)
Diversity reception, diversity earth station, 198
Duplex transmission, full and half, 35, 38

E

EARLY BIRD, 211, 214, 221
Earth stations:
costs of, 219–23, 473–99
for satellite transmissions, 211–36 *passim*
(*see also* Antennae)
Echo suppressors, 224, 485
Edison, Thomas A., 9
EFT, EETS (*see* Electronic Fund Transfer)
Electromagnetic spectrum, 181–82, 406–10
Electronic flip charts, 299
Electronic Fund Transfer (EFT), 41, 239–57, 267–68, 336, 384, 389–90
and automated clearing, 242
CBCTs, 241
and preauthorizations, 242–43
and security, 255–56
standards, 249
SWIFT, 244–46
Electronic mail, 41, 71, 225, 259–69, 536–37
Electronic payroll, 242
Emery Air Freight voice response system, 108–10
Equalizers, 120–21
Error control, in packet system, 586
Error correction, 161
Error detection, 161
in satellite transmissions, 486, 582
Error rate, in Datran, 278
in private millimeter wave system, 419
Exchanges, intelligent, 71–93
Execunet, 274, 276
Executone, Inc., 84

F

Facsimile transmission, 32, 41, 47, 67, 125, 166, 173, 264, 294–95, 297–98, 330, 381, 386, 583
and cable television, 138–39, 148
and digital bit stream, 5, 67, 264, 386, 556–57
by satellite, 218, 235, 264, 583
FCC (U. S. Federal Communications Commission), 14, 135, 138, 141, 182, 184, 192–93, 215, 223–24, 272, 280, 282, 347, 351–56, 360
"open skies" satellite policy, 215, 360
FDMA (*see* Frequency Division Multiple Access)
Federal Reserve Board, 267
Forecasting, technological, 19–23
Foreign attachments, 362–63, 365
Fourier analysis, 51–2, 416, 422
Frame grabber, 148–49
Frequency:
modulation (FM), 415–16
shortage of, 182, 193
Frequency Division Multiple Access (FDMA), 382, 389, 482–84, 575–76
Frequency-division multiplexing, 446 (*see also* Multiplexing)
interfacing (sharing), 431
and lasing, 467
and mobile transceivers (*see* Mobile radio, transceivers)
Full duplex transmission, 278

G

Gauss and Weber, 445
Glass fiber cables, 455–70
Government regulation of telecommunications (*see* FCC)
Grade of service, 37, 40, 68
and channel utilization, 160, 260–61
Graphics, in dialogue, 156–57
Graphnet, 280–83, 371
Ground station (*see* Antennae, satellite; Earth stations)
GTE, 358
satellite, 480
"Guardband," 34
and cable television, 139

H

Hazeltine simultaneous data/TV broadcasting, 172
Helical waveguide, 4, 27, 445, 448–54
Heterodyne detection, 423
Hicap (cellular mobile radio), 192, 202, 206
Hinky Dinky case, 247–49, 251

Holography, 13
"Hot lines", 306–7
Hot potato network (*see* Packet switching networks)
Hybrid services, 353–54, 368–70, 381
  communications, 354
  data processing, 353

I

IBM 3750 Voice and Data Switching System, 75, 84
IBM 7772, 97–100
IBM and SBS satellite, 223–24, 233–36, 280, 484–86, 583
IBM's SNA (System Network Architecture), 539
IBM video education network, 300–301
Idle capacity telephone network, 260–61
Infrared transmission, 422–23
Intelligent terminals, 156, 380
INTELSAT IVA, 111–12, 114, 119, 220–21, 361, 382, 475, 491
Interconnect industry, 350–52, 362–63
ITT COMPAK, 371
ITV, 346

L

Large-scale integration (*see* LSI)
Laser, 4, 13, 339–40, 380, 394–95, 423–27, 458, 463–70, 474, 501
  and bandwidth, 464
  CO_2 laser, 474
  multiplexed, 467
  and satellite transmission, 339–40, 474
  semiconductor, 380
  transmission by, digital and analog, 501
Leased lines, 290, 307
LED (Light Emitting Diode), 423, 461–63
Light pen, 32
Line brokering/sharing, 370–71
Line driver, 78
Loadcoils, 165
Loops, local, 31, 32, 165, 273, 274, 469–70
  alternatives to, 469–70
  and interconnection, 273
LSI, 4, 13, 21, 27–8, 74, 77, 165, 177, 181, 197, 379–80, 385, 465–67, 523, 527
  and cost, 380

M

Magnetic bubbles, 380
Magnetic card typewriter, 295

Magnetic stripe cards, for banking, 244–49
Mail, computer originated, 261, 289–301 *passim*
  electronic, 228, 259–69, 381, 536
  by satellite, 228, 536
Mailgram, 41
Marcon, 170, 180, 584
MARISAT, 226, 296, 301, 343, 360
McFadden Act, 247
MCI, 274–76, 352, 363–65, 371
Message switching, 539–56 *passim* (*see also* Switching; Packet switching networks)
Message switching, corporate, 294
Microcomputer, 4, 13, 71, 313, 380–81, 473, 585
Microprocessor, 28, 322, 380, 386
Microwave, 21, 52, 121, 145, 165, 290, 412, 449–54 (*see also* Datran; MCI)
  absorption, 412 (*see also* Absorption)
  attenuation, 412
  and satellites, 211
  waveguides, 449–54
Millimeter wave, 290, 382, 388, 419
  satellites, 340
Millimeter waveguides, 449–54
Millimeter wave radio, 5, 382, 406–7, 409–11
  attenuation of, 412
Minicomputer, 74, 168, 380–81
Minimum cost routing, 289–301
Mobile radio, 5, 39, 41, 178, 180–209, 301, 408
  cellular system organization, 5
  transceivers, 180–209
Modem, 46, 165, 278, 381–82, 385, 390, 523
  burst, 581, 586
  vs. line driver, 278
Modulation index, 491
Monitoring equipment, 293
Motorola, portable hand telephone, 202–3
Mulsak broadcast satellite, 216, 225, 486, 491
Multidrop lines, 576–77
Multiple access:
  burst, 155–58
  and cable television, 150
  cost, of FDM rising, 47, 53
  of digital signal, 52–3
  and electronic mail, 261–62
  Frequency division (FDM), 34, 47, 53, 150
    and satellite transmission, 486
  of lasers, 467–69
  time division (TDM), 34, 53–5, 150, 160–61, 164, 169
  and switching, 67–8
Multiplexing, 34, 38, 47, 52–5, 165, 169, 261–62
Multiplexor, 502–5

N

NACHA, 242
NASA Record transmission traffic growth, 263, 264
NASA satellites, 178, 225, 309, 340–45, 360–61, 387–89, 477, 486
  ATS-6, 340–45
Network:
  and cable television (*see* Cable TV)
  computerized, 305–11
  corporate, 264–67, 536
  minimizing cost of, 289–301
  new types of, 264–67
Network, switched data, 153–69, 536
Node computers (*see* ARPANET)
Noise-to-signal ratio, 195
North Electric Company communications extender, 91

O

Office of Telecommunications Policy, 136, 182, 183, 351, 370
Optical fibers, 4, 21, 27, 357, 383, 388, 392–93, 395, 445, 454–70, 473
Optical transmission, 382, 383, 388, 409–12, 422–27, 445, 454–70
  and dispersion, 457–60
  and reflective index, 457–60
  and self-focusing fiber, 458–60
Optical waveguide, 445, 454–55
OPTRAN, 423, 425
OTP, 136, 182, 183, 351, 370

P

PABX (PBX in U.S.), 71–93, 122, 290, 297, 301
  computerized, 71–93, 290, 301, 307, 383, 385
  dial codes for, 85–7
  functions of, 75–83, 85–7, 122
  for satellites, 235
Packet control procedures, 543–50
Packet radio, 5, 166, 177–79, 191, 206, 342, 385, 390–91, 585–98
Packet switching networks, 5, 168–69, 279, 381, 386, 534–57 (*see also* Message switching; Packet switching system)
  vs. packet radio, 585
Packet switching system (*see* ARPANET; TELENET)
Paging, radio, 41, 80, 84 (*see also* Radio paging)
PAM, 49–50, 52, 53
Parallelism, 21
PBS, 172
PCM, 12, 46–8, 50–5, 59–60, 161, 166, 261, 359, 382–83, 454, 501–27 (*see also* Transmission)
  advantages of, 520

PCM *(cont.)*
and compression, 523–37
DPCM, 524, 532
and laser transmission, 467–79
and television transmission, 59
and voice channel, 62
and waveguides, 404
PDP-11, 110
Peak-to-average ratio, 155–58, 166
Peak traffic, 259–69
Periphonics, 98, 110
Picturephone, 5, 18, 32, 35, 52, 113, 298, 329, 384, 388, 476, 501–2, 511
and bandwidth, 52, 119
and cost, 52, 117–19, 122, 298, 329
and dialing, 113, 119
and digital dialing, 113
and PAM (*see* PAM)
and PCM, 52, 532–34
and private branch exchange, 122
psychology, 116–18
and slow scan transmission, 125–27, 384
and switching, 119, 122–24
and transmission, 119–22
Pocket calculators, 313, 380
POS (point of sale) terminals, 251
POTS (plain old telephone service), 154–55, 218
Priority preempt signals, 41–2
Priority transmission, 262–63, 292, 306
Privacy, 6
and EFT, 256–57, 381
Pushbutton phones, 95–111

Q

Quantizing, 50–2, 526–27
Quantizing noise, 51, 525–26
QUBE system, 148, 149

R

Radio, CB, 182–85
Radio, data, 171–79, 191–92, 204–6, 207, 406–7, 416–19
Radio, millimeter wave, 409, 411–19
frequencies, 416
Radio, mobile transceivers, 181–209
Radio channels, 405–43
and digital, 415–19
Radio dispatch, 186, 188–91, 194, 335
Radio paging, 41, 188–90, 192, 207, 301, 307, 335, 342

Radio spectrum, 373–76, 405–42 *passim*
and attenuation, 412–13, 453
and data vs. voice, 182, 192
extending use of, 181–82, 191–92, 198, 204–6
FCC channel allocation, 422, 427
frequency management, 182, 192
long wave, 407
medium wave, 407
and satellites, 437–42
scarcity of, 405, 427–39
short waves, 407
Radio telephones, 183, 188–90, 335
cellular organization, 191, 193–203
RCA, satellite:
SATCOM, 211, 214, 382, 490
video voice, 113, 173
Real-time telecommunications, 36, 38–40
mixed with non-real time traffic, 260–61, 264–67
Regenerative digital repeater, 454, 456, 461, 505, 507, 520
Regenerative repeater, 48, 507
Regulation, government (*see* FCC)
Repeater, packet, 597–98
Resource sharing, 283
Rohm PABX, 86–7
Routing, least cost (minimum cost), 82, 86, 90

S

SAGE, 310
SATCOM (RCA), 211, 217
Satellites, 211–36, 339–45, 354–55, 382, 388–89, 473–99
11-14 GHz, 280, 337, 382, 389
antennae, 211–36, 280, 340–45, 360
and cable television, 51, 139, 145–47, 151, 225, 227, 382, 388, 487, 490
capacity of, 27
and cost, 31, 219–23
DAMA (Demand Assignment Multiple Access), 233–34
and data transmission, 224–25, 232, 233–36, 386
degraded signal, 230–32
delay, 223–25
digital transmission, 217–18, 233–36, 259, 474
earth segment, 219–23
and microwave interference, 229–32, 475
multiple access, 217–19, 232–36
NASA, in India, 225 (*see also* NASA satellites)
networks, 267, 360, 386, 486
"open skies" policy (FCC), 215, 360
and packet radio (*see* Packet radio)

Satellites *(cont.)*
  and power generation, 340, 474, 486–87
  radio, 437–42
  relays, 179, 211, 219–23
  space segment, 219–23
  synchronous, 12, 214, 339–45, 389
  TDMA (Time Division Multiple Access), 579–80
  transmissions:
    advantages of, 211–19
    bandwidth, 211, 229–32, 235
    broadcast, 216, 235–36, 341–42, 486–87
      of data, 219–29
    burst, 474
    and DAMA, 233–34
    delay, 223–25
    line control for, 224–25, 232–36, 474
    and microwave interference, 229–32, 475
    to movable transceivers, 177–78, 191, 206
    and multiple access, 217–19, 232–36
    point-to-point, 218
    spot beam, 474, 476–77, 478
    and TDMA, 579–80
    of television, 216, 232
  transponders, 165
Satellite specialized common carriers, 271, 280, 350
Satellite system, corporate, 71, 223–24, 233–36, 264–67, 280, 473, 484–86, 583
Satellite system capacity, 217–23, 475–76
Security, 279, 307, 381
Signal, binary bipolar (T1), 509
Signal-to-noise ratio, 48, 51, 195, 198, 232, 415
  with satellites, 232
Skin effect, 446
Sky lab (ATS-6), 474
Slow scan transmission, 37
  and Picturephone, 125–27, 384
Solar cells, 340, 474
Space division switching, 482
Space shuttle, 219, 228, 389, 477
Specialized common carriers, 271, 273–80, 290, 350–52, 363–65, 381
Spectrum engineering, 435–36
Speech, digitized, 35–6, 536
Speed, of channels, 257–68
Standards, 249, 501–5
Storage, mass on line, 380
Sub-ocean cables, 210–11, 218, 224, 361, 383, 535
Switched CATV, 333–35
Switched data channels, 153–69
  analog, 160
  digital, 153–69 *passim*

Switched data networks, 153–69
Switching, 31, 34
  of burst transmissions, 155–58
  of cellular radio telephone systems, 191, 193–203, 385
  circuit, 168–69, 279
  computerized, 5, 71–93, 386
    of CATV, 333–35
    of circuits, 166–69, 279, 307
    and DAMA (Demand Assignment Multiple Access), 333–34
    distributed, 333
    fast, 166–69, 279, 307, 386
    of hot line, 307
    item, 166–68
    of Picturephone, 122–24
    of satellite network, 233–35, 476–77
    session, 166–68
    space division, 482
    TASI (Time Assigned Speech Interpolation), 35, 224, 535–36
    time division, 559–73 *passim*
  Datran, 274–79
  dynamic allocation of channels, 35, 36, 536
  electromechanical, 72
  fast, 160, 166–69, 279, 307, 386
  and intelligent exchanges, 71–93
  message (*see* Packet switching networks)
  mobile radio signals, 193–206
  vs. multiplexing, 67
  multipoint, 34–5
  by noise-to-signal ratio, 195
  packet, 168–69, 279
  for packet radio (*see* Packet radio)
  point-to-point, 345
  speed of, 160, 166–67, 168
Synchronous transmission, 55

T

T1 carrier (*see* Bell System T1 carrier)
TACSAT, 213
TASI (Time Assigned Speech Interpolation), 35, 224, 535–36
TDMA (Time Division Multiple Access), 382, 389, 482–86, 559–72, 575–83
Telecommunications industry growth, 16–23
Telecommunications links, 4–5, 328–37
  categories of, 27–42 (Box 3.1), 38–9, 328–37
  cost of networks, 289–301
    minimizing, 289–301
    and revenue, 154
  design options, 601–31
  parts of a network, 31

Telecommunications links *(cont.)*
and switched data networks, 153–69
trend to digital links, 259, 294, 376, 379
and wired city, 327–37
Telecommunications regulations, 347–77
Telegraphy, 28–9
channel speeds, 59, 294
TELENET, 177, 281–82, 371, 381, 539–41
Telephone answering machine, 385
Telephone bill, lowering it, 82, 85, 87–90, 154, 289–301 *passim*
Telephone transmission characteristics, 155–58, 159 (Box 10.1)
data needs vs. other, 153–69
Teleprocessing piano, 15
Teletext, 18, 172–73, 176, 207
Television, 17–8, 384, 407, 434, 445
and bandwidth, 29, 37
cable (*see* Cable TV)
capacity, 27, 534–35
color, 52
digital transmission of, 47, 52, 59, 217, 332, 532–35
interactive, 18
satellites, 67, 216, 227, 384, 477
screens, large, 5
University of the Aire (U.K), 216, 390
TELEX-TWX, 345
TELSTAR, 215, 219
Terminals:
and continuous burst transmission, 155–58
forms of, 32–3, 155
in the home, 4–5, 32, 217–18, 313–25, 327–37
intelligent, 327–37
mobile, 5, 41, 181–209 (*see also* CB radio; Mobile radio; Packet radio)
needs, vs. telephone, 154–69
and packet radio (*see* Packet radio)
personal, 218 (*see also* Packet radio)
speeds of, 160
and Touchtone, 32–3, 383
Terrabit store, 380
Tieline, corporate, 40
Time division multiplexing (TDM), 32, 53–5
and lasers, 467
Time division switching, 559–73 (*see also* Switching)
advantages of, 559–60
Time sharing, 13
Touchtone, AT&T, 9, 95–111, 323, 359, 383–84
Transaction II telephone (AT&T), 33
Transceivers, mobile, 181–209

Transmission:
analog or digital?, 45–55, 217–18, 376, 379, 515
in bursts, 36
categories, 38–9 (Box 3.1), 327–37, 336–37
characteristics of data vs. voice, 153–69
of data (*see* Data transmission)
delay (satellite), 224–25
digital, economies of, 47–9
digitizing of, 28, 376, 515
efficiency, 166
by PCM, 46–8, 50–5, 59, 383, 501–23, 525–37
by Picturephone, 119–22, 532–35
satellite, 211–36, 473–99 (*see also* Satellites)
delay, 224–25
speeds, 57–68, 160
of still pictures, 125–27
TDM, 53–4, 383
television, 47, 59, 218, 405–7 (*see also* Television)
voice, 46–7, 217
Transparency of code of terminal, 31, 161, 164–65
Transponder (satellite), 165, 211, 382, 416–17, 482

U

UHF, 177–78, 182, 192, 195, 337, 341–42, 372–73, 385, 390, 408, 445
and data broadcasting, 5, 385, 390
for radio vs. TV, 182, 192
and satellites, 213, 341–42
Utilization, of transmission facilities, 262–63

V

Value added common carriers, 162, 271, 350, 355, 381, 389
Value added networks, 264, 350, 370–71
VHF, 144–45, 337, 341–42, 372, 406–8, 445
and data broadcasting (*see* Data broadcasting)
Video communications network, 300–301
Video conferencing, 384
Videophone, 52, 124, 327
Video telephones, 4, 113–31, 300
Video voice (RCA), 113
Viewdata, 269, 321, 322, 323, 325
Vocoder, 261–62, 268, 527–31
Voice, digitizing of, 48–54 (*see also* Delta modulation)
Voice answerback, 4, 32, 97–100, 269, 315, 383 (*see also* Voice response)
Voicegram, 41, 295, 297, 385
Voice print, 208, 297, 384

Voice response, 95–111, 261
Voice transmission:
  digital bit stream, 46–7, 161–62 (*see also* Digital transmission)
  by satellite, costs of, 219–23
VuSet (AT&T), 33

W

Warner Brothers QUBE system, 148, 149
WATS Box, 90–1
WATS lines, 81, 290, 292, 355
Waveguide, 259, 445–54 (*see also* Helical waveguide; Optical waveguide)
Weather, 405–16
WESTAR (Western Union Satellite), 211–12, 214–16, 225, 227, 301, 339–40, 360, 382, 474, 482, 487, 490, 515
  analog or digital transmission?, 515
Western Union, 11–2, 515
Wired city, 327–37
Wire tapping, and EFT, 255

THE JAMES MARTIN BOOKS ON COMPUTE

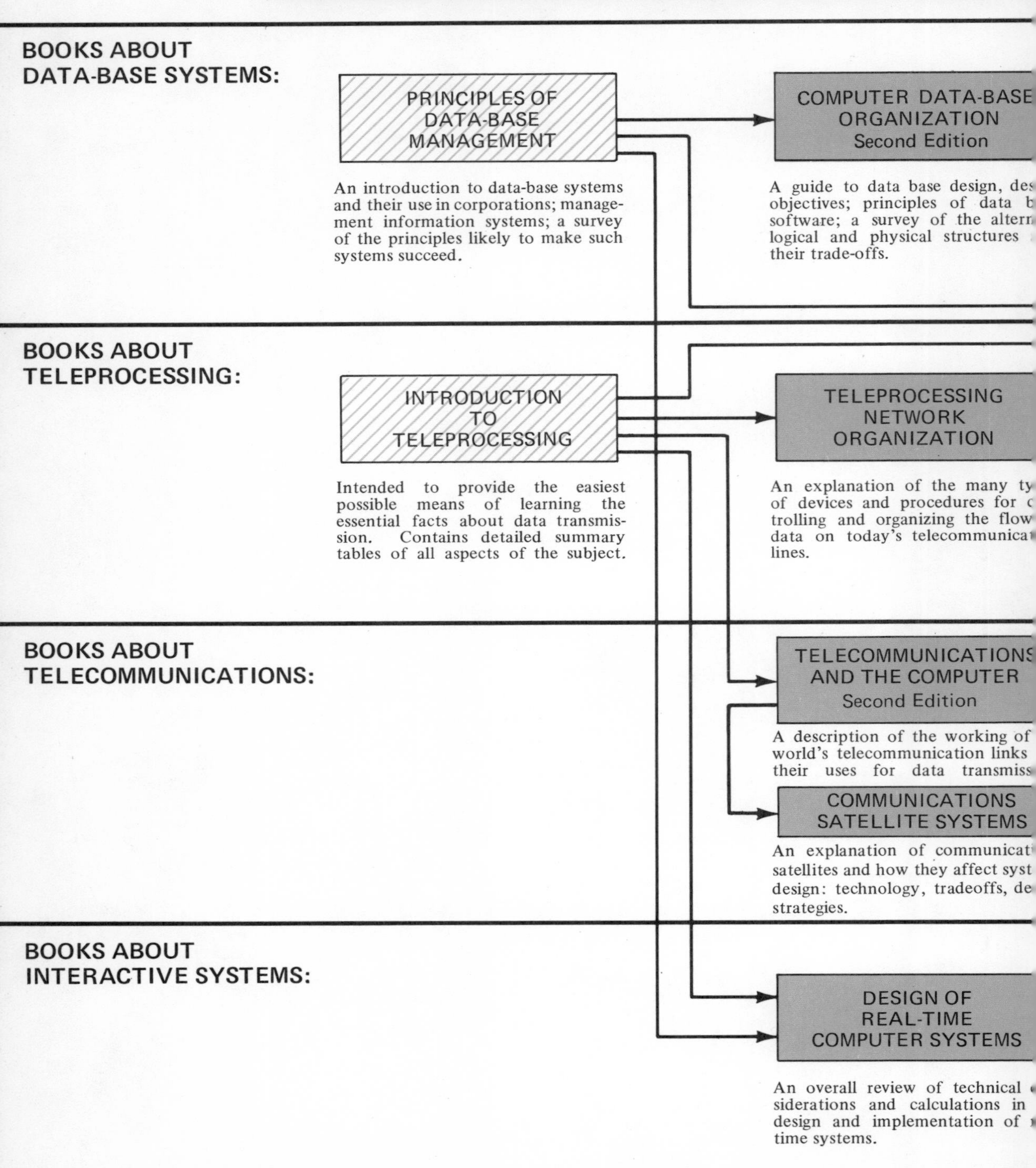

BOOKS ABOUT
DATA-BASE SYSTEMS:
PRINCIPLES OF
DATA-BASE
MANAGEMENT
An introduction to data-base systems
and their use in corporations; manage-
ment information systems; a survey
of the principles likely to make such
systems succeed.
COMPUTER DATA-BASE
ORGANIZATION
Second Edition
A guide to data base design, des
objectives; principles of data b
software; a survey of the altern
logical and physical structures
their trade-offs.
BOOKS ABOUT
TELEPROCESSING:
INTRODUCTION
TO
TELEPROCESSING
Intended to provide the easiest
possible means of learning the
essential facts about data transmis-
sion. Contains detailed summary
tables of all aspects of the subject.
TELEPROCESSING
NETWORK
ORGANIZATION
An explanation of the many ty
of devices and procedures for c
trolling and organizing the flow
data on today's telecommunica
lines.
BOOKS ABOUT
TELECOMMUNICATIONS:
TELECOMMUNICATIONS
AND THE COMPUTER
Second Edition
A description of the working of
world's telecommunication links
their uses for data transmiss
COMMUNICATIONS
SATELLITE SYSTEMS
An explanation of communicat
satellites and how they affect syst
design: technology, tradeoffs, de
strategies.
BOOKS ABOUT
INTERACTIVE SYSTEMS:
DESIGN OF
REAL-TIME
COMPUTER SYSTEMS
An overall review of technical
siderations and calculations in
design and implementation of
time systems.

KEY:

INTRODUCTORY BOOKS

These books are an easy-to-read introduction to the subjects.

DETAIL BOOKS